# Inverse Problems
# in Engineering Mechanics

International Union of Theoretical
and Applied Mechanics

M. Tanaka, H. D. Bui (Eds.)

# Inverse Problems
# in Engineering Mechanics

IUTAM Symposium Tokyo, 1992

Springer-Verlag

Berlin  Heidelberg  New York
London  Paris  Tokyo
Hong Kong  Barcelona  Budapest

Prof. Masataka Tanaka
Department of Mechanical Systems Engineering
Faculty of Engineering
Shinshu University
500 Wakasato, Nagano 380
Japan

Prof. Huy Duong Bui
Laboratoire de Mécanique des Solides
École Polytechnique
91128 Palaiseau
France

ISBN 978-3-642-52441-7      ISBN 978-3-642-52439-4 (eBook)
DOI 10.1007/978-3-642-52439-4

© Springer-Verlag, Berlin Heidelberg 1993
Softcover reprint of the hardcover 1st edition 1993

Typesetting: Camera ready by authors

61/3020-5 4 3 2 1 0 – Printed on acid-free paper.

# Preface

There are many kinds of inverse problems across a wide variety of fields. In general, the inverse problem can be defined as the problem where one should estimate the cause from the result, while the direct problem is concerned with how to obtain the result from the cause. At present in engineering fields, CT scan, ultrasonic techniques and so on can be successfully applied to some of the inverse problems including nondestructive evaluation or testing. On the other hand, different attempts have also been made recently in such a way that the computational software available for the direct problem is applied to the corresponding inverse problem analysis. In most of these computational approaches, the inverse problem is formulated into a parameter identification problem in which the set of parameters corresponding to the lacking data to be estimated should be found by minimizing a suitable cost function. Two main difficulties encountered in the inverse analysis are non-uniqueness and ill-posedness of the inverse solution. The former difficulty should be overcome by selecting one useful solution from the engineering view point. The latter difficulty could be circumvented or overcome from theoretical or mathematical considerations.

To make further advances in the inverse problem research, fruitful discussion is required between researchers in mathematical sience and engineering. It was intended in the present IUTAM Symposium to gather scientists and researchers in engineering mechanics who are working in the inverse problems, in order to exchange the newest research information and hopefully to make a breakthrough in computational and experimental approaches to the inverse problems. The following domains have been the object of presentations and discussions: mathematical and computational aspects of the inverse problems, parameter or system identification, shape determination, sensitivity analysis, optimization, material property characterization, ultrasonic nondestructive testings, elastodynamic inverse problems, thermal inverse problems and other miscellaneous engineering applications. In the Symposium, five general lectures and fiftyeight contributions were presented from attendees participanting from Asia, Europe and North America. The current volume includes most of these presentations, and can provide a state-of-the-art review of the inverse problems in engineering mechanics. As the editors of this topical book we hope that technology transfer of the newest research on inverse problems all over the world will be stimulated and accelerated through this book.

As the organizers of the Symposium we wish to express our cordial thanks to all the members of the Scientific Committee and the Local Organizing Committee formed in the JASCOME. Financial support from the IUTAM is gratefully acknowledged. Co-sponsorship by the following academic societies in Japan and France is heartily appreciated: Acoustical Society of Japan, Japan Society for Aeronautical and Space Sciences, Japan Society for Industrial and Applied Mathematics, Japan Society of Mechanical Engineers, Japanese Society for Nondestructive Inspection, Japan Society of Precision Engineering, Japan Society for Simulation Technology, Japan Society for Technology of Plasticity, Japan Society for Computational Methods in Engineering, Ecole Polytechnique (France) and Electricite de France. Helpful efforts of the JASCOME office as the secretariat of the Symposium should also be appreciated.

Masataka TANAKA, Shinshu University / Japan
Huy Duong BUI, Ecole Polytechnique / France

September 1992

# Symposium Chairmen

Prof. Masataka TANAKA
Department of Mechanical Systems Engineering
Faculty of Engineering
Shinshu University
500 Wakasato, Nagano 380 /Japan


Prof. Huy Duong BUI
Laboratoire de Mecanique des Solides
Ecole Polytechnique
91128 Palaiseau / France

# International Scientific Committee

Prof. J.D. Achenbach
Center for Quality Engineering and Failure Prevention
Northwestern University
Everston, IL 60208-3020 / USA

Prof. A. Bamberger
Institut Francais du Petrole
Preau BP 311 / France

Prof. C.A. Brebbia
Wessex Institute of Technology
University of Portsmouth
Southampton SO4 2AA / UK

Prof. H.D. Bui (Co-Chairman)
Laboratoire de Mecanique des Solides
Ecole Polytechnique
91128 Palaiseau / France

Prof. G.M.L. Gladwell
Faculty of Engineering
University of Waterloo
Waterloo / Canada

Prof. F. Hartmann
FB 14
University of Kassel
D-3500 Kassel / Germany

Prof. S. Kubo
Faculty of Engineering
Osaka University
Suita 565 / Japan

Prof. G. Maier
Institut Polytecnico di Milano
Milan / Italy

Prof. K. Miya
Faculty of Engineering
University of Tokyo
Tokyo 113 / Japan

Prof. J. Orkisz
Institute of Structural Mechanics
Technical University of Cracow
Cracow / Poland

Prof. M. Tanaka (Co-Chairman)
Faculty of Engineering
Shinshu University
Nagano 380 / Japan

Prof. F. Ziegler
Institut fuer Allgemeine Mechnik
Technische Universitaet Wien
Vienna / Austria

# Participants

| | | |
|---|---|---|
| J.D. Achenbach | Northwestern University | USA |
| S. Andrieux | Electricite de France | FRANCE |
| S. Audebert | Electricite de France | FRANCE |
| M. Bonnet | Ecole Polytechnique | FRANCE |
| H.D. Bui | Ecole Polytechnique | FRANCE |
| M. Cahouet | Electricite de France | FRANCE |
| T.H. Chau | Electricite de France | FRANCE |
| Y.M. Chen | State University of New York | USA |
| D.J. Chinn | Delft University of Technology | NETHERLANDS |
| S.K. Datta | University of Connecticut | USA |
| G.S. Dulikravich | Pennsylvania State University | USA |
| G.M.L. Gladwell | University of Waterloo | CANADA |
| M. Hendriks | TNO Building and Construction Research | NETHERLANDS |
| L.S. Koo | Idaho National Engineering Laboratory | USA |
| B.M. Kwak | Korea Advanced Institute of Science and Technology | KOREA |
| A. Ledesma | Technical University of Catalonia | SPAIN |
| M. Link | University of Kassel | GERMANY |
| A. Louis | University of Saarland | GERMANY |
| R.H. Lyon | Massachusetts Institute of Technology | USA |
| T. Mura | Northwestern University | USA |
| H.G. Natke | University of Hannover | GERMANY |
| B. Novotny | Solvak Academy of Sciences | CZECKOSLOVAKIA |
| J. Orkisz | Technical University of Cracow | POLAND |
| C.L. Tan | Carleton University | CANADA |
| A. Vautrin | Ecole des Mines | FRANCE |
| B. Zhu | Shandong University | CHINA |
| | | |
| Y. Akiyama | Unic Corporation | JAPAN |
| S. Aoki | Tokyo Institute of Technology | JAPAN |
| De Cerqueira e Souza | Kyoto University | JAPAN |
| T. Fukui | Fukui University | JAPAN |
| H. Fukuoka | Osaka University | JAPAN |
| K. Fujimata | Kozo Keikaku Engineering Inc. | JAPAN |
| H. Hangai | University of Tokyo | JAPAN |
| K. Harumi | Tokyo University of Information Sciences | JAPAN |
| M. Hirao | Osaka University | JAPAN |
| S. Hirose | Okayama University | JAPAN |
| T. Honma | Hokkaido University | JAPAN |
| Y. Hosoda | Nagoya University | JAPAN |
| Y. Ichikawa | Nagoya University | JAPAN |
| M. Iida | University of Tokyo | JAPAN |

| | | |
|---|---|---|
| H. Imabayashi | Shinko Electric Co., Ltd. | JAPAN |
| H. Imai | University of Tsukuba | JAPAN |
| H. Inoue | Tokyo Institute of Technology | JAPAN |
| Y. Iso | Kyoto University | JAPAN |
| K. Iwamura | Josai University | JAPAN |
| Y. Kanai | University of Industrial Technology | JAPAN |
| H. Kanda | Yanmer Diesel Engine Co., Ltd. | JAPAN |
| Y. Kasuga | Yokohama National University | JAPAN |
| Y. Kagawa | Okayama University | JAPAN |
| H. Kawaguchi | Hokkaido University | JAPAN |
| K. Kishimoto | Tokyo Institute of Technology | JAPAN |
| T. Kitagawa | University of Tsukuba | JAPAN |
| M. Kitahara | Tokai University | JAPAN |
| S. Kobayashi | Kyoto University | JAPAN |
| T. Koike | NTT Advanced Technology Corporation | JAPAN |
| F. Kojima | Osaka Institute of Technology | JAPAN |
| W. Kozukue | Nissan Motor Co., Ltd. | JAPAN |
| S. Kubo | Osaka University | JAPAN |
| M. Kubo | Kyoto University | JAPAN |
| T. Kuroyanagi | | JAPAN |
| K. Matsui | Tokyo Denki University | JAPAN |
| T. Matsumoto | Shinshu Universuty | JAPAN |
| A. Murakami | Kyoto University | JAPAN |
| M. Nakamura | Shinshu Universuty | JAPAN |
| N. Nishimura | Kyoto University | JAPAN |
| N. Noda | Shizuoka University | JAPAN |
| M. Notake | Mitsubishi Research Institute Inc. | JAPAN |
| T. Ohe | Osaka University | JAPAN |
| K. Ohhashi | Kyosera Corporation | JAPAN |
| K. Ohtsuka | Hiroshima-Denki Institute of Technology | JAPAN |
| S. Ohwaki | Kumamoto University | JAPAN |
| S. Oie | Ono Sokki Co., Ltd. | JAPAN |
| M. Okada | Tohoku University | JAPAN |
| T. Okuno | Shimizu Corporation | JAPAN |
| K. Onishi | Science University of Tokyo | JAPAN |
| Y. Otake | Ishikawajima-Harima Heavy Industries Co., Ltd. | JAPAN |
| Y. Otani | Fuji Research Institute Corporation | JAPAN |
| J. Pavo | University of Tokyo | JAPAN |
| M. Sakakihara | Okayama University of Science | JAPAN |
| K. Sato | Kozo Keikaku Engineering Inc. | JAPAN |
| M. Sato | Mitsubishi Electric Corporation | JAPAN |

| T. Sayama | Toyama Industrial Technology Center | JAPAN |
| T. Shimada | University of Marketing & Distribution Science | JAPAN |
| Y. Shinozaki | Kyoto University | JAPAN |
| Y. Suzuki | Aoyama Gakuin Universuty | JAPAN |
| Y. Tada | Kobe University | JAPAN |
| M. Takadoya | Mitsubishi Research Institute Inc. | JAPAN |
| T. Takagi | Tohoku University | JAPAN |
| M. Tanaka | Shinshu University | JAPAN |
| M. Tohyama | Nippon Telegraph and Telephone Corporation | JAPAN |
| N. Tosaka | Nihon University | JAPAN |
| T. Tsuchiya | Okayama University | JAPAN |
| T. Tsuji | Shizuoka University | JAPAN |
| T. Tsukiji | Ashikaga Institute of Technology | JAPAN |
| A. Utani | Nihon University | JAPAN |
| M. Yamano | Sumitomo Metal Industries Co., Ltd. | JAPAN |
| K. Yamazaki | Kanazawa University | JAPAN |
| F. Yano | Obirin Junior College | JAPAN |
| K. Yoda | Mitsubishi Electric Corporation | JAPAN |
| S. Yoshimura | University of Tokyo | JAPAN |

Total number of participants = 100

# List of Contents

## Chapter 3   PARAMETER IDENTIFICATION

## Chapter 4   SHAPE DETERMINATION AND OPTIMIZATION

## Chapter 5   MATERIAL PROPERTY CHARACTERIZATION

## Chapter 6  ELASTODYNAMIC INVERSE PROBLEMS

## Chapter 7  ULTRASONIC NONDESTRUCTIVE EVALUATION

## Chapter 8   THERMAL INVERSE PROBLEMS

## Chapter 9   OTHER ENGINEERING APPLICATIONS

# Chapter 1
# Mathematical Aspects

# On Regularization Methods Within System Identification

*H.G. Natke*

Curt-Risch Institute of the University of Hannover, Germany

## 1 Introduction

Linear and non-linear systems are considered. The term system is used here as a synonym for the technical constructions of mechanical, civil, naval, aeroplane and aerospace engineering. The following is restricted to spatially finite dimensional models.

### 1.1 Motivation.

Theoretical system analysis includes modelling, simulation and prediction, and, of course, optimization. In every case a validated model is needed: system identification ([13], [33]). System identification is the inverse problem of system analysis using measured data. In consequence, the system or a dynamically similar hardware model must be available, and measurements result in erroneous and incomplete data. If the structure of the parametrical model is verified, then the identification is reduced to parameter estimation.

### 1.2 System Identification Problems.

The model structure of a linear system is uncertain with respect to

- the number of degrees of freedom of the bandlimited model,
- the type of damping force.

As far as the type of damping is concerned various methods are available for modelling. However, the prior information of the type of damping is generally poor, although the easiest model should be taken (e.g. viscous type).

Model structure identification of non-linear systems is much more difficult than that of linear systems. This is due to their diversity. Choosing the class of polynomials as the model class then a priori structure identification means the estimation of the maximum order simultaneously with the coefficients [30]; a posteriori structure identification means the verification of the prior (assumed) model structure by statistical tests ([40], [53]). The modelling of hysteretic sytem behaviour is excluded remembering the Stone-Weierstraß theorem. However, Vestroni and Capecchi [50] show how the application to this class of models is performed.

The estimation of physical parameters (inertias, stiffnesses or flexibilities, damping coefficients) often leads to ill-posed problems. A well-known example is deconvolution, and the reader can find a further example in [48] . The characteristics of an ill-posed problem are defined by the three Hadamard conditions, i.e. by violating

1. the existence of solutions (consistency of the problem) or
2. the uniqueness of the solution or
3. the stability (continuity) of the solution with respect to the data.

As a consequence, for the inverse problem of parameter estimation the ill-posedness leading to

- a small radius of convergence for iterative methods (see also [47])
- the uncertainty of convergence, which is assured only in the case a small number of parameters are to be estimated;
- highly correlated parameters are also to be estimated
- ...

It is a disadvantage that the

- measured data are erroneous and
- in general are incomplete.

Counter-measures in connection with unbiased minimum variance estimates are: the optimization of test design ([8], [4]), statistical weighting within the estimation procedure ([13], [7], [10], [11], [28]) and special mathematical manipulations such as the scaling or truncation of decompositions.

## 1.3 Applications of System Identification.

Experimental modal analysis is widely known and accepted ([26], [29], [33]). Here the numerical problems are less serious than in the estimation of the physical parameters. This is due to the possible optimum test design (control and consideration of the effective number of degrees of freedom in the measurements and the non-correlation of eigenfrequencies and modal vectors compared with the often correlated stiffnesses and inertias).

The estimation of physical parameters is often called mathematical model correction, improvement, updating, calibration etc.. The result is a validated mathematical model, that means a model with known confidence and dependent on the used data set with sufficiently small errors ([29], [33], [19], [20], [17], to mention only a few references).

This validated mathematical model will be applied for ([34], [31], [38], [39], [51], [52])

- simulation and prediction

- investigation of the inner structure of the system
- mathematical operations as modal transforms
- modification investigations
- optimization
- system monitoring
- fault (damage) detection, localization and diagnosis, in general, and for
- the knowledge base within an expert system.

As can be seen, system identification is very important in engineering for safety assessment, comfort requirements, and in consequence it plays an important economic role.

## 2 Regularization Approaches

Let us consider the linear equation

$$Af = g, \ A \ : F \longrightarrow W, \ f \epsilon \ F, \ g \epsilon \ W, \tag{1}$$

where $F$ and $W$ are Hilbert domains. Generally $A$ is assumed to be a non-continuously invertible operator, which means that evaluating $f$ for given $A$ and $g$ from Eq.(1) is an ill-posed problem. In the case of the finite dimensional vector spaces $F, W$ one can identifiy the operator $A$ with its matrix representation.

### 2.1 Existence of Solutions.

Generally the mapping of $F$ by $A$ results in a solution domain $R(A) \neq W$: no solution exists for each $g$. A necessary condition for the existence of a solution of Eq.(1) is therefore given by

$$g \in R(A) \ . \tag{2}$$

### 2.2 Uniqueness of Solutions.

Instead of Eq.(1) the defect

$$\|Af - g\| \longrightarrow \text{minimum} \tag{3}$$

is considered. The result is the normal solution

$$\overline{f} := A^+ A f = A^+ g \tag{4}$$

with the generalized inverse operator

$$A^+ : \ R(A) \oplus N(A^*) \subseteq W \longrightarrow F, \tag{5}$$

$A^*$ designates the adjoint operator of $A$, and $N(A^*)$ denotes its nullspace. The generalized solution means

- an enlargement of the solution domain and
- a unique solution with a minimum norm (see e.g. [3]).

This makes sense in mechanics, where the minimum energy principle plays an important role.

## 2.3 Stability of Solutions.

The uniqueness of the solution is one important problem to be solved. However, this is not sufficient, and in general the solution is unstable, since $A^+$ is not continuous, or $A^+$ is continuous but ill-conditioned. Thus, especially when $A$ depends on measurements, small changes of $A$ will cause large changes of the normal solution. Even if $A$ is a regular matrix the normal solution is generally sensititive due to the worse condition of $A$, i.e. $\|A\|\|A^{-1}\| \gg 1$, where $\|\cdot\|$ is some matrix norm. The application of improved algorithms is necessary in order not to worsen the stability. However, they cannot remove the difficulties which are characteristic of the problem.

## 2.4 Adjacent Solutions.

Looking for adjacent solutions by means of a modification of the operator $A^+$ is one way of overcoming the difficulties. This procedure, of course, implies error investigation. Modified operators can be achieved by a coarser parameter topology, by taking into account additional information, by the truncation of series expansions of the operator and, for example, by transformation of the parameters due to adaptive excitations as external conditions. The regularized operator of $A^+$ may be denoted by $T_\gamma = T_\gamma(A)$ . For each real-valued positive regularization parameter $\gamma$ it generally holds true that:

$$\forall\, \gamma > 0 \quad T_\gamma \, : \, W \longrightarrow F \tag{6}$$

in which

$$\lim_{\substack{(\varepsilon, g_\varepsilon) \to (0, g) \\ (\mu, A_\mu) \to (0, A)}} \gamma(\varepsilon, g_\varepsilon, \mu, A_\mu) \; = \; 0 \tag{7}$$

such that with random errors $\|g - g_\varepsilon\| \le \varepsilon$ of the randomly disturbed right side $g_\varepsilon \in W$ and $\|A - A_\mu\| \le \mu$ for the randomly disturbed operator it holds that

$$\lim_{\substack{(\varepsilon, g_\varepsilon) \to (0, g) \\ (\mu, A_\mu) \to (0, A)}} T_{\gamma(\varepsilon, g_\varepsilon, \mu, A_\mu)}(A_\mu) g_\varepsilon \; = \; A^+ g \; . \tag{8}$$

From the triangular inequality follows (principle of discrepancy [23])

$$\|T_\gamma(A_\mu)g_\varepsilon - A^+ g\| \le \|T_\gamma(A_\mu)g_\varepsilon - T_\gamma(A)g\| \; + \; \|(T_\gamma(A) - A^+)g\| \; . \tag{9}$$

The first term describes the total error, the second term the data error and the third term the regularization error. The importance of this inequality, which enables the estimation of the regularization effect, can be seen.

# 3 Stable Solution Methods: A Review

## 3.1 Literature Review.

As far as mathematics are concerned only books on inverse problems will be mentioned here: [3] Baumeister 1987, [21] Louis 1989, [23] Morozov 1984, [12] Deif 1986, [49] Tikhonov, Arsenin 1977, [16] Kuhnert 1976. Survey articles are, for instance, Louis [22] concerning numerical problems of inverse problem and Natke [32] with regard to error localization within spatially discretized models, i.e. within system identification.

Error localization within mathematical models, the fault localization of a system, model correction etc. are applications of inverse problems. The basic principles are presented in [33] . These are the submodel formulations for overcoming the incomplete measurements [35], and the Bayesian approach ([15], [27], [7]), which leads with the assumption of normal distributions for the random variables to the extended weighted least squares with a penalty term. This penalty term permits the inclusion of the results of the prior mathematical model. This has the advantages of:

- being economical (by sometimes using very expensive results from theoretical analysis)
- substituting missing measurements
- restricting the distances of the estimates from the prior parameters
- making the problem convex
- having the effect of regularization (by operator modification) and
- constructing an always convergent iterative (perturbation) procedure.

Mottershead and Foster [24] dealt with the ill-conditioned problem successfully with the aid of a reference model, i.e. from finite element modelling. This approach corresponds to the Bayesian approach, where the difficulty exists how to choose a statistically based weighting. Mottershead and Foster proposed a weighting equal to $\gamma I$ , $I$ is the identy matrix, though that $(\gamma + \sigma_1)/(\gamma + \sigma_N)$ defines the required degree of conditioning with $\sigma$ the singular values of the problem and $N$ the number of measurements.

Sensitivity analysis will help to find out those parameters which are controlled by the measured quantities, and thus it reduces the number of parameters to be considered and gives some additional information about the system and its model. This tool is well-known and is applied in optimization and identification.

Chen and Garba [6] worked with a minimum norm solution, they do not mention the ill-condition of the problem. Agbabian et al. [1] presented a time-domain method using the generalized inverse (pseudo-inverse), where regularization is not mentioned explicitly. However, the authors use additional information from probability functions (detectable variations). Additional information also is used by Lallement and Cogan [18], who take anti-resonances in account. Brandon and Cremona [2] also applied the generalized inverse in connection with the singular value decomposition (SVD) which is stated to have optimal numerical properties. O'Callahan [42] discussed a consequent finite element formulation with a

8

least squares approach (generalized solution), and in order to make the method useful the number of equations are extended by taking additional constraints into consideration.

Ojalvo ([43], [44]) stated that the SVD is one method of regularization which can become costly for models with a high number of degrees of freedom, therefore he described his $\varepsilon$-decomposition for linear algebraic systems of equations. A direct modification of the algebraic operator is performed (without establishing an aim functional) by an eigenvalue shift $\varepsilon \ll \lambda_i$, $i = 1(1)n$, $n$ the order of the system of equations. $\varepsilon$ here plays the role of a regularization parameter which only depends on the operator $A$ (see Eq.(1)). The choice of the regularization parameter here is simplified compared with that of the Tikhonov-Phillips method (see Sect. 3.5). Zhang and Ojalvo [54] have improved the $\varepsilon$-decomposition [43] by taking into account a second modified algebraic operator using the negative eigenvalue shift. The improved solution is then obtained by taking the average of both modified systems $(A + \varepsilon I, \ A - \varepsilon I)$. Zhang stated that the improvement refers to a second order approximation of the corresponding SVD instead of a first order approximation of the simple $\varepsilon$-decomposition. It can be shown that the difference in the two solutions are due to the inverse matrix of $(I - \varepsilon^2 A^{-4})$ that appears. In the $\varepsilon$-decomposition this is approximated by an expansion using the first term, and in the improved method the inverse is taken directly. $A$ is the diagonal matrix of the corresponding non-zero singular values of $A$. Prells [45] has extended the result by Natke [32], mainly concerning the choice of the penalty term within the extended weighted least squares using the sensitivities of the parameters to be estimated with respect to the data errors (see Sect. 3.5). Zhang and Natke [55] applied the component-mode synthesis in addition to the subsystem modelling in order to achieve a two-level correction procedure which means a reduction of the parameters to be estimated in each level. Ben-Haim [4] and Ben-Haim and Prells [5]; Prells, Ben-Haim [46] proposed a quite different method for regularization: selective sensitivity combined with an appropriate excitation, i.e. separating the parameters into subsets with a minimum number, if possible with at least a single parameter each (see Sect. 3.8).

In order to overcome some disadvantages of the input error and output error methods (see Fig. 1), Nalitolela ([25], and see the cited refs.) transforms the dynamic response problem into an eigenvalue problem with additional masses and stiffnesses, and applies updating to this modal model using eigenvalue sensitivities. So far the method has been restricted to undamped systems and investigated for undisturbed simulated measurements.

Natke and Prells [37] have used regularization methods in structure (of the mathematical model) identification of non-linear systems (see Sect. 4).

### 3.2 Series Truncation.

Considering the SVD

$$T_\gamma g = \sum_{\sigma_n > 0} \sigma_n^{-1} F_\gamma(\sigma_n, g) < g, u_n > v_n \qquad (10)$$

with $\sigma_n$ the singular values and $u_n, v_n$ the corresponding left and right vectors respectively of the SVD of $A$. The weighting $F_\gamma$ is a real-valued function which acts on the spectrum of $A^+$ like a filter (low-, high-, band-pass filter). For example, the low-pass filter is defined as

$$F_\gamma(\sigma_n) = \begin{cases} 1 \text{ for } \sigma_n \geq \gamma, \\ 0 \text{ for } \sigma_n < \gamma \end{cases}$$

which truncates the sum with respect to a pre-given regularization parameter $\gamma$ (truncated SVD). It can be shown that this choice is not optimal, although where noisy data with high frequency contents is looked at the filtering makes sense.

## 3.3 Parameter Topology.

One procedure of indirect information enlargement is the reduction of the number of parameters with respect to the available measurements in order to construct an overdetermined system of equations. Subsystem modelling permits this task [35] . All the parameters of the same degree of uncertainty to be estimated are assembled in one correponding matrix, defining one submodel. This can be regarded as a type of uncertainty modell. In the end it is the introduction of a coarser parameter topology. By designating the quadratic matrices of assumed inertia, damping and stiffness as $M$, $C$, $K$ respectively of the prior discrete mathematical model of order $n$ the subsystem modelling will result as

$$M = \sum_{\sigma=1}^{S} M_\sigma, \tag{11}$$

$$C = \sum_{\rho=1}^{R} C_\rho, \tag{12}$$

$$K = \sum_{\iota=1}^{\dot{I}} K_\iota \quad \text{and} \tag{13}$$

$$G = \sum_{\iota=1}^{\dot{I}} G_\iota, \tag{14}$$

with $S + R + \dot{I} \leq 3n^2$ and the flexibility matrix $G := K^{-1}$ . Introducing the (now global) correction factors $a_\sigma, b_\rho, c_\iota, d_\iota$ one obtains the error model

$$M^C = \sum_{\sigma=1}^{S} a_\sigma M_\sigma, \tag{15}$$

$$C^C = \sum_{\rho=1}^{R} b_\rho C_\rho, \tag{16}$$

$$K^C = \sum_{i=1}^{l} c_i K_i \quad \text{and} \tag{17}$$

$$G^C = \sum_{i=1}^{l} d_i G_i . \tag{18}$$

As can be seen, when the correction parameters are all chosen equal to 1 this produces the prior matrices; by choosing each matrix element of $M, C$ and $K$ as one submodel the topology includes a total of $3n^2$ parameters.

The general content of information is one aspect. However, random errors can also be reduced. A dexterous choice of some (at least linearly independent) parameters, i.e. of the submodels, will influence the condition of the problem, which for the weighted least squares can be read from the Hesse matrix [13] . Performing a moderate partition of this kind into subsystems without a priori knowledge of uncertainties generally is unsolved, if one excludes recalculations.

### 3.4 Bayesian Approach.

In general the probability density functions are unknown, therefore it is assumed that the measurement noise and the parameters to be estimated obey normal distributions. It follows that the aim functional of the extended weighted least squares with the parameters introduced in Sect. 3.3 and assembled in the vector $a$, is

$$J(a) = V^*(a) G_v V(a) + (a - a_0)^T G_{a_0}(a - a_0) \tag{19}$$

with

$V$ the finite dimensional residuals
$V^*$ the conjugate complex and transposed vector $V$
$a_0$ the vector corresponding to $a$ due to the prior mathematical model
$(...)^T$ the transposed of $(...)$
$G_v$ and $G_{a_0}$ connectable positive semi-definite weighting matrices.

The residuals can be chosen as equation errors or as partial residuals, as shown in Fig. 1.

Modal partial residuals are, for example, eigenvalue residuals with a relatively poor information content. Non-modal residuals are, for example, stress residuals or acoustic levels.

The properties of this method are already enumerated in Sect. 3.1 . However, it is often difficult to determine the confidence of the prior information $a_0$ as an inverse covariance matrix. Therefore a deterministic method is needed as described in the next section.

### 3.5 Tikhonov- Phillips Method.

Related to Eq.(1) the aim functional is defined as

$$J_\gamma(f) = \|Af - g\|_C^2 + \gamma^2 \|Bf\|_2^2 = (Af - g)^* C(Af - g) + \gamma^2 (Bf)^* Bf \tag{20}$$

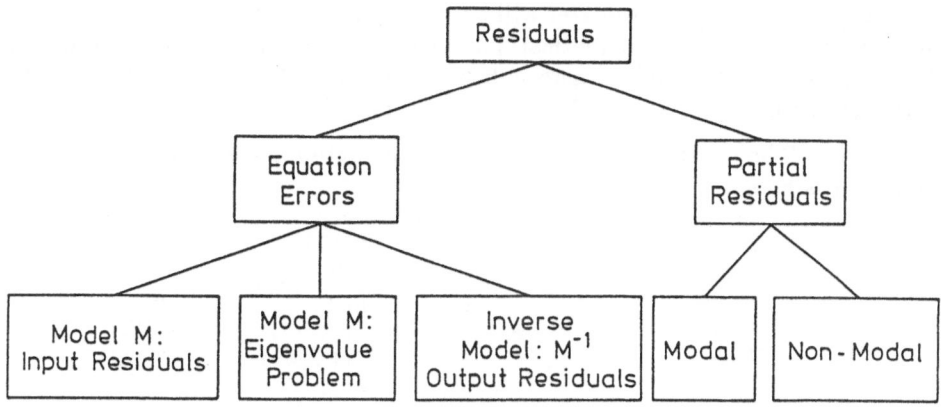

**Fig. 1.** Determination of the residuals in different ways

with the prior knowledge $f_0 = 0$. It follows the generalized solution dependent on the regularization parameter $\gamma$ from equation

$$(A^*CA + \gamma^2 B^*B)f_\gamma = A^*Cg . \tag{21}$$

As can be seen, Eq.(21) can be written without a definition of an aim functional and with various weightings, however the optimum property of the procedure will be missing.

The problem of this method is the choice of the regularization parameter. An a priori choice for $\gamma \neg g_\epsilon$: for example with $B = I, \gamma^2 = O(\frac{\bar{\epsilon}^2}{E^2})$, with $\bar{\epsilon}$ the known level of inaccuracy and $\|f\|_q \leq E, q \geq 1, E > 0$ . Cross-validation can be used (e.g. [3]) as an a posteriori choice. The reader can find further possibilities in the above-mentioned books on ill-posed problems. Additional information, for example of the data errors, serves for the determination of the regularization parameter.

Prells [45] is looking for the sensitivity of the normal solution with respect to the measurement errors. It can be shown that even small additive errors affect the solution domain of the non-disturbed solution operator inducing additional components of its nullspace. In order to reduce this effect the regularization term $\gamma^2 B^*B$ in Eq. (21) is taken as $\gamma^2 W$, with the diagonal weighting matrix $W = \text{diag}(w_i)$. If $W$ is regarded as a measure of the sensitivity of the normal solution with respect to the data, then a large sensitivity corresponds to a small variance (Bayes). The regularization information here is that the sensitive error locations are better known than the less sensitive ones. The weighting matrix elements are determined as follows: the already disturbed measurements are additionally disturbed additively by known mean-free and equally distributed white noise; the resulting generalized solutions $\bar{f}_{dk}, k = 1(1)N$, are taken with the original solution of the ill-conditioned problem $\bar{f}$ to perform

$$W \doteq \frac{1}{N-1} \sum_{i=1}^{N} \text{diag} \left[ (\bar{f}_{di} - \bar{f})(\bar{f}_{di} - \bar{f})^T \right] . \tag{22}$$

The regularization parameter will be determined using error estimates of the measured data. The mathematical foundation can be found in [45] . Link and Zhang [20] present a similar method heuristically based on the sensitivity of the residuals with respect to the design parameters.

*Example 1.* The model of the non-proportionally damped vibrator chain depicted in Fig. 2 is given.

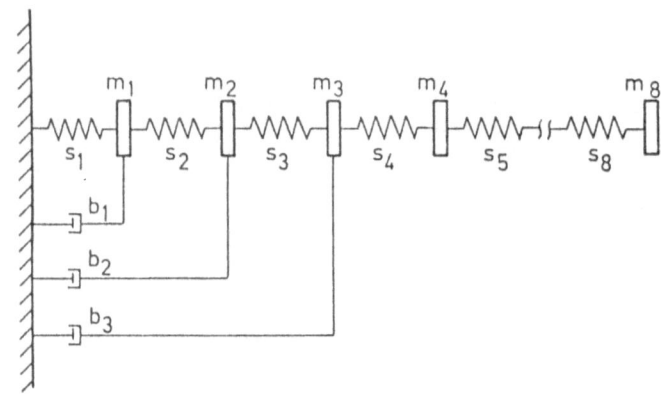

**Fig. 2.** The test model for detection of damage, localized at $m_8, b_2$ and $s_1$

The damage is simulated by a 6 percent increase of mass $m_8$, a 2 percent decrease of damper $b_2$ and a 5 percent increase of the spring stiffness at the boundary $s_1$. Application of the same excitation to the non-disturbed as well as the disturbed model leads to different responses $\tilde{x} \neq x$ . The measurement error is simulated by a uniformly distributed additive random noise of unit variance of 10 percent of the maximum norm of the response matrices. Assuming that the damage has to be localized in each element of the three matrices $[M, C, K] = A$ the problem is to determine $\Delta A$ from the (physical) dimensionless transformed equation $\Delta A \tilde{x} + A \Delta x = 0$, with $\tilde{x} = x + \Delta x$ . The result of error localization is represented in Fig. 3 by the indicator $\mathcal{I}_{r \leq 24}(\Delta A) = \sum_{i \leq 8} \|\Delta A_{ir}\| / \max_{s \leq 24} \sum_{j \leq 8} \|\Delta A_{js}\|$ for the normal solution $\overline{\Delta A}$, for the regularized solution $\Delta A_\gamma$ with respect to the sensitivity matrix $W$ defined above and for the true locations of damage $\overset{o}{\Delta A}$ . The influence of the sensitivity matrix $W$ on the result is controlled by the regularization parameter $\gamma$ which is determined from $\|\Delta A_\gamma \tilde{x} + A \Delta x\|^2 = \delta^2 + h^2 \|\Delta A_\gamma\|^2$ with the additional information of estimated error bounds $\delta, h$ (double of the bounds of the error simulated). The results can be much more improved by additionally taking into account the matrix structures of the chain vibrator.

## 3.6 Mechanical and Mathematical Model Properties.

In order to increase the information for system identification , that means in order to enlarge the number of equations additional to the structured mathe-

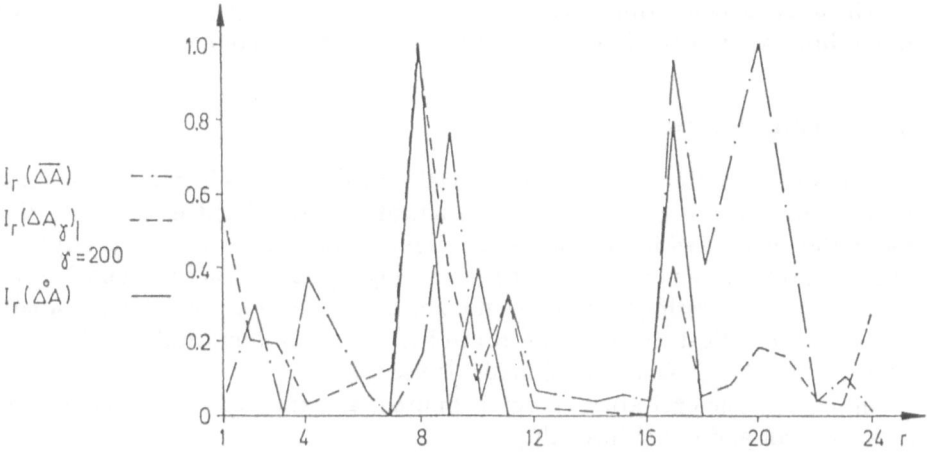

**Fig. 3.** The results of damage detection for normal and regularized solutions

matical model (with real, complex quantities) the following can be taken into account:

- The eigenstructure of the model. This means the generalized orthonormalization including the off-diagonal elements and the spectral decomposition of the physical parameter matrices.
- The large number of points of the frequency response functions: If resonance characteristics are chosen the anti-resonances should also be taken [18] .
- From the point of view of strength load paths should be considered. This can be done in order to apply parameter sensitivities with respect to stresses for regularization.
- Symmetry properties of the physical parameter matrices (Maxwell) which help to reduce the number of parameters of $n(n+1)/2$, with $n$ the order of the matrices.
- The band limitation of the finite-dimensional models. This can be achieved by condensation, balancing, transformation (e.g. incomplete modal transformation ([14], [41], [36]). These manipulations reduce the matrix orders enormously, and consequently the computational errors.
- The band-structure of the matrices resulting from finite element modelling. This consideration will introduce many additional equations due to the prior known (= 0 ) matrix elements.
- Transformation (re-formulation) of the starting equations into a numerically suitable form (measured by the Hessian matrix, for example, see Nalitolela [25]).

Of course, one can also formulate the problem in another way, for example, by first considering the eigenstructure of the system and then taking into account the equation of motion (force adjustment) as additional information.

The additional information, such as equations and inequalities, can be introduced directly or in the form of penalties in the aim functional (Lagrange).

## 3.7 Testing Restrictions.

It is superfluous to mention that additional measurements will enlarge the information. However, in general the measurements are restricted in the sense that not all the nodes used in the mathematical model can be observed at the system. That means that one has to take the measuring equation into consideration. Alternatively, the missing measurements can be substituted by prior knowledge (prior mathematical model). The control matrix is automatically included as information if the mathematical model is used.

In practice the excitation energy is limited, and this can be considered, too. It leads to an additional inequality.

## 3.8 Selective Sensitivity and Adaptive Excitation.

The method is developed for parametric viscously damped models in the frequency domain. It reduces the degree of ill-conditioning of large models in system identification by splitting up the problem into smaller (with respect to the number of parameters to be estimated) subproblems ([5], [46]) .

The equation of motion reads

$$M\ddot{u}(t) + C\dot{u}(t) + Ku(t) = p(t) \tag{23}$$

or Fourier transformed

$$(-\omega^2 M + j\omega C + K)U(j\omega) = P(j\omega) =: S(j\omega)U(j\omega) \tag{24}$$

while designating the corresponding Fourier transformed quantities with capital letters. The frequency response matrix is denoted by

$$F(j\omega) := S^{-1}(j\omega) . \tag{25}$$

Because not all the quantities can be measured, the measuring equation

$$x(t) = Hu(t), \; X(j\omega) = HU(j\omega) = HF(j\omega)P(j\omega) \tag{26}$$

is introduced, showing that only some displacements are measured, for example. The sensitivity of the output with respect to the model parameters $p_\alpha, \alpha \in \mathbb{N}$, is defined as

$$\frac{\partial X(j\omega)}{\partial p_\alpha} = H\frac{\partial F(j\omega)}{\partial p_\alpha}P(j\omega) \tag{27}$$

with

$$-\frac{\partial F(j\omega)}{\partial p_\alpha} = F(j\omega)\frac{\partial S(j\omega)}{\partial p_\alpha}F(j\omega) =: F(j\omega)A_\alpha(j\omega)F(j\omega) . \tag{28}$$

It follows

$$\frac{\partial X(j\omega)}{\partial p_\alpha} = -HF(j\omega)A_\alpha(j\omega)F(j\omega)P(j\omega) \ . \tag{29}$$

Then the sensitivity of the responses with respect to the model parameter $p_\alpha$ is given by

$$S_\alpha[P(j\omega)] := \left(\frac{\partial X(j\omega)}{\partial p_\alpha}\right)^* \frac{\partial X(j\omega)}{\partial p_\alpha} \ . \tag{30}$$

It is assumed that the matrix $H$ is independent of $p_\alpha$ and $\omega$. The selective sensitivity is now defined as follows. Let $\mathcal{J}$ be a proper subset of $\{1, ..., R\}$, $R \in \mathbb{N}$ and $\mathcal{J}'$ the subset of the remaining indices. The system is said to be selectively sensitive to parameters corresponding to $\mathcal{J}$ at frequency $\omega$ if an excitation $P(j\omega)$ exists such that:

$$S_\alpha[P(j\omega)] = \begin{cases} 0 \text{ if } \alpha\epsilon\mathcal{J}' \\ \neq 0 \text{ if } \alpha\epsilon\mathcal{J} \end{cases} \ . \tag{31}$$

Insensitivity is defind in a corresponding way. As can be seen, the requirements are strong, therefore it is proposed to moderate them to the weak selectivity with the requirement of a high sensitivity compared with a sensitivity measure of about zero.

In the above cited references it is shown how the appropriate excitations have to be chosen. However, as is well-known, it can be very difficult to realize such adaptive forcing for pre-determined displacements (see Eq. (31)) .

# 4 Regularization within Structure Identification of Non-linear Systems

The assumption is that the systems are describable by the model class of polynomials. Then the model structure is given by the maximum power. In consequence structure identification is defined by power estimation simultaneously with the estimation of the coefficients. For simplicity, the separation of damping and restoring forces is assumed. Classical identification methods estimate the polynomial coefficients for a pre-given power $N$ dependent on the permitted equation error. Model validation then uses physical and statistical tests [53] . In [30] the economization formula known in approximation theory is taken in addition to the classical coefficients estimation for power estimation from the same measured data set as well. This is performed by statistical multi-hypothesis testing: consideration of the model complexity. The results are encouraging, but the attempt to include additional information, as done in the Bayesian approach, leads to unsatisfactory results [41] . In addition the balanced accuracy has been lacking up to now. And, of course, the regularization of the ill-posedness of the inverse problem remains unsolved.

In order to overcome these problems the Tikhonov-Phillips regularization is applied in connection with the multi-hypothesis testing [37] . In a first approach the economization formula for power estimation is included in the used aim functional as a penalty term, and the regularization parameter is determined by the estimated bound of the measurement errors. In a second approach the regularized normal solution is evaluated and the regularization parameter is determined without using information about the measurement errors explicitly (cross-validation). The results confirm those in [30] .

**Table 1.** Regularized estimates using the weighted cross-validation method

| $N$ | $\overset{\triangle}{\gamma}$ | | $\sigma_a(n)$ | | $\Delta a_1$ | $\Delta a_2$ | $\Delta a_3$ | $\Delta a_4$ |
|---|---|---|---|---|---|---|---|---|
| $s = 10^{-8}$ | | | | | | | | |
| 3 | 3.1 | $10^{-4}$ | 3.4 | $10^{-2}$ | .2386 | .2168 | | |
| 5 | 2.2 | $10^{-10}$ | 4.8 | $10^{-7}$ | .0009 | .0015 | .016 | |
| 7 | 2.2 | $10^{-16}$ | 2.0 | $10^{-9}$ | - | - | - | - |
| $s = .0005$ | | | | | | | | |
| 3 | 3.1 | $10^{-4}$ | 3.4 | $10^{-2}$ | .2386 | .2168 | | |
| 5 | 1.3 | $10^{-7}$ | 4.1 | $10^{-6}$ | .0012 | .0017 | .017 | |
| 7 | 2.9 | $10^{-7}$ | 1.3 | $10^{-5}$ | .0003 | .0007 | .009 | .6 |
| $s = .005$ | | | | | | | | |
| 3 | 3.5 | $10^{-4}$ | 3.4 | $10^{-3}$ | .238 | .2166 | | |
| 5 | 1.3 | $10^{-5}$ | 3.4 | $10^{-4}$ | .0009 | .0002 | .011 | |
| 7 | 5.5 | $10^{-5}$ | 5.5 | $10^{-5}$ | .0283 | .0409 | .413 | 25.0 |
| $s = .01$ | | | | | | | | |
| 3 | 4.4 | $10^{-4}$ | 7.0 | $10^{-2}$ | .239 | .217 | | |
| 5 | 5.4 | $10^{-5}$ | 4.1 | $10^{-3}$ | .0026 | .0026 | .02 | |
| 7 | 2.9 | $10^{-4}$ | 15.14 | | .2386 | .3642 | 3.78 | 231.0 |
| $s = .05$ | | | | | | | | |
| 3 | 4.1 | $10^{-3}$ | 4.0 | $10^{-2}$ | .217 | .2086 | | |
| 5 | 5.7 | $10^{-3}$ | 58.3 | | .5097 | .4452 | .01 | |
| 7 | 3.5 | $10^{-3}$ | 35.6 | | .3814 | .6641 | 5.48 | 479.0 |

*Example 2.* The equally distributed mean-free measurement errors are controlled by the scalar $s \in [0,1]$ . The regularization parameter is estimated with the assumed power $N$ for the iterative procedure.It is designated by $\overset{\triangle}{\gamma}$ . The mean value of the standard deviations of the polynomial coefficients is designed by $\sigma_a(N)$ and the deviations of the coefficient estimates from the exact values are given by $\Delta a_i$ . The resulting values for the pre-given polynomial

$$P_7(x) := x(1 + 0.5x^2 - 0.01x^4 + 0.00001x^6) \qquad (32)$$

are summarized in Table 1 . It can be concluded, that except for working with the exact values, $\gamma$ can be chosen independently of $N$ for the relevant values.

# 5 Conclusions

Inverse problems of modelling (design), system identification and optimization are often ill-posed. Therefore they need regularization. This means enlargement of the solution domain in order to achieve the existence of a unique solution. Generalized solutions are considered here. But this is not sufficient, and especially when dealing with erroneous and incomplete measurements, additional measures have to be introduced which are mainly operator modifications based on additional information.

Various regularization methods are discussed from a mathematical point of view as well as from the applicational point of view. A literature review is presented within system identification and some new results are mentioned.

Applying regularization within system identification is the correct step towards for solving the problems, however, much research still is needed to overcome the difficulties.

# References

1. Agbabian, M.S.; Masri, S.F.; Miller, R.K.; Caughy, T.K.: A System Identification Approach to the Detection of Changes in Structural Parameters; in Natke, Yao 1988, 341-356
2. Brandon, J.A.; Cremona, C.F.: Singular Value Decomposition: Sufficient, But Not Necessary; IMAC '90, 1990, 1376-1380
3. Baumeister, J.: Stable Solution of Inverse Problems, Vieweg Advanced Lectures in Mathematics, Friedr. Vieweg & Sohn, Braunschweig, Wiesbaden, 1987
4. Ben-Haim, Y.: Adaptive Diagnosis of Faults in Elastic Structures by Static Displacement Measurement: The Method of Selective Sensitivity, Mechanical Systems and Signal Processing, Vol.6, No.1, January 1992, 85-96
5. Ben-Haim, Y.; Prells, U.: Selective Sensitivity in the Frequency Domain, Part I: Theory; Mechanical Systems and Signal Processing; to appear in 1992
6. Chen, J.-C.; Garba, J.A.: Structural Damage Assessment Using a System Identification Technique; in Natke, Yao 1988,474-492
7. Cottin, N.: Parameterschätzung mit Hilfe des Bayesschen Ansatzes bei linearen elastomechanischen Systemen; Forschungsbericht, CRI-F-2/1983, Curt-Risch-Institut, Universität Hannover, 1983
8. Cottin, N.: Optimale Versuchsauslegung für die Identifikation elastomechanischer Systeme; Forschungsbericht CRI-F-1/1987, Curt-Risch-Institut, Universität Hannover, 1987
9. Cottin, N.: On the Optimum Experimental Design for the Parametric Identification of Linear Elastomechanical Systems; Proceedings of the European Conference on Structural Dynamics, EURODYN '90, 5.-7. June 1990, Bochum, in: Structural Dynamics, Eds. Krätzig et al., A. A. Balkema, Rotterdam, Brookfield, 1991
10. Cottin, N.;Felgenhauer, H.-P.; Natke, H.G.: On the Parameter Identification of Elastomechanical Systems Using Input and Output Residuals; Ing. Arch. 54(1984), 378-387
11. Cottin, N.; Natke, H. G.: On the Parameter Identification of Elastomechanical Systems Using Weighted Input and Modal Residuals; Ing. Arch. 56 (1986), 106-113.

12. Deif, A.: Sensitivity Analysis in Linear Systems; Springer-Verlag, Berlin, Heidelberg, New York, 1986

13. Eykhoff, P.: System Identification- Parameter and State Estimation; John Wiley & Sons, London, New York, Sydney, Toronto,1974

14. Gawronski, W.; Natke, H.G.: Balancing Linear Systems; Int. J. Sci. 18 (1987), 237-249

15. Isenberg, J.: Progressing from Least Squares to Bayesian Estimation; J.H. Wiggins Co., Redondo Beach, CA, ASME Paper No. 79-WA/DSC-16,1981

16. Kuhnert, F.: Pseudoinverse Matrizen und die Methoden der Regularisierung; Teubner-Texte zur Mathematik, BSB B.G Teubner Verlagsgesellschaft, Leipzig 1976

17. Lallement, G.; Piranda, J., Fillod, R.: Parametric Identification of Conservative Self Adjoint Structures; Proc. Int. Conf. Spacecraft Structures and Mechanical Testing ESA SP-289, ESA/Estec, Nordwijk, The Netherlands, 1989

18. Lallement, G.; Cogan, S.: Reconciliation Between Measured and Calculated Dynamic Behaviors: Enlargement of the Knowledge Space; IMAC X, San Diego, CA, 1992, 487-493

19. Link, M.; Badenhausen, K.: Identification and Dynamic Condensation of Physical System Matrices Using Incomplete Dynamic Response Data; $2^{nd}$ Int. Symposium on Aeroelasticity and Structural Dynamics, April 1-3, 1985; DGLR-Report 85-02, pp. 536-545.

20. Link, M.; Zhang, L.: Experience with Different Procedures for Updating Structural Parameters of Analytical Models Using Test Data; Proc. IMAC X, San Diego, CA, 1992

21. Louis, A.K.: Inverse und schlecht gestellte Probleme; Teubner Studienbücher Mathematik, B.G. Teubner, Stuttgart, 1989

22. Louis, A.K.: Numerik inverser Probleme; GAMM-Mitteilungen 1990, 1, 5-27

23. Morozov, V.A.: Methods for Solving Incorrectly Posed Problems; Springer-Verlag, New York, Berlin, Heidelberg, Tokyo, 1984

24. Mottershead, J.E.; Foster, C.D.: On the Treatment of Ill-conditioning in Spatial Parameter Estimation from Measured Vibration Data; Mechanical Systems and Signal Processing 5,2, 1991, 139-154

25. Nalitolela, N.G.: A New Approach to Update Model Parameters Using the Frequency Response Data; IMAC X, San Diego, CA, 1992, 1267-1273

26. Natke, H.G. (Ed.): Identification of Vibrating Structures; CISM Courses and Lectures No. 272, Springer-Verlag Wien, New York 1982

27. Natke, H.G.: Survey on Parameter Estimation within System Identification Using a priori-knowledge of System Analysis; in: Computational Methods and Measurements (Eds. G.A. Keramidas, C.A. Brebbia), Proc. of the Int. Conference, Washington,DC, July 1982, Springer-Verlag Berlin, Heidelberg, New York, and Computational Mech. Centre, Southampton, 17-27

28. Natke, H.G.: Updating Computational Models in the Frequency Domain Based on Measured Data: A Survey; Probabilistic Engineering Mechanics, 3, 1988, 1, 28-35

29. Natke, H.G. (Ed.): Application of System Identification in Engineering; CISM Courses and Lectures No. 296, Springer-Verlag Wien, New York, 1988

30. Natke, H.G.: Survey on the Identification of Mechanical Systems; in: Road-Vehicle-Systems and Related Mathematics (Ed. H. Neunzert), B.G. Teubner, Stuttgart and Kluwer Academic Publ. 1989, 69-116

31. Natke, H.G.: Identification Approaches in Damage Detection and Diagnosis; IX. Szkoła, Diagnostics '89, Rydzyna, Poland, Aug. 27- Sept. 1, 1989, 99-110

32. Natke, H.G.: Error Localization within Spatially Finite-dimensional Mathematical Models, A Review of Methods and the Application of Regularization Techniques; Computational Mechanics (1991) 8, 153-160

33. Natke, H. G.: Einführung in die Theorie und Praxis der Zeitreihen- und Modalanalyse; Friedr. Vieweg & Sohn, Braunschweig/Wiesbaden, 3. verbesserte und erweiterte Ausgabe 1992

34. Natke, H: G., Cempel, C.: Fault Detection and Localization in Structures: A Discussion, Mech. Syst. and Signal Processing 5(5), 1991, 345-356

35. Natke, H.G.; Collmann, D.; Zimmermann, H.: Beitrag zur Korrektur des Rechenmodells eines elastomechanischen Systems anhand von Versuchsergebnissen; VDI-Berichte Nr. 221, 1974, 23-32

36. Natke, H.G.; Danisch, R.; Delinic, K.: Condensation Methods for the Dynamic Analysis of Large Models; Nuclear Engineering and Design 111 (1989), 265-271

37. Natke, H.G.; Prells, U.: A Contribution to Structure Identification of Nonlinear Mechanical Systems, EUROMECH 280, Oct. 29-31, 1991, Lyon, France

38. Natke, H.G. and Yao, J.T.P. (Eds.): Structural Safety Evaluation Based on System Identification Approaches; Vieweg-Verlag Braunschweig, Wiesbaden, 1988

39. Natke, H.G. and Yao, J.T.P.: System Identification Methods for Fault Detection and Diagnosis; in: Structural Safety and Reliability (Eds. A.H.-S. Ang, M. Shinozuka, G.I. Schuëller), Proc. of ICOSSAR '89, ASCE New York, 1990, 1387-1393

40. Natke, H. G.; Zamirowski, M.: On Methods of Structure Identification for the Class of Polynomials within Mechanical Systems; ZAMM Zeitschrift für Angewandte Mathematik und Mechanik 70(1990) 10,415-420

41. Natke, H. G.; Zhang, J. H.: The Application of Modal Transformation as a Tool for Handling Large Models in Updating Procedures; in: Computers and Experiments in Stress Analysis; Eds.: G. M. Carlomagro, C. A. Brebbia; Proc. $4^{th}$ Int. Conf. on Computational Methods and Experimental Measurements, Capri, Italy, May 1989, Computational Mechanics Publ., Springer-Verlag Berlin, Heidelberg, New York, London, Paris, Tokyo, 1989, 335-346.

42. O'Callahan, J.: Determination of Analytical Model Differences Using Modal Updating; IMAC '90, 1990, 1180-1189

43. Ojalvo, I.U.: Efficient Solution of Ill-conditioned Equations Arising in System Identification; IMAC '90, 1990, 554-558

44. Ojalvo, I.U.: Improved Solution for Ill-conditioned Algebraic Equations by Epsilon Decomposition; AIAA Journal 29 (1991)12, 2274-2277

45. Prells, U.: Regularisierte Modellfehlerlokalisierungen; Forschungsbericht des Curt-Risch-Instituts der Universität Hannover, CRI F-2/91, 1991.

46. Prells, U.; Ben-Haim, Y.: Selective Sensitivity in the Frequency Domain, Part II: Applications; Mechanical Systems and Signal Processing, 1992 to appear

47. Schwetlick, H.: Nichtlineare Parameterschätzung: Modelle, Schätzkriterien und numerische Algorithmen; GAMM-Mitteilungen 1991,2, 13-51

48. Söderström, T.; Stoica, P.: System Identification, Prentice Hall International Series in System and Control Engineering, Series Ed.: M.J. Grimble, Prentice Hall New York, London, Toronto, Sydney, Tokyo, 1989

49. Tikhonov, A.N.; Arsenin, V.Y.: Solution of Ill-posed Problems; Wiley, New York,1977

50. Vestroni, F. and Capecchi, D.: Use of Different State Variables in Nonparametric Identification of Hysteretic Systems; EUROMECH 280, Oct. 29-31,1991, Lyon, France

51. Yao, J. T. P.: Safety and Reliability of Existing Structures; Pitman Advanced Publ. Program, Boston, London, Melbourne, 1985
52. Yao, J. T. P.; Natke, H. G.: Uncertainties in Structural Identification and Control; Proc. Int. Fuzzy Eng. Symposium '91, Yokohama, Japan, Nov. 13-15, 1991, 844-849
53. Zamirowski, M.: Einige zeitdiskrete Parameterschätzmethoden zur Identifikation nichtlinearer mechanischer Systeme, Bericht CRI-F-1/1992 des Curt-Risch-Institutes der Universität Hannover, 1992
54. Zhang, L.; Ojalvo, I.U.: An Improved Epsilon Decomposition Approach for Structural System Identification; IMAC X, San Diego, CA, 1992, 591-594
55. Zhang, J. H.; Natke, H. G.: A Two-level Updating Procedure of the Component Mode Synthesis Model; Mechanical Systems and Signal Processing, 5, 1991, 501-514

This article was processed using the LATEX macro package with LMAMULT style

# New Approaches to the Optimal Regularization

Takashi Kitagawa*    Yosuke Hosoda**

*Institute of Information Sciences and Electronics
University of Tsukuba, Tsukuba 305, Japan

**Department of Information Engineering, Faculty of Engineering
Nagoya Unversity, Nagoya 464, Japan

### Abstract

This paper introduces two new approaches to determine the optimal parameter in the mehod of regularization. One is based on the error analysis made in [4] and [5]. The other is based on, what is called in [2], L-curve, which is formulated and analyzed in [3].

## 1   Introduction

One of the most important problems in approximating the solution of a linear ill-posed problems by the method of regularization resides in the selection of the optimal regularization parameter. we present new two approaches to the optimal regularization.

We consider the ill-conditioned linear systems arising from Fredholm integral equations of the first kind of the form

$$\int_a^b k(s,t)\hat{f}(t)dt = \hat{g}(s), \qquad s_{min} \leq s \leq s_{max}, \tag{1}$$

where $K(s,t)$ and $\hat{g}(s)$ are known $L_2$ functions and $\hat{f}$ is the unknown function in $L_2[a,b]$. This equation is known to be an ill-posed problem in the sense that $\hat{f}$ dose not depend on $\hat{g}$ continuously, namely, any small perturbation in $\hat{g}$ results in arbitrarily large change in $\hat{f}$. Via some discretization process [1, 7], one can reduce (1) to the equation

$$T f = g, \tag{2}$$

with $f = (f_1, f_2, ..., f_n) \in R^n$ , $g = (g_1, g_2, ..., g_m) \in R^m$ and $T : R^n \mapsto R^m$.

The ill-posedness of (1) results from the fact that the operator $\hat{T}$ which is the integral operator in (1) dose not have a bounded inverse, which in turn, implies that the condition number of the matrix T increases rapidly as m and n increase. Consequently, any attempts to solve (2) by a conventional least squares method may produce disastrous results. A number of methods are available to mitigate the effect of this ill-conditioning. Best known of them are the truncation of the singular value decomposition and the method of regularization.

## 2 Optimal regularization

The method of regularization solves the related well-posed problem of minimizing a smoothing functional. In other words:

*For given $g_\Delta = g + \Delta g \in R^m$, find $f = f(\mu, \Delta g) \in R^n$ and $\mu \in [0, \infty)$ for which*

$$\min_{f \in R^m} \{ \|T f - g_\Delta\|^2 + \mu \|f\|^2 \} \tag{3}$$

*is attained.*

The parameter $\mu$ is called the regularization parameter, which controls the tradeoff between the stabilty of the system (3) and the fidelity to the original equation. This technique is known to be very successful in practice, provided that the optimal value of the regularization parameter $\mu$ is determined appropriately [1, 4, 7].
We set, for further use,

$$e(\mu; \Delta g) = T^\dagger g - f(\mu; \Delta g), \tag{4}$$

where $T^\dagger$ denotes the Moore-Penrose generalized inverse of T and $f(\mu; \Delta g)$ represents the minimizer of the smoothing functional (3).

We define the optimal regularization parameter as follows.

**Definition 1** *We call $\mu_o$ the optimal regularization parameter if*

$$\mu_o \in \{\bar{\mu} | \min_{\mu \in [0,\infty)} \|e(\mu)\| = \|e(\bar{\mu})\|\}. \tag{5}$$

Hereafter we may write $f(\mu) = f(\mu; \Delta g)$, etc. for simplicity.

# 3   New approaches to the optimal regularization

We present the following two new approaches to this problem:

1) The first approach is by introducing a function to determin the optimal parameter. The method chooses the value of $\mu$ for which

$$\min_{\mu \in P_\sigma} \zeta(\mu) \quad with \quad \zeta(\mu) = \|\frac{d}{d\xi} f(\mu; \Delta g)\| \tag{6}$$

is attained, where $P_\sigma$ is the set of singular values of $T^t T$ and $\xi = \log \mu$. We monitor the values of the function $\zeta(\mu)$ among the values of $\sigma_i^2$ 's, where $\sigma_i$, $i = 1, 2, ..., n$, are singular values of T. Then we employ the value of $\mu$ which gives the minimum of $\zeta(\mu)$. Namely, one advantage of this method is that the number of the evaluation of the function is at most n. The theoretical aspect which explains why this method works out well is discussed in [4] and the practical numerical algorithm together with some numerical experiments are given in [5].

To estimate $\mu_o$, we introduce two vector valued functions $\tau(\mu)$ and $\eta(\mu)$ which defined by $\tau(\mu) = e(\mu; 0)$ and $\eta(\mu) = e(\mu; 0) - e(\mu; \Delta g)$. We call $\tau(\mu)$ theoretical error vector and $\eta(\mu)$ computational error vector. Then $\|\tau(\mu)\|$ is monotone increasing and $\|\eta(\mu)\|$ monotone decreasing with respect to $\mu > 0$. The outline of the theory to estimate $\mu_o$, developed in [4] is as follows.

i) Since $\|\tau(\mu)\|$ and $\|\eta(\mu)\|$ are monotone increasing and decreasing respectively, there are regions in which $\|\tau(\mu)\| > C\|\eta(\mu)\|$ and $\|\tau(\mu)\| < C\|\eta(\mu)\|$ for some $C > 1$. we call the former the theoretical error dominant region $\Omega_\tau$ and the latter the computation error dominant region $\Omega_\eta$.

ii) We can show that $\|e(\mu)\|$ is is nearly equal to $\|\tau(\mu)\|$ in $\Omega_\tau$ and $\|e(\mu)\|$ is nearly equal to $\|\eta(\mu)\|$ in $\Omega_\eta$ and the monotonicity of $\|\tau(\mu)\|$ and $\|\eta(\mu)\|$ inherits to $\|e(\mu)\|$ in each region. In other words, $\|e(\mu)\|$ is monotone increasing in $\Omega_\tau$, and monotone decreasing in $\Omega_\eta$. Thus the optimal parameter $\mu_o$ lies in the optimal region $\Omega_o \equiv [0, \infty) \backslash (\Omega_\tau \cup \Omega_\eta)$.

iii) Since the optimal parameter $\mu_o$ satisfies

$$\frac{d}{d\mu}\|e(\mu)\|\,|_{\mu=\mu_o} = 0,$$

it seems natural to estimate $\mu_o$ by minimizing some upper bound of

$$\left|\frac{d}{d\mu}\|e(\mu)\|\right|$$

One of the upper bounds for it involves $\|df(\mu)/d\mu\|$.

iv) We introduce a function

$$\zeta : \mu \longrightarrow \left\|\frac{d}{d\xi}f(\mu)\right\|, \quad \xi = \log\mu$$

to estimate the optimal regularization parameter $\mu_o$. Then for any $\mu \in [0, \infty)$, we have

$$\left|\frac{d}{d\xi}\|e(\mu)\|\right| \leq \left\|\frac{d}{d\xi}f(\mu)\right\|$$

We expect that the minimizer of $\zeta$ is close to the optimal regularzation parameter $\mu_o$.

v) Though we cannot assert that the minimizer of $\zeta(\mu)$ coincide with $\mu_o$, under some conditions the minimizer among $\mu \in P_\sigma$, where $P_\sigma$ is the set of singular values of $T^tT$, is in $\bar{\Omega}_o$, the extersion of $\Omega_o$ to the closest singular values of $T^tT$.

We note here that the variable transformation $\xi = \log\mu$ and observing the function $\zeta(\mu)$ among $P_\sigma$ are crucial for this method. A comparison with other conventional methods like generalized crossvalidation is made in [6].

2) The second approach uses the notion of L-curve which is termed by [2] and is defined as the graph of

$$(\|r_\mu^\Delta\|, \|f(\mu)\|) \quad with \quad r_\mu^\Delta = Tf(\mu) - g_\Delta \tag{7}$$

which is parametrized by $\mu$. The name of L-curve comes from the numerical obserbation that the graph (7) has a steep bend in its middle and it looks like L.

It can be shown that for $\mu \in (0, \infty)$ $\|f(\mu)\|$ is monotone decreasing with respect to $\|r_\mu^\Delta\|$ [3]. At the corner of the L-curve, it is heuristically observed that $\|r_\mu^\Delta\| = \|\Delta g\|$

and $\|f(\mu)\| = \|f(0;0)\|$. We note that by the discrepancy principle [1] some upper bound of the error is minimized when $\|r_\mu^\Delta\| = \|\Delta g\|$. From this observation we expect that the corner of the L-curve gives a good estimation for the optimal regularization parameter $\mu_o$.

The maximizer of the curvature of the L-curve is employed as the optimal parameter. The formulatin of this method with numerical examples is given in [3]. The explicit expression of the curvature denoted by $\kappa(\mu)$ using the singular system, which is not simple but still we can compute, is given as follows;

$$\kappa(\mu) = \frac{1}{(\|r_\mu^\Delta\|^2 + \mu^2\|f(\mu)\|^2)^{\frac{3}{2}}} \times$$
$$\left| \frac{\|r_\mu^\Delta\|^2\|f(\mu)\|^2(\Sigma_1(\mu) + 3\mu\Sigma_2(\mu))}{\Sigma_3(\mu)^2} - \mu(\mu\|r_\mu^\Delta\|^2 + \|f(\mu)\|^2) \right| \quad (8)$$

where

$$\Sigma_1(\mu) \equiv \sum_{i=1}^{k} \frac{\sigma_i^2(\sigma_i^2 - 2\mu)}{(\sigma_i^2 + \mu)^4}(u_i, g_\Delta)_m^2 \quad (9)$$

$$\Sigma_2(\mu) \equiv \sum_{i=1}^{k} \frac{\sigma_i^2}{(\sigma_i^2 + \mu)^4}(u_i, g_\Delta)_m^2 \quad (10)$$

$$\Sigma_3(\mu) \equiv \sum_{i=1}^{k} \frac{\sigma_i^2}{(\sigma_i^2 + \mu)^3}(u_i, g_\Delta)_m^2 \quad (11)$$

with $u_i$'s are the left singular vectors of $T$ and $k = rank(T)$. It should be noted here that some intuitive explanation is possible for this method [3, 2], but no rigorous theoretical foundation is available as to why this technique works out well in practice.

# 4   Concluding remarks

It is observed through numerical experiments that the function $\kappa(\mu)$ has a preferable feature of convexity and it looks easy to seek the global minimum as compared to the method (2). This feature, however, is not verified theoretically, because its representation is not simple enough to analyze. While $\zeta(\mu)$ looks wavy and finding the global minimum does not seem easy task. It can be shown that the local extrema are attained at the points of $P_\sigma$ and we can identify the global minimum by monitoring the points [4, 5].

# References

[1] Groetsche, C. W., *The Theory of Tikonov Regularization for Fredholm Integral Equation of the First Kind*, Pitman, Boston, 1984.

[2] Hansen, P. C.,*Analysis of discrete ill-posed problems by means of the L-curve*, preprint 1990.

[3] Hosoda, Y. and Kitagawa, T.,*Optimal regularization for ill-posed problems by means of the L-curve*, Trans. of JSIAM, 2(1992), (in Japanese).

[4] Kitagawa, T.,*A deterministic approach to the optimal regularization - the finite dimensional case -*, Japan J. of Appl. Math. 4(1987), pp.371-379

[5] Kitagawa, T.,*A numerical method to estimate the optimal regularization parameter*, J. of Info. Proc., 11(1988), pp.263-270

[6] Kitagawa, T.,*Methods in estimating the optimal regularization parameters*, in Inverse Problems in Mathematical Engineering, M. Yamaguti et.al. eds., Springer Verlag(1991)pp.

[7] Nashed, M. Z.,*Operator theoretic and computational approaches to ill-posed problems with applications to antenna theory*, IEEE Trans. on Antennas and Propagat.,29(1981), pp.220-231.

# A Method for Solving Inverse Boundary-Value Problems Using Symbolic Computation

Kiyoshi Yoda

Central Research Laboratory, Mitsubishi Electric Corporation
8-1-1 Tsukaguchi-Hommachi, Amagasaki 661, Japan
yoda@ele.crl.melco.co.jp

Summary

An approach to inverse finite-element-analysis has been demonstrated using symbolic computation language, Mathematica. The forward calculation of two-dimensional finite elements is performed with symbolic shape parameters. This results in symbolically-expressed solutions such as potential distributions  as functions of the shape variables. Because the calculated distributions are explicitly given as functions of the shape parameters, the shape optimization can be directly achieved using the resulting symbolic expressions. Preliminary experimental results aiming at shape optimization are shown.

## Introduction

Finite element or boundary element analysis is a powerful method for analyzing complicated systems, which is usually achieved by numerical calculation. Some researchers employed symbolic computation to derive matrices for the following numerical matrix calculations, thereby simplifying the matrix-element derivation[1-2]. On the other hand, a totally-symbolic computation version for solving boundary value problems was proposed[3]. In this case, shape or material parameters were expressed symbolically and the matrix calculation was performed in a symbolic way. An advantage of the totally-symbolic version lies in the ability of direct inverse-problem calculation, and a one-dimensional finite element calculation aiming at material property optimization  was already demonstrated[3]. In this paper, I have extended the previous one-dimensional results to two-dimensional finite element models, and shown a simple example of shape optimization.

## Schematic comparison of the symbolic approach and conventional numerical methods

The idea could be best understood by the following two diagrams. Figure 1 shows a conventional numerical approach to inverse boundary-value problems, where the calculation is initiated by the numerical input parameters such as initial sets of shape or material unknowns; subsequently, output quantities such as potential distributions are numerically calculated by solving matrix equations. Then comparison of the calculated quantities and the target or desired quantities revises the input parameters, and the loop calculation continues until we meet an appropriate convergence.

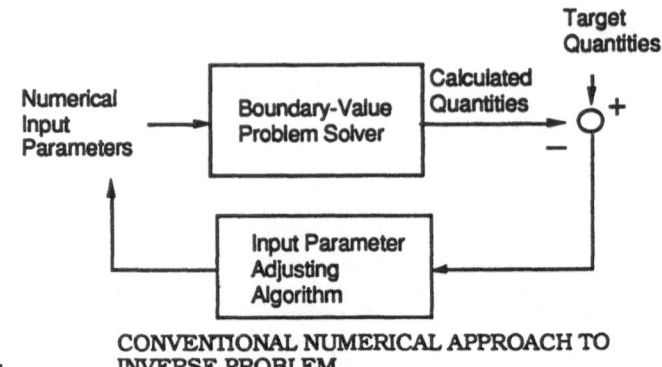

Fig. 1    CONVENTIONAL NUMERICAL APPROACH TO INVERSE PROBLEM

On the other hand, Fig. 2 depicts the proposed symbolic approach to inverse boundary-value problems. In this case, the input parameters, $r_1$, $r_2$,...,$r_n$, are given symbolically; then, the calculated symbolic quantities, $V_k(r_1, r_2,..., r_n)$, are obtained by the symbolic matrix solvers, where k corresponds to each spatial point or nodal number. The direct comparison of the output symbolic expressions $V_k(r_1, r_2,..., r_n)$ and the target quantities $V_{kd}$ leads to the desired unknown parameters, $r_1$, $r_2$,..., $r_n$. This can be done either by the squared residual-function minimization using gradient methods or by the point matching technique using the Newton's or its related methods.

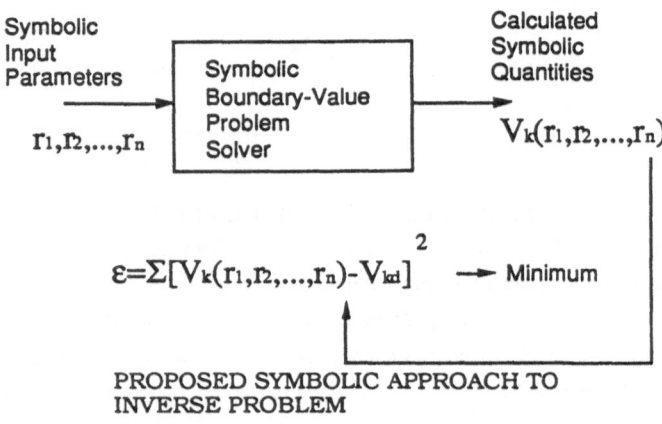

$$\varepsilon = \sum [V_k(r_1, r_2,...,r_n) - V_{kd}]^2 \longrightarrow \text{Minimum}$$

Fig. 2    PROPOSED SYMBOLIC APPROACH TO INVERSE PROBLEM

## Examples of shape optimization

Figure 3 shows a two-dimensional oval-shaped resistor model. The major radius of the oval resistor is symbolically expressed as 'a', and the minor radius is 2, while σ represents the resistor conductivity. A DC voltage is applied to the both ends of the oval major axis.

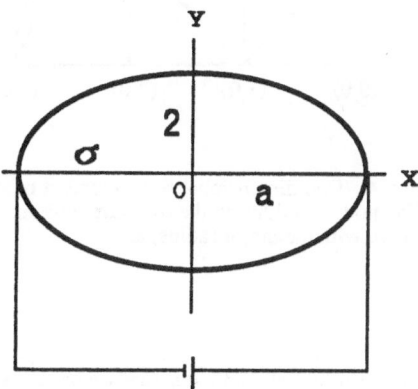

Fig. 3 Two-dimensional thin film resistor model. The major radius, a, of the oval resistor is parametrically given, and the minor radius is 2, while σ represents the resistor conductivity.

Figure 4 shows the corresponding two-dimensional finite element model. The node positions on the oval circumference associatively move with the major radius, a. Only a fourth of the oval has been considered due to its symmetry. The total number of the elements was only 11 for simplicity. Using a symbolic computation language, Mathematica, the forward finite element calculation has been done in a symbolic way. The resulting node voltages are given as functions of the input variable, a. Thus the shape parameter corresponding to each design goal can be solved by using the residual function minimization or the Newton's method. For example, let us set the applied voltage of Fig. 1 to 1V, and the desired node voltage at the point (2,0) to 0.48 V. Using the Newton's method, the resulting major radius to meet the target design was 3.0.

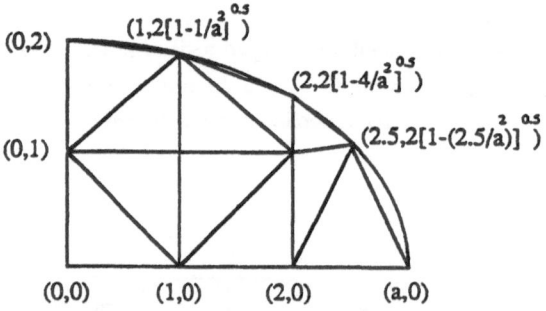

Fig. 4   Two-dimensional finite element model.
The node positions on the oval circumference associatively
move with the major radius, a.

## Discussion

One of the advantages of the proposed method is that we have a general expression to meet each target design. Therefore we can iteratively use the final nonlinear design equation whenever we have different design goals. Another advantage may be the symbolic associativity shown in Fig. 4; in other words, we can add certain symbolic rules to the elements defined. Extensive work on this procedure continues towards several different boundary value problems, including finite element, boundary element and moment methods. The applications also cover identification of unknown internal structures, such as reconstruction of impedance distribution in a conductive body. The details of each work will be published elsewhere.

## Acknowledgment
The author is grateful to Drs. S. Sakabe and T. Yamada for their guidance and encouragement.

## References
1.   W. Luft, J. M. Roesset and J. J. Connor, 'Automatic generation of finite element matrices,' J. Structural Div. ASCE 97, 349-362 (1971)
2.   A. K. Noor and C. M. Andersen, 'Computerized symbolic manipulation in structural mechanics -- progress and potential,' Comput. Structures 10, 95-118 (1977)
3.   K. Yoda,' Forward and inverse boundary value solver using symbolic computation,' Proc. ISEM Nagoya (1992)

# An Application of the Fuzzy Theory for an Ill-Posed Problem

Hitoshi Imai[†]
Akira Sasamoto[††]
Hideo Kawarada[†††]
Makoto Natori[†]

[†] Institute of Information Sciences and Electronics,
University of Tsukuba, Ibaraki 305, Japan
[††] Mechanical Engineering Laboratory, Tukuba , Ibaraki 305 , Japan
[†††] Faculty of Engineering, University of Chiba, Chiba 260, Japan

## 0. ABSTRACT

Ill-posed problems are usually solved by the transformation to minimization problems. They are ill-conditioned, then additional techniques, i.e. regularizations, are adopted to avoid the oscillation. In practical problems they are usually so complicated that it is not easy to adopt such effective techniques. Here two points should be focused on for practical problems. First, high accuracy is not necessary. Second, engineers who have much experience about the problems know how to deal with them qualitatively. These points suggest the validity of flexible minimizers. In this paper the fuzzy theory is introduced to construct such minimizers and it is applied to an ill-posed shape design problem. Numerical results are satisfactory.

## 1. INTRODUCTION

Ill-posed problems are usually solved by the transformation to minimization problems. They are ill-conditioned then additional techniques should be adopted theoretically to avoid the oscillation. A lot of theoretical approaches have been developed[4,13].

However, practical situation is different. Practical problems are nonlinear and very complicated, so it is not easy to choose one suitable technique corresponding to them. Even you can choose such a technique, you might fail in real analysis due to theoretically unpredictable errors. In such a case theory rarely give you any useful information. Even you have enough theoretical knowledge, situation would be same. Moreover, almost analyzers of practical problems are not mathematicians, so they do not have enough theoretical knowledge. In a given term they can hardly solve problems properly from the mathematical view point. However, e.g. in companies, they are obligated to obtain some solutions in a given term.

Here two points should be focused on for practical problems. First, high accuracy is not necessary. Second, engineers who have much experience about the problems know how to deal with them qualitatively.

Thus AI(Artificial Intelligence) approaches or expert systems will be valid for practical problems[5,14]. In such systems the fuzzy theory is often adopted, because it is easy to reflect the experience of experts to the system[10,15].

Here we construct a simple minimizer for an ill-posed shape design problem using the fuzzy theory. In consideration of the practical situation we are assumed to not be specialists in ill-posed problems but know the fact that these problems involve the oscillation phenomenon.

## 2. Ill-Posed Shape Design Problem

Plasma is confined in the vessel by the magnetic field. The shape of the plasma is determined under the balance of the pressure between the plasma and the magnetic field. This magnetic field is determined by the current distribution outside the vessel. The problem of the determination of the plasma shape for the given current distribution becomes a free boundary problem[11]. In a simple MHD model the current distribution is replaced by the vessel shape. In this model the plasma shape is determined by the vessel shape[1,2].

We consider the following inverse problem. In this problem the vessel shape is determined by the plasma shape. This problem is related to the controllability of plasma. It is an initial value problem of the Laplace operator, so it is ill-posed[8].

_Problem1_(Ill-posed Shape Design Problem).

For a fixed boundary $\gamma_d$ and a positive constant $\kappa$, find a Jordan curve $\Gamma=\{(x, y)|\ u(x, y) = \kappa\ \}$ where

$$\Delta u = 0 \quad in \quad \Omega_{\gamma_d}, \tag{1}$$

$$u = 0 \quad on \quad \gamma_d, \tag{2}$$

$$\frac{\partial u}{\partial n} = \frac{4}{l_{\gamma_d}} \quad on \quad \gamma_d. \tag{3}$$

Both $\gamma_d$ and $\Gamma$ are assumed to be symmetric with respect to the both $x$ and $y$ axis. $\Omega_{\gamma_d}$ is the outer region of $\gamma_d$. $n$ is an inner normal unit vector on $\gamma_d$. $l_{\gamma_d}$ is the length of $\gamma_d$ (see Figure 1).

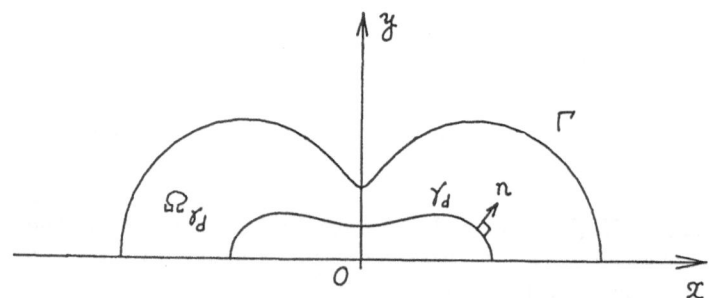

Figure 1. An ill-posed shape design problem.

If you are a specialist in ill-posed problems, then you will transform it to a minimization problem as follows.

*Problem2* (ILL-CONDITIONED SHAPE DESIGN PROBLEM).

For given positive constants $\varepsilon$, $\kappa$ and a given symmetric curve $\gamma_d(\theta)$, find a symmetric Jordan curve $\Gamma$ which is located outside $\gamma_d$ such that

$$J(\Gamma) = \int_0^{2\pi} (\gamma(\Gamma)(\theta) - \gamma_d(\theta))^2 d\theta \le \varepsilon \tag{4}$$

where $\gamma$ is a solution to the following free boundary problem.

*Free Boundary Problem* ( FBP ( $\Gamma$ ) ).

For a fixed boundary $\Gamma$ and a positive constant $\kappa$, find u and a boundary $\gamma$ such that

$$\Delta u = 0 \quad in \quad \Omega_{\Gamma,\gamma}, \tag{5}$$

$$u = 0 \quad on \quad \gamma, \tag{6}$$

$$\frac{\partial u}{\partial n} = \frac{4}{l_\gamma} \quad on \quad \gamma, \tag{7}$$

$$u = \kappa \quad on \quad \Gamma. \tag{8}$$

Here $\Omega_{\Gamma,\gamma}$ is a region between $\gamma$ and $\Gamma$. n is an inner normal unit vector on $\gamma$ (see Figure 2).

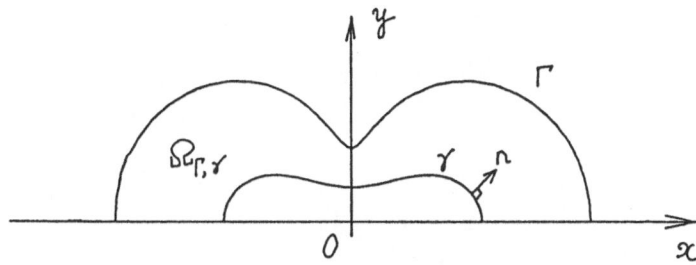

Figure 2. The free boundary problem.

The parameter $\epsilon$ is introduced considering practical situations. Problem2 is an ill-conditioned minimization problem. It is already solved by a mild minimizer with the quadratic approximation to the cost function in eq.(4)[9]. Parameters in this minimizer are so mathematical that it is not easy to control them in the occurrence of the oscillation. In short, this minimizer is not suitable for the interactive systems.

## 3. Our Approach Using Fuzzy Theory

We solve Problem1 along the following qualitative consideration.

(1) Take a vessel shape $\Gamma$, and solve FBP($\Gamma$). If its solution $\gamma$ is located inside $\gamma_d$, this means that the pressure of the magnetic field is too strong, or $\Gamma$ is too near to $\gamma_d$. Then take a new $\Gamma$ further from $\gamma_d$ than the last $\Gamma$ as follows. If the deviation, i.e. difference between the locations of $\gamma_d$ and $\gamma$, is large, then take a new $\Gamma$ largely further.

(2) The position of a part of $\gamma$ is expected to be mainly influenced from the nearest part of $\Gamma$. Then move a part of $\Gamma$ corresponding the nearest deviation.

(3) To avoid the oscillation, restrict the maximum movement of $\Gamma$. Also restrict the movement of $\Gamma$ if the second deviation, i.e. the deviation's deviation, is large.

(4) To enhance the convergence of the movement of $\Gamma$, restrict the movement of $\Gamma$ as in the bisection method if the sign of the deviation. This may be valid to avoid the oscillation.

These consideration is qualitative. Then we use the fuzzy theory to implement them on computers. For the simplicity, it is applied to the considerations (1)-(3). The basic idea is same as in[3]. Here we simply omit membership functions those mean "small" and we add in the reasoning multipliers to control the sensitivity of the reasoning to the second deviations( see Figure 3 ).

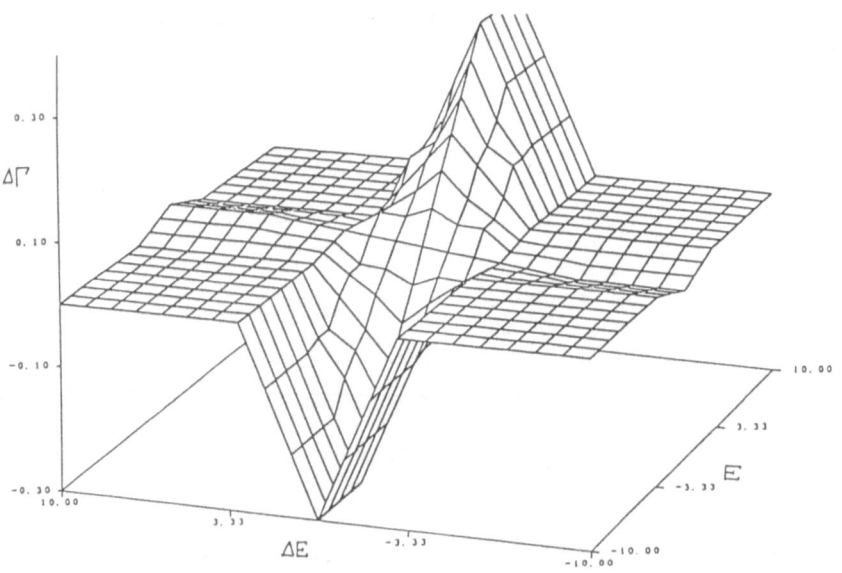

Figure 3. The profile of the output for the deviation($E$) and the second deviation($\Delta E$) :

Note that this approach to Problem1 is different from the usual least square method. The consideration (1) may be supported by the maximum principle.

We control $\Gamma$ by the r coordinates on the equi-$\theta$ lines those are distributed homogeneously around the origin. Then $\Gamma$ is generated in such a way that

$$\Gamma = \{(x,y)|x = R_N(\theta)\cos(\theta), y = R_N(\theta)\sin(\theta),$$
$$R_N(\theta) = P_0 + 2\sum_{m=1}^{\frac{N}{4}-1} P_{2m}\cos(2m\theta) + P_{\frac{N}{2}}\cos(\frac{N}{2}\theta)\}. \tag{9}$$

$\{P_j\}$ are determined by the discrete Fourier transform.

The determination of parameters in the fuzzy control is difficult, if you do not have some good test cases. However, in our system parameters are not so many and their roles are clear, so it was not difficult to determine them. They could be determined after a little trial and error.

Our algorithm is given as follows.

## *Algorithm*

Step 1. Fix parameters on discretization.
Step 2. Set values of parameters on control.
Step 3. Choose suitable starting values $\Gamma^{(0)}$ and $\gamma^{(-1)}$. Set n = 0.
Step 4. Choose $\gamma^{(n-1)}$ as a starting value and solve FBP $(\Gamma^{(n)})$ by a fixed domain method and Powell's method[7,9,12]. Set its solution $\gamma^{(n)}$ and $\Gamma'^{(n)}$.
Step 5. If the deviation is small enough then stop, else move $\Gamma'^{(n)}$ according to the controller and set this boundary $\Gamma^{(n+1)}$. Set n = n + 1 and go to step 4.
Step 6. If the oscillation occurs, then go to step 2.

## 4. NUMERICAL RESULTS

Numerical results are shown in Figures 4-7. Here $N = 24$, but essentially independent parameters of $\Gamma$ are 7 because of the symmetry. Data $\gamma_d$ used here are the solutions to the FBP($\Gamma_r$)[6]. Circles are chosen as starting values $\Gamma^{(0)}$ and $\gamma^{(-1)}$. Numerical results show that in spite of the simplicity the controller goes well.

## 5. CONSEQUENCES

The validity of the fuzzy theory to practical ill-posed problems is discussed and its application to an ill-posed problem is carried out here. A simple minimizer is constructed using the fuzzy theory and it is shown to go well numerically. This means that this approach, i.e. the expert system, is valid for practical ill-posed problems.

36

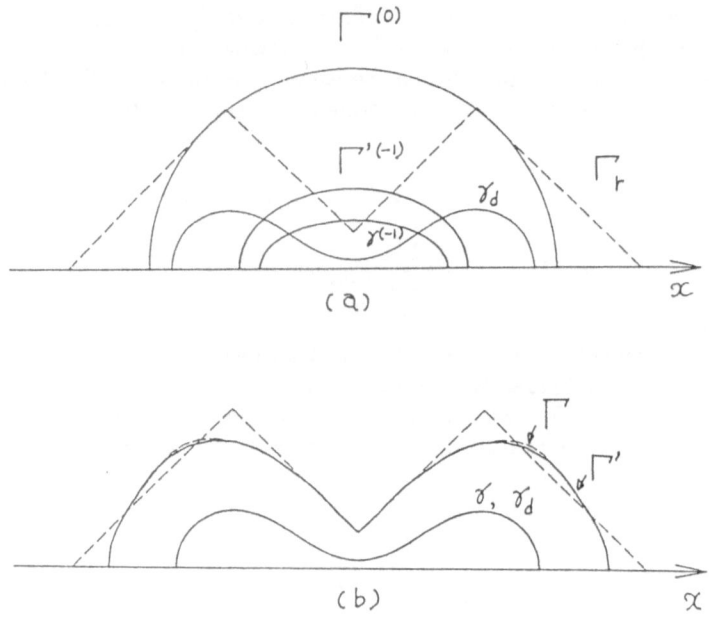

Figure 4. The shape design for $\kappa = 0.2$: (a)initial, (b)final.

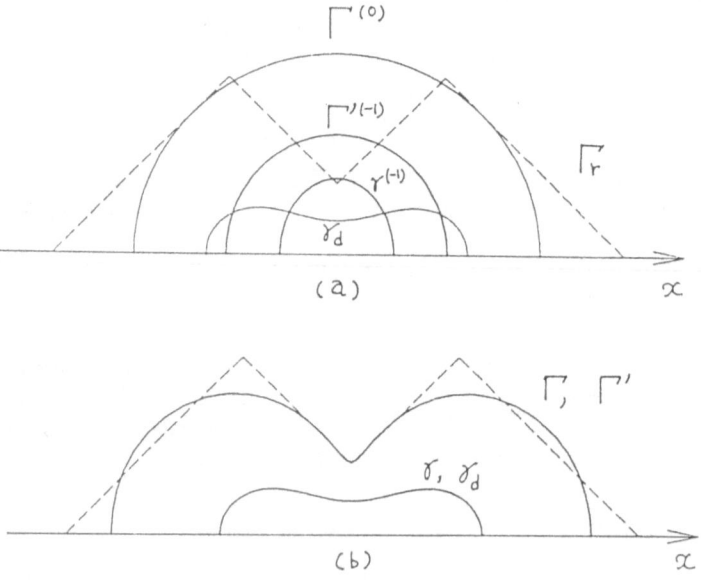

Figure 5. The shape design for $\kappa = 0.35$: (a)initial, (b)final.

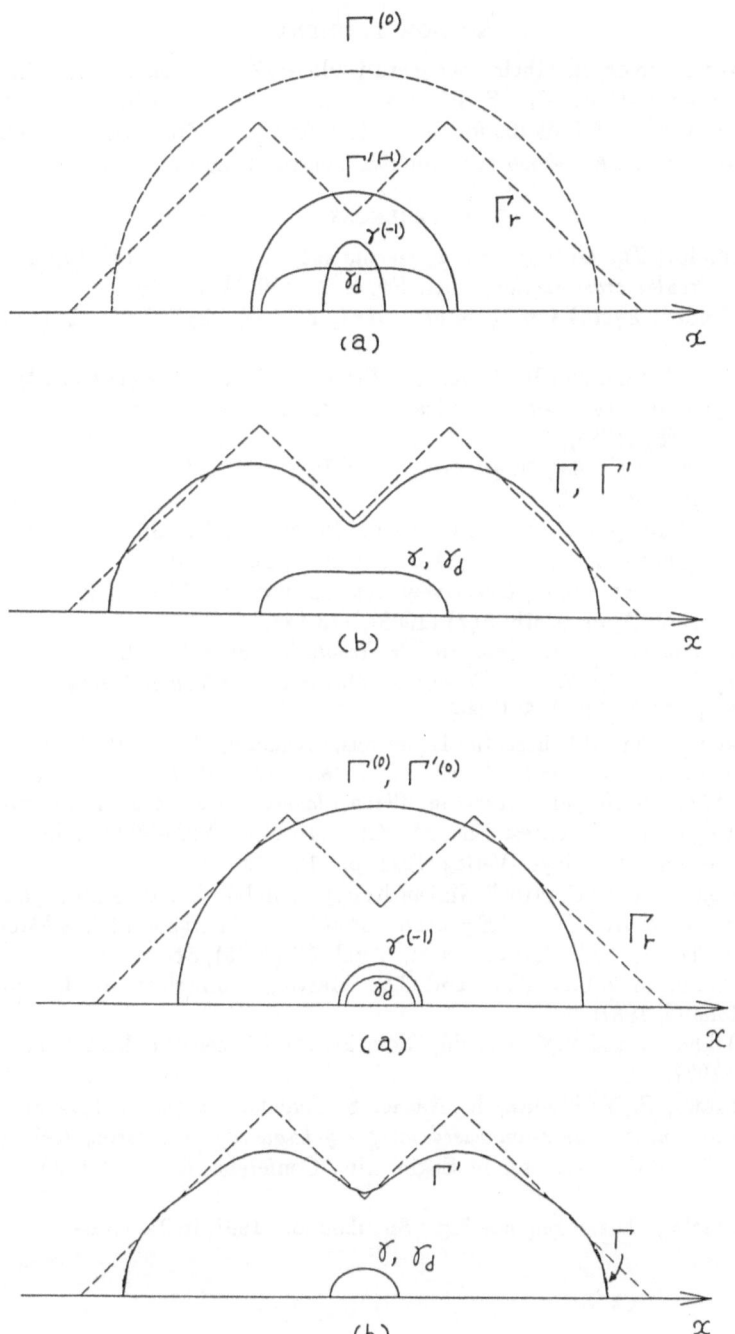

Figure 7. The shape design for $\kappa = 1.0$: (a)initial, (b)final.

## 6. ACKNOWLEDGMENT

The authors wish to express their sincere gratitude to Dr. T. Hanada of the University of Electro-Communications, Dr. S. Hiyama of the University of Tokyo and Dr. T. Kitagawa of University of Tsukuba for the fruitful discussion. Numerical computations were carried out on HITAC M880 at Computer Centre of the University of Tokyo.

## REFERENCES

[1]   A.S. Demidov, *The form of a steady plasma subject to the skin effect in a tokamak with non-circular cross-section*, Nucl. Fusion **15** (1975), 765-768.

[2]   A.S. Demidov, *Equilibrium form of a steady plasma*, Phys. Fluids **21**(6) (1978), 902-904.

[3]   L. -S. Gao, H. Imai and H. Kawarada, *Fuzzy control systems governed by elliptic partial differential equation*, Technical Reports of Mathematical Sciences , Chiba University **5** (8) (1989).

[4]   G. Hammerlin et al.,eds., "Improperly Posed Problems and Their Numerical Treatment.," Birkhauser, 1983.

[5]   T. Hirano, "Computer Analysis of Inverse Problems ( Ed. Japan Society of Mechanical Engineers )," Corona, 1991, in Japanese, pp. 151-169.

[6]   H. Imai and H. Kawarada, *One-Component Asymmetric Plasmas in a Symmetric Vessel*, Japan J. Appl. Math. **5**(2) (1988), 173-186.

[7]   H. Kawarada et al., *An Approximate Resolution of a Free Boundary Problem Appearing in the Equilibrium Plasma by Means of Conformal Mapping*, Japan J. Appl. Math. **6**(3) (1989), 331-340.

[8]   S. Mizohata, "Partial Differential Equations," Iwanami-Shoten, 1981, in Japanese.

[9]   A. Sasamoto, H. Imai and H. Kawarada, *A Practical Method for an Ill-Conditioned Optimal Shape Design of a Vessel in Which Plasma Is Confined*, in "Inverse Problems in Engineering Sciences( Eds. M. Yamaguti et al. ),ICM-90 Satellite Conference Proceedings," Springer-Verlag, 1991, pp. 120-125.

[10]   M. Sugeno, "Fuzzy Control," Nikkan Kougyo Shinbunsha, 1988 ,in Japanese.

[11]   R. Temam, *A Non-Linear Eigenvalue Problem: The Shape at Equilibium of a Confined Plasma*, Arch. Ration. Mecn. Anal. **60** (1975), 51-73.

[12]   J.F. Thompson, Z.U.A. Warsi and C.W. Mastin, "Numerical Grid Generation," North-Holland, 1987.

[13]   A.N. Tikhonov and V.Y. Arsenin, "Solution of Ill-Posed Problems," John Wiley & Sons, 1977.

[14]   G. Yagawa, S. Yoshimura, K. Nakao, S. Soneda, *Automatic two- and three-dimensional mesh generation based on fuzzy knowledge processing technique*, in "Proc.1990 ASME Computer in Engineering Conference,Boston," 1990, pp. 107-114.

[15]   S. Yasunobu, "Fuzzy Engineering," Shoukoudou, 1991, in Japanese.

# Inverse Filtering for Reverberant Transfer Functions

Mikio Tohyama[*],    Richard H. Lyon[**],   and   Tsunehiko Koike[***]

[*] NTT Human Interface Laboratories, #3-9-11, Midoricho, Musashino-shi, Tokyo 180, Japan
[**] Department of Mechanical Engineering, Massachusetts Institute of Technology, U.S.A.
[***] NTT Advanced Technology, Tokyo 167, Japan

## Summary

Inverse filtering is an important issue in sound and vibration control. The phase trend of a transfer function (TF) is predictable; however, inverse filtering for waveform recovery is still a problem because of the non-minimum-phase property of the TF and the local variability of the TF phase. This paper describes pulselike source waveform recovery in a reverberant space. The minimum-phase components of the source waveform can be recovered by inverse filtering the minimum-phase part of the reverberant TF. The effects of smoothing TFs on the waveform recovery and blind dereverberation using the "minimum-phase" complex cepstrum are also discussed.

## Introduction

The relative phase in the response of a dynamic system to its excitation has an essential role in designing filters to "dereverberate" a received signal. Source waveform recovery is particularly useful for machine diagnostics[1] or speech processing[2]. If we could recover features of a time waveform, we could have more detailed information about what is happening during machine operation than can be obtained from the power spectrum only.

The reverberant phase trend is predictable[3] even under reverberant conditions; however, inverse filtering is not a stable process, since the TF generally has non-minimum-phase zeros[2-5], and the TF phase also shows large variations from the theoretical trend[1,6,7]. We describe waveform recovery for pulselike sources by inversely filtering only the minimum-phase components of the TF in a reverberant space[7]. The minimum-phase waveform can be recovered using the inverse filter for the minimum-phase part of the TF. Robust waveform recovery using a smoothed TF[8,9] and blind dereverberation[10,11,12,13] by minimum-phase cepstrum windowing are also described.

## Ideal Inverse Filtering

In a linear system, the recovery of an input signal from the output response records is called inverse filtering. Suppose that $H(\omega)$ is the frequency response of a TF where $\omega$ is the angular frequency. We write the relation between the output signal $Y(\omega)$ and the input signal $X(\omega)$ as

$$Y(\omega) = H(\omega) X(\omega) \tag{1}$$

Thus, if we can obtain perfect records of $H(\omega)$, the input signal $X(\omega)$ is recovered from $Y(\omega)$ as

$$X(\omega) = Y(\omega) / H(\omega) \tag{2a}$$

$$= Y(\omega)\{1/H(\omega)\} \tag{2b}$$

Here we call $1/H(\omega)$ the inverse filter of the TF of the linear system.

However, it is impossible to create a causal and stable inverse filter of a TF that has non-minimum phase zeros[2-4]. In spite of this, Eq.(2a) still holds and the input signal $X(\omega)$ is recovered, even if the inverse filter $1/H(\omega)$ itself is not achievable. This is because, all the non-minimum phase zeros of the denominator $H(\omega)$ will be ideally eliminated by the same zeros of the numerator $Y(\omega)$. This reduction, however, is a virtual process, since it is almost impossible to get a perfect TF. Therefore, we cannot normally expect "perfect zero cancellations" in Eq.(2a). We need a causal and stable inverse filter $1/H(\omega)$.

## Decomposition of a TF into All-pass and Minimum-phase Components

A TF of a non-minimum phase system can always be written on the real frequency axis as the product of a minimum-phase TF and an all-pass TF as shown in Fig.1[1]:

$$H(\omega) = H_{\min \phi} \times H_{\text{all-pass}} \tag{3}$$

We can obtain a stable inverse filter only for the minimum-phase part. If the source waveform is itself minimum-phase, it is completely recovered from the minimum-phase component of the response by the inverse filtering only for the minimum-phase part of the TF. We can disregard the all-pass components of the TF for recovering the minimum phase component of the source signal.

## Minimum Phase Creation and Inverse Filtering

We propose a pulselike waveform recovery by inverse filtering the minimum phase part of the TF. The signal processing procedure is explained in the Appendix[7]. Our "reference" TF is obtained from an impulse-response-like record.

### Minimum Phase Component Extraction and Inverse Filter Creation

The minimum-phase decomposition is generally obtained based on the Hilbert transform[1]; the authors here describe a new method that uses the absolute value of the group delay of the TF[7]. As expressed in the middle part of Fig.1, the non-minimum-phase zeros of the original TF are reflected across the real frequency axis in the minimum-phase part of the TF. The group delay of the TF due to those pairs of zeros is therefore reversed in sign. Our TF has a finite data length; thus, all the TFs have only zeros in the discrete frequency plane[3, 14]. Therefore, after taking the absolute value of the group delay, the newly obtained TF, which corresponds to the middle part of Fig.1 is a minimum-phase TF. We can create a stable inverse filter only for the newly created minimum-phase TF.

### Inverse Filtering and Waveform Recovery

We take a test response at the same receiving point used for the reference TF. Again, we change the sign of the negative group delay part of the test response. After that, the new minimum-phase response is inverted by the inverse filter already created for the minimum-phase part of the reference TF. The minimum-phase components of the source waveform are recovered as the output signal through the inverse filter.

### Non-Minimum Phase Component Recovery

When the source waveform is non-minimum-phase, the non-minimum-phase component is recovered by adding in the phase difference between the all-pass parts of the test response and the reference TF, under the condition that the TF is not greatly changed during the observation period.

## Samples of Recovered Waveforms by Inverse Filtering

### Reference TF Measurements and Creating an Inverse Filter

We carried out a series of measurements of the reverberant responses to a train of pulselike source signals as shown in Fig.2. The interval between the pulselike signals in the train has a sufficiently long duration compared with the reverberation time, so there is very little overlap. We assumed here that an ideal 1/1 octave band-pass filter was used to limit the frequency band of the pulselike signal. The ideal source waveform is shown in Fig.3.

We take one of the responses to the pulselike signals as the reference impulse response $h_1(t)$ at M1. Figure 3 shows only the initial 512 samples of the impulse response record. The total record length for the impulse response completely covers the reverberation time of 1.8s. Figure 4 illustrates the amplitude, phase, and group delay of the TF derived from the impulse response. The minimum-phase part is extracted by taking the absolute value of the group delay.

## Samples of Recovered Waveforms

Figure 5 shows the recovered waveform from the test response at M1 when a pulselike signal is radiated from the loudspeaker into the space. We used the inverse filter already created from the minimum-phase part of the TF at M1. A greatly compressed waveform is extracted from the test response. The difference between the recovered waveform and the ideal source signal is due to slight changes over time of both the TF and the source signal radiated from the loudspeaker.

## Non-Minimum Phase Waveform Recovery

The change of the minimum-phase source signal into the non-minimum-phase waveform is also detectable. Figure 6 is an example of a recovered source waveform which has an all-pass part. The difference from the ideal-source waveform shown in Fig. 3 illustrates a "temporal change" of the source waveform, as we ideally assumed here that the TF is not changed over time.

## Inverse Filtering using a Smoothed TF

Inverse filtering is highly sensitive to the location of the observation point. The phase trend of a TF has a nearly constant group (linear phase) delay under high-modal-overlap conditions[3]. The minimum-phase part of the TF, however, still includes the residual phase variations from the theoretical linear phase[1,8]. The TF phase variation due to the observation locations is difficult to predict, therefore we need "robust" waveform recovery.

The TF variations can be greatly reduced by smoothing processes[6,7, 9]. Figure 7 shows examples of waveforms at M2a and M2b recovered by the inverse filter. This inverse filter was created by applying an exponential time window and the smoothing average in the frequency domain to the TF at M2. Almost the same waveforms can be recovered even at the different observation points. The recovery process seems more "robust" to the unpredictable changes of the TFs when using the "smoothed" inverse filter and smoothing the test response.

## Blind Dereverberation using Complex Cepstrum

Waveform recovery by inverse filtering requires a reference TF. "Blind dereverberation" without the reference TF, however, is possible using the complex cepstrum[1,10,11,12,13]. The complex cepstrum of a response is defined as the Fourier transform of the complex logarithm of the response. We take a reverberant test response, and again divide it into minimum-phase and all-pass parts using the cepstrum[1]. Similar to above, the minimum-phase component of the source waveform is recovered from the minimum-phase part of the test response.

Following Eq.(1), the minimum-phase cepstrum of the response is written as

$$c_{Ymin}(\tau) = c_{Xmin}(\tau) + c_{Hmin}(\tau) . \tag{4}$$

For the waveform recovery in a reverberant space, we can normally assume that the complex cepstrum of the source waveform $c_{Xmin}$ has only "low time" components compared to those of the reverberant TF $c_{Hmin}$[1,10,11,12,13]. Thus, the minimum-phase source waveform can be recovered by "low-time-cepstrum windowing".

The all-pass cepstrum is an odd function, while the minimum-phase cepstrum is zero in the negative time region. The minimum-phase cepstrum is also composed of both the magnitude cepstrum of an even function and the phase cepstrum of an odd function. Thus, the minimum-phase waveform is recovered using these phase and magnitude cepstra of the minimum phase part.

## Samples of Recovered Waveforms by Blind Dereverberation

### *Cepstrum Decomposition of a TF*

We carried out further experiments under the same conditions as shown in Fig. 2. Figure 8 illustrates the all-pass and minimum-phase cepstra of TFs which are obtained at M2a and M2b. The minimum-phase cepstrum in the low time region is not very sensitive to the observation location. Therefore we can expect to get almost the same recovered waveforms at different locations, even if the fine structures of the reverberant TFs are changed.

### *Samples of Recovered Waveforms*

Figure 9 demonstrates waveforms recovered by windowing the low time component (lower than 10 ms) of the minimum-phase cepstrum(Fig.9A) or the total cepstrum(Fig.9B) of the reverberant response. Almost the same pulselike waveforms are extracted from the minimum-phase cepstrum of the responses at different observation points. This recovery process is a blind procedure when TFs are not used; however, if we use the "robust" minimum-phase cepstrum of a TF in the low time region, then we can expect to get a precisely recovered waveform by subtracting the TF cepstrum from the windowed minimum-phase cepstrum of the response.

## Conclusion

We have shown that a pulselike source waveform can be recovered by inverse filtering from the minimum-phase parts of both the TF and a test response. This "minimum-phase inverse filtering" provides a stable inverse filter; however, it is highly sensitive to changes in the TF. Smoothing the TF makes the recovery process less sensitive to the changes in path information over time. We also described blind dereverberation without using TFs. The minimum-phase waveform can be recovered by "low-time windowing" the minimum-phase part of a test response. This recovery process is less sensitive to the observation location. The non-minimum-phase waveform is also recovered using the phase information of the all-pass parts; however, in the case of speech waveform recovery, only the minimum-phase part will be essential because of the lower phase sensitivity of the human auditory system in short time intervals. The authors wish to thank the NTT Computing Center for the calculation using the CRAY-II.

## References
1. Lyon, R. : *Machinery Noise and Diagnostics*    (Butterworths, Boston and London, 1987)
2. Neely, S.; Allen, J.: Invertibility of a Room Response, *J. A.S. A.* **66** (1979) 165 - 169
3. Tohyama, M.; Lyon, R.; Koike, T.: Reverberant Phase In a Room and Zeros in the Complex Frequency Plane, *J. A. S. A.* **89** (1991) 1701-1707
4. Miyoshi, M.; Kaneda, Y.: Inverse Filtering of Room Acoustics, *IEEE* **ASSP-36** (1988) 145-152
5. Wang, H.; Itakura, F.: Recovering of Reverberated Speech using a Narrow Band Envelope Estimation Method, *IEEE Workshop on Signal Processing to Audio and Acoustics*, Session 3-7 (1989)
6. Tohyama, M.; Lyon, R; Koike, T.: "1/f" Phase Fluctuations and Zeros of a Transfer Function in a Multi-degree-of-freedom System, *Int. Conf. Noise and 1/f Fluctuations* (1991) 567-570
7. Tohyama, M.; Lyon, R; Koike, T.: Statistics on Reverberant Transfer Functions, *2nd. Int. Con. Recent Developments in air- and Structure-borne Sound and Vibration*, (1992) 869-876
8. Mourjopoulos, J.: On the Variation and Invertibility of Room Impulse Response Functions, *J. Sound & Vib.* **102** (1985) 217-228
9. Mourjopoulos, J.; Clarkson, P. M.; Hammond, J.K.: A comparative study of least-squares and homomorphic techniques for the inversion of mixed phase signals, *ICASSP'82*, (1982) 1858-1861
10. Yamasaki, Y.; Moriwake, K; Uno, T. ; Ueno, Y. et alli, : An Application of Homomorphic Filtering to Wide-range Audio Signals, *Proc. Autumn Meeting of A.S.J.*, (1982) 315-316
11. Kim, J.; Lyon, R.,: Reducing Transfer Function Variability and Complexity by Cepstrum Windowing, *Proc, Noise-Con '88* , (1988) 493-498
12. Liu, L.: Effect of Reverberation on Group Delay in Acoustical Spaces, *Ms Thesis, Mech. Eng. Dept. MIT*, (1990 Feb.)
13. Sawatari, H.; Yamada. H.; Wang, H., Itakura, F.: The Distribution of Zeros of the Acoustic Transfer Function, *Proc. Spring Meeting of A.S.J.*, (1992) 437-438
14. Tohyama, M.; Lyon, R. : Transfer Function Phase and Truncated Impulse Response, *J.A.S.A.* **86** (1989) 2025-2029

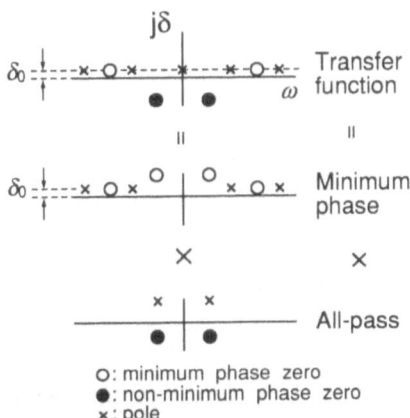

Fig.1 Decomposition of TF into all-pass and minimum-phase components.

o: minimum phase zero
●: non-minimum phase zero
×: pole

Fig.2 Experimental set-up in a reverberant space. The distance between M2a, b is 0.85m.
Reverberation time $T_R$=1.8s at 500Hz

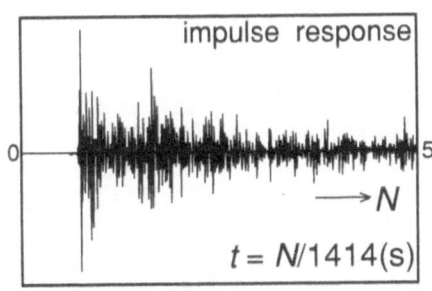

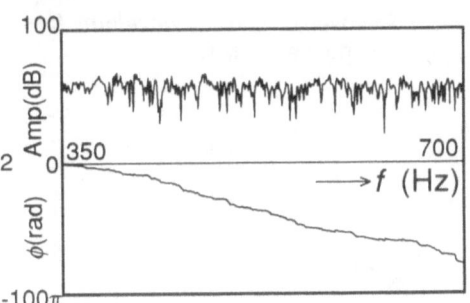

(A) Amplitude and phase

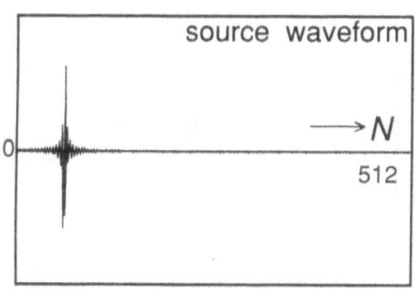

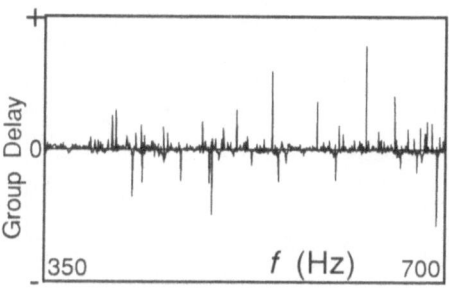

(B) Group delay

Fig.3 Reference impulse response at M1 and ideally band-limited source waveform $t = N/1414(s)$.

Fig.4 The reference TF at M1.

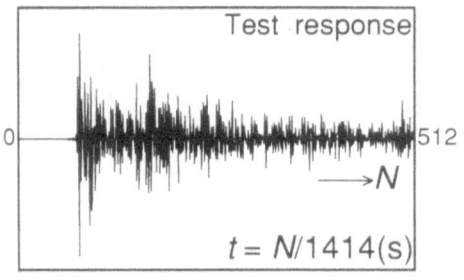

Test response

$t = N/1414(s)$

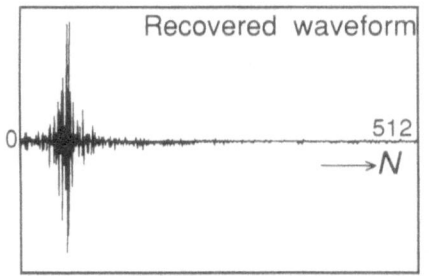

Recovered waveform

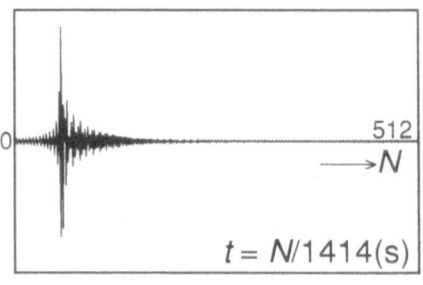

$t = N/1414(s)$

Fig.6 A sample of Recovered Non-minimum-phase waveforms subject to the unchanged TF

Fig.5 Recovered source waveform from a test response at M1.

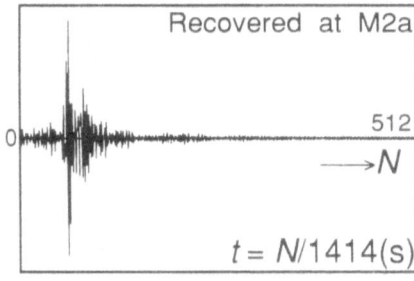

Recovered at M2a

$t = N/1414(s)$

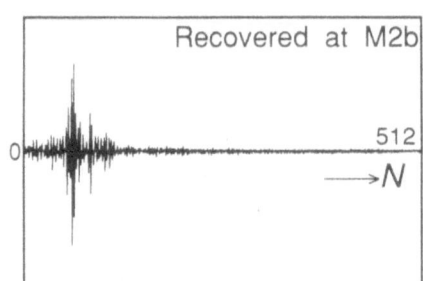

Recovered at M2b

Fig.7 Recovery at M2a and M2b by the inverse filter for the TF at M2. A smoothing average (within 60Hz) and an exponential time window (reduced effective $T_R$ =0.36s) are performed.

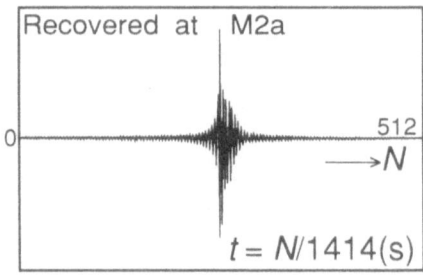

Recovered at M2a

$t = N/1414(s)$

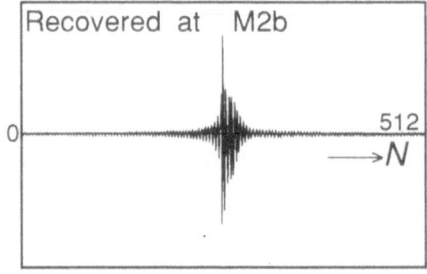

Recovered at M2b

Fig.9A Blind dereverberation at M2a and M2b by "low time windowing" the minimum-phase cepstrum shown in Fig.8c.
Cepstrum window length = ±10ms.

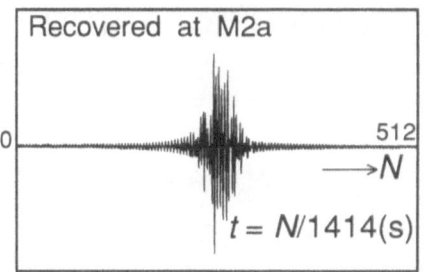

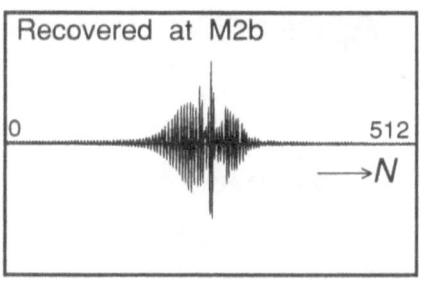

Fig.9B Blind dereverberation at M2a and M2b using the total cepstrum including the all-pass part shown in Fig.8A. Low time cepstrum window length = ±10ms.

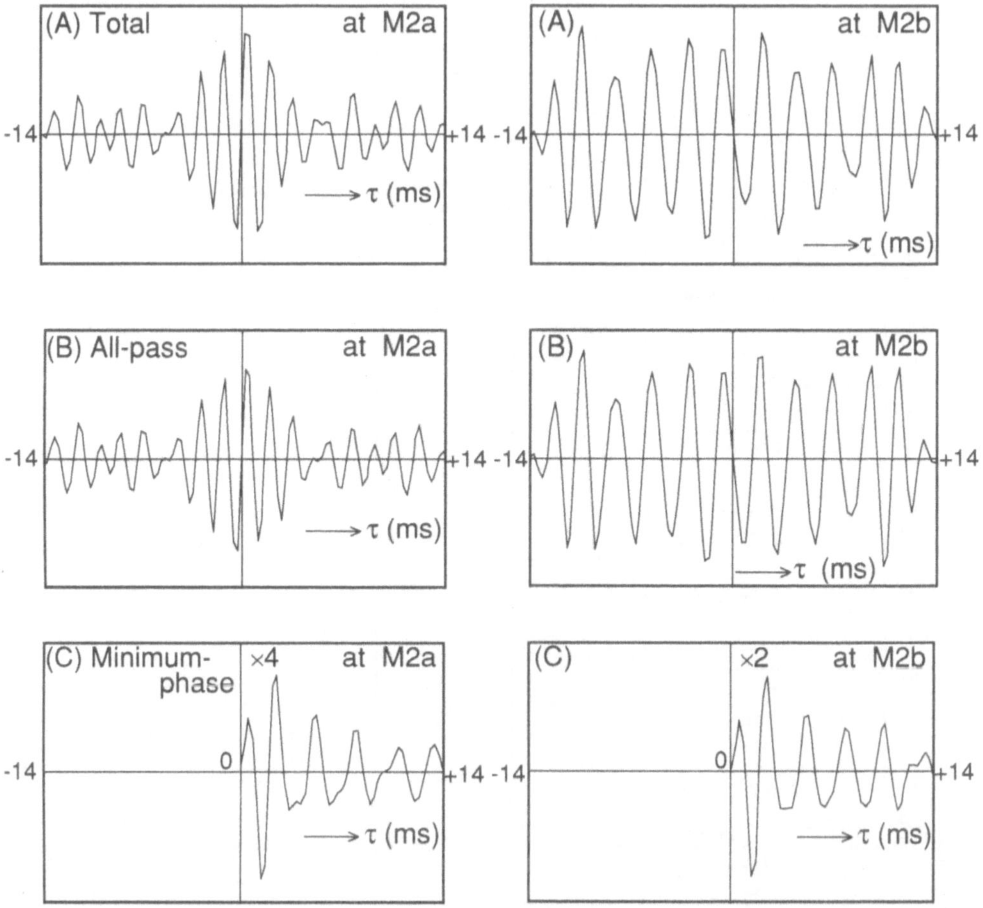

Fig.8 Cepstrum decomposition into minimum-phase part and all-pass component (A) Total cepstrum, (B) All-pass cepstrum, (C) Minimum-phase Cepstrum.

# Appendix: Inverse Filtering Procedure

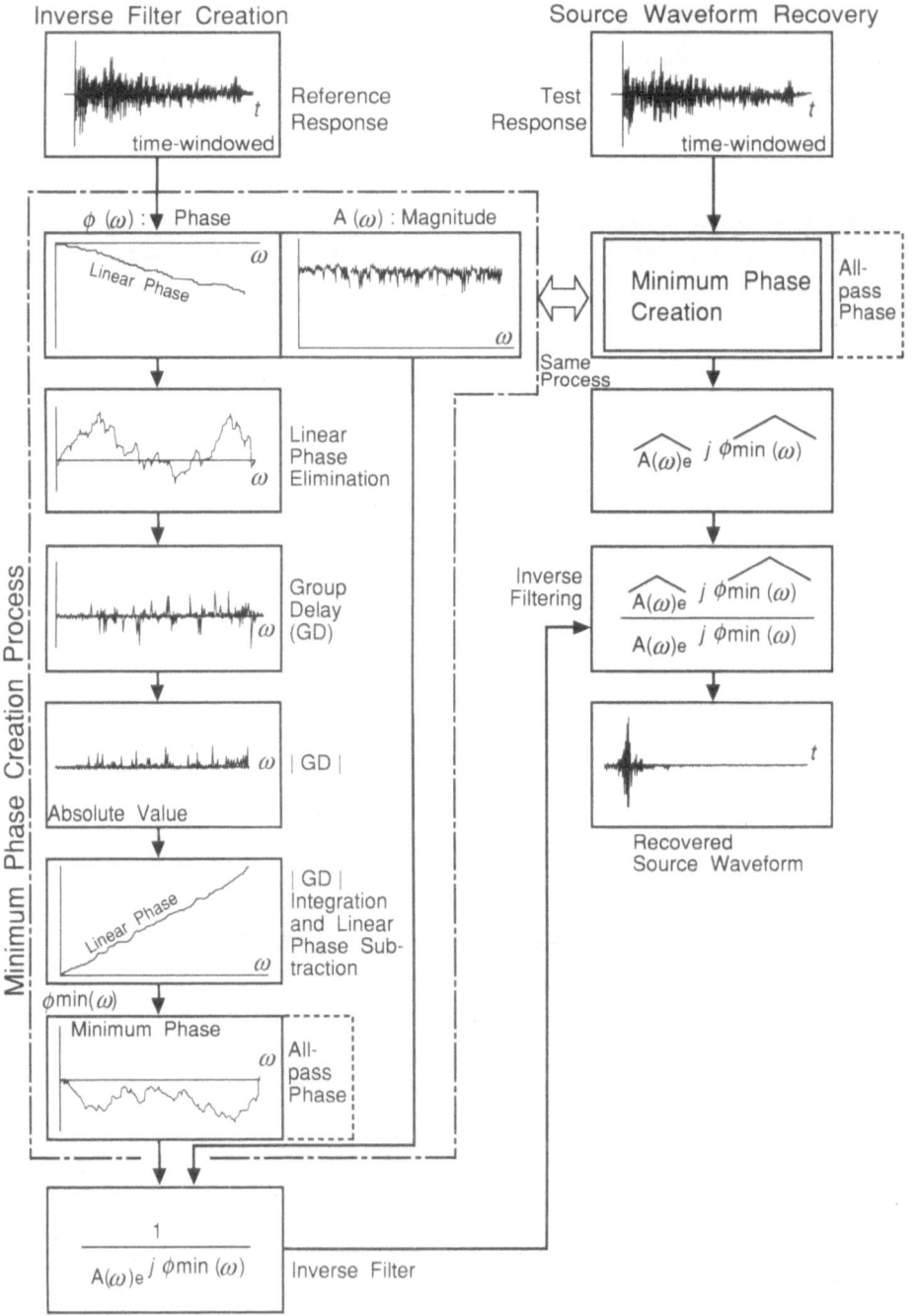

Inverse Filter Creation

Reference Response
time-windowed

Source Waveform Recovery

Test Response
time-windowed

$\phi(\omega)$ : Phase

Linear Phase

$\omega$

$A(\omega)$ : Magnitude

$\omega$

Same Process

Minimum Phase Creation

All-pass Phase

$\widehat{A(\omega)}e^{j\,\widehat{\phi min}(\omega)}$

Linear Phase Elimination

$\omega$

Group Delay (GD)

$\omega$

Inverse Filtering

$\dfrac{\widehat{A(\omega)}e^{j\,\widehat{\phi min}(\omega)}}{A(\omega)e^{j\,\phi min(\omega)}}$

$|GD|$

$\omega$

Absolute Value

$|GD|$ Integration and Linear Phase Sub-traction

Linear Phase

$\omega$

Recovered Source Waveform

$t$

$\phi min(\omega)$

Minimum Phase

$\omega$

All-pass Phase

Minimum Phase Creation Process

$\dfrac{1}{A(\omega)e^{j\,\phi min(\omega)}}$

Inverse Filter

# Chapter 2
# Computational Aspects

# Classification of Inverse Problems Arising in Field Problems and Their Treatments

Shiro KUBO

Department of Mechanical Engineering,
Faculty of Engineering, Osaka University,
2-1, Yamadaoka, Suita, Osaka, 565  JAPAN

## SUMMARY

An overview of inverse problems related to field problems and inverse approaches for their treatments is made. Discussion is made on the classification of inverse problems arising in field problems and it is shown that inverse problems can be classified into domain/boundary inverse problems, governing equation inverse problems, boundary value/initial value inverse problems, force inverse problems, and material properties inverse problems. Examples of inverse analyses conducted by the author's group for dealing with these 5 categories of inverse problems are shown with special emphasis on the applicability of the representative inverse approaches and the regularization schemes.

## INTRODUCTION

The inverse problems are receiving increasing attention in various branches of science and engineering [1-7]. The inverse problems may be regarded as the problems concerning the determination of input or source from output or response, in contrast to direct problems in which output or response is sought from input or source. We can find many inverse problems in computation of fields and relevant research areas.

In this paper a classification of inverse problems arising in analyses of variation of physical quantities are made, and it is stated that there are five kinds of inverse problems. Examples of these inverse problems and their treatments conducted by the author's research group are demonstrated.

## DEFINITION OF INVERSE PROBLEMS

The term "inverse problems" is used in many ways, depending on research areas, and sometimes on researchers. A rational definition of inverse problems can be made by referring to direct problems, which can be considered to be opposite to the inverse problems [6].

Let us consider an analysis of distribution of a field quantity $\phi$ representing physical states of concern. The governing equation is expressed as

$$L(\kappa)\phi = f \tag{1}$$

where $L(\kappa)$ denotes an operator, $\kappa$ and $f$ being material properties and a force defined in domain $\Omega$. For a direct field analysis to be conducted, the information of the following items is indispensable.

(a) Domain $\Omega$ of concern and its boundary $S$.
(b) Governing equations representing the variation of the field quantity $\phi$.
(c) Boundary conditions on the entire or partial boundary of domain $\Omega$ and/or initial conditions.
(d) Forces $f$ acting in domain $\Omega$.
(e) Material properties $\kappa$ involved in the governing equations.

When full information of these 5 items is available in advance, output or response can be determined. The output or response can be calculated by using conventional numerical schemes, such as the finite element method, the boundary element method, and the finite difference method.

If any of the requisites (a) to (e) is lacking, we can not calculate the distribution of field quantity $\phi$. Those problems, which can not be classified as direct problems in the sense mentioned above, can be classified into the inverse problems.

For the field problem, there may be the following inverse problems corresponding to the lack of requisites (a) to (e) for direct analyses.

(a)' Estimation of domain $\Omega$, its boundary $S$ or unknown inner boundary (domain/boundary inverse problems).
(b)' Inference of the governing equations (governing equation inverse problems).
(c)' Estimation of the boundary conditions on the entire or partial boundary and/or estimation of the initial conditions (initial value/boundary value inverse problems).
(d)' Estimation of the forces $f$ acting in $\Omega$ (force inverse problems).
(e)' Estimation of the material properties $\kappa$ defined in $\Omega$ and involved in the governing equations (material properties inverse problems).

Any combination of these inverse problems can be another inverse problem.

The inverse problems are inherently lacking in information as compared with that for direct problems. Additional information is necessary to conduct inverse analyses. output or response can be used as primary information to conduct inverse analyses, in contrast to the direct problem, in which output or response are determined. The information concerning output or response can be obtained by measurements. Subsidiary information expressing physical requirements and a priori information may be used to achieve an effective estimation. In the followings, examples of inverse problems treated by the present author's group for each category are given.

## DOMAIN/BOUNDARY INVERSE PROBLEMS

### Electric Potential CT (Computed Tomography) Method

Electric potential method is used for monitoring crack length. This method is based on the fact that the existence of cracks gives rise to disturbance in electric potential distributions. The location, size and shape of a two- or three-dimensional crack may be identified, if the measured distribution of electric potential is available. This crack identification from electric potential distributions can be recognized as one of the inverse problems of category (a)'. By introducing inverse analysis schemes formulated on the basis of the boundary element method, the present authors proposed the electric potential CT (computed tomography) method to identify the crack location, size and shape [8, 9].

The plane which contains the cracks to be detected is called an incompletely-prescribed boundary $S_o$, since neither the potential $\phi$, nor the flux $q$ can be prescribed on $S_o$. We introduce supplementally an over-prescribed boundary $S_3$, where both $\phi$ and $q$ are given by measuring the value of $\phi$ on some parts of the Neumann boundary $S_2$.

### Inversion Schemes for Electric Potential CT Method

Two inverse analysis schemes, i.e. the inverse boundary integral equation method and the least residual method, were proposed based on the boundary element formulation. The former is formulated by referring to the formulation of the boundary element method.

The variation of the distribution of D.C. electric potential $\phi$ is determined using Laplace's equation. If the entire boundary $S$ is divided into boundary elements, and nodes are introduced in these elements, the boundary integral equation can be written in the form of matrix equation, which relates potential and flux on boundary $S$. This equation can be solved for boundary values on the incompletely-prescribed boundary $S_o$, which contains cracks to be detected. The cracked portions in the plane $S_o$ are identified as flux-free portions in $S_o$. Thus, in the inverse boundary integral equation method the problem of crack identification of category (a)' is reduced to the problem of identification of boundary values, which is one of the inverse problems of category (c)'.

The other inversion scheme, i.e. the least residual method, is based on boundary element potential calculations for assumed cracks. Cracks are assumed, which are expressed by various combinations of the plane $S_o$ containing cracks, crack location, size and shape. Then $S_o$ is separated into a cracked portion and the remaining uncracked portion. If the boundary values of $q$ only are used on the over-prescribed boundary $S_\sigma$, a direct analysis can be made, which gives the value of $\phi$ on the over-prescribed boundary $S_\sigma$. To determine the most plausible crack, the square sum $R$ of residuals is evaluated between the computed potential values $\phi^{(C)}$ and the measured values $\phi^{(M)}$ on $S_\sigma$. The most plausible crack is identified as the crack giving the smallest $R$ value. Thus a quasi-solution is sought in the least residual method.

The present authors discussed the uniqueness of the inverse solution in the crack identification by the electric potential CT method [10, 11]. This discussion is based on the uniqueness of boundary value inverse problems with over-prescribed boundaries. It is found that cracks can be uniquely identified from the electric potential distribution, when the plane $S_o$ containing cracks is known in advance. When $S_o$ is not known, the electric potential distributions under two current application conditions are necessary to determine a single two-dimensional crack embedded in a body. To determine a single three-dimensional crack in an unknown plane, the electric potential distributions under three current application conditions are needed.

### Simulations and Experiments of Crack Identification

Numerical simulations of the crack identification by the inverse boundary integral equation method were made for two-dimensional edge cracks, embedded cracks and plural embedded cracks [8]. This method was also applied to the identification of three-dimensional surface cracks and embedded cracks [12], and it was shown that the inverse boundary integral equation method was applicable to the determination of crack sizes and shapes.

The applicability of the least residual method was examined demonstrated by numerical simulations and experiments [13-18].

As was described in the foregoing, the crack lying in an unknown plane can not be uniquely identified from the potential distribution under only one current application condition. To ensure the uniqueness of the identification, multiple current application method was proposed, in which potential data measured under several current application conditions were processed simultaneously [14]. The experiments demonstrated the usefulness of the multiple current application method. To accomplish efficient identification of the crack by the least residual method, a hierarchical analysis scheme was adopted, in which a gradual refinement of assumed cracks were made.

Experiments were made for determining a three-dimensional surface crack in a steel plate by using the least residual method [16]. The height of the plane containing cracks and the crack shape in the plane were determined from the electric potential distributions.

For efficient identification of the three-dimensional surface crack by the least residual method, a hierarchical inverse analysis scheme was proposed, in which two-dimensional scanning inverse analyses were combined with full three-dimensional inverse analyses. By conducting the two-dimensional inverse analyses for many cross sections, a rough estimate of the height of the plane containing cracks and the crack shape was obtained. Following the two-dimensional analyses, three-dimensional least residual calculations were made to obtain more accurate estimation. Gradual refinement of assumed cracks was also incorporated in the hierarchical analyses.

It was shown that the height of the plane containing cracks and the crack shape can be estimated with good accuracy by the least residual method, even when the potential data are given on the back surface of the crack only.

To achieve a high speed computation, an analytical expression of electric potential distribution reported by Johnson is used in the two-dimensional scanning analysis, and database of electric potential distribution on a three-dimensional cracked body is utilized in the three-dimensional inverse analysis [19, 20]. This idea was applied to identify a crack in a pipe. Numerical simulations and experiments has shown that the proposed scheme is useful for identifying cracks.

The crack identification by the least residual method can be treated as an optimization problem, when the residual $R$ is used as an objective function [21-23]. Several schemes based on the optimization method were proposed. Applicability of the scheme was assessed by numerical simulations and experiments. A hierarchical optimization scheme was proposed to realize a reasonable and high-speed identification.

## GOVERNING EQUATION INVERSE PROBLEMS

Governing equation inverse problems deal with the inference of a differential equation or equations governing the variation of field quantities of the present concern from observations of the field quantities [24, 25]. The estimation of the governing equation can be reduced to the estimation of the order of the differential equation and coefficients involved in the equation. The present authors proposed the local derivatives method, in which the derivatives are determined approximately from observed field quantities using finite difference approximation and the sets of values of these derivative are simultaneously used to identify the coefficients involved in the linear differential equation. This method can be applied to the estimation of nonlinear differential equation.

The local derivatives method was applied to the estimation of second order ordinary differential equation. Numerical simulations were made on the estimation of the governing equation from observed quantities at some points for certain distribution of the field quantity. When the order of the differential equation was known in advance, good estimation of the coefficients could be made from estimated derivative values.

When the order of the differential equation was not known in advance, the order has to be assumed. When the assumed order was lower than the real one, inconsistent estimation resulted for different sets of observed values. When the assumed order was coincident with the real one, consistent estimation could be made for different sets of the observed data. When the assumed order was higher than the real one, inconsistent results were obtained again. Then, the principle of parsimony works well, i.e. when assumed order employed in the estimation was increased gradually and the lowest order giving consistent estimation was adopted as the solution.

When the assumed order of the differential equation was higher than the real one, the estimated governing equation can be written as a linear combination of the original governing equation and its differentiated forms. The successive elimination method was then proposed, in which the highest order term of the estimated governing equations was eliminated. Successive elimination finally yielded consistent equations, which were coincident with the real governing equation. Numerical simulations of the successive elimination method were made and showed the usefulness of the elimination method.

A method was also proposed to reduce the effect of errors involved in the observations. It was found that it was possible to estimate the governing equation from noisy observations.

## INITIAL VALUE/BOUNDARY VALUE INVERSE PROBLEMS

Boundary value inverse problems deal with the estimation of unknown boundary conditions on incompletely prescribed boundaries using over-prescribed boundary value on other boundaries or in the domain. This kind of inverse problems are usually ill-conditioned.

A regularizing scheme using a priori information on the unknown variables was proposed for solving the ill-conditioned boundary value inverse problem [26]. The scheme was based on a multivariable constrained optimization algorithm for determining the most plausible solution satisfying inequality constraints deduced from a priori information available in advance. To demonstrate the applicability of the scheme, it was applied to a boundary value inverse problem with truss-like structures. The nonpositiveness and nonnegativeness of unknown variables were used as the constraint. It was found that the scheme using the constraint was effective in obtaining reasonable estimates of the unknown boundary values and was rather insensitive to errors involved in input data, while the unconstrained scheme was not.

A finite element-based inverse scheme was proposed for boundary value inverse problems, in which no boundary values were known in advance on the incompletely prescribed boundary [27, 28]. This scheme can be applied to the estimation of force and displacement on inaccessible boundary, such as contact region. This boundary value inverse problem is also ill-posed. For the regularization of the problem, function expansion method was introduced. For determining the optimum number of terms in the expansion, estimated error criterion was proposed. The usefulness of the proposed regularization based on the function expansion and the estimated error criterion was shown using numerical simulations of contact force determination.

The Tikhonov regularization method was also introduced in the finite element inverse analysis scheme for regularizing the inverse solution. As stabilizing functionals, norm of forces and displacements, their derivatives, and their second derivatives were employed [28]. For determining the optimum value of the regularizing parameter in the Tikhonov regularization, the estimated error criterion was applied, with error defined as the residual between the estimated and measured displacements on the over-prescribed boundary was used. Numerical simulations of estimation of contact force and displacement demonstrated the usefulness of the Tikhonov regularization method.

Symbolic manipulation was applied to evaluate the ill-posedness of the boundary value inverse problems [29]. It was found that the technique could be applied for evaluating the breakdown of uniqueness and stability of the solution.

## FORCE INVERSE PROBLEMS

A deterministic approach was proposed for identifying the force term in the governing equation [30-32]. For steady-state heat conduction problems, it was shown that a volume integral of the intensity of a heat source multiplied by a harmonic function can be expressed by a boundary integral. This enables us to determine the intensity and the location of the heat source by combining the values of the boundary integrals evaluated using several harmonic functions.

Similar method was also developed for determining the force term governed by linear partial differential equation. The usefulness of the proposed method was demonstrated by numerical simulations of identification of heat source and concentrated body forces.

This method was further generalized for determining force terms in other linear differential governing equations from boundary observations. The adjoint operator $L^*$ and subsidiary function, which satisfied the adjoint governing equation were introduced. Then, it was possible to estimate force terms involved in governing equation expressed in the form of differential equation. The estimation of concentrated mechanical forces acting in a domain was made based on this formulation.

A method was presented for determining initial residual stress fields from redistributed residual stresses measured at several points [33, 34]. Determination of the initial residual stress distribution was effectively achieved by introducing fundamental residual stress distribution functions, which satisfied physical requirements for the residual stress, i.e. the balance of stresses. The least residual criterion was used in the determination. An inverse sensitivity matrix was evaluated to estimate the effect of errors involved in the measurements and to select the best combination of measuring points. Numerical simulations of the determination of initial residual stress fields and the prediction of fatigue crack propagation lives showed the usefulness of the proposed method.

## MATERIAL PROPERTIES INVERSE PROBLEMS

Material properties of individual components of a discrete system are estimated using the response of the system to several sets of external input [35]. Two inversion scheme were proposed for estimating the material properties: [K] matrix method and {C} vector method. The [K] matrix method determined a stiffness matrix of the system for estimating material properties. The {C} vector method was based on the stiffness equation expressed in terms of the {C} vector, whose components consist of parameters of material properties. The usefulness of these methods were shown by numerical simulations. These methods were successfully extended to the determination of material properties in a continuum by employing finite element discretization [36]. Numerical simulations demonstrated the usefulness of the scheme.

## CONCLUSIONS

A brief review of inverse problems was made. Inverse problems were classified into domain/boundary inverse problems, governing equation inverse problems, boundary value/initial value inverse problems, force inverse problems, and material properties inverse problems. Examples of these inverse problems and inverse analysis schemes were demonstrated.

**ACKNOWLEDGEMENTS**

This work was partly supported by the Ministry of Education, Science and Culture under Grant-in-Aid for Co-Operative Research (principal investigator: S. Kubo).

**REFERENCES**

[1] Tikhonov, A.N. and Arsenin, V.Y., Solutions of Ill-Posed Problems, (John Willy & Sons, 1977).

[2] Groetch, C.W., The Theory of Tikhonov Regularization for Fredholm Equations of the First Kind, (Martinus Nijhoff, 1984).

[3] Lavrent'ev, Romonov, V.G. and Shishatshii, S.P., Ill-Posed Problems of Mathematical Physics and Analysis, (Amer. Math. Soc., 1986).

[4] Kubo, S., Inverse Problems, (Baifukan, Tokyo, 1992).

[5] Kubo, S. and Ohji, K., Applications of the Boundary Element Method to Inverse Problems, in: Applications of the Boundary Element Method (Corona, Tokyo, 1987) pp.181-198.

[6] Kubo, S., JSME Int. J., Ser. I 31 (1988) 157.

[7] Kubo, S., Systems, Control and Information 35 (1991) 634.

[8] Ohji, K., Kubo, S. and Sakagami, T., Trans. Japan Soc. Mech. Eng., Ser. A 51 (1985) 1818.

[9] Kubo, S., Sakagami, T. and Ohji, K., in Fracture Mechanics (Current Japanese Materials Research, Vol. 8) (Elsevier, 1991) pp.235-254.

[10] Kubo, S., Sakagami, T. and Ohji, K., Trans. Japan Soc. Mech. Eng., Ser. A 55 (1989) 2316.

[11] Kubo, S., ICM-90 Satellite Conf. Proc., Invere Problems in Engineering Sciences, (Springer, Tokyo, 1991), pp.52-58.

[12] Kubo, S., Sakagami, T. and Ohji, K., Computational Mechanics '86, (Proc. Int. Conf. on Computational Mechanics), (Springer, Berlin, 1986) pp.V-339-344.

[13] Ohji, K., Kubo, S., Sakagami, T. and Hashimoto, T., J. Soc. Mater. Sci., Japan 35 (1986) 942.

[14] Sakagami, T., Kubo, S., Hashimoto, T., Yamawaki, H. and Ohji, K., JSME Int. J., Ser. I 31 (1988) 76.

[15] Kubo, S., Sakagami, T. and Ohji, K., Computational Mechanics '88, (Proc. Int. Conf. on Computational Engineering Science), (Springer, Berlin, 1988), Vol.1 pp.12.i.1-5.

[16] Kubo, S., Sakagami, T., Ohji, K., Hashimoto, T. and Matsumuro, M., Trans. Japan Soc. Mech. Eng., Ser.A 54 (1988) 218.

[17] Sakagami, T., Kubo, S. and Ohji, K., Int. J. Pressure Vessels Piping 44 (1990) 35.

[18] Sakagami, T., Kubo, S., Ohji, K., Yamamoto, K. and Nakatsuka, K., Trans. of Japan Soc. Mech. Eng., Ser. A 56 (1990) 27.

[19] Ohji, K., Kubo, S., Kagoshima, K. and Imajuku, T., Trans. of Japan Soc. Mech. Eng., Ser. A 57 (1991) 2078.

[20] Kubo, S., Ohji, K., Kagoshima, K. and Imajuku, T., Mechanical Behaviour of Materials IV, (Proc. Int. Conf. on Mechanical Behaviour of Materials), (J. Soc. Mater. Sci., Jpn, Kyoto, 1991), Vol. 4, 717.

[21] Sakagami, T., Kubo, S. and Ohji, K., Eng. Analysis with Boundary Elements 7 (1990) 59.

[22] Kubo, S., Ohji, K., Nakatsuka, K. and Fujito, H., Preprint of Japan Soc. Mech. Eng. No. 900-59 (1990) 526.

[23] Kubo, S., Ohji, K. and Nakatsuka, K., Preprint of Japan Soc. Mech. Eng. No. 900-86 (1990) 667.

[24] Kubo, S., Ohji, K. and Shiojiri, A., Trans. of Japan Soc. Mech. Eng., Ser. A 57 (1991) 2250.

[25] Kubo, S., Ohji, K. and Shiojiri, A., Preprint of Japan Soc. Mech. Eng. No. 913-4 (1991) 72.

[26] Kubo, S., Ohji, K. and Matsui, Y., Trans. of Japan Soc. Mech. Eng., Ser. A 57 (1991) 2403.

[27] Kubo, S., Ohji, K. and Ueda, T., Preprint of Japan Soc. Mech. Eng. No. 900-86 (1990) 670.

[28] Kubo, S., Ohji, K. and Hagimoto, K., Preprint of Japan Soc. Mech. Eng. No. 913-4 (1991) 22.

[29] Kubo, S., Ohji, K. and Shiojiri, A., Preprint of Japan Soc. Mech. Eng. No. 913-4 (1991) 69.

[30] Kubo, S., Ohnaka, K. and Ohji, K., Trans. of Japan Soc. Mech. Eng., Ser. A 54 (1988) 1329.

[31] Kubo, S., Ohnaka, K. and Ohji, K., Preprint of Japan Soc. Mech. Eng. No. 890-34 (1989) 49.

[32] Kubo, S. and Ohji, K., Preprint of Japan Soc. Mech. Eng. No. 890-50 (1989) 41.

[33] Kubo, S., Tsuji, M. and Ohji, K., Trans. of Japan Soc. Mech. Eng., Ser. A 54 (1988) 892.

[34] Ohji, K., Kubo, S. and Tsuji, M., Fatigue 90, Vol. IV (1990) 2377.

[35] Kubo, S. and Ohji, K., Trans. of Japan Soc. Mech. Eng., Ser. A 57 (1991) 2257.

[36] Kubo, S., Ohji, K. and Konishi, K., Preprint of Japan Soc. Mech. Eng., (1992).

# A Physically Based Method of Enhancement of Experimental Data Concepts, Formulation and Application to Identification of Residual Stresses

Wojciech Karmowski, Janusz Orkisz

Cracow Institute of Technology
ul. Warszawska 24, 31-155 Kraków, Poland

Summary

A typical way of fitting a curve or a surface to experimental data is based on the least squares technique or on a similar mathematical method. If this is possible such a result is usually compared with a theoretical solution.

A new combined approach presented now, uses simultaneously all the information available for the investigated problem, resulting from both the various experimental measurements as well as the theoretical model. In this way only one , i.e. "the best" solution is obtained. It tries to fit best the experimental data and, at the same time, to satisfy, as much as possible, the requirements of the theory involved.

In terms of mathematics, this approach results in constrained optimization of a functional composed of experimental and theoretical parts.

Presented is the general concept of the approach, and its reference to residual stress analysis, done upon strain gauge measurements.

## 1. Introduction

Following some earlier ideas [1-7,10] presented is a general approach to the enhancement of experimental data by means of use of all information available for a considered problem. The approximation obtained in this way is physically based since all basic physical relations, relevant to the problem, are taken into account. Moreover the results obtained are within the error bounds corresponding to the appropriate physical statistics.

The approach is specially convenient when data is insufficient, uncertain, not uniformly distributed or missing in regions, where measurements are often difficult or even impossible. It then allows, to avoid ill-conditioned problems, and at the same time to improve the quality of the results obtained. It is worth noticing, that the fitted solution is smooth enough, and partially cancels experimental data errors. Its differentiation yields therefore reasonable effects.

In many engineering problems a multidimensional physical field (e.g. stress in mechanics) is needed in each point of the domain considered, but we can measure only a function of the required quantities (e.g. isochromatics). Such information is often not sufficient in order to determine the unknown field. However, taking into account appropriate physical relations that ought to be fulfilled by an unknown field, it is usually possible to obtain all required information.

The main objective of this paper is to present several variants of the general formulation of the problem, following the above mentioned approach. Stress analysis will be used as a particular case. Sample examples of application to the enhancement of strain gauge measurements and moire interferometry data will be also given.

### 2. Formulation

Considered are formulations based on two concepts of analysis called the global and global - local ones. The first one is posed as the constrained optimization problem. Enhanced results are then obtained simultaneously in the whole domain. The global - local analysis also uses measured data from the entire domain, but it provides results only in one required point at a time.

### 2.1. The global method

We may take advantage from all information available about a considered problem and pose the problem in the following general way:

find the stationary point of the functional

$$\Phi = \lambda \Phi^E + (1-\lambda)\Phi^T , \qquad \lambda \in [0,1] , \qquad (1)$$

satisfying the constraints

theoretical (mostly)     $A(\sigma) = 0 ,$     (2)

experimental (mostly)     $B(\sigma) \leq e .$     (3)

Here $\Phi^T(\sigma)$ and $\Phi^E(\sigma)$ are the theoretical and experimental parts of the functional, $\sigma$ is the required solution, $\lambda$ is a scalar weighting factor.

## Experimental requirements
### Functional

As the experimental part of the functional we introduce an averaged global "error norm" as follows

$$\Phi^E(\sigma) = \frac{1}{N} \sum_{n|1}^{N} F\left( \frac{f(\sigma(r_n) - f_n)}{e_n} \right) , \tag{4}$$

Here $\sigma$ presents the required unknown field, $f$ is a measured function of $\sigma$, $f_n$ is its experimental value in the point $r_n$, $e_n$ is an admissible experimental error, N is a number of measurements, $F() = p(0)-p()$ is a data scattering function, defined by means of the probability density function $p()$.

It is worth stressing that results of experiments of various kind e.g. photoelasticity, moire interferometry, strain gauge technique may be combined together, and then the averaged global error is

$$\Phi^E = \frac{1}{I} \sum_{i|1}^{I} \Phi_i^E . \tag{5}$$

The summation is extended over all types of experiments. This is possible because the error norm (4) is dimensionless, and particular types of measurements are relatively normalized by $e_n$ , standing here for a weighting factor. More precise measurements result in a smaller $e_n$ . Consequently the share of such experiment in an error function increases.

### Constraints

The enhanced field $\sigma(r)$ cannot differ too much from experimental data. Thus the constraint $B(\sigma) \leq 0$ are defined first of all as local requirements

$$\left| f(\sigma(r_n)) - f_n \right| \leq e_n \quad , \qquad n=1,2,\ldots,N . \tag{6}$$

Moreover it is useful to also impose an averaged global constraint

$$\Phi^E \leq e_E . \tag{7}$$

In practical analysis both types of constraints (6) and (7), or only one of them, may be used. Evaluation of admissible experimental errors $e_E$

and $e_n$ , n = 1,2,...N should be done taking into account the true statistics of measurements.

## Theoretical requirements

### Functional

When the theoretical part of the functional $\Phi^T$ is concerned, two situations may be distinguished

(i) Theory is known. In mechanics $\Phi^T(\sigma)$ may be represented by one of well known energy functionals that has to be minimized e.g. the total complementary energy of statically admissible stresses.

(ii) Theory is not known. A heuristic principle e.g. requirement of smoothness i.e. minimal average curvature $\kappa(\sigma)$ may then be used.

For a scalar function f we introduce a mean value of the second directional derivative and assume the following definition

$$\kappa^2 = \frac{1}{2\pi} \int_0^{2\pi} \left( \frac{\partial^2 f}{\partial \nu^2} \right)^2 d\varphi \tag{8}$$

of its average curvature in the given point. This definition is objective, with the respect to any rotation of the coordinate system, and may be extended into the case of a tensorial function $\sigma_{ij}$ (e.g. stresses) as to obtain

$$\kappa^2(\sigma_{ij}) = \frac{1}{2\pi} \int_0^{2\pi} \frac{\partial^2 \sigma_{ij}}{\partial \nu^2} \frac{\partial^2 \sigma_{ij}}{\partial \nu^2} d\varphi \tag{9}$$

Finally the theoretical (heuristic) part of the functional, we then assume it to be in the following form

$$\Phi^T = \frac{1}{\Omega} \int_\Omega \kappa^2 d\Omega \tag{10}$$

### Constraints

Theoretical constraints are usually given in the form of equality conditions $A(\sigma) = 0$, e.g. for the complementary energy functional these are the equilibrium equations $\sigma_{ij,j} = 0$ in $\Omega$ and the static boundary conditions $\nu_i \sigma_{ij} = p_i$

### Specific formulations proposed

One of the main disadvantage of the general formulation is (1)-(3) how to establish the weighting factor $\lambda$ i.e. how to determine a reasonable balance between experiment and theory. The following specific formulation are proposed in order to address this problem.

(i) Experimental requirements are removed from the functional $\Phi$ and appear only as constraints (6) and (7). We want to find

$$\min_{\sigma} \Phi^T \tag{11}$$

satisfying the constraints

$$A(\sigma) = 0 \ , \quad \Phi^E \leq e_n \ , \quad \left| f(\sigma(r_n)) - f_n \right|, \ n = 1, 2, \ldots, N \ . \tag{12}$$

(ii) The problem (1)-(3) is divided into the subsequent optimization problems. In the first one we evaluate the family of admissible solutions $\sigma(\lambda)$. In the second one we determine the optional value of $\lambda$ . Thus:

a) we want to find $\sigma(\lambda)$ that yields the stationary value of the functional

$$\Phi = \lambda \Phi^E + (1-\lambda) \Phi^T \tag{13}$$

and satisfy the theoretical equality constraints $A(\sigma) = 0$,

b) we want to find the maximal $\lambda$

$$\max_{\lambda} \lambda \tag{14}$$

satisfying admissible bounds of global and local experimental errors

$$\Phi^E \leq e_E \ , \quad \left| f(\sigma(r_n)) - f_n \right| \leq e_n, \ , \qquad n = 1, 2, \ldots, N \ . \tag{15}$$

(iii) Original formulation (1)-(3) but with $\lambda$ considered as variable throughout the domain. A special technique of local $\lambda$ adjustment was proposed in [2,7] and will not be discussed here. In every of those formulation we have to deal with a constraint optimization problem. However out of the three proposed formulations the second one (ii) seems to be computationally the most effective, especially when the constraints $A(\sigma) = 0$ are linear.

## 2.2 The global - local method

In this method we search for a local surface (curve) that in a neighbourhood of the considered points fits the best the required solution. The concept of the method is based on the local expansion of the searched field function

$$\sigma(r-r_n) = \sum_{i,j|0}^{i+j \leq p} c_{ij} (x-x_n)^i (y-y_n)^j \tag{16}$$

into the Taylor series in a considered point $r_n$. Unknown values of $c_{ij}$ are found by means of the minimization of the global error $\Phi$

$$\min_{c_{ij}} \Phi(r, \sigma(c_{ij})) \tag{17}$$

with equality constraints $A(\sigma(c_{ij})) = 0$ locally enforced. Such a problem has to be solved subsequently in each required point. The inequality constraints $B(\sigma) \leq 0$ still remain to be satisfied, but we consider them later.

In the mathematical sense the global - local method may be viewed as a generalization of some earlier concepts of local approximation [8,9,11]. Some preliminaries of the physically based approach to data handling were earlier done by the authors [5,6,7,10]. Here presented is a new, advanced version of the method.

We consider the global error function (4) in a modified form

$$\Phi(r, \sigma) = \frac{1}{N} \sum_{n|1}^{N} v(\rho_n) \; F\left( \frac{f(\sigma(r_n)) - f_n}{e_n} \right) \tag{18}$$

with a weighting factor $v(\rho_n)$ introduced. This factor depends on the distance $\rho_n = |r - r_n|$ between an analyzed point $r$ and a data point $r_n$.

The proposed weighting factor function is of the form

$$v(\rho) = \left( \rho^2 + \frac{g^4}{\rho^2 + g^2} \right)^{-m} \tag{19}$$

where $g$ is so far a free parameter of approximation and has yet to be determined. Power $m$, is chosen in a way as to cut information from a distance $\rho_n$ when it reaches the approximation error level [1]. Thus e.g. $m = 3$ when $p = 2$.

In the way $v$ is chosen, it serves the rule "longer the distance $\rho_n$ smaller the influence of experimental data on the considered value of field function $\sigma$". Additionally the weighting factor (19) provides

$$v'(0) = v''(0) = v'''(0) = 0 \; . \tag{20}$$

Due to these conditions fitted field function $\sigma$ is smooth enough and is not "pulled" by an experimental point.

Parameter $g$ is responsible for accuracy of approximation. When $g = 0$ we deal with the interpolation. On the other hand when $g$ is infinite we obtain a polynomial approximation in the entire domain (e.g. quadratic for $p = 2$). The appropriate, finite, value of $g$ should be chosen in a way based on the experimental error, in order to satisfy the constraints $B(\sigma) \leq 0$.

It is proposed here to solve that problems in two steps. First we search a solution of the original optimization problem (18), but only for the purpose to determine g . Having found g we return again to the problem (18) and find the final solution.

In the first step all experimental points are then taken into account, however, excluding the measured value in the considered point. In order to determine g , one of the following optimization problems may then be solved:

(i) find

$$\max_{g} g \tag{21}$$

satisfying the constraints $B(\sigma) \leq 0$ i.e.

$$\Phi(r,\sigma) \leq e_{E}, \quad \text{and/or} \quad |f(\sigma(r_n))-f_n| \leq e_n, \qquad n=1,2,\ldots,N \ ,$$

(ii) find

$$\min_{g} \Phi(r,\sigma) \tag{22}$$

The criteria satisfying the constraints $|f(\sigma(r_n))-f_n| \leq e_n$ n=1,2,...,N , are related to the experimental error and compatible with the formulation (1)-(3).

Formulation for strain gauge measurements

In this case the global error function (18) consists of two terms

$$\Phi = \Phi_1 + \alpha \Phi_b \tag{23}$$

The first one is given by

$$\Phi_1(r,\sigma) = \frac{1}{N} \sum_{n|1}^{N} v(\rho_n) \frac{1}{3} \sum_{j|1}^{3} (\sigma_j(r_n) - \sigma_{j(n)})^2 \tag{24}$$

where $\sigma_j = \{\sigma_{xx}, \sigma_{xy}, \sigma_{yy}\}$, $\sigma_{j(n)}$ are "measured stresses" obtained from strain gauge rosettes. The second term is the boundary one

$$\Phi_b(r,\sigma) = \sum_{1|1}^{L} v_1 \frac{1}{2} \left[ (\sigma_1 v_1 + \sigma_3 v_2 - q_1)_1^2 + (\sigma_3 v_1 + \sigma_2 v_2 - q_2)_1^2 \right] \tag{25}$$

Here L is the total number of boundary points, q is loading, and $v$ is the vector normal to the boundary.

In the considered global - local formulation the boundary conditions may be not exactly satisfied. A weighting factor $\alpha > 1$ was assumed, therefore, in order to enforce better fulfillment of the boundary conditions.

On the other hand , as mentioned above, the equilibrium equation are strictly satisfied in advance. We impose then onto the Taylor series expansion of the global error function (18) and reduce, in this way, the number of independent parameters. In the case of quadratic approximation $\sigma_j$ we may have e.g. $3 \cdot 6 - 2 \cdot 3 = 12$ independent unknowns that are found by minimization of (18). We deal then with 12 simultaneous linear equations. Such procedure may be fast and and easily repeated in each subsequent point of domain.

Application of the global - local method to evaluation of residual stresses.

In some technical problems (e.g. in analysis of railroad rails working in service conditions) required is information about distribution of residual stresses in the body considered. This is usually a complex problem. Various experimental methods are tried, therefore, in order to find these stresses. Experiments are usually expensive, and do not yield results of sufficient quality. Therefore an application of enhanced methods of evaluation of measured data is considered now, especially in the case of strain gauge technique and moire interferometry.

The approach formulated above in the general way is verified in numerical experiments where random, but controlled distribution of residual stresses is simulated.

In the first example we deal with 'strain gauges' randomly distributed in the circular domain (Fig. 1). True stresses evaluated from an assumed Airy function, were subjected to artificial randomization. In Fig. 1 presented are results for all three stress components. Compared are their 'true' values (continuous line) and simulated 'experimental' values (dots) with their approximation done by the global - local approach. Three different values of g were used namely zero (interpolation), a large value ( here g = 10) is sufficient to generate the global quadratic approximation), and the optimal value based on formulation (18).

Final remarks

Presented was a new general approach to enhancement of experimental measurements. It uses all available information about the given problem, both experimental and theoretical (or heuristic) nature. Several particular formulations were proposed here in the form of constrained optimization problems. They reflect two different technique used, namely the global approximation and the global - local one. The first technique provides solution at once for the entire considered domain, but it may demand large computer power. The second one yields solution in one point at a time, and may be subsequently applied in any required point. A test example for strain gauge measurements was also presented.

The approach has potentially broad applications. It seems to be specially promising for enhancement of insufficient or uncertain data.

The research is under current development. Further investigations are planned including intensive testing and various applications of the approach, especially in experimental analysis of residual stress.

References

1. Karmowski.W., "Global-local methods of linear problems by use of the physical laws, boundary conditions and experimental data", VII Conf. "Computational Methods in Structure Mechanics, Gdynia May 1985 (in Polish).

2. Karmowski.W., "Ph.D. Thesis", Cracow Institute of Technology, 1989.

3. Karmowski.W, "Physically Based Global - Local Interpretation of Strain Gauge Data", X Polish Conference "Computer Methods in Mechanics", Świnoujście, May 1989.

4. Karmowski.W, "Determination of Stress and Strain Fields by Physically Based Interpretation of Moire Patterns", 20 Convegno Nazionale AIAS", Palermo Sept. 1991.

5. Karmowski.W, Magiera.J, Orkisz.J, "Enhancement of experimental results by constrained minimization", published in "Residual Stress in Rails - Effects on Rail Integrity and Railroad Economics", O.Orringer et al., Kluwer Academic Publishing, Dordrecht, Boston, London, 1992.

6. Karmowski W., Orkisz J., "Fitting of curves and surfaces based on interaction of physical relations and experimental data", Appl. Math. Modelling 7,1983.

7. Karmowski.W, Orkisz.J, "Physically Based Enhanced Analysis of Stresses Using Experimental Data", X Polish Conference "Computer Methods in Mechanics", Świnoujście, May 1989.

8. Liszka.T, 'An Interpolation Method for an Irregular Net of Nodes', Int. J. for Num. Meth. in Engng. 20(1984).

9. Liszka.L, Orkisz.J, "The Finite Difference Method at Arbitrary Irregular Grids and Its Application in Applied Mechanics, "Int. J.Computer & Structures", 11(1980) pp. 83-95.

10. Karmowski.W, Orkisz.J, "A Physically Based Method of Enhancement of Experimental Measurements of Residual Stresses", 5th German - Polish Symposium "Mechanics of Inelastic Solid and Structures", 1990.

11. Shepard.D, 'A two dimensional interpolation function for irregularly spaced data', Proc. 23rd Nat.Conf. A.C.M. (1965), pp. 517-523.

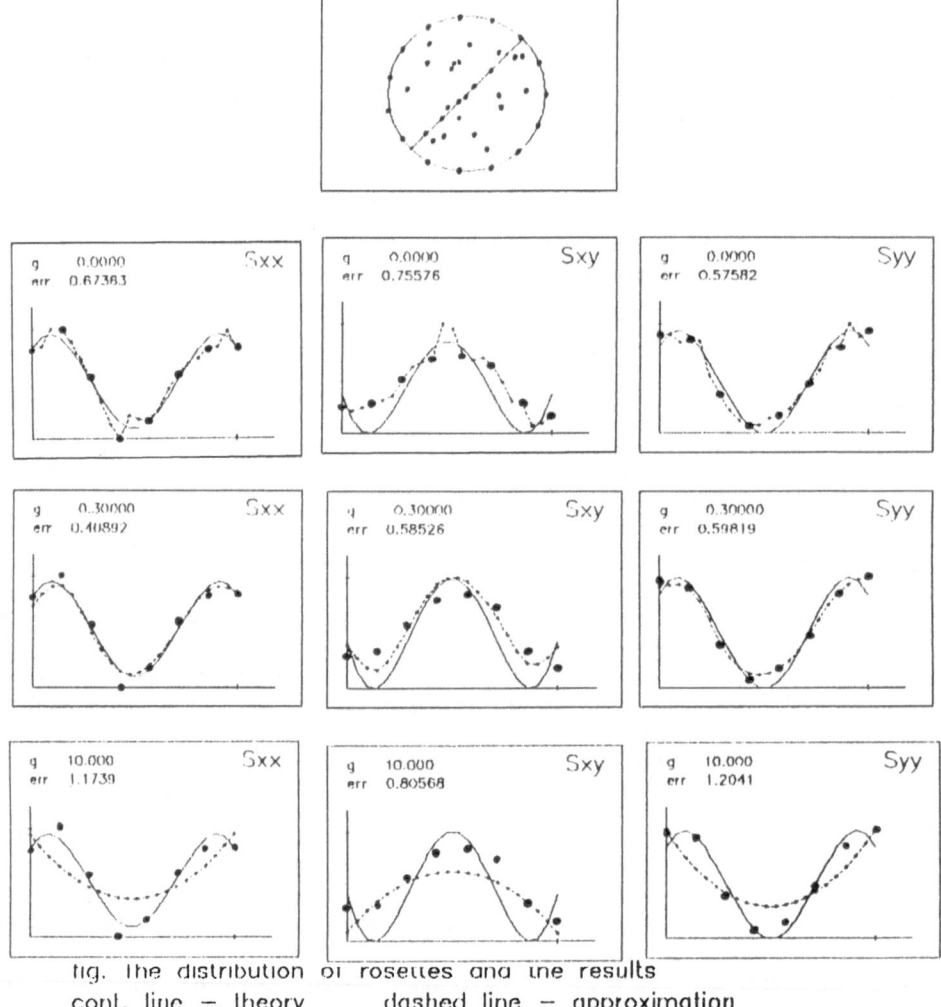

fig. The distribution of rosettes and the results
cont. line — theory          dashed line — approximation

# Inverse Problems for Vibrating Systems in Two and Three Dimensions

G.M.L. Gladwell* and B.R. Zhu**


*   Solid Mechanics Division, Faculty of Engineering
    University of Waterloo, Waterloo, Ontario, Canada N2L 3G1
**  Department of Mathematics, Shandong University,
    Jinan, Shandong, P.R. China 250100

## Abstract

Many vibrating systems involve interactions only between neighbouring parts of the system. Frequently, such systems are analysed by supposing that the mass distribution is lumped at the generalized coordinates. Undamped systems of this type involve a block tri-diagonal stiffness matrix $K$ and a block diagonal inertia matrix $M$. The inverse problem is to reconstruct $K$ and $M$ from frequency response data. The known reconstruction of a Jacobi matrix using the Forsythe algorithm is generalised so that a block Jacobi matrix can be reconstructed from a certain spectral function. This analysis is used to construct the dynamic stiffness matrix $A = L^{-1}KL^{-T}$ ($M = LL^{T}$) from the frequency response; $A$ is determined to within a block diagonal orthogonal matrix. It is shown that $K$ and $M$ can be separated using special information about their forms. A vibrating lattice composed of rods and masses is used as an example.

## Introduction

The simplest type of vibrating system consists of a set of in-line masses $(m_i)_1^n$ connected by linear springs of stiffness $(k_i)_1^n$ and excited by in-line forces $(f_i)_1^n$ applied to the masses, as shown in Figure 1. In the absence of damping the vibration at frequency $\omega$ is governed

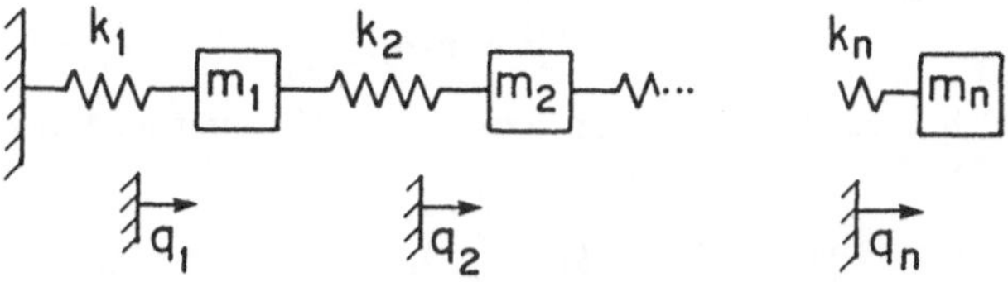

Fig. 1   A set of masses connected by springs.

by the equation

$$(\mathbf{K} - \lambda\mathbf{M})\mathbf{q} = \mathbf{f}, \quad \lambda = \omega^2. \tag{1}$$

If the left hand end of spring $k_1$ is fixed, and the right hand mass $m_n$ is free, then

$$\mathbf{K} = \begin{bmatrix} k_1 + k_2 & -k_2 & & \\ -k_2 & k_2 + k_3 & -k_3 & \\ \cdots & \cdots & \cdots & \\ & & -k_n & k_n \end{bmatrix}, \quad \mathbf{M} = \begin{bmatrix} m_1 & & & \\ & m_2 & & \\ & & \ddots & \\ & & & m_n \end{bmatrix} \tag{2}$$

Write

$$\mathbf{M}^{1/2}\mathbf{q} = \mathbf{x}, \quad \mathbf{A} = \mathbf{M}^{-1/2}\mathbf{K}\mathbf{M}^{-1/2}, \quad \mathbf{M}^{-1/2}\mathbf{f} = \mathbf{g},$$

then equation (1) becomes

$$(\mathbf{A} - \lambda\mathbf{I})\mathbf{x} = \mathbf{g}. \tag{3}$$

$\mathbf{A}$ is a Jacobi matrix, i.e. a tridiagonal matrix with non-zero off-diagonal elements, and so has distinct eigenvalues $(\lambda_i)_1^n$. The solution of (3) is

$$\mathbf{x} = \sum_{i=1}^{n} \frac{(\mathbf{x}^{(i)T} \cdot \mathbf{g})\mathbf{x}^{(i)}}{\lambda_i - \lambda},$$

where $(\mathbf{x}^{(i)})_1^n$ are the normalised eigenvectors of $\mathbf{A}$. In particular therefore, if a unit force is applied to $m_1$, then

$$q_1 = m_1^{-1/2} x_1 = m_1^{-1} \sum_{i=1}^{n} \frac{[x_1^{(i)}]^2}{\lambda_i - \lambda}.$$

Suppose that we know $q_1$ as a function of $\omega$, i.e. of $\lambda$; it has the form

$$q_1 = \sum_{i=1}^{n} \frac{\alpha_i}{\lambda_i - \lambda},$$

where we know $(\alpha_i, \lambda_i)_1^n$. Thus

$$\alpha_i = m_1^{-1}[x_1^{(i)}]^2,$$

so that

$$\sum_{i=1}^{n} \alpha_i = m_1^{-1} \sum_{i=1}^{n} [x_1^{(i)}]^2 = m_1^{-1}. \tag{4}$$

Thus we know $m_1$ and

$$[x_1^{(i)}]^2 = m_1 \alpha_i. \tag{5}$$

Also we know $k_1$ because the static displacement due to a unit load at $m_1$ is

$$\frac{1}{k_1} = \sum_{i=1}^{n} \frac{\alpha_i}{\lambda_i}. \tag{6}$$

The matrix **A** has the form

$$\mathbf{A} = \begin{bmatrix} a_1 & b_1 & & \\ b_1 & a_2 & b_2 & \\ \cdots & \cdots & \cdots & \cdots \\ & & b_{n-1} & a_n \end{bmatrix}.$$

The polynomials $p_i(\lambda)$ defined by $p_0 = 1$, $p_1(\lambda) = \lambda - a_1$,

$$p_i(\lambda) = (\lambda - a_i)p_{i-1}(\lambda) - b_{i-1}^2 p_{i-2}(\lambda),$$

are orthogonal w.r.t the points $(\lambda_i)_1^n$ and the weights $[x_1^{(i)}]^2$, so that knowing the weights and the points from $q_1$, we may construct **A** using Forsythe's algorithm [1]. There is a sign ambiguity in the off-diagonal terms $b_i$; equation (2) shows that we must take them all negative. Boley and Golub [2] review the literature.

The spring system has the property that a *static* force $k_1$ at mass $m_1$ will extend spring 1 by unity; all the other masses and springs will move to the right as a rigid body. Therefore

$$K\{1, 1, \ldots 1\} = \{k_1, 0, \ldots 0\}.$$

This is equivalent to

$$\mathbf{M}^{1/2} A \mathbf{M}^{1/2}\{1, 1, \ldots 1\} = \{k_1, 0, \ldots 0\}$$

or to

$$\mathbf{A}\{m_1^{1/2}, m_2^{1/2}, \ldots m_n^{1/2}\} = \{m_1^{-1/2} k_1, 0, \ldots 0\}. \tag{7}$$

But we know $m_1$ from (4), $k_1$ from (6), and **A**, so that on solving (7) we can find the masses $m_1, m_2, \ldots m_n$. A consistency check on our calculation is given by the requirement that all the $m_i$ be positive. Having found **M** we compute $\mathbf{K} = \mathbf{M}^{1/2} A \mathbf{M}^{1/2}$ and read off the $k_i$ from the entries in **K**.

We now generalise this procedure for a block tridiagonal $\acute{\mathbf{K}}$ and diagonal **M**.

**The spectral function of a block Jacobi matrix**

The full theory is presented in Zhu, Jackson and Chan [3] and Gladwell and Zhu [4]. Let **A** be a symmetric matrix

$$\mathbf{A} = \begin{bmatrix} \mathbf{a}_1 & \mathbf{b}_1 & & \\ \mathbf{b}_1^T & \mathbf{a}_2 & \mathbf{b}_2 & \\ & & \mathbf{b}_{n-1}^T & \mathbf{a}_n \end{bmatrix}$$

where $\mathbf{a}_i$, $\mathbf{b}_i$ are real $r \times r$ matrices, $\mathbf{a}_i = \mathbf{a}_i^T$, $\mathbf{b}_i$ is non-singular. For any real $\lambda$ define

$$\phi(\lambda) = \{\phi_1(\lambda), \phi_2(\lambda), \ldots \phi_n(\lambda)\}$$

where $\phi_i(\lambda)$ is $r \times r$ and $\phi_1(\lambda) = I_r = I$, and

$$A\phi = \begin{bmatrix} a_1 & b_1 & & \\ b_1^T & a_2 & b_2 & \\ \cdots & \cdots & & \cdots \\ & & b_{n-1}^T & a_n \end{bmatrix} \begin{bmatrix} \phi_1 \\ \phi_2 \\ \vdots \\ \phi_n \end{bmatrix} = \lambda \begin{bmatrix} \phi_1 \\ \phi_2 \\ \vdots \\ \phi_n \end{bmatrix} + \begin{bmatrix} 0 \\ 0 \\ \vdots \\ \Phi_n(\lambda) \end{bmatrix}.$$

Thus the $\phi_i(\lambda)$ satisfy the recurrence

$$b_{i-1}^T \phi_{i-1}(\lambda) + (a_i - \lambda I)\phi_i(\lambda) + b_i \phi_{i+1}(\lambda) = 0 \tag{8}$$

where $i = 1, 2, ... n - 1$, $\phi_0(\lambda) = 0$, $\phi_1(\lambda) = I$, and

$$\Phi_n(\lambda) = b_{n-1}^T \phi_{n-1}(\lambda) + (a_n - \lambda I)\phi_n(\lambda).$$

In general $\Phi_n(\lambda) \neq 0$. However, when $\lambda$ is an eigenvalue of $A$, $\Phi_n(\lambda)$ is singular. Let $\lambda_k$ be an eigenvalue of multiplicity $r_k$, then the homogeneous equation

$$\Phi_n(\lambda_k)\gamma = 0$$

will have $r_k \leq r$ linearly independent solutions $\gamma^l(\lambda_k)$, $l = 1, 2, ... r_k$. Each will give an eigenvector of $A$, namely

$$x^l(\lambda_k) = \phi(\lambda_k)\gamma^l(\lambda_k). \tag{9}$$

We normalise the eigenvectors so that

$$(x^l(\lambda_i))^T(x^m(\lambda_j)) = \delta_{lm}\delta_{ij} \tag{10}$$

and define the spectral function of $A$ as

$$\rho(\lambda) = \sum_{\lambda_k < \lambda} \sum_{l=1}^{r_k} \gamma^l(\lambda_k)\gamma^{lT}(\lambda_k).$$

The $r \times r$ matrix function $\rho(\lambda)$ is a sum of rank-one matrices, since each term is a column vector multiplied by its transpose. The completeness of the eigenvectors of a symmetric matrix ensures that

a) $\mathrm{rank}(\rho(\lambda_k + 0) - \rho(\lambda_k - 0)) = r_k$

b) $\sum_k r_k = rn = N$.

Thus $\rho(\lambda)$ is a matrix function with monotonically increasing rank, and may thus be used to define a matrix inner product. If $f(\lambda)$, $g(\lambda)$ are two real $r \times r$ matrix functions, we define their inner product as

$$(f, g) = \sum_k f(\lambda_k) \sum_l \gamma^l(\lambda_k)\gamma^{lT}(\lambda_k)g^T(\lambda_k)$$

which we abbreviate to

$$(f, g) = \sum_k f(\lambda)\gamma(\lambda_k)\gamma^T(\lambda_k)g^T(\lambda_k).$$

We now show that, with this definition, the matrix functions $\phi_i(\lambda)$ are orthogonal in the sense that

$$(\phi_i, \phi_j) = \delta_{ij}\mathbf{I}. \tag{11}$$

We proceed as follows. The matrix $\mathbf{A}$ has $N$ eigenvector $\mathbf{x}^l(\lambda_k)$ given by equation (9). Assemble these $N$ column vectors into an $N \times N$ matrix $\mathbf{X}$. The orthonormality conditions (10) imply that $\mathbf{X}^T\mathbf{X} = \mathbf{I}_N$, but therefore

$$\mathbf{XX}^T = \mathbf{I}_N. \tag{12}$$

But since $\mathbf{x}^l(\lambda_k) = \phi(\lambda_k)\gamma^l(\lambda_k)$, equation (12) may be written, with proper attention to multiple eigenvalues, as

$$[\phi(\lambda_1)\gamma(\lambda_1), ...\phi(\lambda_N)\gamma(\lambda_N)]\{\gamma^T(\lambda_1)\phi^T(\lambda_1), ...\gamma^T(\lambda_N)\phi^T(\lambda_N)\} = \mathbf{I}_N.$$

Thus

$$\sum_k \phi(\lambda_k)\gamma(\lambda_k)\gamma^T(\lambda_k)\phi^T(\lambda_k) = \mathbf{I}_N.$$

But this is equivalent to

$$\sum_k \begin{bmatrix} \phi_1(\lambda_k) \\ \vdots \\ \phi_n(\lambda_k) \end{bmatrix} \gamma(\lambda_k)\gamma^T(\lambda_k)[\phi_1^T(\lambda_k), ...\phi_n^T(\lambda_k)] = \begin{bmatrix} I & & & \\ & I & & \\ & & \ddots & \\ & & & I \end{bmatrix},$$

so that, on equating terms on both sides we find

$$\sum_k \phi_i(\lambda_k)\gamma(\lambda_k)\gamma^T(\lambda_k)\phi_j^T(\lambda_k) = \delta_{ij}\mathbf{I},$$

which is equivalent to (10).

We note that $\mathbf{x}_1^l(\lambda_k)$, the first $r$ components of $\mathbf{x}^l(\lambda_k)$, are given by

$$\mathbf{x}_1^l(\lambda_k) = \phi_1(\lambda_k)\gamma^l(\lambda_k) = \gamma^l(\lambda_k),$$

so that we can write the spectral function as

$$\rho(\lambda) = \sum_{\lambda_k < \lambda} \sum_{l=1}^{r_k} \mathbf{x}_1^l(\lambda_k)\mathbf{x}_1^{lT}(\lambda_k). \tag{13}$$

**Construction of a block Jacobi matrix**

Equation (8) shows that

$$(\mathbf{a}_1 - \lambda\mathbf{I})\phi_1(\lambda) + \mathbf{b}_1\phi_2(\lambda) = 0.$$

Taking the inner product with $\phi_1(\lambda) \equiv I$, and noting that

$$(\phi_1, \phi_1) \equiv \| \phi_1 \|^2 = \mathbf{I}, \quad (\phi_1, \phi_2) = 0$$

we find

$$\mathbf{a}_1 = (\lambda\boldsymbol{\phi}_1, \boldsymbol{\phi}_1) = \sum_{\lambda_k < \lambda} \sum_{l=1}^{r_k} \lambda_k \mathbf{x}_1^l(\lambda_k)\mathbf{x}_1^{lT}(\lambda_k).$$

Now

$$\mathbf{b}_1\boldsymbol{\phi}_2(\lambda) = -(\mathbf{a}_1 - \lambda\mathbf{I})\boldsymbol{\phi}_1(\lambda), \tag{14}$$

so that

$$\mathbf{b}_1\mathbf{b}_1^T = \parallel \mathbf{a}_1 - \lambda\mathbf{I} \parallel^2 = \mathbf{B}_1.$$

Since $\mathbf{b}_1$ is, by assumption, non-singular, $\mathbf{B}_1$ will be positive definite. There is then a unique factorisation $\mathbf{B}_1 = \mathbf{U}_1^T\mathbf{U}_1$, where $\mathbf{U}_1$ is an upper triangular matrix with positive diagonal entries. Thus

$$\mathbf{b}_1\mathbf{b}_1^T = \mathbf{U}_1^T\mathbf{U}_1, \quad \mathbf{b}_1 = \mathbf{U}_1^T\mathbf{Q}_2^T,$$

where $\mathbf{Q}_2$ is an orthogonal matrix. Equation (14) now becomes

$$\boldsymbol{\phi}_2(\lambda) = -\mathbf{Q}_2\{\mathbf{U}_1^{-T}(\mathbf{a}_1 - \lambda\mathbf{I})\mathbf{I}\} = \mathbf{Q}_2\boldsymbol{\psi}_2(\lambda),$$

where $\boldsymbol{\psi}_2(\lambda)$ is known. Proceeding in this way we find

$$\mathbf{A} = \mathbf{Q}\bar{\mathbf{A}}\mathbf{Q}^T,$$

where $\mathbf{Q} = diag(I, \mathbf{Q}_2, ...\mathbf{Q}_n)$ is an unknown orthogonal matrix and

$$\bar{\mathbf{A}} = \begin{bmatrix} \mathbf{a}_1 & \mathbf{U}_1^T & & \\ \mathbf{U}_1 & \bar{\mathbf{a}}_2 & U_2^T & \\ \cdots & \cdots & \cdots & \cdots \\ & & \mathbf{U}_{n-1} & \bar{a}_n \end{bmatrix},$$

where the $\bar{a}_i$ and $\mathbf{U}_i$ are known. This construction can be carried out iff all the $\mathbf{b}_k$, and hence the $\mathbf{U}_k$ are non-singular. In the next section we examine this condition.

## Fully connected systems

Consider a linear, conservative, anchored vibrating system with stiffness matrix $\mathbf{K}$ and mass matrix $\mathbf{M}$ excited by generalised forces $\mathbf{f}\cos\omega t$ at its $N = nr$ generalised coordinates $\mathbf{q}$. The equation governing the vibration with frequency $\omega$ is

$$(\mathbf{K} - \lambda\mathbf{M})\mathbf{q} = \mathbf{f}, \quad \lambda = \omega^2. \tag{15}$$

The static response is governed by $\mathbf{Kq} = \mathbf{f}$. Suppose $\mathbf{K}$ has block-Jacobi form, then

$$\begin{bmatrix} \mathbf{c}_1 & \mathbf{d}_1 & & & & \\ \mathbf{d}_1^T & \mathbf{c}_2 & & & & \\ & & \mathbf{c}_k & \mathbf{d}_k & & \\ & & \mathbf{d}_k^T & \mathbf{c}_{k+1} & & \\ & & & & & \mathbf{d}_n \\ & & & & \mathbf{d}_{n-1}^T & \mathbf{c}_n \end{bmatrix} \begin{bmatrix} \mathbf{q}_1 \\ \mathbf{q}_2 \\ \mathbf{q}_k \\ \\ \\ \mathbf{q}_n \end{bmatrix} = \begin{bmatrix} \mathbf{f}_1 \\ \mathbf{f}_2 \\ \vdots \\ \\ \\ \mathbf{f}_n \end{bmatrix}.$$

Consider $r$ sets of forces $\mathbf{f}^{(1)}, ...\mathbf{f}^{(r)}$ applied to the first $r$ coordinates, then $f_i^{(j)} = 0$ if $i > 1$. Choose the $r$ column vectors $f_1^{(j)}$, $j = 1, 2, ...r$ to be the columns of the unit matrix $\mathbf{I}_r \equiv \mathbf{I}$, then the corresponding displacement vectors $\mathbf{q}^{(j)}$ satisfy

$$\mathbf{K}[\mathbf{q}^{(1)}, \mathbf{q}^{(2)}, ...\mathbf{q}^{(r)}] = \{I, 0, ... 0\}$$

which we will write as

$$\mathbf{K}\mathbf{q} = \mathbf{I}$$

where now the $N \times r$ matrix $\mathbf{q}$ has been partitioned

$$\mathbf{q} = [\mathbf{q}^{(1)}, \mathbf{q}^{(2)}, ...\mathbf{q}^{(r)}] = \{q_1, q_2, ...q_n\}$$

i.e. as a column of $n$ $r \times r$ matrices $q_i$.

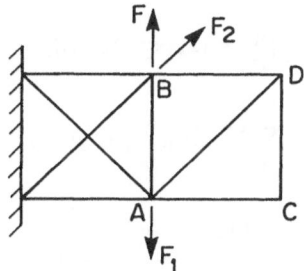

Fig. 2    This truss is not fully connected.

We may now prove *All $q_i$ are non-singular iff all $(d_k)_1^{n-1}$ are non-singular.* We may interpret this by relating it to the flow of information from the level of application of the forces, level 1, to the level where the displacement are measured. Full information reaches level $k$, i.e. $rank(\mathbf{q}_k) = r$, iff $d_1, d_2, ..., d_{k-1}$ are all non-singular. We call a system in which $(d_k)_1^{n-1}$ are non-singular a *fully connected system.*

Fig. 2 shows a plane truss which is not fully connected. We may apply a vertical load at $A$ and loads at $B$ so that $A$ is not deflected at all, while $B$ is deflected vertically. In this case neither $C$ nor $D$ will be deflected at all. The force at level 1 has rank 1, while the deflection at level 2 has rank 0. A diagonal bar $BC$ must be inserted to make the truss fully connected.

**The frequency response gives the spectral function**

We analyse the block system as we did the scalar one. Suppose

$$\mathbf{M} = diag(\mathbf{M}_1, \mathbf{M}_2, ... \mathbf{M}_n)$$

where $\mathbf{M}_i = \mathbf{L}_i \mathbf{L}_i^T$. Equation (15) may be written

$$(\mathbf{A} - \lambda_N \mathbf{I})\mathbf{x} = \mathbf{g} \tag{16}$$

where

$$\mathbf{A} = \mathbf{L}^{-1} \mathbf{K} \mathbf{L}^{-T}, \quad \mathbf{x} = \mathbf{L}^T \mathbf{g}, \quad \mathbf{g} = \mathbf{L}^{-1} \mathbf{f}$$

and $\mathbf{L} = diag(\mathbf{L}_1, \mathbf{L}_2, ... \mathbf{L}_n)$. The solution of (16) gives

$$\mathbf{q}_1 = \mathbf{L}_1^{-T} \left( \sum_k \sum_{l=1}^{r_k} \frac{\mathbf{x}_1^l(\lambda_k)\mathbf{x}_1^{lT}(\lambda_k)}{\lambda_k - \lambda} \right) \mathbf{L}_1^{-1}.$$

In the (idealised) measurement process we have

$$\mathbf{q}_1 = \sum_k \frac{\alpha_k}{\lambda_k - \lambda}. \tag{17}$$

Thus

$$\mathbf{L}_1^{-T} \{ \sum_{l=1}^{r_k} \mathbf{x}_1^l(\lambda_k)\mathbf{x}_1^{lT}(\lambda_k) \} \mathbf{L}_1^{-1} = \alpha_k$$

but

$$\sum_k \sum_{l=1}^{r_k} \mathbf{x}_1^l(\lambda_k)\mathbf{x}_1^{lT}(\lambda_k) = \mathbf{I}_r$$

so that

$$\mathbf{L}_1^{-T} \mathbf{L}_1^{-1} = \mathbf{m}_1^{-1} = \sum_{k=1}^n \alpha_k. \tag{18}$$

This yields a necessary condition on the data, namely that $\sum_k \alpha_k$ be positive definite. If this condition is fulfilled, then we can find a unique $\mathbf{L}_1$ with positive diagonal entries such that (18) holds. Then

$$\sum_{l=1}^{r_k} \mathbf{x}_1^l(\lambda_k)\mathbf{x}_1^{lT}(\lambda_k) = \mathbf{L}_1^T \alpha_k \mathbf{L}_1 \tag{19}$$

which yields the spectral function. The static component of (17) is

$$\mathbf{q}_1 = \sum_k \frac{\alpha_k}{\lambda_k}. \tag{20}$$

This is the flexibility matrix for displacements at level one due to loads at level one.

**Separation of the stiffness and mass matrices**

The general case, in which $\mathbf{M}$ is block diagonal, is discussed in [4]; here we consider the simpler case in which $\mathbf{M}$ is diagonal, and study an example. The lattice shown in

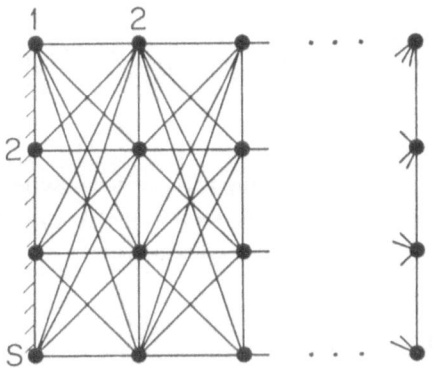

Fig. 3   A two-dimensional system of masses and rods.

Fig. 3 has $n$ levels, and vibrates in its own plane. The masses at each level are connected to all the masses at the neighbouring levels by linearly elastic rods. There are $r = 2s$ degress of freedom at each level, $N = rn$ in all. The stiffness matrix is block tridiagonal and the mass matrix is diagonal, of the form

$$\mathbf{M} = diag(m_1, m_1, m_2, m_2, \ldots \ m_{N/2}, m_{N/2}).$$

We proceed as follows. We apply sinusoidal loads at each mass on the first level, both horizontally and vertically. The static component of this forcing yields the flexibility submatrix for the displacements at level one due to static loads at level one, according to equation (20). The dynamic component of the response yields the masses at level one, according to equation (18), and the spectral function, from equation (19).

We now construct $\mathbf{A} = \mathbf{M}^{-1/2}\mathbf{K}\mathbf{M}^{-1/2}$, and we can do this uniquely because, knowing the form of the off-diagonal blocks $b_i$, we can deduce $b_i$ from $b_i b_i^T$.

Knowing the flexibility matrix for the first level we can determine what static loads must be applied at level one to move that level, and all the remaining levels, one unit to the right; note that such a rigid body movement is independent of the stiffness to the right of level one. Thus we know the solution of the equation

$$\mathbf{K}\{1, 0, 1, 0, \ldots 1, 0\} = \{\mathbf{f}_1, 0, 0 \ldots 0\}$$

in the sense that we know the $r \times 1$ vector $\mathbf{f}_1$, but we do not know $\mathbf{K}$. Now this equation is equivalent to

$$\mathbf{M}^{1/2}\mathbf{A}\mathbf{M}^{1/2}\{1, 0, 1, 0, \ldots 1, 0\} = \{\mathbf{f}_1, 0, \ldots 0\}$$

which can be written

$$A\{m_1^{1/2}, 0, m_2^{1/2}, 0, \dots m_N^{1/2}, 0\} = \{m_1^{1/2}f_1, 0, \dots 0\} \tag{21}$$

where on the right

$$m_1 = diag(m_1, m_1, m_2, m_2, \dots m_s, m_s).$$

We know $m_1$, $f_1$ and $A$, so that, on Solving (21), we can determine the remaining masses.

Note that there are 2 stages where we assume that the measurements are exact and they do correspond to the system that we have modelled:

i) in deducing $b_i$ from $b_i b_i^T$

ii) in deducing the $m_i$ from equation (21).

These stages thus provide consistency checks for the data.

## Conclusion

We have shown that a vibrating system that may be modelled as *slices* or *levels* with equal numbers of degrees of freedom, connected just to neighbouring levels, may be identified by using a block matrix analogue of the procedure used for reconstructing a tri-diagonal matrix. The procedure is valid only for fully connected systems, although it is clear that it may be applied also to identify fully connected *parts* of a system which is not necessarily fully connected.

The analyses may be applied to the identification of skeletal frameworks of the type used in traditional buildings, i.e. rectangular grids of beams and columns, fixed at the base, free at the top. The analyses presupposes a lumped mass model. This means that the mass and moment of inertia must be lumped at the joints, not distributed as in consistent mass finite element models. Such a structure will be fully connected if the beams and columns are rigidly connected. Again it may be shown that the off-diagonal matrices $b_i$ can be determined from the squares $b_i b_i^T$.

The analysis described here is just the first step in solving inverse problems of this type. It needs to be supplemented by investigations of the stability of the inversion algorithms and of the effects of inaccurate data. This research is under way.

## References

1. Forsythe, G.E.: Generation and use of orthogonal polynomials for data fitting with a digital computer, *J. SIAM*, **5** (1957) 74-88.

2. Boley, B.; Golub, G.H.: A survey of matrix inverse eigenvalue problems, *Inverse Problems* **3** (1987) 595-622.

3. Zhu, B.R.; Jackson, K.R.; Chan, R.P.K.: Inverse spectrum problems for the Jacobi matrices, *to be published*.

4. Gladwell, G.M.L.; Zhu, B.R.: Inverse problems for multi-dimensional vibrating systems, *to be published*.

# Inverse Spectrum Problems for Block Jacobi Matrix

*Benren Zhu\*, K.R. Jackson† and R.P.K. Chan‡*

\*Department of Computer Science
University of Waterloo, Waterloo
Ontario, N2L 3G


†Computer Science Department
University of Toronto, Toronto
Ontario, M5S 1A4


‡Department of Mathematics and Statistics
University of Auckland
New Zealand

## ABSTRACT

By establishing the Spectrum (matrix) function for the Block Jacobi Matrix theorems of existence and uniqueness for the inverse problem and algorithms for its solution are obtained. The study takes into account all possible multiple-eigenvalue cases that are very difficult to deal with by other means.

Keywords: block matrix, inverse problem, Gelfand-Levitan equation
AMS(MOS) Subject Classification: 15A18, 15A33

## 1. Introduction

There are extensive papers [1-5] on inverse eigenvalue problems for Jacobi matrices, but only a few papers [4] on block or banded matrices which arise more often in practice. In these papers most work is restricted to the case of simple-eigenvalue only. As this is often not the case in practice. Further study of such problems involving multiple-eigenvalue is urgently needed.

In this paper we study the Jacobi matrix with entries as $r \times r$ matrices of the form:

$$
A = \begin{pmatrix}
a_1 & b_1 & & & & \\
b^*_1 & a_2 & b_2 & & & \\
& \cdot & \cdot & \cdot & & \\
& & \cdot & \cdot & \cdot & \\
& & & & & b_{n-1} \\
& & & & b^*_{n-1} & a_n
\end{pmatrix}
\tag{1}
$$

$a_j, b_j$; $r \times r$ matrices on C; $a_j = a^*_j$ ; all $b_j$ invertible

We denote the set of all $r \times r$ matrices on C as a ring F. The main points we must bear in mind are

(1)  for $a, b \in F$, then $a\,b \neq b\,a$ in general;

(2)  for $a \in F$ and $a \neq 0$, then $a^{-1}$ may not exist in general.

First we need to establish operations in F, linear vector space and operation in the space, and functions and vector-function on F. Then in Section 2 we will apply the Fourier theory to the matrix A, this fresh approach will turns out to be very useful. In Section 3. a general inverse spectrum problem and the main theorems of the problem will be given. Two algorithms based on G-L theory are formulated, and another algorithm based on orthogonalization is given in Section 4. Finally a brief description of the applications and numerical testing will be presented in the last section. The main advantage of our treatment is that our conclusion takes into account all possible multiple-eigenvalue cases that are diffcult to deal with by other means.

Beside the well known operations such as multiplication, scalar multiplication, scalar multiplication and addition, we recall the conjugate operation '\*' in F:

(1)    for $a \in F$,  then $a^* = \bar{a}^T$, $(a^*)^* = a$; $\qquad\qquad$ (2)

(2)    for $a, b \in F$, then $(ab)^* = b^* a^*$;

(3)    for $a \in F$ then $a^* a$ and $aa^*$ are semipositive and selfadjoint;

(4)    $a = 0 \Leftrightarrow a^* a = aa^* = 0$.

Let H denote a linear space of 'n-dimensional column vectors' on F, that is

$$H = \{\, f = (f_1, f_2, \ldots, f_n)^T, \text{ all } f_K \in F \}.$$

And denote $H^*$ its conjugate space, that is

$$H^* = \{f^* = (f^*_1, f^*_2, \ldots, f^*_n), \text{ all } f_K \in F\}.$$

Then we define the 'inner product' of $f, g \in H$ as follows:

(1)    $(f, g) = f^* g = f^*_1 g_1 + f^*_2 g_2 + \cdots + f^*_n g_n \in F$;    (3)

(2)    $(f, g) = (g, f)^*$;

(3)    $(f, f) = f^* f \geq 0$;

(4)    $f = 0 \Leftrightarrow (f, f) = 0$.

Denote by 0 and I the zero element and identity element respectively of either F or H or some matrices without causing confusion.

Since $(f, g)$ is matrix in F , then it is natural to regard $a^*a$ $(aa^*)$ in (2) and $(f, f)$ in (3) be certain matrix measurement instead of a usual norm. This is some thing interesting beyond our problem.

## 2. Fourier Theory for a Block Jaccobi Matrix

First we introduce an eigenfunction $\phi(\lambda)$, $\lambda \in (-\infty, +\infty)$ for (1) as follows:

$$\phi(\lambda) = (\phi_1(\lambda), \phi_2(\lambda), \ldots, \phi_n(\lambda))^T, \ \phi(\lambda) = I,$$    (4)

$$A\phi(\lambda) = \lambda\phi(\lambda) + R(\lambda), \quad R(\lambda) = (0, 0, .., 0, \Phi_n(\lambda))^T$$

where $\Phi_n(\lambda) = b^*_{n-1} \phi_{n-1}(\lambda) + (a_n - \lambda)\phi_n(\lambda)$. In general $\Phi_n(\lambda) \neq 0$, unless when $\lambda$ is an eigenvalue, in which case $\det |\Phi_n(\lambda)| = 0$ and the following homogenous system $\Phi_n(\lambda)\gamma = 0$ has nontrival solution $\gamma = \gamma(\lambda)$.

Let $\lambda_k$ be an eigenvalue with multiplicity $s_k$, then the homogenous system possesses $s_k$ linearly independent solutions $\gamma^\ell(\lambda_k)$, $\ell = 1, 2, .., s_k \leq r$. Furthermore we normalize $\phi(\lambda_k) \gamma^\ell(\lambda_k)$ as follows:

$$(\phi\gamma^\ell)^* (\phi\gamma^k)_{\lambda=\lambda_k} = \gamma^{\ell*} \phi^* \phi\gamma^k = \delta_{k\ell}I, \quad \delta_{kk} = 1; \quad \delta_{k\ell} = 0 \text{ for } k \neq \ell$$    (5)

If we take into account the multiplicity, then we may omit the superscript of $\gamma$, and just simply write $\gamma(\lambda_k)$.

**Definition:** The spectrum function $\rho(\lambda)$ of the block matrix (1) is defined by

$$\rho(\lambda) = \sum_{\lambda_k < \lambda} \gamma(\lambda_k) \gamma^*(\lambda_k), \ \lambda \in (-\infty, +\infty)$$

By the definition $\rho(\lambda)$ is a nondecreasing and semipositive function:

(1)    for $\lambda, \mu \in (-\infty, +\infty)$ and $\lambda < \mu$, then $\rho(\lambda) \leq \rho(\mu)$;

(2)    $\rho(\lambda) = \rho^*(\lambda) \geq 0$.

**Remark 1.** It is easy to check that for matrix (1) the spectrum function is uniquely determined and assumption (A2) must be valid.

Denote by $\hat{F}\rho$ is the set of all functions of $\lambda$ in F with measure $d\rho(\lambda)$, then we define

(1)    $a(\lambda) \overset{\cdot\rho}{=} b(\lambda) \Leftrightarrow$ for all $c(\lambda) \in \hat{F}_\rho$, $\int a(\lambda)\, d\rho\,(\lambda)\, c(\lambda) = \int b(\lambda)\, d\rho(\lambda)\, c(\lambda)$;    (6)

(2)    $a(\lambda) \overset{\rho\cdot}{=} b(\lambda) \Leftrightarrow$ for all $c(\lambda) \in \hat{F}_\rho$, $\int c(\lambda)\, d\rho(\lambda)\, a(\lambda) = \int c(\lambda)\, d\rho(\lambda)\, b(\lambda)$;

(3)    $a(\lambda) \overset{\cdot\rho}{=} b(\lambda) \Leftrightarrow a^*(\lambda) \overset{\rho\cdot}{=} b^*(\lambda)$.

The same definition are used for vectors, if $\hat{H}\rho$ denote the space of vector functions of $\lambda$, then $\phi(\lambda) \in \hat{H}\rho$ and in the above sense we have

$$A\phi(\lambda) \overset{\cdot\rho}{=} \lambda\phi(\lambda) \qquad (7)$$

For each $f \in H$ we define its Fourier transform with respect to $\phi(\lambda)$ by

$$\hat{f}(\lambda) = \phi^*(\lambda) f \in \hat{F}_\rho \qquad (8)$$

and the inverse Fourier transform by

$$f = \int_{-\infty}^{+\infty} \phi(\lambda)\, d\rho(\lambda)\, \hat{f}(\lambda) \qquad (8')$$

By direct calculating it is easy to derive (8') from (8). Since

$$f = \int_{-\infty}^{+\infty} \phi(\lambda)\, d\rho(\lambda)\, \hat{f}(\lambda) = \int_{-\infty}^{+\infty} \phi\, d\rho\, \phi^* f, \quad \text{for all } f \in H, \qquad (9)$$

So we have

$$I = \int_{-\infty}^{+\infty} \phi(\lambda)\, d\rho\, \phi^*(\lambda) \qquad (10)$$

Notice that if all the $b_j$'s are invertible, then for $\phi(\lambda) = (\phi_1, \phi_2, .., \phi_n)$ $\phi_k(\lambda)$ is just a polynomial in $\lambda$ of order $k-1$, i.e.

$$\phi_k(\lambda) = b_{k-1}^{-1} \cdots b_1^{-1} \lambda^{k-1} + \cdots = (b_1 b_2 \cdots b_{k-1})^{-1} \lambda^{k-1} + \cdots, \; I = \phi_1(\lambda) \qquad (11)$$

and it is easy to see that the transforms defined by (8) and (8') is an one to one correspondence. Furthermore

$$\hat{f}(\lambda) = \int_{-\infty}^{+\infty} \phi^*(\lambda)\, \phi(\mu)\, d\rho(\mu)\, \hat{f}(\mu) = \int_{-\infty}^{+\infty} \delta_\rho(\lambda, \mu)\, d\rho(\mu)\, \hat{f}(\mu), \text{ for all } \hat{f} \in \hat{H}_\rho \qquad (12)$$

So we denote

$\phi^*(\lambda)\, \phi(\mu) = \delta_\rho(\lambda, \mu)$ — the Dirac delta function with resp. to $\rho(\lambda)$. Now we reach a generalized form of Parsaval Equality:

$$f^* g = \int\limits_{-\infty}^{+\infty} \hat{f}^*(\lambda) \, d\rho(\lambda) \, \hat{g}(\lambda), \text{ for all } f, g \in H,$$

or in particular

$$f^* f = \int\limits_{-\infty}^{+\infty} \hat{f}^*(\lambda) \, d\rho(\lambda) \hat{f}(\lambda), \text{ for all } f \in H \tag{13}$$

## 3. Inverse Eigenvalue Problems

In this section we will formulate the inverse eigenvalue problem and present the main result for a block Jacobi matrix. First we need the following two assuptions (A1) and (A2):

(1)    all $b_k$, $k = 1, 2 ..., n-1$ are invertible;    (A1)

(2)    $b_k$'s are fully determined by $a_1, \ldots, a_k$ and $b_1, \ldots, b_{k-1}$.

In practice $b_k$ may be given, or $b_k$ may only depend on $a_k$. For the completeness of the spectrum function $\rho(\lambda)$ we also assume that

(1)    $\text{rank} \{\rho(\lambda_k + 0) - \rho(\lambda_k - 0)\} = r_k \leq r$, $r_k$—multiplicity of $\lambda_k$    (A2)

(2)    $\sum\limits_k r_k = r.n$.

It is easy to see for a given block Jacobi matrix (1) assumption (A2) must be fulfilled.

**Theorem 1.** Under (A1), (A2) the matrix A of the form (1) is uniquely determined by the spectrum function $\rho(\lambda)$.

**Proof:** Our proof of this theorem is constructive, first from (7) we have

$$\phi^* A = (A\phi)^* = \lambda \phi^* + R^*(\lambda),$$

i.e.

$$(\phi_1^*(\lambda), \ldots, \phi_n^*(\lambda)) A = (\lambda \phi_1^*, \lambda, \phi_2^*, \ldots, \lambda \phi_n^* + \Phi^*(\lambda))$$

this implies

$$\phi_1^*(\lambda) a_1 + \phi_2^*(\lambda) b_1^* = \lambda \phi_1^*(\lambda)$$
$$\phi_1^*(\lambda) b_1 + \phi_2^*(\lambda) a_2 + \phi_3^*(\lambda) b_2^* = \lambda \phi_2^*(\lambda) \tag{14}$$

$$\ldots\ldots$$

$$\phi_{n-1}^*(\lambda) b_{n-1} + \phi_n^*(\lambda) a_n = \lambda \phi_n^*(\lambda) + \Phi_n^*(\lambda) \overset{\rho.}{=} \lambda \phi_n^*.$$

Moreover relation (10) implies

$$\int\limits_{-\infty}^{+\infty} \phi_k(\lambda) \, d\rho(\lambda) \, \phi_j^*(\lambda) = I \delta_{kj}, \; \delta_{kk} = 1, \; \delta_{kj} = 0 \text{ for } k \neq j.$$

Therefore multiplying the equation of (14) from left by $\phi_1(\lambda)$ and using (14) gives

$$a_1 = \int \lambda \phi_1(\lambda) \, d\rho \, \phi_1^*(\lambda), \; \phi_1(\lambda) = \phi_1^*(\lambda) = I$$

now by $(A_1)$ $b_1$ can be found and $\phi_2(\lambda) = b_1^{-1} \, (\lambda - a_1), \ldots$ If $a_1, a_2, \ldots, a_{k-1}$ ; $b_1, b_2, \cdots, b_{k-1}$ and $\phi_1, \phi_2, \ldots, \phi_k$ are known, then from the k-th equality of (14), we can find as follows

$$a_k = \int_{-\infty}^{\infty} \lambda \phi_k(\lambda) \, d\rho \, \phi_k^*(\lambda), \; b_k, \text{ and } \phi_{k+1}(\lambda), \; k = 2, 3, \ldots,$$

by induction the proof is completed.

The proof of the theorem is constructive, but it does not provide a practical algorithm for numerical computation, this is because the construction is numerically unstable for large n (see de Boor and Golub [2] for $r = 1$). Interestingly we have the following theorem:

**Theorem 2.** If A is a block Jacobi matrix of form (1), where all $b_k$ 's are nonsingular, then under (A2) all $a_k$'s and $b_k b_k^*$'s are uniquely determined by the spectrum function $\rho(\lambda)$.

**Proof.** There is only a little difference in the proof. First we have

$$a_1 = \int \lambda \phi_1 \, d\rho \, \phi_1^* \text{ and } b_1 = \int \lambda \, \phi_1 \, d\rho \, \phi_2^*, \; \phi_1 = I$$

then we have

$$b_1 b_1^* = \int \lambda \phi_1 \, d\rho (b_1 \phi_2)^* = \int \lambda \, \phi_1 \, d\rho ((\lambda - a_1) \phi_1)^*$$

Similarly,

$$a_k = \int \lambda \phi_k \, d\rho \, \phi_k^*$$

$$b_k b_k^* = \int \lambda \, \phi_k \, d\rho \, \Phi_k^*(\lambda), \; \Phi_k(\lambda) = b_{k-1}^* \, \phi_{k-1} + (\lambda - a_k) \phi_k$$

$$k = 1, 2, \ldots, n-1, \; a_n = \int \lambda \, \phi_n \, d\rho \, \phi_n^*.$$

The proof is completed.

**Remark 2.** Since $b_k b_k^*$ is unique, then by QL factorization method we may suppose $b_k$ 's are low triangular matrices, and A becomes band matrix.

**Remark 3.** We will show in the next section that assumption (A2) makes all $b_k$ 's invertible(nonsingular).

## 4. Some Algorithms for Solving the Problem

Let A be a given block Jacobi matrix, $B = A + D$, where $D = \{d_1, \ldots, d_n\}$ is a block diagonal matrix, and let $\rho_0(\lambda)$ and $\rho(\lambda)$ be given as the spectrum functions of A and B respectively, then we are required to determine D.

For this purpose we introduce two block matrices U and V as follows

$$\tilde{U} = \int\limits_{-\infty}^{+\infty} \phi^0(\lambda)\, d\rho(\lambda)\, \phi^*(\lambda), \quad \tilde{V} = \int\limits_{-\infty}^{+\infty} \phi(\lambda)\, d\rho_0(\lambda)\, \phi^{0*}(\lambda) \tag{15}$$

where $\phi^0(\lambda)$ and $\phi(\lambda)$ are eigenfunctions of A and B respectively. It is easy to see that

$$\phi^0(\lambda) = \tilde{U}\, \phi(\lambda), \qquad \phi(\lambda) = \tilde{V}\, \phi^0(\lambda) \tag{16}$$

$$A\, \tilde{U} = \tilde{U}\, B + R_1, \qquad B\, \tilde{V} = \tilde{V}\, A + R_2 \tag{17}$$

where $R_1 = \int R^0(\lambda)\, d\rho(\lambda)\, \phi^*(\lambda)$, $R_2^* = \int \phi^0(\lambda)\, d\rho_0(\lambda)\, R^*(\lambda)$, and $R^0$, $R$ defined in (4), moreover entries of $R_1$ and $R_2$ are 0's except the last row. Noticing the structure of A and B we know

$$\tilde{U} = I + U, \quad U = \{U_{kj}\}, \quad U_{kj} = 0, \quad \text{for } k \leq j \tag{18}$$

$$\tilde{V} = I + V, \quad V = \{V_{kj}\}, \quad V_{kj} = 0, \quad \text{for } k \leq j$$

letting $\rho(\lambda) = \rho_0(\lambda) + \sigma(\lambda)$, we introduce a Hermitian matrix F as follows:

$$F = \int\limits_{-\infty}^{+\infty} \phi^0(\lambda)\, d\sigma(\lambda)\, \phi^{0*}(\lambda) = -I + \int\limits_{-\infty}^{+\infty} \phi^0(\lambda)\, d\rho\, \phi^{0*}(\lambda) \tag{19}$$

From (16), (18) and (19) we have the following relations:

$$(I+U)\,(I+V) = I, \quad (I+V)\,(I+U) = I \tag{20}$$

$$I + F = \int\limits_{-\infty}^{+\infty} (I+U)\,\phi\, d\rho\, \phi^*\,(I+U)^* = (I+U)\,(I+U)^*$$

It follows that

$$(I+V)\,(I+U)\,(I+U^*) = (I+U^*) = (I+V)\,(I+F)$$

i.e.

$$V + F + VF = U^*, \tag{21}$$

this is equivalent to the following systems:

$$V_{kj} + F_{kj} + \sum_{\ell=1}^{k-1} V_{k\ell} F_{\ell k} = 0, \quad 1 \leq j < k \,(\text{Gel'fand–Levitan})$$

$$U_{kj}^* = F_{jk} + \sum_{\ell=1}^{j-1} V_{j\ell} F_{\ell k}, \quad 1 \leq j < k$$

From (20) and (17) we have

$$V_{kk-1} = -U_{kk-1}, \quad d_1 = -b_1 V_2, \quad d_k = -b_k V_{k+1,k} + V_{kk-1} b_{k-1}, \quad k = 2, \ldots, n-1 \qquad (22)$$

Therefore the basic steps of the algorithm are:

(1)   From $\rho(\lambda)$, $\rho_0(\lambda)$ construct F;

(2)   solve linear systems (G−L) for $k = 2, 3, \ldots$

(3)   determine $d_k$'s, for $k = 1, 2, \ldots, n-1$.

An extra work is needed to compute $d_n$.

Another way to solve this problem is to apply Cholesky Factorization to matrix I+F as in (20), we leave it to the readers.

From the proof of theorem 1 we also can formulate an algorithm based on orthogonalization procedure. In fact if we denote

$$<a(\lambda), b(\lambda)> = \int_{-\infty}^{+\infty} a(\lambda) \, d\rho \; \overset{*}{b}(\lambda), \quad a, b \in \hat{F}_\rho \qquad (23)$$

then we have

$$<\phi_k(\lambda), \phi_\ell(\lambda)> = \delta_{k\ell} \, I, \quad k, \ell = 1, 2, .., n$$

So $\{\phi_k(\lambda)\}$ are normolized-orthogonal polynomials, let $\{P_k(\lambda)\}$

$$<P_k(\lambda), P_\ell(\lambda)> = 0, \quad 1 \le \ell < k, \quad P_1(\lambda) = I$$

with the form $P_k(\lambda) = \lambda^{k-1} + p_1 \lambda^{k-2} + \ldots + p_{k-1}, \quad p_\ell \in F$.

If all $b_k$'s are nonsingular, then it is easy to see that

(1)   $P_{k+1}(\lambda) = \lambda P_k(\lambda) + \alpha_k P_k(\lambda) + \beta_k P_{k-1}(\lambda), \quad P_1(\lambda) = I.$

$\alpha_k <P_k, P_k> = -<\lambda P_k, P_k>,$

$\beta_k <P_{k-1}, P_{k-1}> = -<P_k, \lambda P_{k-1}> = -<P_k, P_k>;$

(2)   $P_k(\lambda) = (b_1 \cdots b_{k-1}) \phi_k(\lambda)$, and

$<P_k, P_k> = b_1 \cdots b_{k-1} (b_1 \cdots b_{k-1})^* > 0,$

So $\alpha_k, \beta_k$ are uniquely determined;

(3)   $a_k = -(b_1 \cdots b_{k-1})^{-1} \alpha_k (b_1 ..., b_{k-1})$

$b_{k-1} b_{k-1}^* = -(b_1 \cdots b_{k-2})^{-1} \beta_k (b_1 \cdots b_{k-2}),$

So $a_k, b_k b_k^*$ are uniquely determined.

We now show an important theorem which insures all $<P_k, P_k>$ ($b_k$'s) invertible.

**Theorem 3.** Assumption (A2) insures all $<P_k, P_k>$'s ($b_k$'s) invertible.

**Proof.** If $<P_1, P_1>, \ldots, <P_{k-1}, P_{k-1}> > 0$, but $<P_k, P_k>$ singular, i.e. $\det <P_k, P_k> = 0$, then for the following block $k \times k$ matrix

$$\Lambda = \begin{pmatrix} <1,1> & <1,\lambda> & & <1,\lambda^{k-1}> \\ <\lambda,1> & <\lambda,\lambda> & \cdots & <\lambda,\lambda^{k-1}> \\ \cdots & \cdots & \cdots & \cdots \\ <\lambda^{k-1},1> & <\lambda^{k-1},\lambda> & \cdots & <\lambda^{k-1},\lambda^{k-1}> \end{pmatrix}, \quad \Lambda Q = 0$$

the homogenous system has nontrival solution $Q = (q_1, q_2, \ldots, q_k)^T$, it implies

$$<Q(\lambda), Q(\lambda)> = 0, \quad Q(\lambda) = q_k + q_{k-1} + \ldots + q_1 \lambda^{k-1};$$

i.e. $\det |Q(\lambda_j)| = 0, j = 1, 2, \ldots, nr$. This means $\det Q(\lambda)$ must be a polynomial of order $nr$, it is impossible for $k \leq n$.

**Remark 4.** The Theorem 3 also shows the solvability of the inverse problem under assumption (A2).

## 5. Application Aspects

The problem presented in this paper is related to those of so-called impulse response problems in many applications such as geophysics, nondestructive testing, identification and acoustics etc. A typical example of such problem can be described as follows: Consider a discrete mechanical system . Matrix B is regarded as its stiffness and $B = A + D$ with A given and D to be determined. The motion of the system can be described as a solution of the following ODE's system:

$$u''(t) + Bu(t) = 0, \quad u(0) = u_0, u'(0) = 0 \tag{23}$$

Suppose B is positive, i.e. $B > 0$, then we know that

$$B = \int_{-\infty}^{+\infty} \lambda \phi(\lambda) d\rho(\lambda) \phi^*(\lambda) = \int_0^{+\infty} \lambda \phi(\lambda) d\rho(\lambda) \phi^*(\lambda)$$

where $\rho = 0$ for $\lambda < 0$. Let $\hat{u}(\lambda, t) = \phi^*(\lambda) u(t)$ is the Fourier transform of $u(t)$, then

$$u(t) = \int_0^{+\infty} (\cos \sqrt{\lambda} t) \phi(\lambda) d\rho(\lambda) \phi^*(\lambda) u_0 \tag{24}$$

For a very special initial value $u = (I, 0, \ldots, 0)^T =$ the impulse, we can then record the response to the impulse at the first component of $u(t)$, that is a set of data:

$$u_1(t) = \int_0^{+\infty} \cos \sqrt{\lambda} t \; \phi_1(\lambda) d\rho(\lambda) \phi_1^*(\lambda) = \int_0^{+\infty} \cos \sqrt{\lambda} t \; d\rho(\lambda) \tag{25}$$

Therefore $\rho(\lambda)$ can be found by an inverse Fourier transform of (25). The next step then would be to use the above algorithm to find D.

For example the figure below represents a discrete mechanical system with 20 nodes, so we have a $5 \times 5$ block Jacobi matrix, and $\rho(\lambda)$ will be $4 \times 4$ as $u_1(t)$.

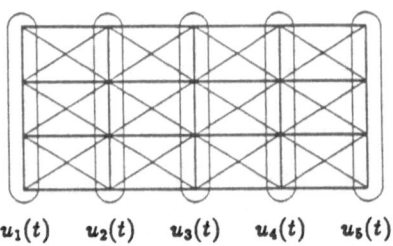

$$u_1(t) \quad u_2(t) \quad u_3(t) \quad u_4(t) \quad u_5(t)$$

Figure 1: A discrete mechanical system with twenty nodes.

This work also provides some insight into the question - "can one hear the shape of a drum"? [6].

## Acknowledgement

This work was started during Dr. Zhu's visit to Universities of Toronto and Waterloo. The authors would like to thank NSF(China), NSERC(Canada) and ITRC(Information Technology Research Centre, Canada) for their support. Dr. Zhu's special thanks should go to the Scientific Computation Group of Computer Science Department of the University of Waterloo for their hospitality and support to his visit. We would also like to thank Mrs. Z. Kaszas for her help in processing this paper.

## References

[1].   F.W. Biegler-Konig, 1981, Lin.Alg.Appl. 40, pp. 79-84.

[2]   de Boor and G.H. Golub, 1978, Lin.Alg.Appl. 21, pp. 245-266.

[3]   D.L. Boley and G.H. Golub, 1977, Lecture Notes in Math. Springer, Berlin.

[4]   D.L. Boley and G.H. Golub, 1987, Inverse Problem 3, pp. 95-622.

[5]   G.M.L. Gladwell, 1984, 1986, Proc. R.Soc.A. 393, pp. 277-295.
      Q.J. Mech.Appl.Math. 39, pp. 297-307.

[6]   M.H. Protter, 1987, SIAM Rev. Vol. 29, No.2.

*Dr. B. Zhu's permanent address is:
Mathematics Department
Shandong University
Jinan, Shandong 250100
P.R. China

# Parametric Correction of Finite Element Models Using Modal Tests

P. LADEVEZE, M. REYNIER, D. NEDJAR

Laboratoire de Mécanique et Technologie
E.N.S. Cachan / C.N.R.S./Université Paris VI
61, Av. du Président Wilson
94235 CACHAN Cedex - France

## Summary
This paper concerns the computational aspects of a the new correction method for the control and the refinement of a given finite element model using modal tests results, when finite element models represent complex structures.The updating strategy is iterative and each iteration is divided into two stages : first, the local error measure on the constitutive relation is used to locate the most erroneous areas, second, the mass and stiffness matrices are corrected. Some examples illustrate this strategy.

## Introduction

A recent tendency in engineering is to reduce the number of tests and to prefer numerical simulations even for complex structures ; the tests are used to validate and verify the model. Tuning problems appear for many free-vibration industrial problems where the results obtained with finite element models are not too far from the observed experimental results. Nevertheless, differences can be observed : they proceed from erroneous estimation of the parameters describing the mass and stiffness properties. These errors are frequent in the modelling of joining sub-structures. The strategy , established by [LAD 89] [REY 90], is based on the concept of the error on the constitutive relation, and on the hypothesis that the experimental eigenvalues have generally a satisfactory quality. We present particularly in this paper the computational aspects of the new parametric correction method that uses the global error on the constitutive relation for the whole structure to quantify both the *mass* and *stiffness* matrices.

Different methods have been proposed to solve tuning problems : Baruch [ BAR 82], Berman [BER 71], Chen [CHE 89], Caesar [CAE 84], Ewins [EWI 88], Link [LIN 87] construct directly the corrected mass and stiffness matrix. Collins [COL 74] and Zhang [ZHA 87] apply a sensitivity analysis. Other approaches based on the residues computation of equilibrium equations have been described by Cottin [COT 84] and Berger [BER 89].

Our strategy is an iterative parametric strategy where each iteration needs two stages : first, the local error measure on the constitutive relation is used to locate the most erroneous areas, and second, the corrections of the involved structural parameters are computed. The iterative process is stopped, when the global error computed for the whole structure and all the experimental modes is less than an accuracy $\varepsilon_0$ fixed by the user.

## The Error Measure on the Constitutive Relation

The data of the tuning problem are : the finite element model, the experimental eigenvalues $\lambda_i$, i$\in$ [1, q] for the q "measured" modes, the experimental eigenshapes $\Pi U_i$, where $\Pi$ is the projection operator indicating that the experimental eigenshape is partly measured.

The starting point is to assume that the quality of the experimental eigenvalues is good enough, i.e., the model has to give the experimental eigenvalue $\lambda$.

Then, let $\Omega$ be a bounded subset with the boundary $\partial\Omega$ corresponding to the structure. To specify the boundary conditions, let us consider two complementary subsets $\partial_1\Omega$ and $\partial_2\Omega$. Let the displacement field be given on $\partial_1\Omega$ and the stress field be given on $\partial_2\Omega$. And let be $U = \{U', U' | \partial_1\Omega=0, U'$ regular$\}$, we then search to construct the complete associated eigenshapes by solving for each given $\lambda$ the following problem $P_\lambda^1$ :

> To find a couple ( $U, \sigma$ ) where $U$ is a displacement field and $\sigma$ a stress field , such as :
>
> - $U$ satisfies the kinematic constraints, $U \in U$          (1)
>
> - ( $U, \sigma$ ) satisfies the equilibrium equation :
>
> $$\forall\, U^* \in U \qquad \int_\Omega Tr(\sigma\varepsilon(U^*))d\Omega = \lambda \int_\Omega \rho\, U.U^* d\Omega \quad (\rho \text{ density}) \qquad (2)$$
>
> - and such as ( $U, \sigma$ ) verifies the constitutive relation $\sigma = K\,\varepsilon(U)$.

$K$ is Hooke's operator and $\varepsilon$ the strain operator, (1) and (2) mean ( $U, \sigma$ ) is an *admissible* couple. Let us introduce : $A_d = \left\{ (U', \sigma') , \quad U' \text{ regular, } U' | \partial_1\Omega = 0, \text{ and } ( U', \sigma') \right.$

verifying $\forall\, U^* \in U \left. \int_\Omega Tr(\sigma'\varepsilon(U^*))d\Omega = \lambda \int_\Omega \rho\, U'\, U^* d\Omega \quad (\rho \text{ density}) \right\}$ .

An equivalent problem is to search, *for the given experimental eigenvalue $\lambda$*, the admissible couple (U, $\sigma$) where U is a displacement field and $\sigma$ a stress field, such as they minimise on $A_d$ the error measure on the constitutive relation :

$$J : ( U', \sigma' ) \dashrightarrow J (U', \sigma') = \| \sigma' - K\varepsilon(U') \|^2 \text{ with } \| \sigma' \|^2 = \int_\Omega Tr (\sigma' K^{-1}\sigma') \, d\Omega$$

We obtain a displacement approach associating to the stress field $\sigma'$ the displacement field $V' \in U$ solution of the following elastic problem :

$$\forall\, U^* \in U \qquad \int_\Omega Tr(K\varepsilon(V') - \sigma')\varepsilon(U^*) \, d\Omega = 0$$

Let us introduce $\mathbf{B_d} = \Big\{ (U', V') ,\ U' \text{and } V' \text{ regular, } U'|_{\partial_1 \Omega} = 0,\ V'|_{\partial_1 \Omega} = 0,\ (U', V')$

verifying$\forall\ U^* \in U \int_\Omega Tr(K\varepsilon(V')\varepsilon(U^*))d\Omega = \lambda \int_\Omega \rho\ U\ U^* d\Omega$ ($\rho$ density)$\Big\}$.

The displacement approach consists then in the finding the couple $( U, V ) \in \mathbf{B_d}$ such as they minimise the error on the constitutive relation measured by :

$$\Big| J : ( U', V' ) \dashrightarrow J ( U', V' ) = \| U'-V' \|^2 \text{ with } \| U \|^2 = \int_\Omega Tr(K\varepsilon(U)\varepsilon(U))d\Omega$$

And to take the measures into account, we propose finally to minimize the **global modified error** on the constitutive relation solving the following problem $P_\lambda^2$

$$\Big| \begin{array}{l} \text{To find } (U,V) \in \mathbf{B_d} \text{ such as they minimize} \\ E : (U',V') \dashrightarrow E(U',V') = \| U'-V' \|^2 + \dfrac{r}{1-r} \| \Pi U' - \underline{\Pi U} \|^2 \end{array}$$

with $\| U' \|^2 = \int_\Omega Tr(K\varepsilon(U')\varepsilon(U'))d\Omega$ where $\underline{\Pi U}$ is the measured part of the $\lambda$-associated eigenshape.

Remarks :

- r is a scalar expressing the confidence in the quality of the experimental shapes, a current value is 1/2. For very noisy measures, low values should be chosen for r , r=0.1 for example.

-We take into account the modes orthogonality : we arrange the experimental modes following the ascending values of E(U,V), and we impose that U is orthogonal with the preceding generalised displacements. To introduce these constraints, we use Lagrange multipliers which become supplementary variables.

## The Error Indicators - The Localization Method

The error measure on the constitutive relation, computed for the whole structure and for the all q measured modes $\mathcal{E}^q = \left( \sum\limits_{i=1}^{q} \dfrac{\|U_i-V_i\|^2}{\frac{1}{2}\left(\|U_i\|^2+\|V_i\|^2\right)} \right)^{1/2}$ , measures the model correctness.

We define the following indicators $\mathcal{E}^{q\ 2}_{(s)}$ and $\eta^{q\ 2}_{(s)}$ for each sub-structure (s),

$$\mathcal{E}^{q\ 2}_{(s)} = \sum\limits_{k=1}^{q} \mathcal{E}^2_{k(s)} \text{ and } \eta^{q\ 2}_{(s)} = \dfrac{\| U - V \|^2_{(s)}}{1/2\left(\| U \|^2_{(s)}+\| V \|^2_{(s)}\right)}.$$

with $\mathcal{E}^2_{k(s)} = \left( \dfrac{\| U_k - V_k \|^2_{(s)}}{1/2\left(\| U_k \|^2+\| V_k \|^2\right)} \right)$ and $\| U \|^2_{(s)} = \int_S Tr(K\varepsilon(U)\varepsilon(U))\ ds$

These indicators allow to locate the areas that should be corrected as a matter of priority.

## Updating of Finite Element Models

We discretize the problem $P_\lambda^2$, and let u and v be the nodal values of U and V. For a free vibration problem without damping, the F.E. model is characterized by the symmetric mass matrix **M** and stiffness matrix **K** dimension in which **n** is the number of degrees of freedom. The m first eigenvalues and their associated eigenmodes have been computed ( m<< n ).

We associate the admissible couple (u , v) with the theoretical model and with the experimental eigenmode ( $\lambda$ , $\underline{\pi u}$ ) such as (u, v) minimises the error on the constitutive relation.

Let be $\mathbf{U} = \{ u' , u' | \partial_1 \Omega = 0 \}$, we write the following problem $P_\lambda^3$ :

> To find the displacement fields ( u , v ) $\in$ **U** minimizing :
> $$e^2 : ( u', v' ) \text{-----> } e^2( u', v' ) = \| u' - v' \|^2 + \frac{r}{1-r} \| \pi u' - \underline{\pi u} \|^2$$
> with the constraint $\mathbf{K} v' = \lambda^2 \mathbf{M} u'$

<u>Remarks</u> : The choice of $\| .. \|^2$ is minor, we use Guyan's reduction $\mathbf{K_r}$ of the stiffness matrix on the measured degrees of freedom $\| \pi u' - \underline{\pi u} \|^2 = (\pi u' - \underline{\pi u} )^t \mathbf{K_r} ( \pi u' - \underline{\pi u} )$.

## Correction of the Mass and Stiffness Errors

For the tuning iteration t, the stiffness matrix is expressed by $\mathbf{K_t} = \mathbf{K_{t-1}} + \Delta K(p_t)$ and the mass matrix by $\mathbf{M_t} = \mathbf{M_{t-1}} + \Delta M(p_t)$ with $\mathbf{M_0} = M$ and $\mathbf{K_0} = K$ . Then, the correction problem $P_c$ is written for the iteration t :

> Let it be that $\mathbf{P_t} = \{ p_t,$ such as they ensure the properties of $\mathbf{K_t}$ and $\underline{M_t}.\}$
> To find $p_t$ minimizing on $\mathbf{P_t}$,
> $$e^2 : p' \text{---> } e^2(p') = \sum_{i=1}^{q} \{ \| u_i - v_i \|^2 + \frac{r}{1-r} \| \pi u_i - \underline{\pi u_i} \|^2 \} \text{ with } \mathbf{K_t} v_i = \lambda_i \mathbf{M_t} u_i$$

Taking the equilibrium equations into account u-v = Q u, $P_c$ becomes:

> Find ( $u_i$, $p_t$ ) $u_i \in U$, $p_t \in P_t$ , such as they minimize
> $$F : u', p' \text{---> } F( u'_i, p') = \sum_{i=1}^{q} \| Q(p')u'_i \|^2 + \frac{r}{1-r} \| \pi u'_i - \underline{\pi u_i} \|^2$$

Then $F( u_i , p') = \sum_{k=1}^{q} (- ( \frac{r}{1-r} ) (\pi u_i)^t \mathbf{K} (\pi u_i - \underline{\pi u_i})$ , finally we have to solve :

> To find $p_t$ minimizing on $\mathbf{P_t}$ : $\qquad H : p' \text{---> } H( p') = F ( u_i, p')$

For each iteration t, the correction problem is non-linear but the number of variables is very low.

## Numerical Implementation.

### Reduced computational base

For the localization stage, the linear problem $P_\lambda^3$ has a dimension n, the number of degrees of freedom of the finite element model, but this dimension can be notably lower, because a reduced modal base is a priori enough to describe the fields u and v without the fineness that should be given using a complete modal base. Let us consider $u = \Phi_t a$ where $\Phi_t$ is the m-truncated modal matrix used in iteration t. We obtain then an approximated problem of $P_\lambda^3$ to solve. For the first iteration we use $\Phi_0$, a m-truncated modal base of the given finite element model.

The first m eigenvalues and eigenvectors are computed ($\lambda_i, x_i$), $i \in (1, m)$, $1 \leq i \leq m <<$

$$n \quad u = \Phi_0 a \qquad \text{or} \qquad u = \sum_{i=1}^{m} a_i x_i \quad \text{then the equilibrium equation gives} \quad v = \sum_{i=1}^{m} \frac{\lambda}{\lambda_i} a_i x_i$$

### The approximated location problem

We write the approximated problem :          Find $a_i$ such as they minimise

$$E = \sum_{i=1}^{m} a_i^2 \left( \frac{\lambda}{\lambda_i} - 1 \right)^2 + \frac{r}{1-r} \sum_{i=1}^{m} \sum_{j=1}^{m} (\pi a_i x_i - \pi \underline{u})^t \, K_r \, (\pi a_j x_j - \pi \underline{u})$$

This problem admits a unique solution if zero is the solution of the following expression :

$$\sum_{i=1}^{m} a_i^2 \left( \frac{\lambda}{\lambda_i} - 1 \right)^2 + \frac{r}{1-r} \sum_{i=1}^{m} \sum_{j=1}^{m} a_i a_j .(\pi x_i)^t \, K_r \, (\pi x_j) = 0$$

When $\lambda$ is not an eigenvalue, the solution is unique. If not, it is enough that the number of measurement points forbid the modes such as $\pi u = 0$. The resolution leads to a linear system. Then the computation of u and v consists in solving a small linear problem.

### The approximated correction problem

The same approximation is made for the correction problem. But, after each correction stage, we construct a new m-truncated modal base taking the corrections $\Delta K(p_t)$, $\Delta M(p_t)$ into account.

Improvement of the reduced base. Using a p-truncated modal base (p>m, for the proposed examples, we choose p=2m) we compute m new eigenvalues and associated eigenvectors by means of an iterative process successively determining $X_t = X_{t-1} + \Delta X_t$ from the equilibrium

equation $\left( \underline{K}_{t-1} + \Delta K(p_t) \right) (X_{t-1} + \Delta X_t) = \lambda_{t-1} \left( \underline{M}_{t-1} + \Delta M(p_t) \right) (X_{t-1} + \Delta X_t)$

which gives $\Delta X_t$. $\lambda_t$ is then computed using Rayleigh's ratio written with the matrices $\underline{K}_t = \underline{K}_{t-1} + \Delta K(p_t)$ and $\underline{M}_t = \underline{M}_{t-1} + \Delta M(p_t)$. We then obtain a "re-actualized" m-truncated base allowing the improvement of the u- and v-field description.

Approximated correction problem. The equilibrium equation (iteration t+1) can be written for the experimental mode k :

$$v_k - u_k = Q_t u_k \quad \text{with} \quad Q_t = [\lambda_k \Phi_t [\Phi_t^T [\underline{K}_t + \Delta K(p_{t+1})] \Phi_t]^{-1} \Phi_t^T [\underline{M}_t + \Delta M(p_{t+1})] - \text{Id}]$$

the dimension of which is m .In the case when the base improvement is made with a good accuracy, we consider the normality properties: $X_i^t \underline{K}_t X_j = \delta_{ij}$ , then

$$v_k - u_k = \Phi_t \left( \lambda_k [\Phi_t^T \Delta K(p_{t+1}) \Phi_t + \text{Id}]^{-1} [\Phi_t^T \Delta M(p_{t+1}) \Phi_t + [1/\Lambda_i]] - \text{Id} \right) a_k$$

where $a_k$ is the colum of the u-components on the reduced base.

Let us consider $v_k - u_k = \Phi_t G a_k$ where $\Phi_t$ has the dimension $(n, m)$ and $G(\Delta K(p_{t+1})$, $\Delta M(p_{t+1})$ ) has the dimension $(m,m)$. We solve then the following correction problem :

To find $p_{t+1} \in P_{t+1}$ such as they minimise :

$$f^2: (a_k, p') \to f^2(a_k, p') = \sum_{i=1}^{q} a_k^T (\Phi_t G)^T K_t \Phi_t G a_k + \frac{r}{1-r}(\pi \Phi_t a_k - \pi u)^T K r_t (\pi \Phi_t a_k - \pi u)$$

$a_k$ is the solution of $\left( (\Phi_t G)^T K_t \Phi_t G + \frac{r}{1-r} (\pi \Phi_t)^T K r_t \pi \Phi_t \right) a_k = \frac{r}{1-r} (\pi \Phi_t)^T K r_t \pi u$

where $a_k$ is a function of $p_{t+1}$. The correction problem is then solved by means of a conjugate gradient algorithm, which needs less than ten iterations to compute the structural corrections $p_t$.

## Examples

We propose two examples figure 1 : the first and the 2nd benchmark of the A.G.11-Group for Aeronautical Research and Technology in Europ, Action Group on Parametric Updating of Finite Element Models Using Experimental Simulations] [GAR] where the test-structures are plane truss-structures discretized of sample beams with lumped mass distribution. The measured data are simulated ( eventually with noise ) by perturbing the geometrical parameters or the elasticity characteristics, and the modal parameters are recomputed with a finite element program. The modifications on the structure have been made on the elements in the localized areas.

The first truss-structure is a plane <u>free</u> truss-structure where only stiffness errors have been simulated. The 2nd is a plane <u>embedded</u> truss-structure where mass and stiffness errors have been made. The location of measured degrees of freedom is denoted on each example in the figures, the dimensions and the properties are also depicted.

<u>Free plane truss-structure</u>. The stiffness errors are localised as show in figure 1 on 8 elements. The modal base used for v-describing contains 30 eigenvectors, the number of measured d.o.f. is 78 as depicted in figure 1. And only 5 measured modes have been given.

<u>Iteration 1</u>: the localization method detects 5 erroneous elements figure 1.1 with $\eta^2_{s\ max} = .159$. The distance between the initial F.E. model previsions and the experimental data is expressed by :

| modes number | 1 | 2 | 3 | 4 | 5 |
|---|---|---|---|---|---|
| $\Delta\omega / \omega$ (%) | 2,93 | 1,53 | 1,17 | 1,42 | 1,19 |

The first correction process supplies a corrected F.E. model such as :

| modes number | 1 | 2 | 3 | 4 | 5 |
|---|---|---|---|---|---|
| $\Delta\omega / \omega$ (%) | 0,08 | 0,04 | 0,01 | 0,02 | 0,10 |

<u>Iteration 2</u> : we find one supplementary erroneous element with $\eta^2_{s\ max} = .051$. The stiffness correction supplies then a F.E. model , which is close by the experimental model.

| modes number | 1 | 2 | 3 | 4 | 5 |
|---|---|---|---|---|---|
| $\Delta\omega / \omega$ (%) | 0,03 | 0,01 | 0,01 | 0,00 | 0,1 |

If we continue the localisation process, we detect three erroneous elements with $\eta^2{}_{s\,max}$ =002. This error level is very low and the localization of the edge errors become difficult with only 5 measured modes, the localization method detect one element which does not contains stiffness error, but it is contiguous to an erroneous element. Let us remark also, that the contribution of the remaining errors is not important for the dynamic behaviour of the 5 first modes.

Embedded plane truss-structure. (2nd benchmark of the G.A.R.T.EUR) .The localization of measured degrees of freedom is denoted figure 2, the dimensions and the properties are also depicted.The initial difference between the computed and the experimental (simulated) eigenvalues are:

| modes number | 1 | 2 | 3 | 4 | 5 |
|---|---|---|---|---|---|
| $\Delta\omega$ / $\omega$ (%) | -124,4 | - 8,15 | - 13,25 | - 11,8 | -3,35 |

We use a truncated modal base ( m = 30 computed for the given initial finite element model ) to describe the displacement fields, and 5 experimental eigenvalues (q=5) are only given with 78 "measured" components of the associated eigenvectors.

Iteration 1 :The first localization stage allows to detect the elements number : 6, 7, 17, 18, 19, 20, 21 (figure 2.1). The global error is equal to 0.54 and the model needs an improvement process. The computed correction proposes the following modifications after 2 iterations of the conjugate gradient algorithm.

| N° éléments | 6 | 7 | 17 | 18 | 19 | 20 | 21 |
|---|---|---|---|---|---|---|---|
| $\Delta I$ / I (%) | -90 | -90 | -40.3 | -88.8 | -90 | -90 | -90 |

The computed corrections on the areas ( S ) can be neglected. The global staying error is equal to .097 (figure 2.1) and 82% of the initial error has been corrected.

Using the proposed "re-actualized" m-truncated modal base, which take the first computed corrections into account, we have the following improved answer for the elements 17 and 18 :

| for the element number 17, | $\Delta I/I$= -80% | and for the element 18, | $\Delta I/I$=-89%. |
|---|---|---|---|

Iteration 2 : The localization stage (figure 2.2) recognizes the erroneous elements number : 19, 52, 67, 70, 71 . The correction process computes neglectable corrections for the elements 19 and 67 and we have the following corrections for the others :

| N° éléments | 52 | 70 | 71 |
|---|---|---|---|
| $\Delta S$ / S (%) | + 90% | +110% | +104% |

Then, the remaining difference between the computed and the experimental behavior becomes :

| modes number | 1 | 2 | 3 | 4 | 5 | iter |
|---|---|---|---|---|---|---|
| $\Delta\omega$ / $\omega$ (%) | -124.4 | -3.15 | -13.25 | -11.8 | -3.35 | |
| $\Delta\omega$ / $\omega$ (%) | -4.4 | -2.2 | -0.15 | -1.35 | -2.6 | 1 |
| $\Delta\omega$ / $\omega$ (%) | -1.8 | 0.15 | -0.75 | -0.2 | -0.5 | 2 |

The final global modified error attains 0.019 figure 2 correcting also 97% of the initial global modified error. Generally, as soon as we reach a small value of the error, we have to stop the tuning process.The initial model may be completely corrected when we describe the solution fields by means of a complete modal base and if we have all the experimental eigenmodes at our disposal. We correct 99% of the global error takingwith m=70 and 10 experimental modes, and :

| N° éléments | 23 | 24 | 76 | 77 |
|---|---|---|---|---|
| ΔI/I % | +107 | +96 | +80 | +76 |

## Conclusions

In this paper we have presented the numerical implementation of the new parametric correction method associated with the localization method based upon the error on the constitutive relation.This correction process computes the structural parameters correcting the *stiffness and mass* errors. Let us remark that the prealable localization stage allows to handle only the erroneous areas, that is why the ill-posed character of the problem is not prejudicial to the computing of structural parameters. The correction problem has a dimension *limited to the number of parameters describing the areas to modify.* Moreover, the numerical implementation has been made easy to carry out using improved truncated modal bases.

## Références

[BAR 82] Baruch, M.,1982," Optimal correction for mass matrix and stiffness matrix using measuring modes", A.I.A.A. 82-4265 ,1982.

[BER 71] Berman, A., Flannelly, o,,1971, " Theory of incomplete models of dynamics structures ", A.I.A.A. journal, Vol 9/8 , pp 1481-1487

[BER 89] Berger, H., Barthe, L., Ohayon, R., April 1989, "Parametric Updating of a Finite Element Model from Experimental Modal Characteristics", Proc.European Forum on Aeroelasticity and Structural Dynamics 1989, Aachan.

[CHE 89] Chen, J.C., Garba, J.A., 1989, " Analytical Model Improvement Using Modal Test Results", AIAA Paper 79-0831, Vol. 18, N°6, April 1989.

[CAE 84] Caesar, B.,1984,"Correlation and update of dynamic mathematical models ", 4th F.E. world congress, Interlaken,Sept. 1984

[COL 74] Collins, J.D., Hart, G.C., Hasselman, J.K. and Kennedy, B.,1974, " Statical identification of structures ",A.I.A.A. journal, Vol 12/2 pp 185-190

[COT 84] Cottin, N.,Felgenhauer, H.P., Natke, H.G., 1984, " On the parameter identification of elastomechanical systems using input and output residuals ", Ingenieur Archiv 54 ,pp 378-387, Spring-Verlag.

[EWI 88] Ewins, D.J., He, J,1988, "A review of the error matrix method for structural dynamic modes comparison ", Proc. Int. Conf. on Spacecraft Structures Testing, ESA-SP-289.

[GAR] GARTEUR-AG11 Group for Aeronautical Research and Technology in Europ, Action Group on Parametric Updating of Finite Element Models Using Experimental Simulations.

[LAD 89] Ladeveze, P., Reynier, M. , "A Localization Method of Stiffness Errors for the Adjustment of F.E. Models", Vibrations Analysis Techniques and Applications, Chap. "F.E. Modeling and Analysis, pp 350-355, A.S.M.E.publishers.

[LIN 87] Link, M., Weilend, M., Barragan, J.M., 1987, "Direct Physical Matrix Identification as Compared to Phase Resonance Testing - An assesment Based on Practical Application", Proc;, IMAC, London.

[REY 90] Reynier, M., 1990, " Sur le controle de modélisations éléments finis : recalage à partir d'essais dynamiques", These de doctorat de l'Université PARIS VI.

[ZHA 88] Zhang, W.,Lallement, G., Fillod, R., Piranda, J., 1988, " Parametric identification of conservative self-adjoint structures, Proc. Int. Conf. on Spacecraft Structures Testing, ESA-SP-289.

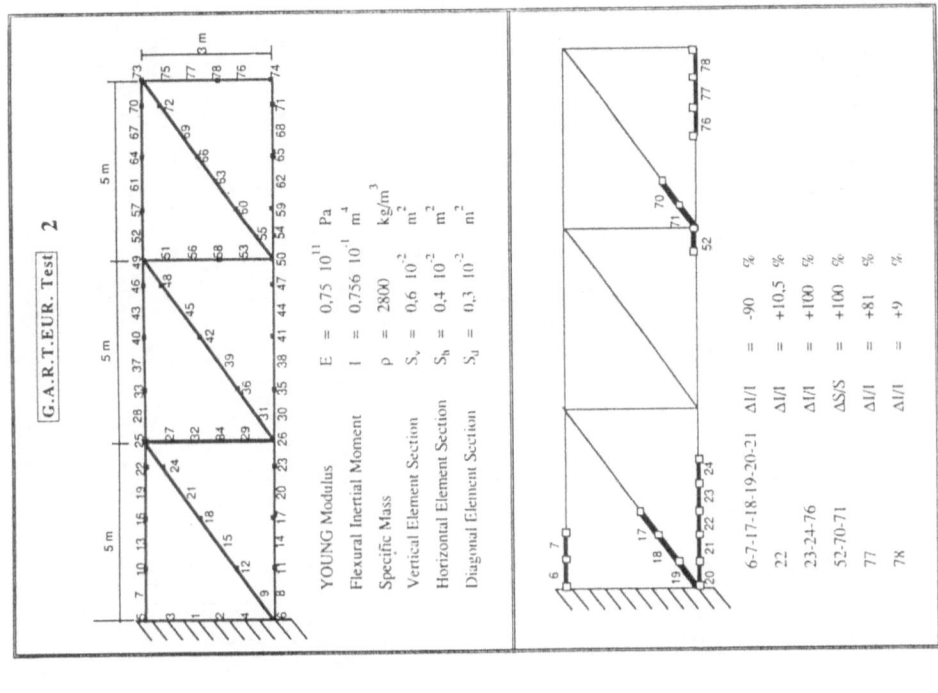

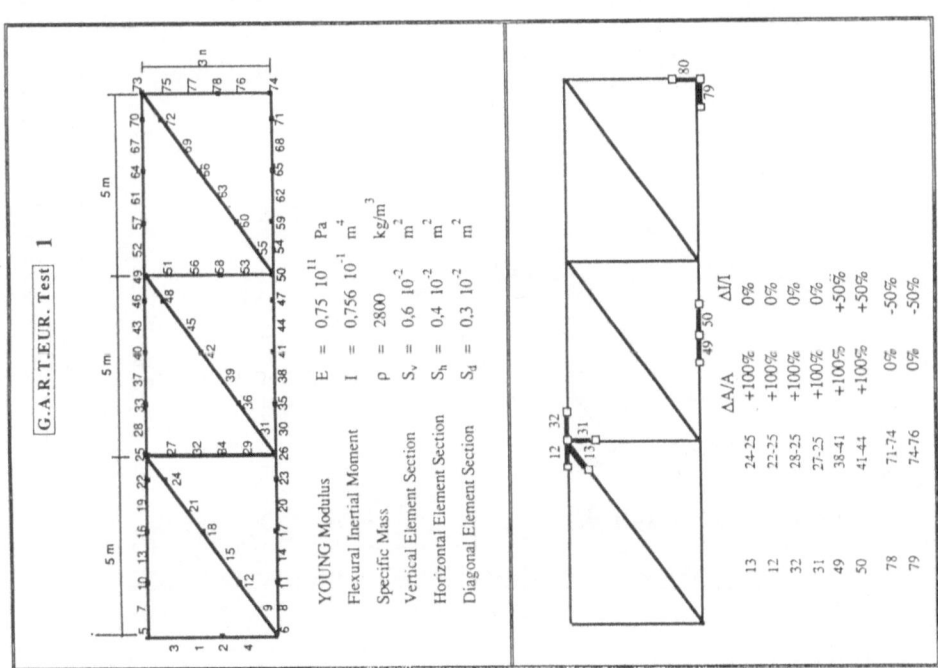

**FIGURE 1**

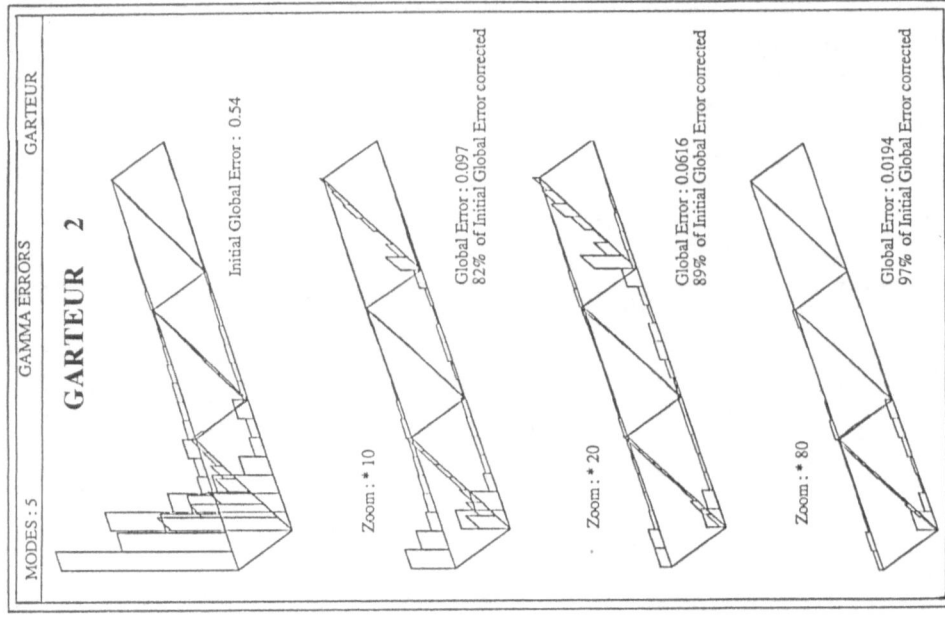

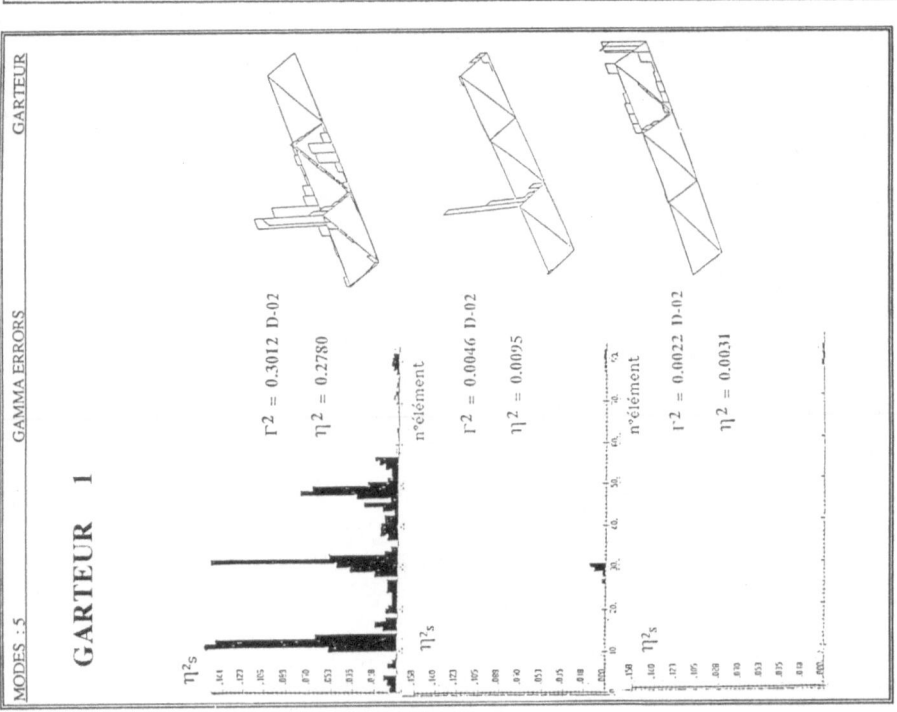

**FIGURE 2**  **Tuning Iterations**

# Structural Defect Synthesis Using Noise-Frequency Response Characteristics for Coupled Acoustic-Structural System by Fuzzy Inference Method

W. Kozukue[*] and I. Hagiwara[*]

[*] Vehicle Research Laboratory, Nissan Motor Co., Ltd
1 Natsushima-cho, Yokosuka 237, Japan

Abstract

Structural identification is the technology which is inevitable for design of complicated strucutures. Since the structure of a vehicle is very complicated, sometimes it happens that the analysis value and experimental value of physical quantities such as noise level do not agree. In such cases the so-called inverse problem is implemented, by which the design variables are assumed properly and the analysis model is modified using the experimental value. However it has the possibility that the model obtained do not have physical meaning, namely the design variables which have essentially the true values are changed improperly. Formerly we tried to investigate the positions which is the cause of the disagreement of the experimental value and analysis value (we call these structural defects) by using vibration data, but it was very difficult to carry out. In this paper the authors will introduce noise data to this field for the first time and try to investigate strucural defects by using the frequency characteristics of noise and sensitivity coefficient of the coupled structural-acoustic system. Moreover we reduce the number of design variables by using Fuzzy inference theory for normalized sensitivity coefficient of noise.

## Introduction

Since the noise and vibration characteristics is one of the basic qualities of vehicles most deeply concerned with the car salability, the analysis technique is strongly desired to yield the proper structure. Therefore the analysis for the characteristics of the original is achieved at first and then the investigation for making it appropriate is implemented. Since the structure of vehicle is complicated, it is not easy to get the agreement between analysis and experiment. The reason for this is as follows : the stiffness of adhesive glass is not clear, it is not possible to make detailed analysis model to reproduce bead or embos set up for taking measures for strength and vibration in relation to calculation time and working volume for model preparation, it is not possible to compute the effect of spot welding, the plate thickness varies widely in the casting product such as engine block, the characteristics of lubricating part is not clear, etc. , namely, various factors can be considered. Therefore the so-called inverse problem has been implemented, which sets up the design variables properly and does identification to modify the model using experimental value. However, it has the possibility to change the intrinsically true value and the model obtained do not have physical meaning. So , the one of the authors has proposed 'certainity coefficient' to find out the reason for the disagreement between experimental value and analysis value [1] . In that paper it was represented that if the number of freedoms and modes is equal it is possiblle to find out the design variables to be changed to some extent in accuracy. However for the large structure such as a vehicle the number of measuring point becomes numerous and it is essentially difficult to measure the same number of mode precisely.

Generally speaking, it is required to devide into finer meshes for vehicle vibration analysis model than those for vehicle noise analysis model. Therefore also in the analysis of coupled structural -acoustic system it is general that the meshes of structural analysis model are much finer than those of acoustic analysis model. In this paper instead of using vibration data of structure we try to do structural identification based on frequency characteristics of noise and sensitivities at the obsevation point. If there is a defect in the structure there should be the clear difference between experimental value and analysis value of noise at some frequencies for some observation points. It is considered that the differnce results from the large sensitivities of defects for the frequencies at those obserbation point. Therefore we calculate the sensitivity coefficients of design variables for noise of some obserbation point and consider the design variables for which the value of those sensitivities is large to be a defect. In this case it is not easy to infer to what extent it might be a defect. Therfore in this paper we try to utilize the Fuzzy inference theory. It is advantageous to set the observation point near the plane on which it is known that a defect exists. There are already some examples that the identification of noise source or structure have been implemented by using BEM. However by using BEM the accuracy near structure is inferior and the strctural vibration based on the new mode superposition method truncating the higher and/or lower mode proposed by one of the authors and the BEM sound field coupling analysis method has not been achieved yet, so in this paper we use FEM for calculating sound filed .

Analysis Method

As we already have described the analysis method [2] , only the outline will be explained in this paper. For simplicity, if we omit damping term, the finite element equation to be treated becomes as follows.

$$M\ddot{u} + Ku = f \tag{1}$$

where $M$ and $K$ represent the mass matrix and the stiffness matrix of the system, respectively.

$$M = \begin{bmatrix} M_{ss} & 0 \\ M_{as} & M_{aa} \end{bmatrix} \qquad K = \begin{bmatrix} K_{ss} & K_{sa} \\ 0 & K_{aa} \end{bmatrix}$$

$$\tag{2}$$

$$u^T = \{u_s^T, u_a^T\} \qquad f^T = \{f_s^T, f_a^T\}$$

In the above equation $u_s$ is displacement vector of structure, $u_a$ is sound pressure level vector, $f_s$ is excitation force acting on structure, $f_a$ is sound pressure of sournd source, $M_{ss}$, $K_{ss}$ are mass matrix and stiffness matrix of structure, $M_{aa}$, $K_{aa}$ are mass matrix and stiffness matrix of acoustic field, and $M_{as}$, $K_{sa}$ are coupling matrix, respectively. Here "s" and "a" represent subscripts for strucural system and acoustic system, respectively , and ¨ represents second derivatives of time. In the following we describe the outline of the method of modal frequency response sensitivity analysis used in this paper for coupled structural-acoustic system based on the sensitivities of eigen-pair.

The authors have obtained the relationship between left eigenvector $\overline{\phi}_i$ and right eigenvector $\phi_i$ as shown in equation (3).

$$\overline{\phi}_i^T = \left\{\phi_{si}^T, \frac{1}{\lambda_i}\phi_{ai}^T\right\} \qquad \text{(for } \lambda_i \neq 0) \tag{3}$$

Here, $\lambda_i$ is the eigenvalue of coupled system and $\phi_{si}$ , $\phi_{ai}$ represent eigenvector of structural and acoustic system, respectively.

*Modal Sensitivity Analysis of Coupled Structural-acoustic System*

The eigenvalue sensitivity $\lambda_j'$ of coupled system about design variable $\alpha_k$ is represented as follows.

$$\lambda_j' = \overline{\phi}_j^T (K' - \lambda_j M')\phi_j \tag{4}$$

Here $K'$, $M'$ are derivatives of $K$, $M$ about design variable $\alpha_k$, respectively. For the eigenvalue problem of equation (1) if we do partial differentiation about design variable $\alpha_k$.

$$(K - \lambda_j M)\phi_j' = b_j \tag{5}$$

where $\phi_j'$ is the sensitivity of right eigenvector for design variable $\alpha_k$ and $b_j = \lambda_j' M\phi_j - (K' - \lambda_j M')\phi_j$ . $\mu$ is the shift value which is arbitarily changed and $X_j$ is the solution of the following equation.

$$(K - \mu M)X_j = b_j \tag{6}$$

By using $X_j$, the sensitivity of right eigenvector for design variable $\alpha_k$ is represented as follows.

$$\phi_j' = X_j + \sum_{i=m}^{n} \phi_j C_{ij} \tag{7}$$

where, $m \geq 1$ and $n \leq N$. ( Here $N$ is the total number of degree of freedom of coupled system.) For $C_{ij}$ the following equation is obtained.

$$C_{ij} = \begin{cases} \dfrac{\lambda_j - \mu}{\lambda_i - \mu} \dfrac{1}{\lambda_j - \lambda_i} \overline{\phi_i}^T (K' - \lambda_j M')\phi_j & \text{(for } i=j) \\[2mm] -\dfrac{1}{2}\overline{\phi_j}^T M'\phi_j & \text{(for } i=j) \end{cases} \tag{8}$$

In the literature it is described that by choosing the shift value $\mu$ this method is reduced to Fox method、Wang method、Nelson method, respectively. Generally the value of $\mu$ may be positive value in a wide range. Moreover by inplementing error analysis we have determined the proper range of shift value $\mu$ .

*Modal Frequency response Sensitivity Analysis of Coupled Structural - acoustic System*
The authors have obtained the two formulation based on the method of using eigen pair sensitivity and the direct differentiation of frequency response. Here we describe the outline of the former used in this report. Sensitivity of modal frequency response for design variable $\alpha$ (Hereafter、for simplicity, we represent $\alpha_k$ as $\alpha$ .) is wriitten as differential form、$\Delta U/\Delta \alpha$ . Here、$\Delta U$ is the change occurred in modal frequency response by perturbation $\Delta \alpha$ . If $U$ is the modal frequency response of original system and $U_1$ is the modal frequency response of perturbed system , $\Delta U = U_1 - U$ .

$$U = \sum_{i=1}^{n} \frac{\phi_i \overline{\phi_i}^T F}{\lambda_i - \Omega^2} \qquad U_1 = \sum_{i=1}^{n} \frac{\phi_{1i} \overline{\phi_{1i}}^T F}{\lambda_{1i} - \Omega^2} \tag{9}$$

$$\lambda_{1i} = \lambda_i + \Delta \alpha \lambda_i' \qquad \phi_{1i} = \phi_i + \Delta \alpha \phi_i'$$

Here、$\overline{\phi_{1i}}$ is obtained by using equation (3) from $\phi_{1i}$. Since $\lambda_i'$ and $\phi_i'$ are caluculated by modal sensitivity analysis by the methood described above, the sensitivity of modal frequency response can be obtained by using those values from equation (9).

## Analysis Model and the Method of Searching Defects

In figure 1 the box model for analysis is presented. The box consists of panels of steel plates and the thickness of plates is uniformly 0.4 cm. The structure model comprises 98 node points and 96 rectangular elements (CQUAD4). The acoustic model comprises 125 node points and 64 solid elements (CHEXA). We assume that the results of FEM analysis based on this model are the analysis values and that the experimental values are the values obtained from the analysis of model in which the thickness of the hatching part on the bottom plate is reduced 20% as shown in Figure 2. Here, for simplicity, we assume that it is known that defects are in the bottom plane for searching. The observation points are four points A, B, C, D which are on the plane 30cm up to the bottom plate.

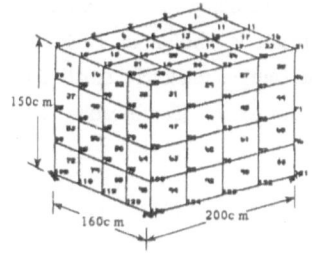

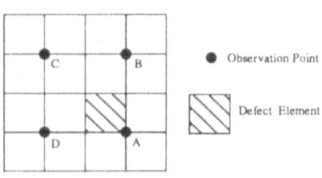

Figure 2    Bottom Plane of Box Model

Figure 1    Box Model for Analysis

## Analysis Results

For the box model in Figure 1 we reduce the degrees of freedom of total system to 70 by using lower 53 eigen-modes in structural syatem and lower 17 eigen-modes in acoustic sytem. At first we compare the analysis and experimental value of noise at the four observation points, A, B, C, D. Since at the observation point C there is no significant difference between the both value, we show the both response at 3 observation points (A, B, D) where the difference between analysis and experimental value exist from 80 Hz to 100 Hz as shown in figure 3 to figure 5. In figure 3 and figure 4 the both response at observation point A and D at which the difference is remarkable at the structural resonance frequency 96 Hz. However at the observation point B the difference remains small at 96 Hz. Since the observation point A, B, D is near the bottom plane including a defect, the sound pressure level at that point are easily affected by the vibration directly below. Therefore it is considered that the possibility for a defect existing directly below the observation point B is small.

To confirm the above qualitative inference result quantitatively we calculate the sensitivity coefficients of plate thickness of bottom elements for noise. The result is represented in Table 1. From this we can say that the sensitivity coefficient of a defect at observation point B is small and those at oberbvation point A, D is large. So for point A we calculate the sensitivity coefficient of plate thickness of each element on the bottom plane for noise at frequency points where there are the sound pressure difference. As shown in Table 2, at these frequency points we assume the set of design variables to be Fuzzy set and the normalized sensitivity coefficients to be membership function. From these we try to reduce the design variables of a defect and the results that the design variables 6 and 11 are most probably the defect is obtained. As described above by using the program which the authors have developed for coupled structural-acoustic systems, we can show the possibility that the identification or defect search become easier by using noise data than by using vibration data so far

Finally to confirm this rsult for structural eigenmode of frequency 96 Hz we calculate the difference of amplitude between the model including defect and the model without defect in the bobbom plane of the box. The result is shown in Table 3. From this table we can se that the difference at the defect element is relatively small compared to other parts, so we conclude that it is quite difficult to do identification by using vibration data.

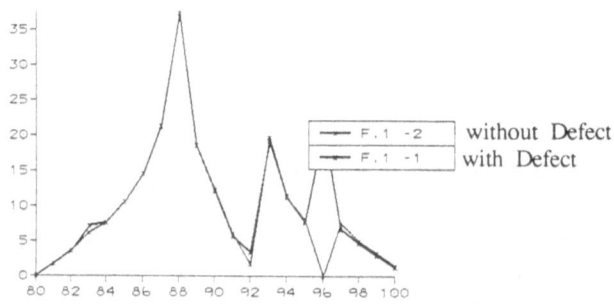

Figure 3   Frequency Response of Analysis and Experiment
at Observation Point A

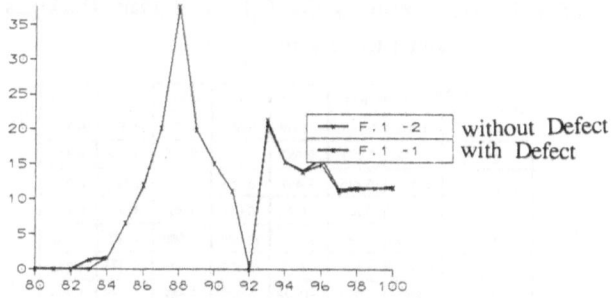

Figure 4  Frequency Response of Analysis and Experiment
at Observation Point B

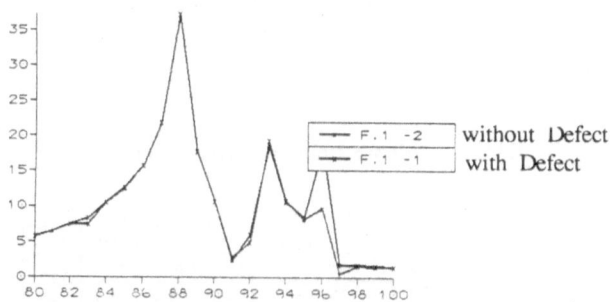

Figure 5  Frequency Response of Analysis and Experiment
at Observation Point D

## Table 1  The Sensitivity Coefficients of Plate Thickness of Bottom Elements for Noise

| Frequency(Hz) / Design Variables | 83.0 | 97.0 | 120.0 | 147.0 | 168.0 | 170.0 | 174.0 |
|---|---|---|---|---|---|---|---|
| 1 | 0.463 | 0.753 | -10.30 | -0.552 | -0.998 | -0.605 | 0.358 |
| 2 | 0.542 | 0.271 | -9.951 | -3.052 | -0.919 | -0.608 | 0.441 |
| 3 | 0.498 | 0.411 | -7.417 | -3.153 | -0.921 | -0.682 | 0.361 |
| 4 | 0.494 | 0.981 | -10.19 | -0.627 | -0.984 | -0.736 | 0.269 |
| 5 | 0.669 | 1.140 | -8.121 | -2.479 | -3.969 | -3.389 | 1.292 |
| 6 | 0.681 | 1.549 | -14.75 | -5.158 | -3.476 | -3.171 | 1.334 |
| 7 | 0.656 | 1.234 | -12.76 | -5.286 | -3.482 | -3.145 | 1.295 |
| 8 | 0.692 | 1.208 | -7.330 | -2.569 | -3.990 | -3.287 | 1.275 |
| 9 | 0.603 | 1.179 | -8.346 | -2.569 | -4.040 | -3.298 | 1.238 |
| 10 | 0.620 | 1.237 | -14.09 | -2.589 | -3.524 | -3.124 | 1.257 |
| 11 ★) | 0.665 | 1.551 | -13.40 | -5.285 | -3.492 | -3.183 | 1.344 |
| 12 | 0.633 | 1.136 | -7.223 | -5.198 | -3.996 | -3.462 | 1.282 |
| 13 | 0.453 | 0.938 | -10.44 | -2.510 | -1.004 | -0.763 | 0.239 |
| 14 | 0.489 | 0.403 | -8.010 | -0.628 | -0.943 | -0.719 | 0.327 |
| 15 | 0.533 | 0.282 | -9.350 | -3.073 | -0.942 | -0.599 | 0.450 |
| 16 | 0.492 | 0.774 | -10.39 | -0.564 | -1.004 | -0.708 | 0.329 |

★)Defect Element

## Table 2  Membership Function of Fuzzy Sets

| Frequency(Hz) / Design Variables | Set 1 83.0 | Set 2 97.0 | Set 3 120.0 | Set 4 147.0 | Set 5 168.0 | Set 6 170.0 | Set 7 174.0 |
|---|---|---|---|---|---|---|---|
| 1 | 0.669 | 0.486 | 0.698 | 0.104 | 0.247 | 0.175 | 0.266 |
| 2 | 0.783 | 0.175 | 0.675 | 0.577 | 0.225 | 0.176 | 0.328 |
| 3 | 0.720 | 0.265 | 0.503 | 0.596 | 0.228 | 0.197 | 0.269 |
| 4 | 0.714 | 0.633 | 0.691 | 0.119 | 0.244 | 0.213 | 0.200 |
| 5 | 0.967 | 0.735 | 0.551 | 0.467 | 0.982 | 0.979 | 0.961 |
| ⑥ | 0.984 | 0.999 | 1.0 | 0.976 | 0.860 | 0.916 | 0.993 |
| 7 | 0.948 | 0.796 | 0.865 | 1.0 | 0.862 | 0.908 | 0.964 |
| 8 | 1.0 | 0.779 | 0.497 | 0.486 | 0.988 | 0.949 | 0.949 |
| 9 | 0.871 | 0.761 | 0.566 | 0.486 | 1.0 | 0.953 | 0.921 |
| 10 | 0.896 | 0.798 | 0.955 | 0.490 | 0.872 | 0.902 | 0.935 |
| ⑪ ★) | 0.961 | 1.0 | 0.908 | 0.999 | 0.864 | 0.919 | 1.0 |
| 12 | 0.915 | 0.733 | 0.490 | 0.983 | 0.989 | 1.0 | 0.954 |
| 13 | 0.655 | 0.605 | 0.708 | 0.475 | 0.249 | 0.220 | 0.178 |
| 14 | 0.707 | 0.260 | 0.543 | 0.119 | 0.233 | 0.208 | 0.243 |
| 15 | 0.770 | 0.182 | 0.634 | 0.581 | 0.233 | 0.173 | 0.335 |
| 16 | 0.711 | 0.499 | 0.704 | 0.107 | 0.249 | 0.205 | 0.245 |

★) Defect Element
◯ Candidates of Defect

Table 3  The Difference of Strucural Eigen Mode between the Model Including Defect and the Model without Defect

| Node # | without Defect | with Defect | Difference | Node # | without Defect | with Defect | Difference |
|---|---|---|---|---|---|---|---|
| 1 | 8.02E-5 | 8.95E-5 | 9.25E-6 | (14) | 1.94E-4 | 1.93E-4 | -9.28E-7 |
| 2 | 2.73E-5 | 3.59E-5 | 8.50E-6 | 15 | -1.41E-4 | -1.40E-4 | 8.62E-7 |
| 3 | 4.31E-5 | 4.69E-5 | 3.84E-6 | 16 | -7.96E-5 | -7.94E-5 | 2.16E-7 |
| 4 | 5.90-E-5 | 6.11E-5 | 2.03E-6 | 17 | 1.09E-4 | 1.10E-4 | 8.85E-7 |
| 5 | -5.30E-5 | -5.61E-5 | -3.14E-6 | (18) | -1.63E-4 | -1.64E-4 | -5.69E-7 |
| 6 | 3.65E-5 | 3.69E-5 | 4.41E-7 | (19) | 1.46E-4 | 1.46E-4 | -3.35E-8 |
| 7 | -5.34E-6 | -8.39E-6 | -3.05E-6 | 20 | -1.64E-4 | -1.64E-4 | 1.70E-7 |
| 8 | -4.63E-5 | -4.49E-5 | 1.42E-6 | 21 | 1.11E-4 | 1.10E-4 | -6.58E-7 |
| 9 | 1.06E-4 | 1.07E-4 | 6.08E-7 | 22 | -7.85E-5 | -7.94E-5 | -9.05E-7 |
| 10 | 1.88E-4 | 1.88E-4 | 3.72E-7 | 23 | -1.39E-4 | -1.40E-4 | -9.24E-7 |
| 11 | 6.11E-5 | 6.27E-5 | 1.57E-6 | 24 | 2.11E-4 | 2.12E-4 | 1.02E-6 |
| 12 | 9.46E-5 | 9.51E-5 | 5.75E-7 | 25 | -1.33E-4 | -1.32E-4 | 1.03E-6 |
| (13) | -9.17E-5 | -9.14E-5 | 3.28E-7 | | | | |

## Conclusions

So far the searching for defects or the identification were implemented by using vibration data. In this paper we tried to apply the new analysis method of coupled structural-acoustic system and the method of Fuzzy inference theory by using noise frequency response characteristics. From these results it became possible to reduce the design variables which have the strong possibility for defects. Hereafter we will try to increase the number of observation points and defects and to use the noise value at high frequency or at the obserevation point on the plane nearer to the bottom plane. Further we will try to identify defects by using optimization analysis by using reduced design variables.

## References

1. Hagiwara, I. and Nagabuchi, K., "Sensitivity Analysis for Structural Frequency Response Characteristics (1 st Report Proposition of new Frequency Response Sensitivity Analysis and Comparison between Some Different Methods)", Trans. Jpn. Soc. Mech. (in Japanese), Vol.54, No.497, C (1987), p.124.
2. Hagiwara, I. ,Ma, Z., Arai, A., and Nagabuchi, K., "Technical Development of Eigenmode Sensitivity Analysis for Coupled Acoustic-Structural Systems", Trans. Jpn. Soc. Mech. (in Japanese), Vol.56, No.527, C (1990), p.60.

# Inverse Problem Solution in Non-Destructive Pavement Diagnostics – Computational Aspects

B. Novotny

Institute of Construction and Architecture
Slovak Academy of Sciences
Dubravska cesta 9, 842 20 Bratislava

Summary

Determination of pavement layer moduli from measured surface deformation is performed by the iteration process based on capability to calculate deformation derivatives with respect to elastic moduli. A priori information on the processed data is used to stabilize process of solution in the case of the greater number of considered layers. The applicability of obtained results is conditioned also by the accuracy with which the imperfect bonding of layers is accounted for.

Formulation of the problem

In the field of pavement design it is felt that the determination of material characteristics of pavement construction layers from the measured deformation of the pavement surface may provide a valuable information on pavement serviceability [1,2]. Different versions of so-called "moduli back calculation" computer programs are now widely used (leaving the most sophisticated part of the inverse problem solution to the skill of the user). In [2], the EFROMD computer program (based on Rosenbrock's search algorithm) has been used to deduce information on modulus-temperature relationship for asphalt layer. In the present paper, the inverse problem for a layered halfspce will be solved using search method based on deformation derivatives with respect to moduli, which should result in better efficiency of the solution procedure.

The scheme of the problem is shown in Fig. 1. Pavement is composed of $n$ elastic, isotropic and homogeneous layers with Young moduli $E_i$, Poisson's ratios $\nu_i$ and constant thicknesses $h_i$. This structure is supported by the proper elastic, isotropic and homogeneous halfspace with material characteristics $E_{n+1}$, $\nu_{n+1}$. The applied load is uniformely distributed with

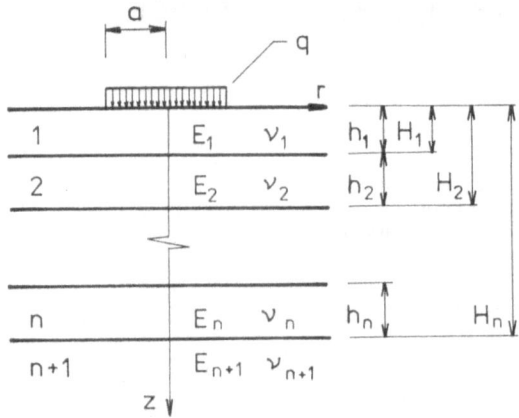

Fig. 1  Scheme of a layered halfspace.

intensity $q$ over circular area having radius  $a$. The mathematical problem is rotationally symmetrical and for the surface deformation we have [3]

$$\chi(E,v,h,C;r) = \int_0^\infty \Omega(E,v,h,C;\xi)\ \Xi(a,q;\xi,r)\ d\xi\ ,$$  (1)

where by  $\chi$  we denote the complex information

$$\chi(E,v,h,C;r) = \left\{ \begin{array}{l} w(r) \\ r^{-1}\ dw(r)/dr\ ; \\ d^2w(r)/dr^2 \end{array} \right.$$  (2)

on the surface deformation:  deflection and both principal curvatures (small deformation theory being applied). The term  $\Xi(a,q;\xi,r)$  contains all the data concerning external loading and the measurement setting, while the term $\Omega(E,v,h,C;\xi)$  comprises the influence of material  $\{E,v\}$  and geometrical $\{h\}$  properties of the pavement structure. By writing  $C$  in the list of parameters we express the dependency of pavement deformation on the quality of layer bonding - usually the complete adhesion is considered. In the final part of the paper, few aspects of the imperfect bonding will be discussed.

After applying external load, the surface deformations $\hat{\chi}(\hat{r})$ are measured in $m$ locations $r = \hat{r}_k$ $(k=1,2,\ldots,m)$ and our problem here is to determine pavement system properties in such a way that the error functional

$$\mathcal{F}(E,\ldots) = \sum_{k=1}^{m} \zeta_k \left[ \chi(E,\ldots;\hat{r}_k) - \hat{\chi}(\hat{r}_k) \right]^2, \tag{3}$$

attains its minimum. The weighting coefficients $\zeta_k$ are of the type

$$\zeta_k = z_k \, \hat{\chi}(\hat{r}_k)^{-2} \, ; \qquad 0 < z_k \leq 1 \, . \tag{4}$$

Before solving this complex problem, it is advisable to acquire as much information on the problem properties as possible. In Figs. 2 to 5, the direct problem results on sensitivity of the pavement N5 to the system parameters are shown (pavement central deflection is $0.06204cm$, sensitivity values are given in $[0.01cm]$). The following conclusions may be drawn:

(a) the material characteristics of subsoil $E_{n+1}$, $\nu_{n+1}$ affect the surface deformations in a sufficiently greater manner than properties of the all other layers, especially in points far away from the loaded area;

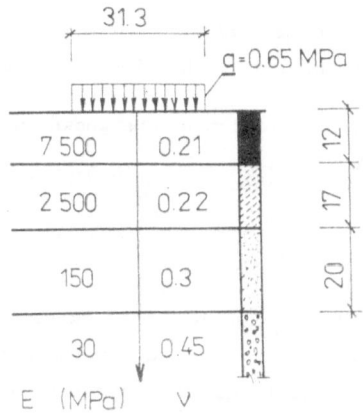

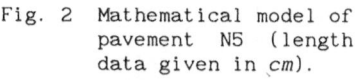

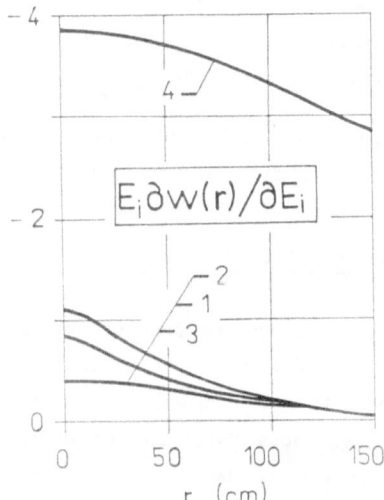

Fig. 2  Mathematical model of pavement N5 (length data given in $cm$).

Fig. 3  Sensitivity analysis : surface deflections vs. elastic moduli $E_i$ of pavement layers.

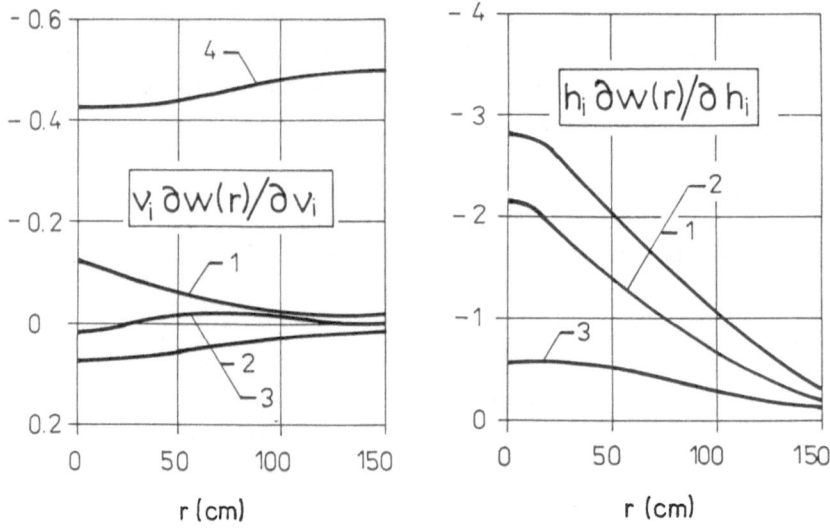

Fig. 4  Sensitivity of the N5 pavement   Fig. 5  Influence of the layer
        deflections to the Poisson's            thicknesses $h_i$.
        ratios.

(b) the effect of Poisson's ratios $\nu_i$ is of a lesser order that of elastic
    moduli;

(c) thicknesses $h_i$ of layers affect the surface deformations in a similar
    way as layer moduli, this points out difficulties of the given problem.

Here we assume that thicknesses $h_i$ and Poisson's ratios $\nu_i$ are known and
that full bonding is observed at all interfaces.


Method of solution


The solution to the problem of functional (3) minimization is constructed
along the same lines as in classical literature, e.g. [3]. Our approach is
based on two techniques, the first expresses the fact that deformations $\chi$
are homogeneous functions of order -1 with respect to the elastic moduli
of the layers, the second one uses Taylor expansions of $\chi(E; r)$

$\underline{\text{T1} [ t \to u ] }:$ $\qquad E_i^{(u)} = E_i^{(t)} / \Lambda_t ,$

$$\Lambda_t = \left[ \sum_{k=1}^{m} \zeta_k \, \chi(E^{(t)}; \hat{r}_k) \, \hat{\chi}(\hat{r}_k) \right] \left[ \sum_{k=1}^{m} \zeta_k \, \hat{\chi}(E^{(t)}; \hat{r}_k)^2 \right]^{-1} \qquad (5)$$

$\underline{\text{T2} [ t \to u ] }:$ $\qquad E_i^{(u)} = E_i^{(t)} + \lambda_t \, \Delta E_i^{(t)} ,$ $\qquad\qquad (6)$

$\underline{\text{direction search}}$ :

$$\Delta E^{(t)} : \qquad \sum_{k=1}^{n+1} \left\{ \sum_{j=1}^{m} \zeta_j \, \tau_{ij}^{(t)} \, \tau_{kj}^{(t)} \right\} \Delta E_k^{(t)} =$$

$$= \sum_{j=1}^{m} \zeta_j \, \tau_{ij}^{(t)} \, [ \hat{\chi}(\hat{r}_j) - \chi(E^{(t)}; \hat{r}_j)] ;$$

$\underline{\text{minimum search}}$ : $\qquad d\mathcal{F}( E^{(t)} + \lambda_{t,min} \, \Delta E^{(t)})/d\lambda = 0 ,$

$$\lambda_t = \min (\lambda_{t,min} ; \lambda_{c,i})$$

the $\lambda_{c,i}$ values following from the basic constraints $E_{i,min} \le E_i \le E_{i,max}$
that prevent solution to enter "forbidden" area. For the typical pavement
structure shown in Fig. 2, each step of the iteration process consists of
two substeps:

$\qquad\qquad\qquad\qquad\qquad\qquad\qquad\qquad\qquad\qquad\qquad\qquad\qquad\qquad (7)$

---

$(\alpha)$ $\quad \underline{\text{T1} [(t = s) \to (u = s + 1/2)]}$

$(\beta)$ $\quad \underline{\text{T2} [(t = s + 1/2) \to (u = s + 1)]}$ : $\quad \tau_{ij}^{(t)} = \partial\chi(E^{(t)}; \hat{r}_j)/\partial E_i$

---

Detailed analysis of this iteration process has been performed in [5] with
the conclusion that on the basis of numerical experiments it is possible to
optimize the overall experimental setting of the diagnostics procedure as

concerns the type of measured quantities and the choice of measurement locations.

When the structure of the layered system is more complicated, the iteration process (7) shows preferrence for the "zig-zag" type of solution. Let us consider the case with following input data (given in *cm* and *MPa*)

$$n = 8; \quad m = 19; \quad a = 15.65, \quad q = 0.65; \tag{8}$$
$$\{v: \ 8*0.35, \ 0.45\}; \quad \{h: \ 3*5., \ 2*10., \ 2*15., \ 20.\};$$
$$\{E^{(0)}: \ 11000, 9500, 8000, 2500, 1500, 550, 300, 90, 5\};$$
$$\{E^*: \ 6000, 5500, 5000, 1000, 800, 300, 200, 80, 25\};$$

and let the "measured" data be computed from

$$\hat{\chi}(\hat{r}_k) = \chi(E^*; \hat{r}_k) , \tag{9}$$

so thus $E^*$ represents the "exact" solution of the inverse problem. Slightly modifying constraints of the problem, three quite different "equivalent" sets of elastic moduli can be found:

(a) $\{E : \ 6013, \ 5483, \ 5007, \ 1001, \ 795.8, \ 301.5, \ 199.1, \ 80.1, \ 25.0\}$

(&) $\{E : \ 6202, \ 5256, \ 4635, \ \underline{1302, \ 324.2, \ 1225, \ 10.90, \ 307.0}, \ 25.39\}$

(c) $\{E : \ \underline{4485, \ 9629, \ 3550, \ 1195}, \ 798.7, \ 300.4, \ 199.7, \ 80.2, \ 25.0\}$

To avoid such situations, the a priori information on the solution quality should be used to stabilize the solution procedure [6]. To achieve this (and to ensure uniqueness of the "reasonable" solution), we impose additional constraints that supplement already set basic constraints (6)

$$E_i^{(u)} \geq \rho_i \ E_{i+1}^{(u)} . \tag{10}$$

Also, the constraint activity indicators $R_i^{(t)}$ are introduced as

$$R_i^{(t)} = \begin{cases} 0, & \text{if} \quad E_i^{(t)} > \rho_i \ E_{i+1}^{(t)} \\ 1, & \text{if} \quad E_i^{(t)} = \rho_i \ E_{i+1}^{(t)} \end{cases} \tag{11}$$

Now, the iteration process (7) is changed so that each step is performed in three substeps

(12)

---

($\alpha$)   T1 $[(t = s) \rightarrow (u = s + 1/2)]$

($\beta$)   T2 $[(t = s + 1/2) \rightarrow (u = s + 3/4)]$

$$\tau_{ij}^{(t)} = (1 - R_i^{(t)}) \sum_{k=1}^{i} \eta_{k,i}^{(t)} \, \partial\chi(E^{(t)};\hat{r}_j)/\partial E_i \; ;$$

$$\eta_{k,i}^{(t)} = \prod_{l=k}^{i-1} \rho_l \, R_l^{(t)}; \qquad \eta_{i,i}^{(t)} = 1 \; ;$$

$$\Delta E_p^{(t)} = \sum_{j=p+1}^{n+1} (1 - R_j^{(t)}) \, \eta_{p,j}^{(t)} \, \Delta E_j^{(t)};$$

$$\lambda_t = \min_i \; ( \lambda_{t,\min}; \lambda_{v,i} );$$

($\gamma$)   T2 $[(t = s + 3/4) \rightarrow (u = s + 1)]$

$$\tau_{ij}^{(t)} = S_{i,i-1}^{(t,u)} \, \partial\chi(E_j^{(t)};\hat{r}_j)/\partial E_i$$

$$S_{i,i-1}^{(t,u)} = 1 \quad \text{if} \; \left[ (R_{i-1}^{(t)} + R_i^{(t)}) > 0 \; \wedge \; \Delta E_i^{(u)} > \rho_i \, \Delta E_{i+1}^{(u)} \right]$$

$$\qquad\qquad\quad 0 \quad \text{otherwise}$$

---

Here, the values $\lambda_{v,i}$ follow from the conditions (10)

$$\lambda_{v,i} = \left( E_i^{(t)} - \rho_i \, E_{i+1}^{(t)} \right) / \left( \rho_i \, \Delta E_{i+1}^{(t)} - \Delta E_i^{(t)} \right) , \qquad (13)$$

and the substep ($\gamma$) should be solved iteratively, since the values $S_{i,i-1}^{(t,u)}$ are guessed in advance and than corrected, if necessary.

When applying (12) to our test problem (8) with

$$\{\rho : 4*1., 1.2, 1., 1.5, 2.\}$$

we can follow the run of the iteration process in Tab. 1. It is seen that

118

| Appr. | Interface No. | | | | | | | |
|---|---|---|---|---|---|---|---|---|
| | 1 | 2 | 3 | 4 | 5 | 6 | 7 | 8 |
| 1 | $0_↘$ | $0_↘$ | $0_↘$ | $0_↘$ | $0_↘$ | $0_↘$ | $0_↘$ | $0_↘$ |
| 2 | $0_↘$ | $0_↘$ | $0_↘$ | $0_↘$ | $0_↘$ | $0_↘$ | 1 | $0_↘$ |
| 3 | 1 | $0_↘$ | $0_↘$ | $0_↘$ | $0_↘$ | $0_↘$ | ↓1 | $0_↘$ |
| 4 | ↓1 | $0_↘$ | $0_↘$ | 1 | $0_↘$ | $0_↘$ | ↓1 | $0_↘$ |
| 5 | ↓1 | $0_↘$ | $0_↘$ | ↓1 | $0_↘$ | $0_↘$ | ↓1 | $0_↘$ |
| 6 | ↓1 | $0_↘$ | $0_↘$ | ↓1 | $0_↘$ | $0_↘$ | ↓1 | $0_↘$ |
| 7 | ↓$0_↘$ | $0_↘$ | $0_↘$ | ↓$0_↘$ | $0_↘$ | $0_↘$ | ↓$0_↘$ | $0_↘$ |
| 8 | 0 | 0 | 0 | 0 | 0 | 0 | 0 | 0 |

Tab. 1  Run of iterations (12), values of $R_i$ presented.

the third substep of the iteration process (12) has been performed at the fifth attempt only.

The alternative to imposing constraints of the type (10) is the reconstruction of the initial approximation $E^{(0)}$ based on a priori information that the deflections in remote points are mostly affected by the subsoil material properties

$$w_{hom}(\hat{r};\ E_{hom} = E_{n+1}^{(0)}) \approx \hat{w}(\hat{r}) \tag{14}$$

for $\hat{r} \sim 10\ a$ .

The best way is to combine iteration process (12) with the initial approximation improvement (13).

## Improper bonding of layers

At the end of the paper, the problem of imperfect bonding of layers will be briefly discussed. According to the model of a *fictive interlayer* the slip between layers adjacent to the interface $z = H_i$ is expressed in the form

$$(1 - g_i)\ {}^{(i)}\Delta u(r) = g_i\ C_i\ {}^{(i)}t(r) \tag{15}$$

where

$$\begin{aligned} {}^{(i)}\Delta u(r) &= {}^{(i+1)}u_r(r, z=H_i) - {}^{(i)}u_r(r, z=H_i) \\ {}^{(i)}t(r) &= {}^{(i)}\tau_{rz}(r, z=H_i) , \end{aligned} \tag{16}$$

denotes the interface slip and shearing stress, respectively. The interaction parameter $g_i$ takes values from the interval $(0, 1)$, $g_i = 0$ when the bonding of layers is perfect and $g_i = 1$ in the case of a frictionless slipping. The parameter $C_i$ is regarded as a material characteristic of the interface and the manner of its dependence on $g_i$ is governed by the choice of the interface control quantity ${}^{(i)}U$ . Here we take

$$ {}^{(i)}U = 1 - {}^{(i+1)}\varepsilon_r(r=0, z=H_i) / {}^{(i)}\varepsilon_r(r=0, z=H_i) \tag{17}$$

with $\varepsilon_r$ being the radial deformation. The *fundamental cases of interaction (FCI)* are defined as the cases having either full bonding or frictionless slip prescribed at all the imperfect interfaces of the given system (in what follows $q$ such interfaces are assumed). In all these cases the values $^{(1,K)}U$ of control quantities are computed and the following requirements are set

$$^{(1)}U = \sum_K g_K \, ^{(1,K)}U \; ; \qquad g_K = \prod_{i=1}^{q} [( 1 - k_i )( 1 - g_i ) + k_i \, g_i], \tag{18}$$

which form the conditions to determine $C_i$ The multiindex $K$ is used here to distinguish different *FCI* (index $i$ refers to $i$-th imperfect interface)

$$K = (k_1, k_2, \ldots, k_q) \; ; \qquad
\begin{aligned}
k_i &= 0 & \text{for} & \quad \text{complete adhesion} \\
& \phantom{=} 1 & & \quad \text{frictionless slip} \, .
\end{aligned} \tag{19}$$

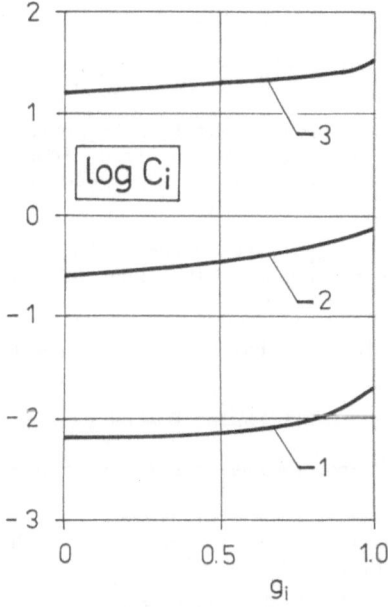

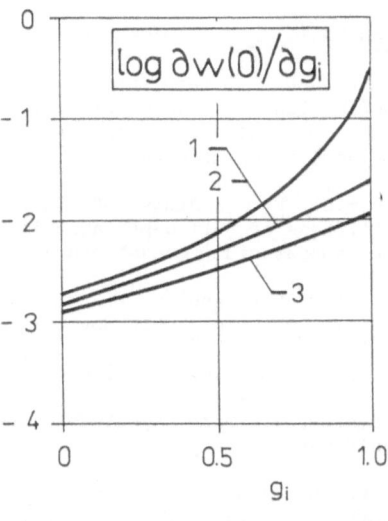

Fig. 6  Pavement N5, material characteristic $C$ as function of interaction parameter $g$ .

Fig. 7. Effect of changes in values of interaction parameters $g$ on surface central deflection.

To illustrate the meaning of these relations, the N5 pavement from Fig. 2 has been subjected to the imperfect bonding analysis and the results are presented in Figs. 6 and 7. The individual interfaces are subsequently regarded as imperfectly bonded ($q = 1$) and Fig. 6 shows that the location of the improperly acting interface affects significantly the values of interface material characteristics $C_i$ and also indicates that the mutual interface interaction may be regarded as a second-order effect. Fig. 7 proves that the effect of improper bonding has a decisive impact on the accuracy of the pavement behaviour predictions and should, therefore, receive proper consideration.

As concerns inverse problem analysis, two remarks at least should be made:

(a)   the decomposition approach is to be applied when minimizing error functional with imperfect bonding effects (what kind of problems is to expected in accomplishing this task can be seen e.g. in [7])

(b)   the type of applied "system exciting" loading should be reconsidered, the horizontally acting load components should be also incorporated, since the imperfectly bonded systems are naturally more sensitive to this particular kind of loading.

References

1.   Irwin, L.H.: Determination of pavement layer moduli from surface deflection data for pavement performance evaluation. In: *Proc. Fourth Int. Conf. Struct. Design of Asph. Pavem.*, vol **1** (1977) 831 - 840, The University of Michigan, Ann Arbor.

2.   Schmidt, B.: A study of the relationship between temperature and stiffness of full depth asphalt pavements. *Report* 215 (1989) Statens Vejlaboratorium, Copenhagen.

3.   Novotny, B.; Hanuska, A.: Theory of layered halfspace (in Slovak). Bratislava: Veda Publ. House (1983).

4.   Cea, J.: Optimisation. Theorie et algorithmes. Paris: Dumond (1971).

5.   Novotny, B.: Some aspects of inverse problem analysis for a layered halfspace (to appear in *Comp. Mech.*).

6.   Tikhonov, A.N.; Goncharovsky, A.V. (eds): Ill-Posed Problems in the Natural Sciences. Moscow: Mir Publ. (1987).

7.   Pasquetti, R.; Le Niliot, C. : Boundary element approach for inverse heat conduction problems: application to a bidimensional transient numerical experiment. *Num. Heat Transfer*, Part B, **20** (1991) 169-189.

# Inverse Formulation for Incompressible Viscous Two-Dimensional and Axisymmetric Flow Problems

Tetsuhiro TSUKIJI

Department of Mechanical Engineering,
Ashikaga Institute of Technology,
268-1, Oomae-cho, Ashikaga-shi, Tochigi,
326 JAPAN

Summary

The inverse problem of an incompressible viscous axisymmetric flow is formulated using the orthogonal curvilinear coordinate system, including a two-dimensional flow. The Stokes stream function or stream function and the coordinate which is constant along the orthogonal trajectories of streamlines or stream tubes are the independent variables. The Navier-Stokes equations and the vorticity transport equation are transformed into the orthogonal coordinate frame. The axisymmetric laminar jet leaving a Poiseuille tube for an inviscid outer fluid phase is solved numerically using parabolic differential equations which are simplified by a boundary-layer-type analysis under the influence of both gravity and surface tension.

## Introduction

There have been many studies of inverse problems in a fluid engineering. The coordinate systems which has the stream function for a two-dimensional flow or Stokes stream function for an axisymmetric flow as an independent variable are sometimes formulated to solve inverse problems. The inverse formulation can be applied to the steady flow problem with unknown shape of the physical boundary.

In potential two-dimensional and axisymmetric flow problems, orthogonal and non-orthogonal coordinate systems were used considering the stream function or Stokes stream function as one of the independent variables[1-13].

For the inverse problems of a viscous two-dimensional flow, the complete equations have been formulated by orthogonal curvilinear coordinate system which has the stream function as the independent variables[14-17]. The two-dimensional boundary-layer equations expressed by the cartesian coordinates were transformed into the form of a generalized heat conduction equation by changing the independent variables into the stream function and the distance in the main-flow direction. The transformation yielded the non-orthogonal coordinate system[18,19].

The coordinate system called the Protean coordinate system for a viscous axisymmetric flow was developed by using the stream function and the distance in axial direction as the independent variables[20]. Essentially, the coordinate transformation is an extension of the above two-dimensional case[18,19]. It is expected that the accuracy of the finite-difference solutions using the Protean coordinate system is not satisfactory in the flow fields where the radius of curvature of a streamline is small, because the Protean coordinate system is non-orthogonal.

In the present study, the inverse problem of an incompressible viscous axisymmetric flow is formulated using the orthogonal curvilinear coordinate system, including a two-dimensional flow. The advantage of the present inverse formulation is that the problem of a viscous flow can be treated easily even though the shape of the physical boundary is unknown. The Stokes stream function or stream function and the new coordinate which is constant along the orthogonal trajectories of streamlines or stream tubes are the independent variables. In potential flows, the new coordinate reduces to the velocity potential. The Navier-Stokes equations and the vorticity transport equation are transformed into the new coordinate frame. The axisymmetric laminar jet leaving a Poiseuille tube for an inviscid outer fluid phase is solved numerically using parabolic differential equations[21] which are simplified by a boundary-layer-type analysis under the influence of both gravity and surface tension. The prediction of the velocity distributions and shape of the free surface profile is undertaken. The results obtained by the present method were found to agree very closely with experimental results[22], and the method used was found to be very useful for investigating the behavior of an axisymmetric laminar jet of incompressible viscous fluid.

## Development of the mathematical inverse formulation
The coordinate system for a two-dimensional and an axisymmetric flow is shown in Fig.1. The quantities are non-dimensionalized with respect the

shows the dimensional quantity. In the case of the axisymmetric flow, the coordinates $x$ and $y$ are the axial coordinate and radial coordinate, respectively, and the plane including a symmetric axis is shown. The coordinate $\phi$ is in the streamwise direction and $\psi$ is orthogonal to the streamlines or stream tubes. The coordinate $\psi$ also denotes the stream function or Stokes stream function.

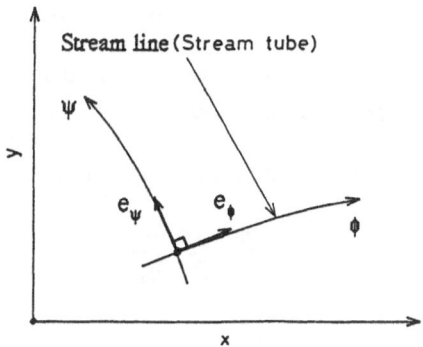

Fig.1 Coordinate system

The stream function $\psi$ is defined by Eqs.(1).

$$v_x = \frac{1}{y^n} \frac{\partial \psi}{\partial y}, \quad v_y = -\frac{1}{y^n} \frac{\partial \psi}{\partial x} \tag{1}$$

where $v_x$ and $v_y$ are the component of dimensionless velocity in the direction $x$ and $y$, respectively, and $n$ is zero for the two-dimensional flow and unit for the axisymmetric flow.

The function $\phi$ is defined by the following equations:

$$\frac{\partial \phi}{\partial x} = \beta v_x, \quad \frac{\partial \phi}{\partial y} = \beta v_y \tag{2}$$

where $\beta$ is a factor and a function of $x$ and $y$.

Equations (1) and (2) satisfy the condition in which the $\phi$-lines perpendicularly intersect the $\psi$-lines. Namely, the streamlines are the lines $\psi$ = constant, and the orthogonal trajectories to the streamlines are the line $\phi$ = constant. In the potential flows, $\phi$ reduces to the velocity potential and $\beta$

= 1.  On the other hand, if $\phi = \phi\,(x,y)$ and $\Psi = \Psi(x,y)$, then $x$ and $y$ must be functions of $\phi$ and $\Psi$, such as that[1],[23]

$$\frac{\partial x}{\partial \phi} = -\frac{1}{J}\frac{\partial \Psi}{\partial y}, \quad \frac{\partial x}{\partial \psi} = \frac{1}{J}\frac{\partial \phi}{\partial y}$$

$$\frac{\partial y}{\partial \phi} = \frac{1}{J}\frac{\partial \Psi}{\partial x}, \quad \frac{\partial y}{\partial \psi} = -\frac{1}{J}\frac{\partial \phi}{\partial x}$$

(3)

in which $J$ is the Jacobian determinant

$$J = \frac{\partial \phi}{\partial y}\frac{\partial \psi}{\partial x} - \frac{\partial \phi}{\partial x}\frac{\partial \psi}{\partial y}$$

(4)

Substituting Eqs.(3) into the equations which are obtained by eliminating the variables $v_x$ and $v_y$ from Eqs. (1) and (2), respectively, gives

$$-\frac{1}{y^n}\frac{\partial y}{\partial \phi} = \frac{1}{\beta}\frac{\partial x}{\partial \psi}$$

(5)

$$\frac{1}{y^n}\frac{\partial x}{\partial \phi} = \frac{1}{\beta}\frac{\partial y}{\partial \psi}$$

(6)

Since the dimensionless velocity $q$ is expressed as $\sqrt{v_x^2 + v_y^2}$, the substitution of Eqs.(1)-(4) into the relation ( $q = \sqrt{v_x^2 + v_y^2}$ ) yields

$$q = \frac{1}{y^n\sqrt{\left(\frac{\partial x}{\partial \psi}\right)^2 + \left(\frac{\partial y}{\partial \psi}\right)^2}}$$

(7)

Transforming the Navier-Stokes equations and the vorticity transport equation into the $\phi$ - $\Psi$ coordinate frame

The Navier -Stokes equations are written in the following vector form, for viscous flow:

$$(q \cdot \nabla)\,q = -\nabla p + K + \frac{1}{Re}\nabla^2 q$$

(8)

where $q, K, p$ and $Re$ are the dimensionless velocity vector ( $= q\,e_\phi$ ), the dimensionless body force, the dimensionless pressure and Reynolds number( $= L*U* / v*$, where $v*$ is the kinematic viscosity of fluid) , respectively. $\nabla$

denotes a dimensionless vector differential operator which is defined by

$$\nabla \equiv \mathbf{e}_\phi \beta q \frac{\partial}{\partial \phi} + \mathbf{e}_\psi y^n q \frac{\partial}{\partial \psi} + n \, \mathbf{e}_\theta \frac{1}{y^n} \frac{\partial}{\partial \theta} \tag{9}$$

where $\mathbf{e}_\phi$, $\mathbf{e}_\psi$ and $\mathbf{e}_\theta$ are the dimensionless unit vectors in the directions $\phi$, $\psi$ and $\theta$, respectively. $\theta$ is the azimuthal coordinate variable for the axisymmetric flow and the coordinate variable perpendicular to $\phi$ - $\psi$ plane for the two-dimensional flow.

The vorticity vector $\Omega$ ( $= \omega \, \mathbf{e}_\theta$ , where $\omega$ is the dimensionless vorticity) is defined by

$$\Omega = \nabla \times q = \frac{y^n q^2}{\beta} \frac{\partial \beta}{\partial \psi} \mathbf{e}_\theta \tag{10}$$

where $\nabla^2 q = - \nabla \times \Omega$ for incompressible flow. Substituting Eqs.(9) and (10) into Eq.(8) and rearranging by using the relation of $q = q \, \mathbf{e}_\phi$ , the equations of motion in the $\phi$ - and $\psi$ - directions can be expressed in the following form:

$$\beta q^2 \frac{\partial q}{\partial \phi} = - \beta q \frac{\partial p}{\partial \phi} + K_\phi - \frac{q}{R_e} \frac{\partial}{\partial \psi} ( y^n \, \omega) \tag{11}$$

$$\frac{y^n q^2}{\beta} \frac{\partial}{\partial \psi} (\beta q) = - y^n q \frac{\partial p}{\partial \psi} + K_\psi + \frac{\beta q}{R_e} \frac{1}{y^n} \frac{\partial}{\partial \phi} ( y^n \, \omega) \tag{12}$$

where $K_\phi$ , and $K_\psi$ are the components of the dimensionless body force which act upon a unit mass in the directions $\phi$ and $\psi$, respectively.

Taking the curl of each term of Eq.(8) to eliminate the pressure term, the vorticity transport equation, which is written in the following dimensionless form, is obtained.

$$- n \, \Omega \cdot \nabla \, q + q \cdot \nabla \Omega = \nabla \times K + \frac{1}{Re} \nabla^2 \Omega \tag{13}$$

Taking the $\mathbf{e}_\theta$ component out of Eq.(13), we finally obtain

$$\beta q^2 \frac{\partial \omega}{\partial \phi} - n \frac{\beta q^2 \, \omega}{y^n} \frac{\partial y}{\partial \phi} = \beta q \frac{\partial K_\psi}{\partial \phi} - y^n q \frac{\partial K_\phi}{\partial \psi} - \frac{\beta}{y^n} K_\psi \frac{\partial}{\partial \phi} (y^n q)$$

$$+ \frac{y^n}{\beta} K_\phi \frac{\partial}{\partial \psi} (\beta q) + \frac{1}{Re} [\beta q^2 \frac{\partial \beta}{\partial \phi} \frac{\partial \omega}{\partial \phi} + \beta^2 q^2 \frac{\partial^2 \omega}{\partial \phi^2}$$

$$+ \left(2ny^nq^2 \frac{\partial y}{\partial \psi} - \frac{y^{2n}q^2}{\beta} \frac{\partial \beta}{\partial \psi}\right) \frac{\partial \omega}{\partial \psi} + y^{2n}q^2 \frac{\partial^2 \omega}{\partial \psi^2} - n \frac{\omega}{y^{2n}}]$$

$$(14)$$

Under the influence of gravity as the body force, the combination of Eqs.(1)-(4) leads to Eqs.(15):

$$K_\phi = \frac{y^n q}{F_r} \frac{\partial y}{\partial \psi}, \qquad K_\psi = \frac{y^n q}{F_r} \frac{\partial x}{\partial \psi} \qquad (15)$$

where $F_r$ is the Froude number, defined as:

$$F_r = \frac{U^{*2}}{L^* g^*} \qquad (16)$$

where $g^*$ is the acceleration of gravity. There are two ways to choose some dependent variables for solving the equations mentioned in this chapter. One way is that Eqs.(5)-(7), (11) and (12) can be solved by choosing $x$, $y$, $p,q$ and $\beta$ as the dependent variable after eliminating $\omega$ by substituting Eq.(10) into Eqs.(11) and (12). The other is that Eqs.(5)-(7), (10) and (14) can be solved by choosing $x$, $y$, $q$, $\beta$ and $\omega$ as the dependent variable.

## Application of the Governing Equations to the Axisymmetric Jet issuing from a circular tube[24]

The dependent variables, $x$, $y$, $q$, $\beta$ and $\omega$ are chosen in order to solve Eqs.(5)-(7), (10) and (14). The numerical marching integration from an upstream station to a downstream one is employed to obtain finite-difference solutions to simplified forms of the partial differential equations (vorticity transport equation, etc.), converting the elliptic partial differential equations to parabolic equations and, thus, formulating an initial-value problem.

The streamlines calculated are shown in Fig.2. The tube diameter is chosen as a reference length $L^*$ and the average velocity in the tube exit section( $x = 0$ ) as a reference velocity $U^*$. The axisymmetric axis is the line at $y = 0$. $We$ is Weber number which is expressed by :

$$We = \frac{\rho^* U^{*2} L^*}{\sigma^*} \qquad (17)$$

where $\rho^*$ is the density of fluid and $\sigma^*$ is the surface tension. In Fig.3 , the free surface profiles calculated by the present method are illustrated by the continuous lines. The corresponding experimental results[22] are indicated by the broken lines. We find that our numerical solutions are in very close agreement with the experimental results. There is a small difference between the calculated results and the corresponding experimental results near the tube outlet in Case 4, because the radius of curvature of the streamline is small. In Cases 1, 2, 3 and 4, the calculated contraction of jets is greater than the experimental one near the tube outlet. Duda and Vrentas[20] showed the similar tendencies in their investigation. The experimental contraction of jets, however, is greater than the calculated one in the downstream region. The difference between the calculated and experimental contraction in the downstream region, which seems to be caused by the accumulation of integral error, increases gradually as the distance $x$ increases, but it is very small.

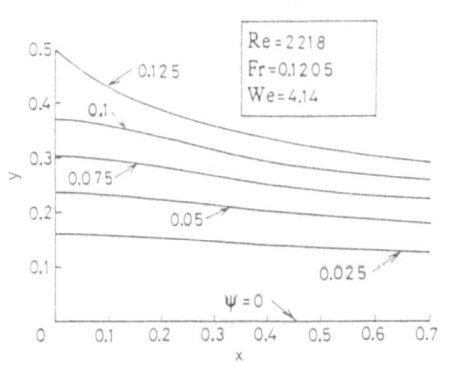

Fig.2  Streamlines

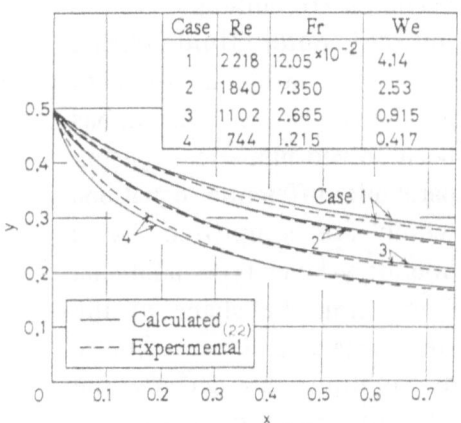

Fig.3  Free surface profiles
        of numerical results
        and experimental ones

128

The axial velocity distributions at cross sections of jets are shown in Fig.4. The axial velocities increase over a cross section of a jet until a uniform velocity distribution is attained. The results are similar to the analytical results of the two-dimensional laminar jet[25]. The axial velocity distributions calculated by the present method are found to agree very closely with experimental results[22] measured by a laser Doppler velocimeter.

Conclusions

Assuming an incompressible and two-dimensional or axisymmetric flow, the complete equations which describe a viscous flow have been formulated for a inverse problem using the orthogonal coordinate system ($\phi$, $\psi$), where $\phi$ is the coordinate of the flow direction and $\psi$ is the stream function or Stokes stream function. The axisymmetric laminar jet leaving a Poiseuille tube for an inviscid outer fluid phase, under the influence of both gravity and surface tension, has been solved numerically using parabolic differential equations which are simplified by a boundary-layer-type analysis. Comparing the shape of the free surface and the velocity distribution obtained by the present method with the experimental results, the present method proved to be very useful for investigating the behavior of an axisymmetric laminar jet of incompressible viscous fluid.

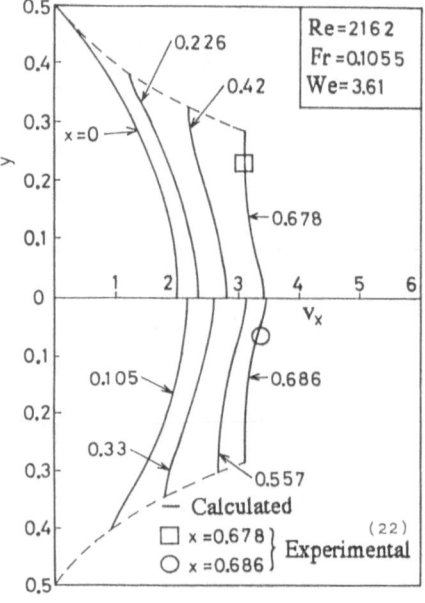

Fig.4  Axial velocity distribution
of a jet  issuing from
a circular tube

References
1. Jeppson,R.W.: Inverse Formulation and Finite Difference Solution for Flow from a Circular Orifice, *J. Fluid Mech.*,**40,1**(1970)215-223.
2. Yang,T. ; Nelson, C.D.: Griffith Diffusers, *Trans. ASME, J. Fluid Eng.*, **101-4**(1979)473-447.
3. Sugiyama, H.: A Numerical Study of Gas- Particle Supersonic Flow Past Blunt Bodies (The Case of Axisymmetric Flow), *Trans. Jpn. Soc. Mech. Eng.*, (in Japanese), **50-449,** B(1984)185-192.
4. Ahmed, N.M.A. ; Myring,D. F.: An Inverse Method for the Design of Axisymmetric Optimal Diffusers, *Int. J. for Numerical Method in Engineering*, **22-2**(1986)377-394.
5. Brown, E.F. ; Eskandarian, A.: Nozzle Design Using an Iterative Dirichlet Approach, *Int. J. for Numerical Methods in Engineering*, **22-2**(1986)481-494.
6. Chen, N. X. ; Zhang, F. X.: A Generalized Numerical Method for Solving Direct Inverse and Hybrid Problems of Blade Cascade Flow by Using Streamline-Co-Ordinate Equation, *ASME paper* **No.87-GT-29**.
7. Tsukiji, T. ; Takahashi, K.: A Two-Dimensional Reattaching Jet issuing from theMetering Orifice of Spool Valves ( Numerical Simulation by a Discrete-Vortex Method), *Trans. Jpn. Soc. Mech. Eng.*,(in Japanese), **54-507**,B(1988)3076-3081.
8. Schmidt, E. ; Berger, P.: Inverse Design of Supercritical Nozzles and Cascades, *Int. J. for Numerical Methods in Engineering*, **22-2**(1986)417-432.
9. Schmidt, E.: Computation of Supercritical Compressor and Turbine Cascades with a Design Method for Transonic Flows, *Trans. ASME, J. Engineering for Power*, **102**(1980)68-74.
10. Zannetti, L.: A Natural Formulation for the Solution of Two-Dimensional or Axisymmetric Inverse Problems, *Int. J. for Numerical Methods in Engineering*, **22-2**(1986)451-463.
11. Sugiyama, H. : A Numerical Study of Gas-Particle Supersonic Flow Past Blunt Bodies ( The Case of Two-Dimensional Flow ), *Trans. Jpn. Soc. Mech. Eng.*,(in Japanese), **50-460**,B(1984)3084-3090.
12. Stanitz, J. D.: Design of two-dimensional channels with prescribed velocity distributions along the channel walls, *NACA* Rep.No.**1115**(1953)153-192.
13. Stanitz, J. D.; Sheldrake, L. J.: Application of a channel design method to high-solidity cascades and tests of an impulse cascade with $90°$ of turning, *NACA* Rep. **1116**(1953)193-211.
14. Takahashi,K.: A Numerical Analysis of Flow Using Streamline Coordinates (The Case of Two-Dimensional Steady Incompressible Flow), *Bull. JSME*, **25-209**(1982)1696-1702.
15. Finnigan, J. J.: A Streamline Coordinate System for Distorted Two-Dimensional Shear Flows, *J. Fluid Mech.*, **130**(1983)241-258.

16. Tsukiji, T. ;  Takahashi, K.: The Numerical Solution of the Jet Issuing from a Skew-Symmetrical Orifice Using a Streamline  Coordinate System, *Trans. Jpn. Soc. Mech. Eng.*,( in Japanese ),**49-445**,B(1983)1832-1839.
17. Takahashi, K. ; Tsukiji, T.: Numerical Analysis of a Laminar Jet Using a Streamline Coordinate System, *Trans. Canadian Soc. Mech. Eng.*, **9-3**(1985)165-170.
18. Von Mises, R. V.: Bemerkungen zur Hydrodynamik, *Zeitschrift Fur Angewandte Mathematik und Mechanik*, **7-6**(1927)425-431.
19. Dubreil-Jacotin, M. L.: Sur la Determination Rigoureuse des ondes Permanentes Periodiques Dampleur Finie, *Journ. de Math.* , **13**(1934)217-291.
20. Duda, J. L. ; Vrentas, J. S.: Fluid Mechanics of Laminar Liquid Jets, *Chem. Eng. Sci.*, **22**(1967)855-869.
21. Patankar, S. V. ; Spalding, D. B. : A Calculation Procedure for Heat, Mass and Momentum Transfer in Three-Dimensional Parabolic Flows, *Int. J. Heat Mass Transf.*, **15**(1972)1787-1806.
22. Yamaguchi, R. ; Takahashi, K.: Flow Pattern Near the Outlet of a Straight Long Circular Tube (1st Report, Experimental Study of Velocity Profile), *Bull. JSME*, **23-185**(1980)1798-1805.
23. Jeffrey, A: *Mathematics for Engineers and Scientists,* (1969)247, Barnes & Noble, Inc.
24. Tsukiji, T.; Takahashi, K.: Numerical Analysis of an Axisymmetric  Jet Using a Streamline Coordinate System, *JSME International Journal*, **30-267**(1987)1406-1413.
25. Miyake, H., Mukai, E. and Iemoto, Y., Two-Dimensional Laminar Liquid Jets, *Trans. Jpn. Soc. Mech. Eng.*, (in Japanese)**45-390**, B(1979)180-187.

# Chapter 3
# Parameter Identification

# Requirements for the Structure of Analytical Models Used for Parameter Identification

M. Link

Prof. of Light Weight Structures
University of Kassel, Moenchebergstr. 7,
D-3500 Kassel, Germany

## 1. Introduction

In order to assess the success and the robustness of procedures for updating elastomechanical parameters of discrete (finite element) models using experimental data investigations in recent years (e. g. refs. 1 - 7) have mainly been directed with respect to the influence of measurement noise, the type and the amount of experimental data like natural frequencies and modes and also with respect to different types of residuals used in the objective function to be minimized and subject to the constraint equations given by possible upper and lower parameter bounds and by linear relations between the parameters.

In all these investigations the structure of the initial model to be updated in terms of the type and the number of finite elements was assumed to be fixed before the updating was started. The unknown correction parameters are generally introduced by developing the analytical system matrices $S$ (stiffness, mass or damping matrix) into a series

$$S = S_A + \sum_i S_i \, \triangle p_i \quad (i = 1, 2, \ldots P)$$

where $S_i = \partial S_A / \partial p_i$ represents a correction matrix (submatrix) defining the location and the source of the unknown modelling error.

Therefore the success of the parameter identification is governed by the skill of the analyst to assume (1) the structure of the initial model, (2) the source and (3) the location of the modelling error. Since these assumptions are not unique in practical applications the solutions will also not be unique although all of them fulfill the mathematical minimisation criteria, i. e. the updated models will reproduce the experimental input data. For example, using a physical design parameter like a bending stiffness to update a dicretisation error caused by a coarse finite element mesh would *not be consistent* with the real error source and would therefore destroy the physical meaning of the design parameter although the updated model will better reproduce the test data.

With two exemplary beam structures we first will illustrate the influence of different types of model structure errors on the results of the updating process. Finally we will discuss indicator functions which shall help to indicate the best choice for the initial model structure and discuss related requirements.

134

## 2. Updating Methods

To ilustrate the influence of the initial modelling assumptions we have applied different updating theories. Their basic formulations are repeated here for completeness (see also [6]).
Starting point for parameter updating is the assumption that the initial analytical stiffness and mass matrices can be updated according to

$$K = K_A + \Sigma \, \alpha_i \, K_i \qquad (1a)$$
$$M = M_A + \Sigma \, \beta_j \, M_j \qquad (1b)$$

$\left[\alpha_i \, \beta_j\right] = \left[p_k\right]$ = unknown correction (design) parameters, ( i = 1, 2..I, j= 1,2..J)

$k = 1,.. \, P = I + J$ = no. of correction parameters
$K_A, M_A$ = analytical (initial) stiffness and mass matrix
$K_i, M_j$ = assumed correction substructure matrices (elements or element groups) *defining source and location of modelling error* (error model).

With this assumption it is possible to derive a bunch of objective functions

$$J(p_k) = \Delta Z^T \, \Delta Z + p^T W_p \, p \longrightarrow \text{min} \qquad (2)$$

$\Delta Z (p_k) = W \Delta Z_U$ = weighted residual vector

($W$, $W_p$ = weighting matrices)

whose minimisation allows to determine the design parameters (A comprehensive collection is presented in ref. [2]). The second term in eq. (2) is used to constrain the variation of the correction parameters p. (Note: W=0 leads to $p = 0$, i.e. no change of the correction factors). Very frequently the differences between the experimental (index M) and the updated analytical (finite element) modal data are used as residual vectors:

$$\Delta Z_L = \lambda_M - \lambda \qquad (3a)$$

$\lambda = \omega^2 := (\lambda_1 ... \lambda_r ... \lambda_R)$ = Eigenvalues of *updated* model

$$\Delta Z_y = y_M - y_c \qquad (3b)$$

$y_c := (y_{c1} ... y_{cr} ... y_{cR})$ = modes of *updated* model related to the reduced set of measured DOF's

$\lambda_M, y_M$ = corresponding measured eigenvalues and modes ( r = 1, 2 ... R = no. of measured modes)

In the present paper we will apply additionally the error given by weighted force residuals [1], [5], [6]

$$\Delta Z_R = W \, \hat{K}_c(\omega_M, p) \, y_M \qquad (3c)$$

with $\hat{K}_c (\omega_M) = -\omega_M^2 \, M_c + K_c$

= updated dynamic stiffness matrix dynamically condensed to $N_M < N$ measured DOF's and the weighting matrix

$$\mathbf{W} = \hat{\mathbf{K}}_{AC}^{-1} = \left(-\omega_F^2 \mathbf{M}_{Ac} + \mathbf{K}_{Ac}\right)^{-1} = \sum_i \mathbf{y}_{Ai} \mathbf{y}_{Ai}^T / (\lambda_{Ai} - \lambda_F) \tag{3d}$$

$\hat{\mathbf{K}}_{Ac}$ = condensed dynamic stiffness matrix of initial model (index A)

$$\lambda_F = \omega_F^2; \quad \omega_F := (\omega_{F1} \ \dots \ \omega_{FR})$$

The weighting matrix $\mathbf{W}$ can be interpreted as a dynamic filter, which transforms the force residuals to response residuals, with given filter frequencies $\omega_F$.

The condensed matrices in the above equations are obtained by first indroducing eqs. (1) into the eigenequation

$$(-\omega_M^2 \mathbf{M} + \mathbf{K}) \mathbf{y} = \hat{\mathbf{K}} \mathbf{y} = 0 \tag{4}$$

and by partitioning the matrices and the eigenvectors with respect to the measured (index M) and unmeasured DOFs (index U):
The vector $\mathbf{y}_u(\omega)$ contains the *unknown* eigenvector components at the $N_u = N - N_M$ unmeasured DOF's and can be expressed as a function of the correction parameters and the measured vector $\mathbf{y}_M$:

$$\mathbf{y}_U = \mathbf{T}(\omega_M, p) \, \mathbf{y}_M = -\hat{\mathbf{K}}_{UU}^{-1} \, \hat{\mathbf{K}}_{UM} \, \mathbf{y}_M \tag{5}$$

which yields the transformation

$$\mathbf{y} = \begin{bmatrix} \mathbf{y}_M \\ \mathbf{y}_U \end{bmatrix} = \begin{bmatrix} \mathbf{I} \\ \mathbf{T}(\omega_M, p) \end{bmatrix} \mathbf{y}_M = \overset{*}{\mathbf{T}}(\omega_M, p) \, \mathbf{y}_M \tag{6}$$

used to condense the system matrices in the equation of motion (4) to the measured DOF's leading to

$$\mathbf{K}_c = \overset{*}{\mathbf{T}}^T (\mathbf{K}_A + \sum_i \alpha_i \mathbf{K}_i) \overset{*}{\mathbf{T}} = \mathbf{K}_{AC} + \sum_i \alpha_i \mathbf{K}_c^i \tag{7a}$$

$$\mathbf{M}_c = \overset{*}{\mathbf{T}}^T (\mathbf{M}_A + \sum_j \beta_j \mathbf{M}_j) \overset{*}{\mathbf{T}} = \mathbf{M}_{AC} + \sum_j \beta_j \mathbf{M}_c^j \tag{7b}$$

with the frequency dependent matrices

$\mathbf{M}_{AC} = \overset{*}{\mathbf{T}}^T \mathbf{M}_A \overset{*}{\mathbf{T}}$
= condensed initial mass matrix

$\mathbf{K}_{AC} = \overset{*}{\mathbf{T}}^T \mathbf{K}_A \overset{*}{\mathbf{T}}$
= condensed initial stiffness matrix

$\hat{\mathbf{K}}_{AC} = -\omega_M^2 \mathbf{M}_{AC} + \mathbf{K}_{AC}$
= condensed initial dynamic stiffness matrix

$\mathbf{K}_c^i = \overset{*}{\mathbf{T}}^T \mathbf{K}_i \overset{*}{\mathbf{T}}$
= condensed stiffness correction matrix

$\mathbf{M}_c^j = \overset{*}{\mathbf{T}}^T \mathbf{M}_j \overset{*}{\mathbf{T}}$
= condensed mass correction matrix

The minimisation in eq. (2) with respect to the unknown design parameters $p_k = [\alpha_i \; \beta_j]$ results in a non-linear problem which is solved iteratively using the first step of the Taylor series expansion

$$\Delta Z = G \, \Delta p - \Delta Z^0 \tag{8a}$$

where the terms of the gradient matrix are obtained from

$$G = \left[ \frac{\partial \Delta Z}{\partial p_K} \right] = \left[ \cdots \frac{\partial \Delta Z}{\partial \alpha_i} \cdots \frac{\partial \Delta Z}{\partial \beta_j} \cdots \right] \tag{8b}$$

Neglecting derivatives $\partial \overset{*}{T} / \partial p$ we obtain for the response residual method from eqs. (3c) and (7)

$$G \rightarrow G_R(\omega_M) = W \, [\cdots K_c^i \; \cdots \; -\omega_M^2 M_c^j \; \cdots] \; y_M \tag{8c}$$

$$\Delta Z_R^0 = - W \, \hat{K}_c(\omega_M, p) \; y_M \tag{8d}$$

Eq. (8a) is set up by $R * N_M$ equations where R is the no. of eigenfrequencies and $N_M$ the no. of measured DOF's.

The gradient matrix for the eigenvalue residuals is obtained from

$$G \rightarrow G_L = \left[ \cdots \frac{\partial \lambda}{\partial \alpha_i} \cdots \frac{\partial \lambda}{\partial \beta_j} \cdots \right] \tag{9a}$$

with

$$\frac{\partial \lambda}{\partial \alpha_i} = y_c^T \, K_c^i \, y_c = \gamma^i$$

$$\frac{\partial \lambda}{\partial \beta_j} = - \lambda \, y_c^T M_c^j \, y_c = - \lambda \, \mu^j$$

$$\Delta Z_L^0 = \lambda^M - \lambda \tag{9b}$$

$\lambda = \omega^2$, $y_c$ = modal data of condensed updated system

The gradient matrix for the mode shape residuals are obtained from

$$G \rightarrow G_y = \left[ \cdots \frac{\partial y_c}{\partial \alpha_i} \cdots \frac{\partial y_c}{\partial \beta_j} \cdots \right] \tag{10a}$$

with

$$\frac{\partial y_c}{\partial \alpha_i} = -\sum_s y_{cs} \frac{\gamma_s^i}{\lambda_s - \lambda} \quad \text{for } \lambda_s \neq \lambda \quad , \quad \frac{\partial y_c}{\partial \beta_j} = \sum_s y_{cs} \frac{\lambda \mu_s^j}{\lambda_s - \lambda} \quad \text{for } \lambda_s \neq \lambda$$

$$\frac{\partial y_c}{\partial \alpha_i} = 0 \quad \text{for } \lambda_s = \lambda \quad , \quad \frac{\partial y_c}{\partial \beta_j} = \frac{-1}{2} \mu_s^j y_c \quad \text{for } \lambda_s = \lambda$$

where $\quad \gamma_s^i = y_{sc}^T \, K^i y_c \quad$ and $\quad \mu_s^j = y_{sc}^T \, M^j y_c$

$$\Delta Z_y^0 = y_M - y_c \tag{10b}$$

In the present paper we applied 2 different combinations

RRU: response equation residuals (index R), eq. (8c,d), combined with eigenvalue residuals (index L), eq. (9).
EVU: eigenvector residuals (index y), eq. (10), combined with eigenvalue residuals, eq. (9)

For all two methods additional diagonal weighting matrices $W_I$ and $W_L$ are introduced for the residuals leading to

$$\Delta Z = \left[ \begin{array}{c} W_I \, \Delta Z_I \\ W_L \Delta Z_L \end{array} \right]$$

$$G = \left[ \begin{array}{c} W_I \, G_I \\ W_L G_L \end{array} \right]$$

$$\Delta Z^0 = \left[ \begin{array}{c} W_I \, \Delta Z_I^0 \\ W_L \Delta Z_L^0 \end{array} \right] \qquad (I=R,y) \tag{11}$$

It is an advantage that the eigenvalue residuals need only be taken into account when there is no problem with pairing the measured and the calculated eigenvalues. No pairing problem exists for the response equation residuals (I=R).

The minimisation (2) (applied at each iteration step $\nu$) leads to the least squares solution for the correction parameter changes.

$$\Delta p^{(\nu)} = (G^T G + W_p)^{-1} G^T \Delta Z^0 \; (\nu-1) \tag{12}$$

The correction parameters, and the system matrices are updated in each iteration step $\nu$ according to

$$\alpha_i^{(\nu)} = \alpha_i^{(\nu-1)} + \Delta \alpha_i^{(\nu)} \tag{13a}$$

$$\beta_j^{(\nu)} = \beta_j^{(\nu-1)} + \Delta \beta_j^{(\nu)} \tag{13b}$$

$$K^{(\nu)} = K_A + \sum_i \alpha_i^{(\nu)} K_i = K^{(\nu-1)} + \sum_i \Delta \alpha_i^{(\nu)} K_i \tag{14a}$$

$$M^{(\nu)} = M_A + \sum_j \beta_j^{(\nu)} M_j = M^{(\nu-1)} + \sum_j \Delta \beta_j^{(\nu)} M_j \tag{14b}$$

### 3. Modelling errors

It can be seen from the basic assumption in eqs. (1) that the solution depends on the contents of the correction matrices

$$K_i = \frac{\partial K}{\partial \alpha_i} \quad \text{and} \quad M_j = \frac{\partial M}{\partial \beta_j}$$

and of the initial model matrices $K_A$ and $M_A$. It will be our task now to investigate the influence of the assumptions contained in eqs. (1). There are two essentially different types of assumptions. The first is characterized by the *model structure* used in $K_A$ and $M_A$ and the second by the type and the location of the correction parameters in the second terms in eqs. (1) which we call the *error model*.

The following model structure errors may occur in practice. These errors are defined with respect to what we call the true or the test model which, of course, is unknown when real test data are used for updating. Analytical test models may be used for test data simulation.

(a) Discretisation errors:
FE-mesh density (insufficient h-convergence), shape function order (insufficient p-convergence), inconsistent shape functions for stiffness and mass matrix representation, e.g. when lumped mass approach is used. In [8] a procedure for excluding the discretisation errors is proposed.

(b) Energy errors:
e. g. due to neglecting shear deformation and / or rotational kinetic energy.

(c) Boundary conditions:
elastic boundaries, missing degrees of freedom at boundaries.

(d) Element connectivity:
resulting in local stiffness and mass distribution errors (this type of error is generally avoided by using graphic mesh pre-processing).

(e) Non-linearity errors:
resulting when linear models are used for non-linear structures.

The *error model* contains the assumptions for the type, the location and the number of the correction parameters like: local design parameters of elements or element groups (bending, shear and extensional stiffness, translational and rotational inertia, joint stiffness, individual matrix elements, spring stiffness), distribution of material properties (E-modulus, densitiy).

We will not discuss here the influence of the test data errors on the updating results in detail (see e. g. [1] - [7] ) but we will always have to account for unavoidable random and systematic errors and the incompleteness of test data with respect to the number of measured modes and measured degrees of freedom.

## 4. Influence of model structure errors

We will now demonstrate the influence of different systematic model structure errors on parameter updating results introduced by the discrepancies between the test and the analysis model. These errors may be interpreted as a systematic discrepancy between test and analysis excluding a treatment by statistical analysis. We will show that any discrepancies between test and analysis model with respect to their inner structure and/or the applied error model will destroy the physical significance of the physical correction parameters after updating.
At first we use the simple cantilever beam shown in fig. 1.

(a) Discretisation errors:
The simulated test data are created with the 18-element discretisation and the physical property data given in fig. 1a ( test model ). The coarser 6-element discretisation ( analysis model ) of fig. 1b is used to introduce discretisation errors affecting the first six eigenfrequencies according to curve "DE" in fig. 2.
Additionally we have introduced the parameter perturbations indicated in fig. 1b to generate the analytical model. This analysis model therefore contains both the

discretisation errors as well as the parametric errors allowing to assess the influence of the discretisation errors on the updating results.

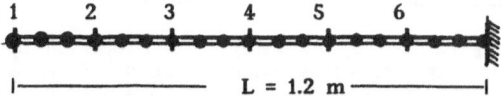

Fig. 1 a: 18-element test model with 6 measurement DOF's

Young's modulus E = 7E9 N/m$^2$
Mass distribution $\mu$ = 2.711 kg/m
Bending moment of inertia I = 2.12E-8 m$^4$
Eigenfrequencies $f_i$ [Hz] = [ 2.875  18.02  50.45  98.87
163.47  244.23  341.35  454.8]

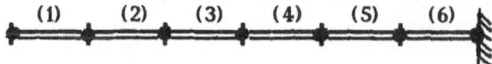

Fig. 1b: 6-element analysis model with the following perturbations on the bending moments of inertia to generate the erroneous analysis model: $I_{(2)}$ = 20 % , $I_{(4)}$ = 30 % , $I_{(5)}$ = 10 %.

Fig. 1 : Test and analysis models for discetisation error study

The initial deviations of the test eigenfrequencies with respect to the initial model frequencies are shown in curve "AM" of fig. 2. For updating we have used the RRU–method (weighted response equation residuals) and the EVU (eigenvector residuals) both in combination with eigenfrequency residuals together with six lateral measurement DOF's and six measured modes.
Six parameters $\alpha_i$ (i = 1, . . . 6)) corresponding to the moments of inertia of the six beam elements were open for updating. These parameters are consistent with the real error source and locations ( exact error model ).
The eigenfrequency deviations and the correction factors 1 + $\alpha_i$ after updating are presented in figs. 2 and 3, curves "EVU" and "RRU". It should be noted that the eigenfrequencies no. 7 and 8 ( not used for updating) have not improved. This result limits the application range of the updated model to the frequency range of the first six active modes. Correction factors are different for both methods.

(b) Error model inconsistent with respect to shear deformations of test model

This type of inconsistency was generated by inclusion of shear deformation effects in the 6 element beam. With the shear parameter $EI/GA_s L^2 = 0.04$ ( $GA_s$ = shear stiffness) the test model then has the following eigenfrequencies:

$f_{test}$ = $\left[2.873\ 17.93\ 49.94\ 97.42\ 160.5\ 235.04\ 377.1\ 507.78\right]$

The same analysis model (6 elements, perturbations on the bending stiffness parameters $I_{(2)}$, $I_{(4)}$, $I_{(6)}$ and 6 correction parameters corresponding to $I_{(1)}$ – $I_{(6)}$) as in case (a) has been used. Fig. 4 shows the frequency deviations of the test model without shear effects (curve "SE") and the updated model frequencies with methods RRU and EVU. The correction factors 1 + $\alpha_i$ are presented in fig. 5.

The factors show opposite tendency for the two methods. With the RRU method the two non-active frequencies no. 7 and 8 have also been improved.

(c) Error model inconsistent with boundary condition of test model

This type of inconsistency was generated by inclusion of a rotational spring in the 6-element beam instead of the rigid clamping condition used before. With the spring stiffness $k_f$ = 6.0 E3 the test model then has the following eigenfrequencies:

$$f_{test} = \begin{bmatrix} 2.76 & 17.37 & 48.80 & 96.23 & 160.66 & 241.45 & 371.45 & 506.65 \end{bmatrix}$$

The same analysis model (6 elements, perturbations on the bending stiffness parameters $I_{(2)}$, $I_{(4)}$, $I_{(6)}$ and 6 correction parameters corresponding to $I_{(1)}$ – $I_{(6)}$) as in case (a) and (b) was used. The spring stiffness parameter $k_f$ was *not* used for updating which means that in this case the selected parameter set is *not consistent* with the real error source and location. This situation occurs in real applications when neither the location nor the source of the physical design parameter errors are known a priori. Fig. 6 shows the frequency deviations of the test model with rigid spring (curve "RS") and the updated model frequencies with method EVU . The correction factors are presented in fig. 7. In particular it can be noticed that the bending stiffness parameters of the beam are altered in order to compensate for the boundary flexibility which is not consistent with reality.

The following conclusions may be drawn from these curves:

· The active eigenfrequencies (those which have been used for updating) have improved,
· the eigenfrequencies *beyond* the active frequencies have <u>not</u> improved except for the RRU- method which gave an improvement of the non-active eigenfrequencies in cases (a) and (b),
· the correction factors are *not unique*, they depend on the type of residual used in the updating procedure,
· the deviations (not shown here) between test modes and updated modes are extremely small for the active modes,
· the correction parameters converged after few iteration steps
· the curves do not allow to answer the question of the best update model.

The last question however is crucial in practical applications and makes it necessary to find criteria to select the best parameter estimate.

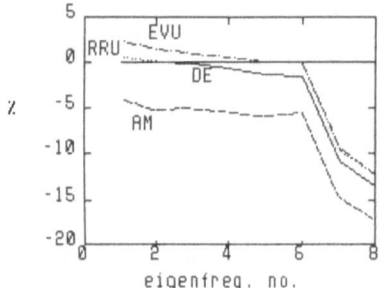

EVU = eigenvector update
RRU = response residual update
DE  = systematic discretisation errors only
AM  = analysis model with discretisation and parameter errors

Fig. 2 : Eigenfrequency deviations w. r. to test frequencies, case (a)

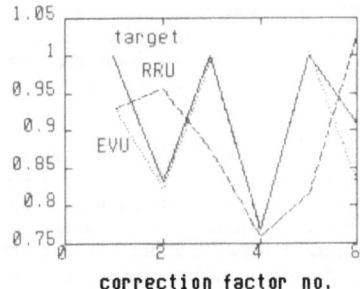

**Fig. 3** : Correction factors, case (a)

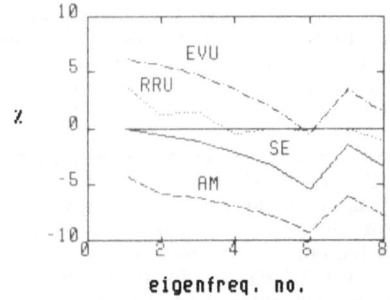

**Fig. 4** : Eigenfrequency deviations w. r.
to test frequencies, case (b)

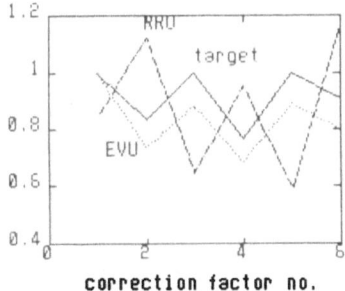

**Fig. 5** : Correction factors, case (b)

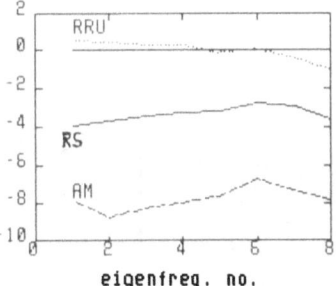

**Fig. 6** : Eigenfrequency deviations w. r.
to test frequencies, case (c)

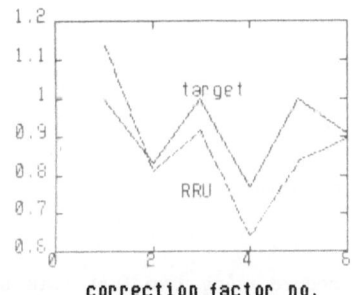

**Fig. 7** : Correction factors, case (c)

In the next example we used real test data of the laboratory test structure shown in fig. 8. It consists of 5 cantilever steel beams carrying a tip mass and coupled by 6 arch springs. With this structure the effect of two popular types of mass representation was investigated, the lumped mass and the consistent mass representation.

Due to the concentrated tip masses the structure could be modelled as a discrete 5-DOF system shown in Fig. 9, where the masses are lumped at the 5 DOF's resulting in a diagonal mass matrix. The alternative approach was to use a consistent mass matrix for each arch and each beam obtained from reducing a multielement mass matrix by static condensation (Guyan reduction) to the connection DOF's at the beam tips. Instead of a diagonal matrix we then get a mass matrix with two non-zero off-diagonal rows. Since the same measured masses have been used for both representations and the same stiffness matrix has been used for the updating examples the resulting deviations between the update results are only caused by the different structures of the mass matrices.

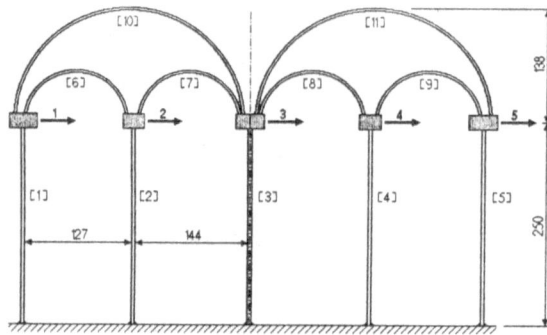

Fig. 8 : Laboratory test structure

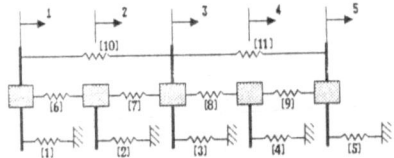

Fig. 9 : Discrete 5-DOF model of laboratory test structure

The updating was performed with the RRU-procedure using two test data sets with different numbers of measured DOF's and modes:

set 5,5: DOF's 1-5 and modes 1-5 (complete set),
set 4,3: DOF's 1-4 and modes 1-3 (incomplete set).

Using the incomplete set with only 3 modes for updating allows to asses the accuracy of the non-active modes 4 and 5 of the updated models. All 11 spring stiffness parameters of the discrete model in fig. 9 were open for updating. Mass matrix parameters were kept unchanged.

The eigenfrequency deviations of the initial models (curves "ini") in figs. 10 and 11 show, that the initial models with the inconsistent diagonal mass matrix is more accurate than the consistent mass model. Updating with the complete 5,5 data sets yields a nearly perfect fit of both updated models to the test frequencies (curves set 55, with symbol *). (The same was found for the mode shapes which are not shown here). Using the incomplete test data set 4,3 the results are characterized by perfect update of the active first three frequencies, however a considerable improvement of the non-active frequencies 4 and 5 was only obtained with the consistent mass representation only.

The lumped mass updated model frequencies even became worse than the initial model frequencies.
The result for the correction parameters $1 + \alpha_i$ (i=1,...,11) are presented in figs. 12 and 13.
Curves for set 5,5 and set 4,3 indicate that the non-uniqueness of the parameters with respect to the completeness of the data set is much less marked in fig. 12 than in fig. 13 for the lumped mass model. Comparing the results in figs. 12 and 13 underlines the non-uniqueness of the identified parameters with respect to the selected structure of the mass model.

With these and the previous results from the academic beam examples in mind we have to discuss the problem of finding criteria to select the best parameter estimate. It must be emphazised that with the above examples we have concentrated on systematic test/analysis discrepancies and did not use statistical weighting matrices. In particular the weighting matrix $W_p$ (Bayesian estimation) in eq. (2) which is responsible for constraining the parameter variations was not used since in the presence of systematic errors estimates on the parameter uncertainties cannot be established. Any deterministic choice of $W_p$ would lead to any answer and thereby increase the number of possible solutions. In order to constrain the number of possible solutions we find it necessary to discuss first the scope of the estimated models from the viewpoint of its subsequent utilization in mechanical, aerospace or civil engineering structural systems.

These models should have the ability of improved prediction of structural response with respect

(1) to other loading conditions than those used in the test,
(2) to design modifications,
(3) to coupled substructure analysis and
(4) to the influence of structural damage.

For all these areas it is important to determine the no. of active modes, i.e. the minimum number and the type of modes necessary for accurate predictions. For substructure coupled analysis it is additionally important to improve not only the active modes but also the interface flexibility which is influenced by the higher modes. These considerations lead to the obvious conclusion, that it makes no sense to use an updated model predicting only $n_{act}$ modes accurately when $n >$ $n_{act}$ modes are necessary for the purposes mentioned above. As shown in fig. 11 for the inconsistent mass model it may happen that the non-active mode predictions are worse than for the initial model. On the other hand when the number of active modes is high enough for the individual application the corresponding updated model may be interpreted as a special purpose model with limited validity. In the ideal case the updated model will also improve the predictions in the frequency range *beyond* the active range, which seems to be a good and reliable criterion to check the suitability of the initial model structure. This behavior was demonstrated in fig. 10, where the non-active frequencies (set 4,3) of the updated model were improved for the consistent mass only but not for the lumped mass model.

Another criterion is the sensitivity of the identified parameters with respect to the type of residuals and the amount of test data used in the objective function to be minimized.

Any initial model structure should be prefered to another when the resulting parameter sensitivity is smaller. From comparison of the parameter sensitivities in figs. 12 and 13 with respect to test sets 5,5 and 4,3 one would again prefer the consistent mass representation. Looking at the high parameter sensitivities at the cantilever beam example in figs. 3 and 5 one would reject all the initial model structures used there.

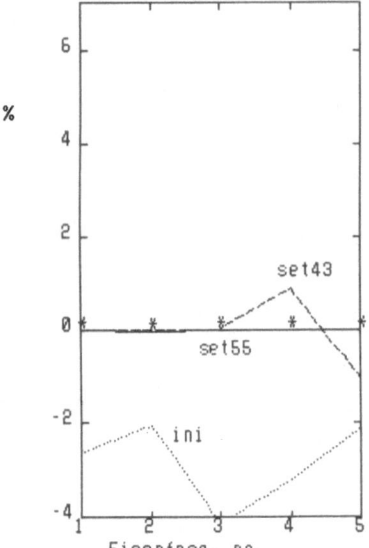 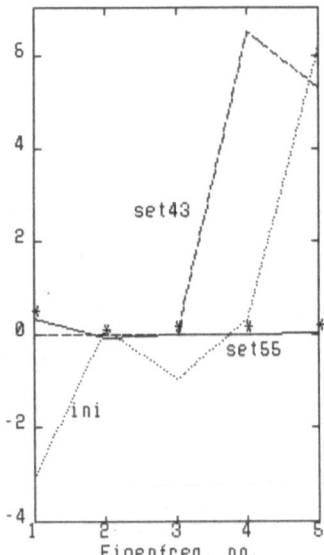

Fig. 10: Frequency errors before and after updating, consistent masses

Fig. 11: Frequency errors before and after updating, lumped masses

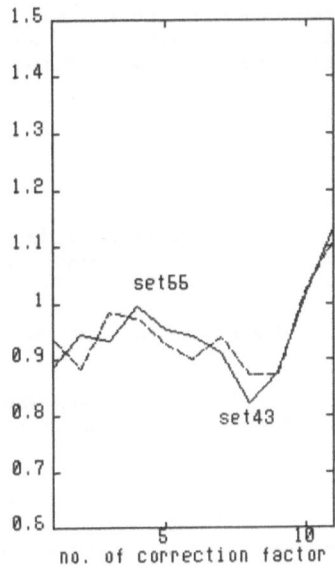

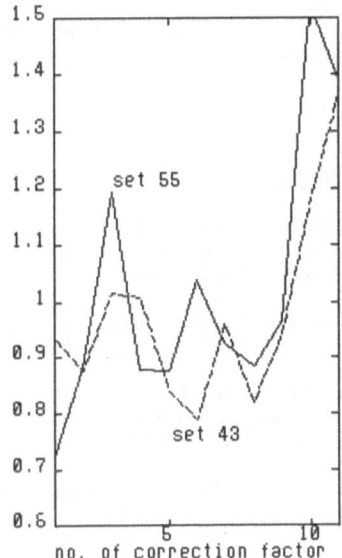

Fig. 12: Correction factors after update
with consistent mass

Fig. 13: Correction factors after update
with lumped mass

## 5. Conclusions

Updating of physical design parameters like bending stiffnesses for minimizing test/analysis discrepancies leads to non-unique solutions which depend on the assumptions contained in the structure of the analysis model, on the amount and the accuracy of the test data, on the type of residuals in the objective function as well as on the error model in terms of type, numbers and location of correction parameters. However, the aim of retaining the physical meaning of the estimated parameters requires that the parameters are <u>invariant</u> with respect to these influences. Each solution may nevertheless produce an adequate fit of the experimental data used for parameter updating. It is therefore necessary to *reduce* the degree of non-uniqueness although unique solutions cannot be expected because of the non-linear nature of the minimisation procedure.

In general several model structures, objective functions and error models should be choosen.

The selection of the best suited model structure is governed by
· the specified use of the identified model,
· its ability to predict test results which have not been used in the updating process,
· the sensitivity of the identified parameters with respect to the choice of the model structure.

As already stated in [7] the assessment whether the identified model is adequate for the specified use will not be possible without the exercise of engineering judgement. Studies similar to those presented here can be helpful for this assessment.

## References

1.  Cottin N., Felgenhauser H.P., Natke H.G.: On the parameter Identification of Elastomechanical Systems Using Input and Output Residuals. Ing. Archiv (54), Springer Verlag, 1984.

2.  Natke H.G.: Einführung in die Theorie und Praxis der Zeitreihen- und Modalanalyse, Vieweg Verlag, Braunschweig, 1988.

3.  Lallement G., Piranda J., Fillod R.: Parametric Identification of Conservative Self Adjoint Structures, Proc. Int. Conf. Spacecraft structures and Mechanical Testing ESA SP-289 , ESA / Estec, Noordwijk, The Netherlands , 1989.

4.  Caesar B., Eckert L., Hoppe A.: Design Parameter Update of Dynamic Mathematical Models in Presence of Test Noise and Mode Pairing Problems. Int. Conf. on Spacecraft Structures and Mech. testing. European Space Agency ESA/ESTEC (1991).

5.  Link M.: Localisation of Errors in Computational Models Using Dynamic Test Data. Proc. European Conference on Structural Dynamics, EURODYN 90. Bochum, Germany, June 1990, Elsivier Publ.

6.  Link M., Zhang L.: Experience with Different Procedures for Updating Structural Parameters of Analytical Models Using Test Data. Proc. IMAC X, San Diego, 1992.

7.  Hasselman T.K.: A Perspective on Dynamic Model Verification. Modal Testing and Model Refinement (ASME, New York, NY), AMO-Vol. 59, 1983, Winter Annual Meeting of the ASME, Boston, MA, 1983.

8.  Mottershead J.E., Weixun Shao: Correction of Joint Stiffnesses and Constraints for Finite Element Models in Structural Dynamics. J. Appl. Mechanics, 1992.

# Inverse Problem in Mechanics of Structures: a New Approach Based on Displacement Field Processing

M. Grédiac, C. Toukourou and A. Vautrin

Ecole des Mines de Saint-Etienne, Département Mécanique et Matériaux
158, cours Fauriel, 42023 Saint-Etienne cedex 2, France

Summary
The experimental determination of the mechanical properties of structures is of prime importance to assess and improve their performances. This problem becomes critical when heterogeneous or anisotropic materials are concerned because of their variable and adjustable characteristics. The present paper addresses a method to identify the elastic characteristics of structures based on displacement field processing. The main idea that governs the approach is to depart from the usual experimental procedures, that restrict their interest to homogeneous strain fields, and to regard heterogeneous strain fields. Assuming a relevant constitutive law, the inverse problem is to compute at best its parameters from the strain field over the specimen surface. The paper describes the theoretical framework of the approach and two examples are discussed.

Introduction

Designing optimized structures can be regarded in terms of cost, weight, geometry or material capabilities like stiffness or strength. Unfortunately, no general rule is available to design at best any part of structure subjected to any given loading. Expert systems associated with finite element analysis and computed aided design only provides optimized shapes and geometries in some particular cases. Heavy computations are often required even if the materials are chosen a priori. This problem becomes critical for composite materials and structures since the design procedure must then incorporate two prominent new features : heterogeneity and anisotropy.

**Heterogeneity** means that the designer can associate several materials to fulfil any prescribed constraint. For instance, a sandwich beam consists of two thin stiff skins and a light core. The resulting beam is usually a compromise between global stiffness and light weight.and its design can be formulated as an optimization problem to achieve the beam with minimum weight that will meet prescribed constraints on stiffness and strength. This particular problem has already received considerable attention at different levels. For instance, for a given cross section geometry, one can consider the core density as well as the skin and core thicknesses as structural parameters that can be optimized for a given simple loading. This problem evolves into a far more complicated one when the cross section shape can be changed too. Core and face thicknesses are tailored to produce a progressive change in stiffness along the length of the beam. Such an approach allows a better design in term of optimal weight but it becomes obvious that the optimization procedure is now far more complicated. First, a stiffness evolution and minimum weight constraint being given, it is not evident to obtain a unique solution for the core and skin thickness evolutions. Second, the optimized arrangement can involve more than two materials, each of them being used for its particular mechanical properties in a part of the beam. The design of the cross section of a ski ([1] for instance) outlines that many different materials are currently in use in such a structure. The final arrangement is the result of a step by step empirical analysis. The global stiffness of the beam is achieved using the traditional rules of the beam theory and a suited experimental procedure is absolutely necessary to check the final design afterwards. This simple example clearly points out that a relevant technique for global measurement of stiffness is essential when an optimal design procedure is being set up.

**Anisotropy** provides the designer with direction dependent mechanical properties. Such a feature is attractive for the optimization procedure : one can design structures which are reinforced in the direction of the loading. However, the design at best in terms of mechanical properties as well as in terms of cost is not correctly carried out up to now, and the advantages of such materials are therefore not completely taken into account. Two main reasons can be emphasized at that level. The first one is that usual rational design rules are not suited to such media. The second one deals with the usual testing procedures that cannot provide the designer with accurate and relevant informations on the variable structure mechanical parameters. In fact, characterizing new materials must involve new experimental techniques as well as new methods for data processing.

Both features described above show that direct measurements on structures must be included in any optimization procedure. Analysis is used to predict the performances and suited experiments are necessary to check the final result. Two basic approaches can be distinguished to assess mechanical characteristics from experimental tests :

- performing tests that produce **uniform** strain fields. The usual procedure is then to collect one or more **local informations** and to relate them with the measured loading through well known analytic formula and assumptions. Such an approach is commonly used when usual tensile or bending tests are carried out. Unfortunately, it is not suitable for the practical characterization of variable mechanical properties required in an optimization procedure.

- performing tests directly on parts of structures. In this latter case, the strain field becomes obviously **heterogeneous** and cannot be easily determined as an analytical function of the loading. If a suited experimental procedure is available to measure the strain or displacement fields, such tests will clearly provide more complete informations to carry out optimization procedures. The aim of this paper is to describe an approach that allows the treatment of this kind of particular tests. The data processing is based on **global informations** on the strain or displacement fields onto the loaded structure. It is not necessary to know the explicit relationship between the strain at each point of the stressed structure and the applied loading. On the other hand, it will be shown that relationships between global mechanical informations and applied loading allow the identification of the unknown parameters.

## Formulation of the inverse problem

The assessment of the strain or stress field of a stressed structure is one of the main goal of experimental mechanics. This type of problem (called herein direct problem) can be defined as follows, limiting the problem to the elastic case (figure 1-a):
- applied loading $F_i$ are known;
- stiffnesses $M_i$ that govern the elastic constitutive law of the structure are known;
- displacement, strain and stress fields $u_i$ $\varepsilon_{ij}$ $\sigma_{ij}$ are unknown.

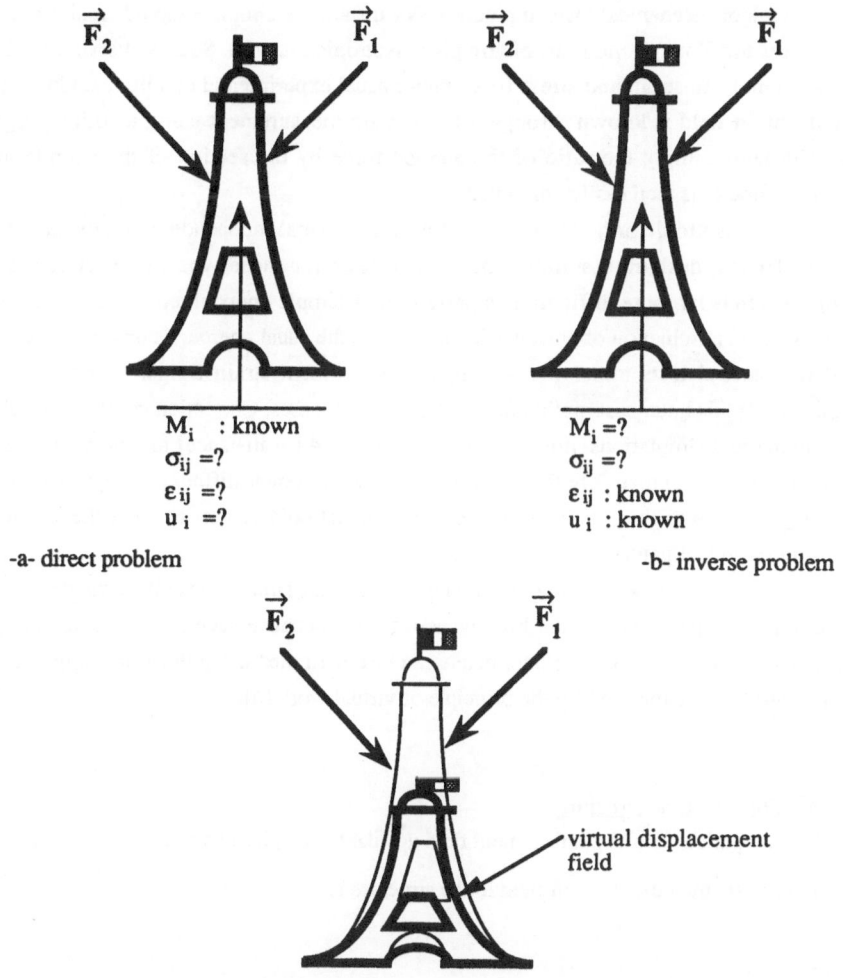

$M_i$ : known
$\sigma_{ij}$ =?
$\varepsilon_{ij}$ =?
$u_i$ =?

-a- direct problem

$M_i$ =?
$\sigma_{ij}$ =?
$\varepsilon_{ij}$ : known
$u_i$ : known

-b- inverse problem

virtual displacement field

-c- example of compatible virtual displacement field

Fig. 1 Loaded structure

The problem of the determination of the set of parameters that governs the constitutive law can be regarded as an inverse problem of the preceding one (figure 1-b) :
  - applied loading $F_i$ are known;
  - displacement and strain fields, $u_i$ and $\varepsilon_{ij}$, are determined over the whole structure through a suitable experimental technique;
  - parameters $M_i$ that govern the constitutive law of the structure are unknown;
  - stress field $\sigma_{ij}$ is therefore unknown.

This second type of problem is obviously easily solved when the strain field is uniform, that is the case for the major mechanical tests in practice. For instance, a simple uniaxial tensile test can be used to assess the Young's modulus of any given isotropic material. Such a characterization test gives rise to uniform strain and stress fields under usual experimental conditions. Consequently the whole strain field is known through a local strain measurement using a strain gauge. The stress field being simply the ratio of the applied force by the section of the sample and the unknown parameter is then easily computed.

In the case of anisotropic or heterogeneous structures, several independent parameters are to be identified. Hence, designing suitable mechanical tests for these type of structures is a real challenge which is far more difficult than performing a simple uniaxial tensile test. For instance, concerning the determination of anisotropic parameters, the usual approach consists of performing several mechanical tests under the assumption of uniform strain fields. It must be clearly emphasized that this latter assumption is not verified in all cases. Correction coefficients, obtained through numerical simulations, are often to be used [2-3-4 for instance] to determine the actual value of the stress - and therefore the actual value of the unknown stiffness - at the location of the strain gauge bonded to the specimen. Moreover, one test only cannot provide the whole set of unknown independent parameters.

On the other hand, it can be easily shown that heterogeneous strain fields allow the determination of more unknown parameters than homogeneous ones because several stiffnesses are jointly involved in the strain field [5]. The parameters can be determined using the global approach of the structure equilibrium expressed by the principle of virtual work [6].


Global equilibrium of the structure

Consider a compatible virtual displacement field $u_i^*$, like the displacement in figure 1-c. The virtual work produced by the virtual strain field in the structure is

$$U_{int}^* = \int_V \sigma_{ij} \varepsilon_{ij}^* \, dv \tag{1}$$

The applied forces to the structure surface S, which are kept constant during the virtual displacement, produce an external virtual work

$$U_{ext}^{*} = \sum_{i=1}^{n} \vec{F_i} . \vec{u_i}$$ (2)

where "." is the scalar product between vectors. Using the principle of virtual work, one can deduce

$$\int_{V} \sigma_{ij} \varepsilon_{ij}^{*} \, dv = \sum_{i=1}^{n} \vec{F_i} . \vec{u_i^{*}}$$ (3)

Each compatible virtual displacement field gives rise to a linear relationship, where the stiffnesses $M_i$ to be identified are the unknowns. A suited choice of the virtual fields, of the structure geometry as well as of the loading leads to several independent linear equations that can be solved to obtain the unknown stiffnesses. Classical optimization procedures can be used to enhance the accuracy of the final results.

It is clear that the accuracy of the approach strongly relies on the choice of the virtual fields and the whole mathematical processing of the experimental data. The identification method must obviously have a reasonable stability towards the scatter of the data. This stability has to be carefully checked through numerical simulations of the whole identification process prior to any actual experimental attempt. The question to determine which virtual fields should be the most relevant in given experimental conditions has to be numerically examined since it is necessary to be able to characterize and discuss the accuracy of the approach afterwards. Finally, numerical simulations have to be performed to contour the experimental conditions to be respected, in particular they have to take into account the experimental scatter.

Consider now two examples The first one deals with a thin **anisotropic** plate whose six independent bending stiffnesses are the unknown parameters that have to be identified. In this case, the final expression of the internal virtual work is the following [5]

$$U^*_{int} = \frac{4}{h^2} \int_S (\varepsilon^*_{xx}, \varepsilon^*_{yy}, 2\varepsilon^*_{xy}) \begin{pmatrix} D_{11} & D_{12} & D_{16} \\ D_{12} & D_{22} & D_{26} \\ D_{16} & D_{26} & D_{66} \end{pmatrix} \begin{pmatrix} \varepsilon_{xx} \\ \varepsilon_{yy} \\ 2\varepsilon_{xy} \end{pmatrix} ds \qquad (4)$$

where h is the thickness, S is the surface and the $D_{ij}$ are the unknown bending stiffnesses of the plate. Note that the compliance matrix [d], such that $[d] = \frac{h^3}{12} [D]^{-1}$ , could then be introduced to provide more simple expressions.

The second example concerns a beam whose stiffness is variable along its length. In this case, the internal virtual work is

$$U^*_{int} = \int_0^1 D(x) \frac{\partial^2 w}{\partial x^2} \frac{\partial^2 w^*}{\partial x^2} dx \qquad (5)$$

where l is the length, D(x) the bending stiffness that depends of the abscissa x, w the deflection and $w^*$ the compatible virtual deflection of the beam.

Equation (1) describes the global equilibrium of the structure and equations (4) and (5) clearly show that the unknown parameters are the solution of a set of equations built up with different virtual displacement fields. The inverse problem is now to determine these unknown stiffnesses through the use of a relevant choice of the virtual fields as well as of the testing configuration.

Choice of the virtual fields

Every different virtual field gives rise to a new relationship. Only a few examples will only be provided below to emphasize the relevancy of the approach. For instance, consider the first example associated with the following virtual field

$$u^*(x,y) = kxy \qquad (6)$$

where k is any real number.

If the plate is circular (diameter R) and subjected to one single force applied at $(R\sqrt{2}, R\sqrt{2})$ (figure 2-a), one can easily show [7] that three out of the six unknown parameters are directly obtained. Field (6) leads to :

$$d_{66} = \frac{-h^2}{3FR^2} \int_S \varepsilon_{xy} \, ds \qquad (7)$$

In fact, each virtual field can be considered as a filter that enhances the contribution of one or more stiffnesses or compliances. For instance, the virtual field of equation (6) points out the shear behaviour of the structure. The stress field in the loaded plate is heterogeneous and theoretically unknown, but the virtual work produced by virtual field (6) is expressed as a function one of the unknown compliance. Other fields and tests can be used [7]. For instance, the two following fields

$$u^*(x,y) = kx(x-R\sqrt{2}) \qquad \text{and} \qquad u^*(x,y) = ky(y-R\sqrt{2}) \qquad (8)$$

allow the determination of $d_{16}$ and $d_{16}$

$$d_{16} = \frac{-h^2}{6FR^2} \int_S \varepsilon_{xx} \, ds \qquad \text{and} \qquad d_{26} = \frac{-h^2}{6FR^2} \int_S \varepsilon_{yy} \, ds \qquad (9)$$

The three remaining unknown compliances are found with other virtual displacement fields associated with a three-point bending test carried out on the same specimen [5].

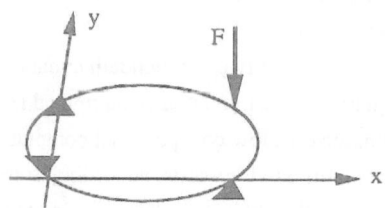

a- testing configuration

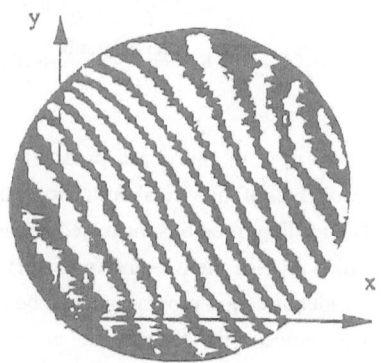

b- example of displacement contour on the top surface of the plate

Fig. 2 Twisting test on a circular plate

The strains are to be measured all over the surface of the plate using an optical technique that should lead to an acceptable data accuracy and that should not be too difficult to master. The thickness of the plate being constant, the surface strains are proportional to the plate curvatures, i.e. the first derivatives of the plate slopes. The field of the slopes is assessed using a moiré optical device. Figure 2-b shows an example of the slope field along direction y in the loading case of figure 2-a. As can be seen, fringes are neither linear nor equidistant. Consequently the surface strain field is heterogeneous.

The second inverse problem aims at determining the function $D(x)$ of the real beam that is considered. $D(x)$ will be approximated by the evolution of the bending stiffness of a beam model which is a priori defined by the designer, taking into account the specific features of the real structure. The beam model could be step by step adjusted as the identification process will be going on. It is composed of several elements of different lengths whose bending stiffnesses are linear functions of the beam longitudinal abscissa (figure 3). The question is to determine the value of the bending stiffness at the beginning ($D_i$) and at the end ($D_{i+1}$) of each element $i + 1$ of the model to approach the real beam at best. These unknowns are identified assuming that the real and the model beams have the same deflections under a given loading, that is a three-point bending test in the studied case. The deflections (or the strains) of the real beam are experimentally measured at a limited number of points.

Two independent virtual fields have been used through the study. Both of them generate a linear virtual deflection along the beam except on the element [a,b] which will be identified and where the following fourth and fifth degree independent polynomial functions are chosen (figure 4)

$$w_1^*(x) = \frac{x^4}{12} + x^3 \frac{a+b}{6} - x^2 \frac{ab}{2} + dx + c \qquad\qquad a<x<b \qquad\qquad (10)$$

$$w_2^*(x) = \frac{x^5}{20} + x^4 \frac{a+b}{12} - x^3 \frac{ab}{6} + d'x + c' \qquad\qquad a<x<b \qquad\qquad (11)$$

where d, c, d' and c' are parameters which are determined in such a way that the two first derivatives of the virtual deflections are continue at $x = a$ and $x = b$.

The loading force F is measured and equations (3) and (5) give a set of two independent equations which provide the two unknown stiffnesses. Integral in equation (5) is numerically computed and the identification programme can be run in less than one minute on a low cost personal computer. The choice of the beam model and the sharing of it can be modified afterwards, according to the profile of the curve which is identified. Important variations in the stiffness curve may incite to work with smaller beam parts or to try other sharings. It must be emphasized that this local identification highly improves the stability of the method since only sets of two equations with two unknowns have to be solved; in particular there is no restriction to the number of beam elements. Various numerical simulations have been performed to check the true performances of the method under different assumptions on experimental random errors, on real beam bending stiffness and model beam number of elements.

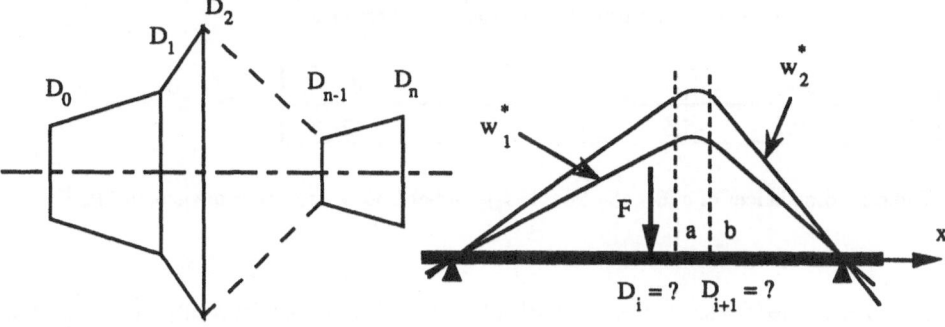

Fig. 3 Beam model to be identified.                Fig. 4 Virtual fields $w_1^*$ and $w_2^*$.

Influence of experimental errors. Stability of the method

The influence of experimental errors on the identified stiffnesses $D_{ij}$ of an anisotropic circular plate has been regarded from a statistical point of view. Full details can be found in [7]. Three different tests have been simulated with a finite element programme. The nodes of the meshed plate are considered as simulated experimental points and an error provided by a Gaussian random number generator routine is added with the exact location of the nodes. Fifty simulations have been carried out and the standard deviation of each stiffness has been computed. It has been observed that the sensitivity to experimental errors depends on the stiffness and on the anisotropy of the material. In all cases, the method can be considered as very stable. For instance, a graphite/epoxy unidirectional circular plate has been considered. The diameter is 150 mm and the standard deviation of the error on the location of the nodes is 1.33 mm. The maximum standard deviation of the stiffness is 2.1 % of the stiffness for $D_{12}$. Roughly speaking, this stability is due to the fact that the present data treatment is equivalent to an average of the strain all over the tested plate. Hence errors made on different experimental points balance themselves.

Results

Several experimental tests have been carried out on different types of anisotropic circular plates of 150 mm in diameter [7]. For instance, table 1 gives the six unknown bending compliances of a quasi isotropic laminate in the 0 deg-frame. Such a plate is characterized by isotropic in-plane compliances and completely anisotropic bending compliances.

Integrals of equations (7) and (9) are deduced from the slope field through a suited data treatment : the strain field is first numerically computed deriving the measured slope field and then integrated on the whole specimen surface using a finite element programme.

| $d_{11}$ | $d_{22}$ | $d_{12}$ | $d_{66}$ | $d_{16}$ | $d_{16}$ |
|------|------|------|------|------|------|
| 10.6 | 38.2 | -5.6 | 97.7 | -6.5 | -21.3 |

Table 1 Compliances of a $[0_3, 45_3, 90_3, 135_3]_S$ carbon/epoxy stacking sequence, in TPa$^{-1}$

It must be pointed out that stiffnesses in table 1 could not be obtained from only one specimen and from a uniform strain fields. The ability of heterogeneous displacement field processing stand therefore an important improvement.

The procedure for non constant beam bending stiffness determination described above has been tested to a preliminary example of beam shown in figure 5. The beam has been subjected to a three point bending test and the strain measured by 14 gauges regularly bonded to the beam.

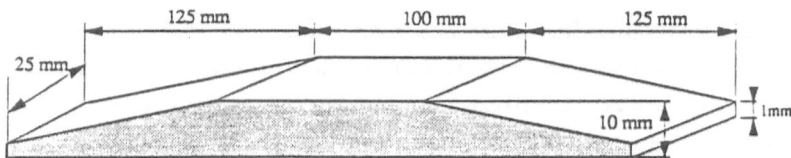

Fig 5 Tested real beam (in PMMA).

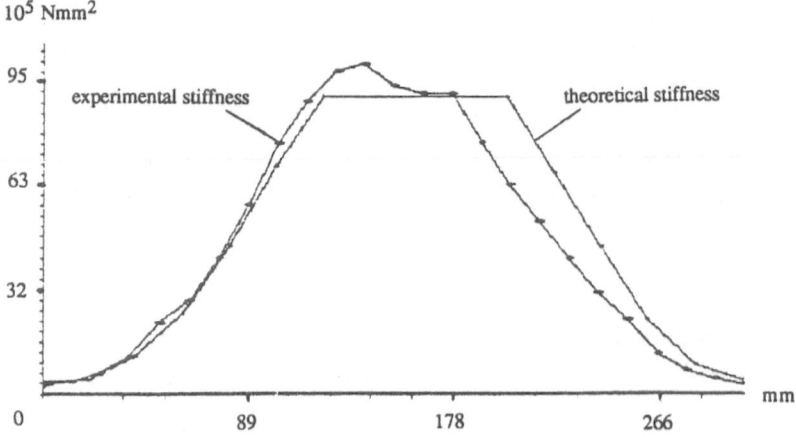

Fig 6 Theoretical and experimental curves

Both theoretical and identified stiffnesses are reported in figure 6 with a beam model sheared into 24 parts of equal length. The apparent discrepancy between identified and theoretical is quite reasonable taking into account that no particular optimizing procedure has been performed.

Conclusion

This paper addresses a general approach to determine any structural parameters in the case of two-dimension structures. Since it is necessary to deal with heterogeneous strain fields in practice, it is shown that processing the whole fields can lead to powerful methods to identify model parameters. Field measurement techniques have to be used. Optical techniques, such as moiré for instance, can easily provide displacement or strain mapping, moreover their practice is now user-friendly owing to the enhancement of personal computer facilities and image processing treatment. The mechanical identification of structures is a challenge of prime importance to set up rational design procedure to use at best the fascinating tailorable properties of new materials.

References
1. Gibson L. J.; Ashby M. F. : *Cellular solids*, Pergamon Press (1988).
2. Pagano N. J.; Halpin J. C. : Influence of end constraint in the testing of anisotropic bodies, *Journal of Composites Materials*, 21, 1(1968), 18-31.
3. Pindera M. J.; Herakovich C. T.: Shear characterisation of unidirectional composites with the off-axis tension test, *Experimental Mechanics*, 26, 1(1986), 103-112.
4. Walrath D. E.; Adams D. F. : The Iosipescu shear test as applied to composite materials, *Experimental Mechanics*, 23, 3(1983), 105-109.
5. Grédiac M.; Vautrin, A.: A new method for determination of bending rigidities of thin anisotropic plates, *Journal of Applied Mechanics*, 57, December 1990, 964-968.
6. Dyme C. L.; Shames I. H. : *Solid mechanics, a variational approach,* international student edition (1973).
7. Grédiac M.; Vautrin, A.: Mechanical characterization of anisotropic plates. Part II : experiments and results, submitted to the *Journal of Applied Mechanics*.
8. Tsai S. W.; *Introduction to composite materials*, Technomic (1980).

# New Method for Determining Contact Pressure Distributions by Using Caustic Images

K.Sato[*] and T.Yamaya[**]

[*] Faculty of Engineering, Chiba University
1–33 Yayoi–cho, Inage–ku, Chiba 263, Japan
[**] Japan Air System
Toranomon 37 Mori Bldg., 5–1 Toranomon 3–chome, Minato–ku, Tokyo 105, Japan

Summary
The method of caustics was applied to determining the contact pressure distribution in the contact between a plate and a pad. A formula based on the caustic theory and an experimental procedure for determining the contact pressure were described. By coinciding the caustic image profile obtained through calculation with the image profile obtained from experiments, the contact pressure distribution could be evaluated. The evaluated contact pressure was compared with the result obtained using a boundary element method, and discussed.

Introduction

The method of caustics has been utilized for obtaining stress intensity factors of cracked plates subjected to static force or dynamic force. It is also known that the method of caustic is applicable to contact problems. Theocaris and Razem [1] have studied the caustic theories for determining the magnitude of concentrated load applied on the edge of a semi–infinite plate and the pressure distribution on the deformed boundary of a semi–infinite plate subject-ed to pressure on a finite area. Sato and Sato [2] have reported the theory for determining the contact load in the Hertzian contact by introducing the contact width modification and exact initial curve into Theocaris' theory, and also reported the experimental results. In this paper, a new technique using the method of caustics for determining the pressure distribution in con-tact between an elastic plate and an elastic pad has been studied.

Theory of caustic formation

When the converged light ray is impinged around the contact interface as shown in Fig.1, its transmitted right ray creates caustic images on a reference screen which is placed at the dis-tance $z_0$ from the specimen plate. Fig.2a shows caustic images created from a semi–infinite plate subjected to contact pressure induced with a pad. The caustic images in this case are divided into two categories: one is the *actual caustic image* and another is the *pseudo–caustic image*. The actual caustic image is like a semi–circle and the pseudo–caustic image is like a bow shaped curve. The former is created by the light rays passed through the position around two contact edges and the latter is created by the light rays passed along the deformed bound-ary. In this contact condition the actual caustic image and two pseudo–caustic images are simultaneously created on the screen. Because the size and shape of the caustic images depends on the contact pressure distribution, by measuring that of the caustic image we can estimate the contact pressure distribution.

The locus of the points on the plate where the light rays passed to create the caustic images is termed as the *initial curve*. Fig.2b shows the initial curve corresponded to the caustic images in Fig.2a. The contact pressure distribution affects the coordinates of the initial curves and the shapes of the caustic images.

The each point on the initial curve must satisfy the following relation with a stress function of $\Phi(z)$;

$$\frac{4\,|C|}{m}\left|\frac{d^2\Phi(z)}{dz^2}\right| = 1. \tag{1}$$

The coordinates of the caustic images created on the screen, $W(x,y)$, are given by substituting the points on the initial curve,

$$W = m\left(z - \frac{4\,C}{m}\frac{\overline{d\Phi(z)}}{dz}\right) \tag{2}$$

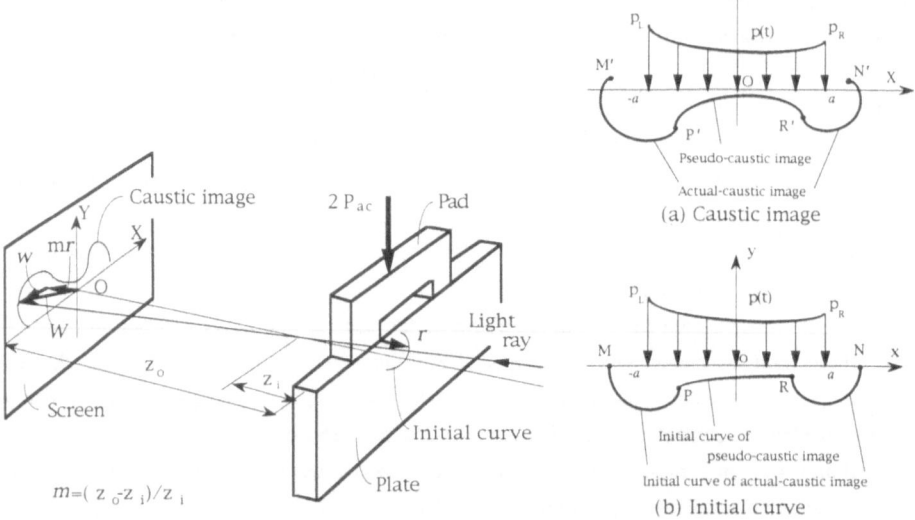

(a) Caustic image

(b) Initial curve

$m = (z_o - z_i)/z_i$

Fig.1 Schematic explanation for caustic image formation and experimental arrangement.

Fig.2 Caustic image and the corresponding initial curves for the plate subjected to distributed contact pressure.

with

$$m = (z_0 - z_i)/z_i \tag{3}$$

and

$$C = z_0 c_t d, \tag{4}$$

where $z_0$ is the distance between the plate and the screen, $z_i$ is the distance between the plate and the focus of the converged light, $c_t$ is an optical constant, $d$ is the thickness of the specimen plate, and $m$ is the optical magnification factor for the experimental set–up used.

When the contact pressure $p(t)$ distributes on the contact width $2a$, as shown in Fig.3, the stress function, $\Phi(z)$, is expressed by

$$\Phi(z) = \frac{1}{2\pi i} \int_{-a}^{a} \frac{p(t)}{t-z} dt, \tag{5}$$

where $t$ is a coordinate along the contact interface and limited to in $2a$, i.e., $(-a \leq t \leq a)$.
The contact pressure distribution, $p(t)$, was assumed as a 6–degree power series; that is

$$p(t) = s_0 + s_1 t + s_2 t^2 + s_3 t^3 + \cdots + s_N t^6. \tag{6}$$

Therefore the stress function, $\Phi(z)$, is given by Eq.(7).

$$\Phi(z) = \frac{1}{2\pi i} [s_0 E + s_1 \{2a + zE\} + s_2 \{2az + z^3 E\} + s_3 \{(2/3)a^3 + 2az^2 + z^3 E\}$$
$$+ s_4 \{(2/3)a^3 z + 2az^3 + z^4 E\} + s_5 \{(2/5)a^5 + (2/3)a^3 z^3 + 2az^4 + z^5 E\}$$
$$+ s_6 \{(2/5)a^5 z + (2/3)a^3 z^3 + 2az^5 + z^6 E\}] \tag{7}$$

where

$$E = ln|(z-a)/(z+a)| \tag{8}$$

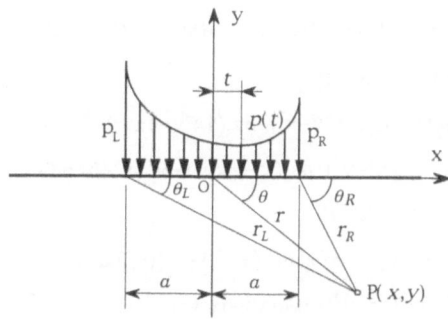

Fig.3 Geometry for the plate subjected to arbitrary distributed contact pressure.

## Actual caustic image

The actual caustic image is created by stress concentration at the contact edge and its shape is like a semi–circle that is similar to the image created when a concentrated load is applied. The coordinates of the initial curve and the actual caustic image are obtained through substituting Eq.(5) into Eqs.(1) and (2). Because calculations are not easy, an interactive symbolic algebra program "MACSYMA" was utilized with an engineering workstation.

As the result after integrating Eq.(5), the stress function, $\Phi(z)$, was expressed as the following relation:

$$\Phi'(z)=A(x,y)+iB(x,y) \tag{9}$$

Functions $A(x,y)$ and $B(x,y)$ are given by the following very long complicated expressions:

$$
\begin{aligned}
A =\frac{1}{2\pi} [&H\{s_1 v+s_2 xv+s_3(3x^2v-v^3)+s_4(4x^3v-4xv^3)\\
&+s_5(5x^4v-10x^2v^3-v^5)+s_6(6x^5v-20x^3v^3-6xv^5)\}\\
&-I\{s_0+s_1 x+s_2(x^2-v^2)+s_3(x^3-3xv^2)+s_4(x^4-6x^2v^2+v^4)\\
&+s_5(x^5-10x^3v^2+5xv^4)+s_6(x^6-15x^4v^2+15x^2v^4+v^6)\}\\
&-J\{s_1+2s_2 x+3s_3(x^2-v^2)+4s_4(x^3-3xv^2)+5s_5(x^4-6x^2v^2+v^4)\\
&+6s_6(x^5-10x^3v^2+5xv^4)\}\\
&+L\{2s_2 v+6s_3 xv+4s_4(3x^2v-v^3)+5s_5(4x^3v-4xv^3)+6s_6(5x^4v-10x^2v^3-v^5)\}\\
&+4s_3 av+12s_4 axv+s_5\{(4/3)a^3v+8a(3x^2v-v^3)\}+s_6\{4a^3xv+10a(4x^3v-4xv^3)\}] \quad (10)
\end{aligned}
$$

$$
\begin{aligned}
B =\frac{1}{2\pi} [&H\{s_0+s_1 x+s_2(x^2-v^2)+s_3(x^3-3xv^2)+s_4(x^4-6x^2v^2+v^4)\\
&+s_5(x^5-10x^3v^2+5xv^4)+s_6(x^6-15x^4v^2+15x^2v^4+v^6)\}\\
&+I\{s_1 v+s_2 xv+s_3(3x^2v-v^3)+s_4(4x^3v-4xv^3)+s_5(5x^4v-10x^2v^3-v^5)\\
&+s_6(6x^5v-20x^3v^3-6xv^5)\}\\
&-J\{2s_2 v+6s_3 xv+4s_4(3x^2v-v^3)+5s_5(4x^3v-4xv^3)+6s_6(5x^4v-10x^2v^3-v^5)\\
&+L\{s_1+2s_2 x+3s_3(x^2-v^2)+4s_4(x^3-3xv^2)+5s_5(x^4-6x^2v^2+v^4)\\
&+6s_6(x^5-10x^3v^2+5xv^4)\}\\
&+2s_2 a+4s_3 ax+s_4\{(2/3)a^3+6a(x^2-v^2)\}+s_5\{(4/3)a^3x+8a(x^3-3xv^2)\}\\
&+s_6\{(2/5)a^5+2a^3(x^2-v^2)+10a(x^4-6x^2v^2+v^4)\}] \quad (11)
\end{aligned}
$$

where
$$H=2a(x^2-a^2-v^2)/\{(x^2-a^2-v^2)^2+4x^2v^2\} \tag{12}$$
$$I=4axv/\{(x^2-a^2-v^2)^2+4x^2v^2\} \tag{13}$$

$$J=tan^{-1}\{2av/(x^2+a^2-v^2)\} \tag{14}$$

$$L=ln|\{(x^2+a^2-v^2)^2+4x^2v^2\}^{(-1/2)}/\{(x+a)^2+v^2\}| \tag{15}$$

By using Eq.(9) into Eq.(2), the coordinates of the caustic image, $W_1$, are given by

$$W_1=X_1+iX_1 \tag{16}$$

where

$$X_1=mx+4CA(x,y)$$
$$Y_1=my-4CB(x,y). \tag{17}$$

The actual caustic image is defined by introducing the coordinates on the initial curve, which satisfies Eq.(1), into Eq.(16). The initial curve was numerically determined by using the regular false method in a personal computer during executing the computer program for estimating the contact pressure distribution.

Pseudo–caustic image

The initial curve of the pseudo–caustic image corresponds to the deformed boundary profile of the contacted interface [1]. The deflection of the interface, $v(x)$, is expressed by Eq.(18).

$$v(x)= - i \ \frac{1+\kappa}{2\mu} \ \phi(x) \tag{18}$$

with

$$\kappa = (\lambda+3\mu)/(\lambda+\mu) \tag{19}$$

and

$$\phi(z)=\int \Phi(z)dz, \tag{20}$$

where $\lambda$ and $\mu$ are Lame's constants.

Therefore the initial curve of the pseudo–caustic image is given by Eq.(21).

$$\zeta = x + i \ v(x) \tag{21}$$

with

$$v(x)=\frac{1}{2\pi i}[s_0\{xF-aG+2a\}+s_1\{(x^2/2)F-(a^2/2)F+ax\}+s_2\{(x^3/3)F-(a^3/3)G+(2/3)ax^2\}$$
$$+s_3\{(x^4/4)F-(a^4/4)F+(1/2)ax^3+(1/6)a^3x\}+s_4\{(x^5/5)F-(a^5/5)G+(2/5)ax^4+(2/15)a^3x^2\}$$
$$+s_5\{(x^6/6)F-(a^6/6)F+(1/15)a^5x+(1/9)a^3x^3+(1/3)ax^5\}$$
$$+s_6\{(x^7/7)F-(a^7/7)G+(2/7)ax^6+(2/21)a^3x^4+(2/35)a^5x^2\} \tag{22}$$

where

$$F=ln|(x-a)/(x+a)| \tag{23}$$

$$G=ln|(x^2-a^2)| \tag{24}$$

Therefore, the coordinates of the pseudo–caustic image on the screen, $W_2$, were given as Eq.(25), or Eq.(26).

$$W_2 = m \ (\zeta-\frac{4 C}{m} \ \frac{\overline{d\Phi(\zeta)}}{d\zeta} ) \tag{25}$$

$$= X_2 + iY_2 \tag{26}$$

where

$$X_2 = mx + 4CA(x,v) \tag{27}$$
$$Y_2 = mv - 4CB(x,v).$$

<u>Computer Program</u>

A computer program based on the above theory of caustics was developed by a BASIC code. By executing this program using characteristic coordinates of the caustic images, as shown in Fig.4, the contact pressure distribution was automatically obtained after some iterations. The algorithm used in the program is as follows:

The contact pressure distribution was firstly assumed as the linear distribution along the inter-face with two end values of $p_R$ and $p_L$, shown in Fig.5. The values of $p_R$ and $p_L$ were determined by using the formula which is derived by Theocaris and Razem [3] under the assumption that the contact length $2a$ is sufficiently large and $p_R$ and $p_L$ are constant. Its form is:

$$p_{R,L} = \frac{\pi(X_{1R,1L}^2 + Y_{1R,1L}^2)}{4Cm} \tag{28}$$

where $X_{1R,1L}$ and $Y_{1R,1L}$ are coordinates of the actual caustic image at the direction of angle of 45 degree for the right-hand edge and of 135 degree for the left-hand edge, respectively.

In order to make the calculated pseudo-caustic image coincide more effectively with the experimental pseudo-caustic image, two caustic image profiles were approximated as the polynomial expressions of $f(x)$ and $g(x)$ by using the method of least squares (Fig.5). More-over, the contact interface creating the pseudo-caustic image was divided into five regions for executing the iteration automatically, shown in Fig.6. In each region, there are three represen-tative points for the caustic image profiles. The value of contact pressure at the coordinate,

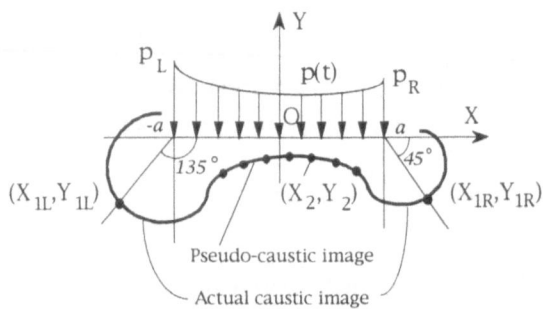

Fig.4 Characteristic coordinates of the caustic images.

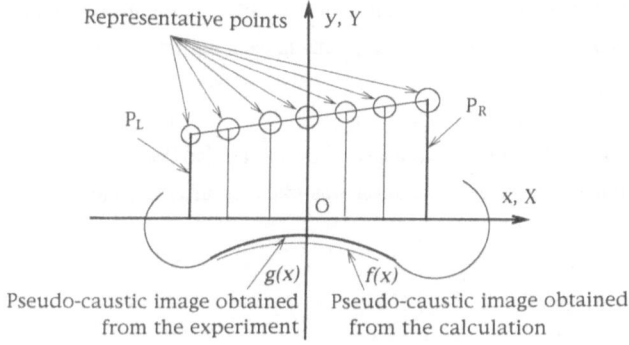

Fig.5 A linear contact pressure distribution assumed at the first step.

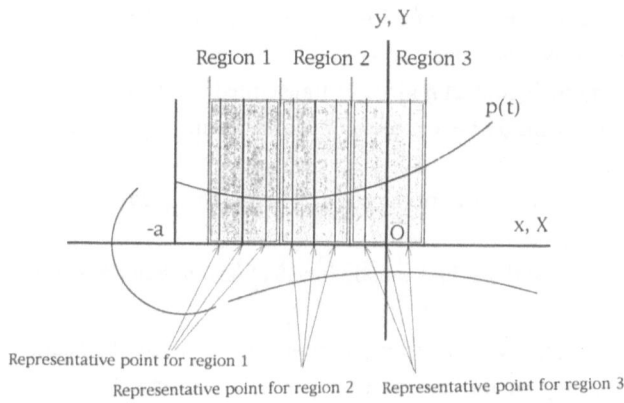

Fig.6 Contact region divided for automatic iteration procedure.

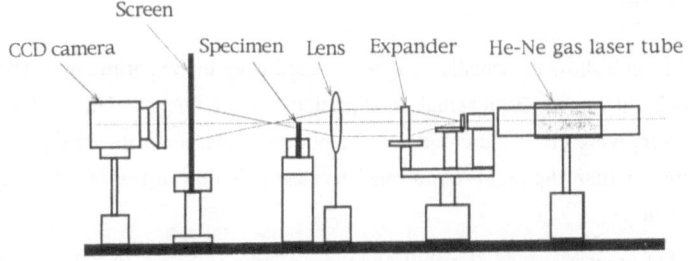

Fig.7 Optical system used in experiments.

corresponded to the representative point which has the maximum difference in the values of $f(x)$ and $g(x)$ among three representative points in each region, was corrected in each iteration step. A few percent of the value of contact pressure was corrected in each step, and such a correction was continued to converge in a given error limit between $f(x)$ and $g(x)$. A similar correction with respect to the actual caustic image was also done. By using this program, the approximate contact pressure distribution was obtained after iterations not more than twenty steps.

## Experimental procedure

Experiments were carried out to examine the theory and computer program developed. Fig.7 shows the experimental set-up used. The PMMA plate specimen of W40xL100xT(2, 3, 4 and 5) mm$^3$ was pressed by a bridge-type pad made of the same PMMA plate of 5mm in thickness, which has two contact feet of 5 mm in length. The values of $z_i$ and $z_o$ are 123mm and 940 mm, respectively. The value of $c_t$ is $-1.03 \times 10^{-4} mm^2/N$. When a He-Ne gas laser ray was impinged around their contact region, its transmitted ray created the caustic images on the screen. The caustic images were recorded by using an image processing unit.

The characteristic coordinates of the caustic images to measure are listed below (shown in Fig.4) :
  1) the coordinates of the actual caustic image at an angle of 45 degree direction from the contact edge;
  2) the coordinates for seven to twelve points of the pseudo-caustic image.
By executing the program on a personal computer, the contact pressure distribution calculated was drawn in its display at each iteration step.

## Experimental results

Photographs in Fig.8 show the caustic images obtained from the experiments with the PMMA plates of 2mm in thickness. Both actual- and pseudo-caustic images, which are similar to the theoretical shape, were clearly created. In the photographs, the left-hand pseudo-caustic image was greater than the right-hand one, because it is due to the fact that the pad was loaded at its center.

The evaluated contact pressure distributions were shown in Fig.9, together with the result obtained by BE analysis. The BE analysis was conducted under using a mesh model whose shape and dimensions are the same as the specimen excluding the thickness and under plane stress condition. All results coincided with in the middle contact area, but did not near contact

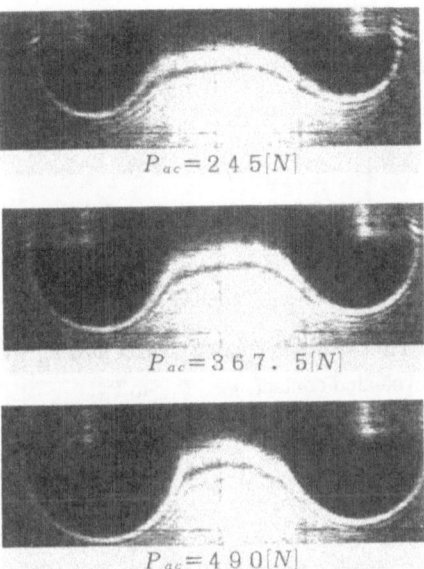

Fig.8 Caustic images obtained from the experiments with the PMMA plate.

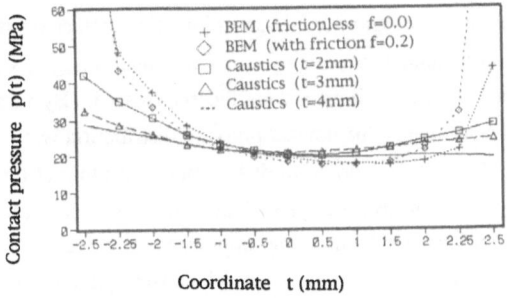

Fig.9 Evaluated contact pressure distribution and the result obtained by BE analysis.

edges. It was considered that this disagreement is due to the effect of yielding of the material at the contact edge. It was also considered, moreover, that the variation in contact pressure distribution with thickness is due to the effect of breaking away from the plane stress condition, which is a basis of the method of caustics.

The applied contact load can be estimated by integrating the contact pressure distribution along the contact interface. The results were shown in Fig.10, using the ratio of the experimental value to the actual value of applied load, $P_{exp}/P_{ac}$. The ratios at low applied load were approximately unity, but the ratios decreased with increasing applied load. This result seems to be due to the effects above.

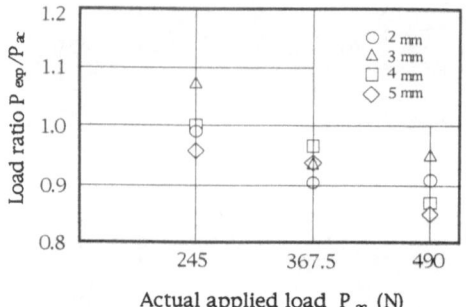

Fig.10  Plots between applied contact load $P_{ac}$ and the ratio of
evaluated contact  load $P_{exp}$ to $P_{ac}$.

## Conclusions

A new technique for determining the contact pressure distribution was proposed.  Basic
expressions for the initial curve and caustic image created under the contact pressure distribution of the polynomial power series form were derived.  Based on such the theory of caustics
a fully automatic  computer program for determining  the contact pressure and contact load
was developed and experiments were carried out. The caustic images, which consist of actual
caustic image and pseudo–caustic image, were clearly created.  By measuring characteristic
coordinates of the caustic images, the contact pressure distribution was obtained through some
iterated calculations.  The  contact pressure distribution excepting at the contact edge agreed
with the result obtained by the BE analysis. The contact pressure obtained by the caustic
method depended on specimen thickness and applied load. It seems to be effect of material
yielding and breaking away from the prerequisite plane stress condition.

## References

1. Theocaris, P.S.; Razem, C.I.: Deformed boundaries determined by the method of caustics,
*Journal of Strain Analysis*, **12** (1977), 223–232.

2. Sato, K.; Sato, M.: Measurement of contact load by the method of caustics, *Proc. the IV
Conference of Asian–Pacific Congress on Strength Evaluation (APCS–91)*, **II** (1991), 1443–
1448.

3. Theocaris, P.S.; Razem, C.I.: The end–values of distributed loads in half planes by caustics, *Journal of Applied Mechanics*, **45** (1978), 313–319.

# Estimation of Impact Force by an Inverse Analysis

Hirotsugu Inoue*, Kikuo Kishimoto*, Toshikazu Shibuya* and Takashi Koizumi*

* Department of Mechanical Engineering for Production,
  Tokyo Institute of Technology,
  2-12-1, O-okayama, Meguro-ku, Tokyo 152, Japan

Summary

An inverse problem of impact force is studied, in which an impact force acting on a body of arbitrary shape is estimated from responses induced at any point of the body. The noncausal Wiener filtering theory and the singular value decomposition (SVD) technique are employed for improving the accuracy of estimation. The present method is applied to the measurement of the impact force acting on a GFRP plate and its effectiveness is examined.

## 1. Introduction

The impact problem of materials and structures is widely studied for evaluating strength, safety and reliability of engineering products; for example, the impact resistance of high-tech materials is given attention in recent years. Accurate evaluation of impact force acting on a body is a fundamental subject for investigating the impact phenomena both analytically and experimentally. However, measurement of the impact force is difficult in contrast with that of the static force since the impact force is transmitted through the impactor in the form of stress waves. This is especially the case when the impactor is complicated in shape or structure.

To overcome this difficulty, a method based on the idea of inverse analysis is given attention and widely studied in recent years[1–19]. The authors have also studied a method for experimentally evaluating the impact force acting on a body of arbitrary shape, which is to estimate the impact force from measured response of the impactor by means of deconvolution[20–25]. Since the deconvolution of experimental data is one of the inverse problems, care must be taken against the ill-posedness of the problem in order to obtain acceptable estimates in practical applications. Unfortunately, as far as the authors know, there are few studies considering the ill-posedness of the inverse problem of the impact force. As for this subject, the authors have presented a method utilizing the noncausal Wiener filtering theory.

However, even if the noncausal Wiener filtering technique is employed, the estimate of the impact force might be still insufficient when, for example, the impactor is extremely complicated in structure. In the present study, a method based on the least squares method is presented for further improvement of the estimation accuracy of the impact force by means of inverse analysis. In addition, the method is applied to the measurement of impact force acting on a GFRP plate and its applicability is discussed.

## 2. Inverse Problem of Impact Force

### 2.1 Basic Scheme

Let $f(t)$ denote an impact force acting on a body of arbitrary shape and let $e(t)$ a response of the body induced by the impact force, for example, strain response at a certain point away from the point of impact. If we consider a linear system whose input and output are $f(t)$ and $e(t)$, respectively, then the input-output relationship is expressed by the convolution integral, namely

$$e(t) = \int_0^t h(t - \tau) f(\tau) d\tau, \tag{1}$$

where $h(t)$ is the impulse response function of the system and it is assumed that $f(t) = e(t) = h(t) = 0$ at $t < 0$. The Fourier transform of Eq.(1) is given by

$$E(\omega) = H(\omega)F(\omega), \tag{2}$$

where the capital letters denote the Fourier transforms of corresponding functions and $H(\omega)$ is called the transfer function. In general, the impact force can be hardly measured directly while the response can be measured rather easily, for example, by using a strain gage. Therefore, if the transfer function has been identified in advance, the impact force can be estimated from the measured response by solving Eq.(2) for $F(\omega)$, that is the deconvolution. The transfer function is identified by conducting a calibration such that the impact force and the response are measured simultaneously by means of a certain special method. For example, the impact force induced by longitudinal impact of a slender rod can be measured using strain gages. The Fourier transformation and inversion can be computed efficiently by utilizing the fast Fourier transform (FFT) algorithm. When applying the FFT algorithm, use of an exponential window function is known to be adequate for minimizing the error of the numerical transformation and inversion[26].

### 2.2 Application of the Noncausal Wiener Filtering Theory

As is well known, the deconvolution is one of the ill-posed problems[27] so that direct application of the above scheme to practical problems is insufficient to provide a satisfactory estimate of the impact force because of the existence of the measurement noise. To overcome this insufficiency, the authors have presented a method utilizing the noncausal Wiener filtering theory and demonstrated its effectiveness through both numerical simulation[25] and experiment[24]. The method provides the best linear estimate of the impact force in the sense that the mean square error becomes minimum. Since the details of the method have been reported previously, a brief description will be given here.

Let $x_k(t)$ and $y_k(t)$ ($k = 1, 2, \cdots, K$) be $K$ sets of impact force and response data obtained by conducting the calibration $K$-times. In usual, these data contain measurement noises. According to the noncausal Wiener filtering theory, as shown in the previous paper[25], the optimal transfer function for the inverse analysis can be identified by

$$\hat{H}(\omega) = \frac{\displaystyle\sum_{k=1}^{K} Y_k^*(\omega)Y_k(\omega)}{\displaystyle\sum_{k=1}^{K} Y_k^*(\omega)X_k(\omega)}, \tag{3}$$

where $X_k(\omega)$ and $Y_k(\omega)$ are the finite Fourier transforms of $x_k(t)$ and $y_k(t)$, respectively, and * denotes the complex conjugate.

## 2.3 Application of the Least Squares Method

There might be a case when the response is much smaller than the noise level of the measurement system, that is, when the transfer function is very small in magnitude. In such a case, the estimate of the impact force would be erroneous and unacceptable because little information about the impact force is contained in the measured response. Even if the response at one location of the impactor is very small, however, the response at other locations might be sufficiently larger in magnitude than the noise level. Therefore, if the impact force is estimated from responses measured at several locations, the estimation accuracy would be improved. In the present study, we employ the least squares method for estimating the impact force from the responses at several locations.

Let $e_i(t)$, $i = 1, \cdots, n$, denote the responses at $n$ locations of the impactor and let

$$\boldsymbol{E} = \{E_1(\omega), E_2(\omega), \cdots, E_n(\omega)\}^T, \tag{4}$$
$$\boldsymbol{H} = \{H_1(\omega), H_2(\omega), \cdots, H_n(\omega)\}^T, \tag{5}$$

where $H_i(\omega)$ denotes the transfer function corresponding to each locations. Then the following equation is obtained from Eq.(2):

$$\boldsymbol{E} = \boldsymbol{H}F(\omega). \tag{6}$$

In order to derive the least squares solution $F(\omega)$ of Eq.(6), we adopt the singular value decomposition (SVD) technique[28].

The singular value decomposition of the matrix $\boldsymbol{H}$ is given by

$$\boldsymbol{H} = \boldsymbol{U\Sigma V}^H, \tag{7}$$

where $\boldsymbol{U}$ is an $n$-by-$n$ hermitian matrix and $\boldsymbol{V}$ is a 1-by-1 hermitian matrix, that is a scalar in fact. In addition,

$$\boldsymbol{\Sigma} = \{\sigma_1, 0, \cdots, 0\}^T, \tag{8}$$

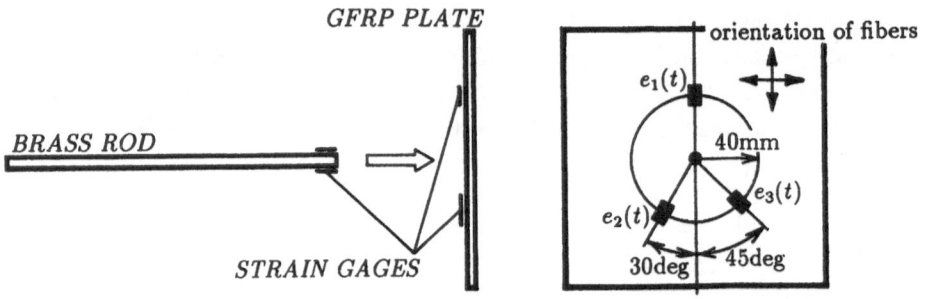

Figure 1: Impact of a GFRP plate with a brass rod.

where $\sigma_1$ is a positive constant called the singular value. According to the theory of the singular value decomposition, the least squares solution of Eq.(6) is given by

$$\tilde{F}(\omega) = \boldsymbol{H}^+ \boldsymbol{E}, \tag{9}$$

where $\boldsymbol{H}^+$ is the Moore-Penrose generalized inverse of $\boldsymbol{H}$ given by

$$\boldsymbol{H}^+ = \boldsymbol{V}\boldsymbol{\Sigma}^+\boldsymbol{U}^H, \tag{10}$$

$$\boldsymbol{\Sigma}^+ = \{\frac{1}{\sigma_1}, 0, \cdots, 0\}. \tag{11}$$

Equation (9) provides a unique least squares solution such that both $||\boldsymbol{E} - \boldsymbol{H}\tilde{F}(\omega)||_2$ and $||\tilde{F}(\omega)||_2$ are minimal.

## 3. Impact Force Acting on GFRP Plate

### 3.1 Experimental Setup

We applied the present method to the measurement of the impact force acting on a GFRP plate. The GFRP plate ($165 \times 165 \times 1.8$ mm) was fabricated using four sheets of woven roving (Nittobo, WR570C-100CS) and vinylester type epoxy resin (Showa Highpolymer, Ripoxy R-802) by means of the hand lay-up method. The plate was supported almost freely using flexible strings and subjected to transverse impact at the center. The impactors used were a brass rod (10-mm diameter by 1-m long) and a PMMA ball (30-mm diameter). The impact force was estimated from the strain responses measured at three locations using strain gages (Kyowa Electronic Instruments, KFRP-2-120-C1-3) as shown in Fig.1. The procedures for the measurement are summarized as follows:

1. The calibration experiment is conducted; that is, the plate is impacted with a brass rod which is instrumented with semiconductor strain gages (Kyowa Electronic In-

struments, KSP-2-E4) for measuring the impact force. Then both the impact force and the strain responses of the plate are measured.

2. The transfer functions are identified by Eq.(3) using the data obtained in the first step.

3. The plate is impacted with the brass rod or the PMMA ball and then the strain responses of the plate are measured.

4. The impact force is estimated by Eq.(9) using the transfer functions obtained in the second step and the strain responses obtained in the third step.

The strain gages were connected to a DC strain amplifier and the data was captured by a digital recorder (Autonics, APC-204). The data processing was performed using a personal computer (NEC, PC-9801). The sampling rate was 5 $\mu$s and the length of the data was 2048 words.

### 3.2 Result of the Calibration
The calibration was conducted by freely dropping the brass rod from the height of 50 mm, which means that the impact velocity was 0.99 m/s. The impact force measured by the strain gage attached to the rod is shown in Fig.2 and the strain responses of the plate are shown in Fig.3. It is seen from Fig.2 that the duration of the impact is about 4 ms. The strain responses at the three locations are slightly different from each other. Ten sets of data similar to Figs.2 and 3 were collected by conducting the calibration for ten times and then the transfer functions were identified by Eq.(3).

### 3.3 Impact with Rod
The plate was impacted again in the same manner as in the calibration and the strain responses of the plate was measured. The impact force at this time was estimated from the measured strain responses using the transfer functions identified above.

At first, the impact force was estimated by directly applying the basic scheme. The transfer functions $H_1(\omega)$, $H_2(\omega)$ and $H_3(\omega)$ were identified from one set of the calibration data by setting $K = 1$ in Eq.(3) and then the impact force was estimated from each of the strain response data $e_1(t)$, $e_2(t)$ and $e_3(t)$. The results are shown in Fig.4. Although the impact force must be estimated as shown in Fig.2, the estimates are extremely noisy and unacceptable. This is due to the ill-posedness of the inverse problem, namely, the measurement noise contained in the data has been expanded through the inverse analysis. Therefore, it is clear that consideration of the ill-posedness is necessary for obtaining satisfactory estimates.

174

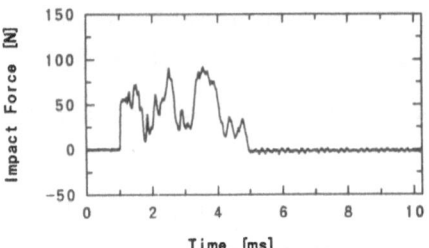

Figure 2: Impact force of the rod in the calibration.

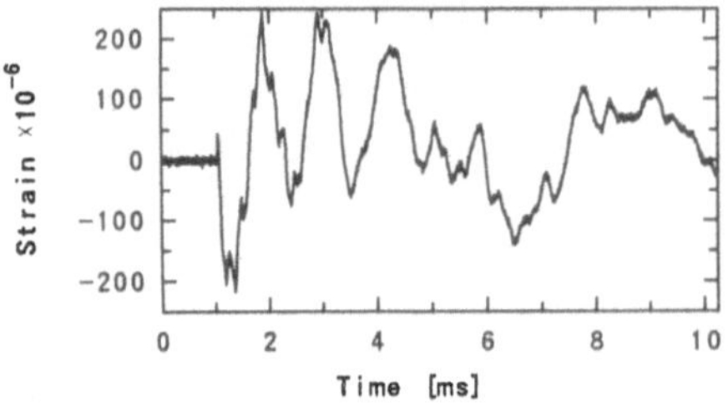

(a) $e_1(t)$

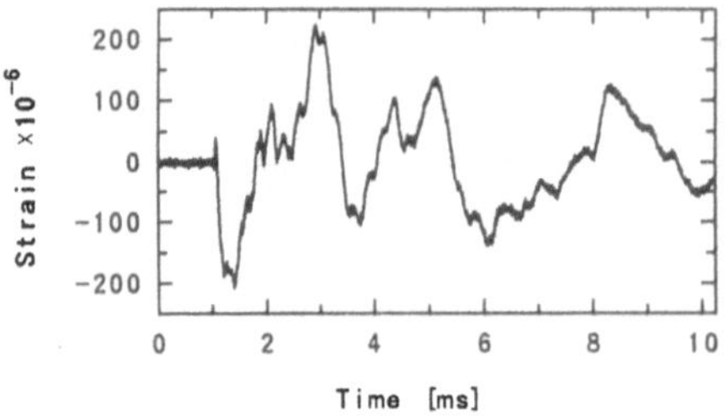

(b) $e_2(t)$

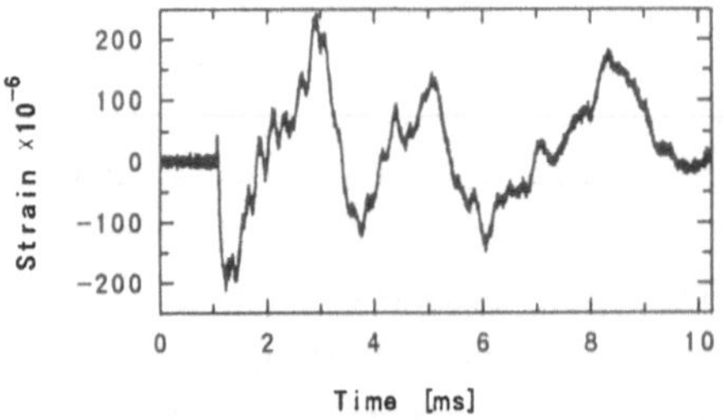

(c) $e_3(t)$

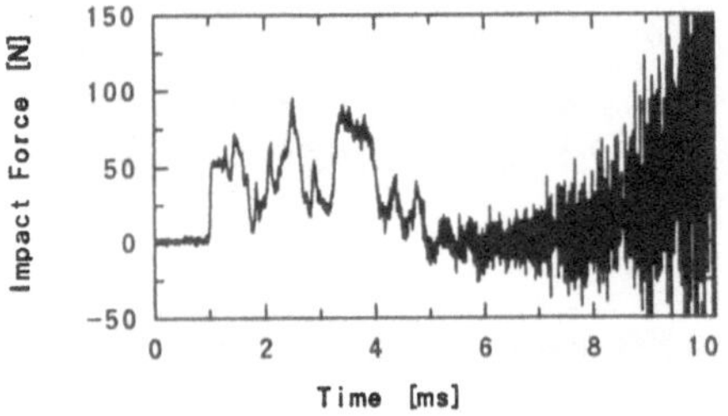

(a) Estimated from $e_1(t)$

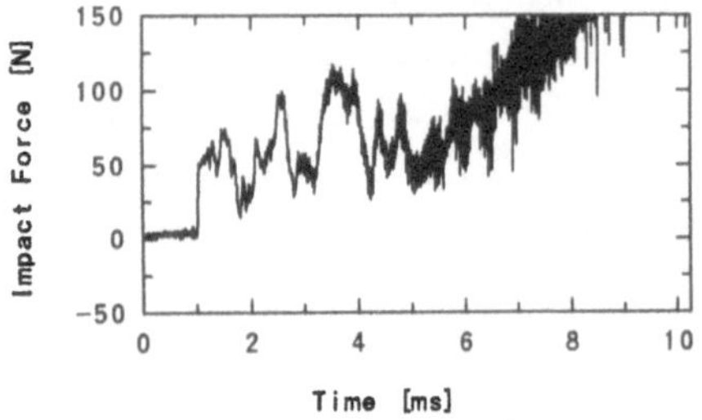

(b) Estimated from $e_2(t)$

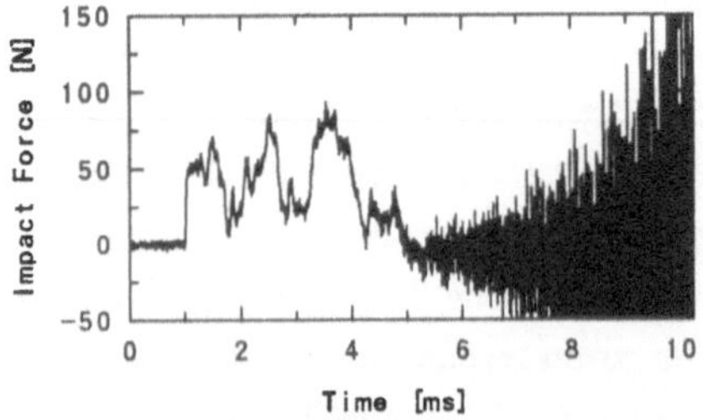

(c) Estimated from $e_3(t)$

Figure 3: Strain response induced in the GFRP plate in the calibration.

Figure 4: Impact force of the rod estimated by directly applying the basic scheme.

Next, the same impact force was estimated by applying the noncausal Wiener filtering technique. The transfer functions were identified from ten sets of the calibration data setting $K = 10$ in Eq.(3) and then the impact force was estimated from each of the strain response data. The results are shown in Fig.5. Comparing Fig.5 to Fig.4, it is seen that the estimation accuracy has been significantly improved by employing the noncausal Wiener filtering technique. However, the estimates obtained from $e_2(t)$ and $e_3(t)$ are still unstable and not agree very well with the direct measurement shown in Fig.2.

For further improvement of the estimation accuracy, the least squares method was applied to the present problem. The transfer functions and the strain responses used in the last example were used for estimating the impact force by Eq.(9). The result is shown in Fig.6. It is seen that the estimate coincides with the direct measurement shown in Fig.2 better than any of the estimates shown in Fig.5. As a result, the present method based on the noncausal Wiener filtering technique and the least squares method has been shown to be effective for the inverse analysis of the impact force.

3.4 Impact with Ball

As an example of application of the present method, the impact force was measured when the GFRP plate was impacted with the PMMA ball. The ball was freely dropped from the height of 150 mm to the center of the plate. The strain responses of the plate were measured using the strain gages.

Using the transfer functions obtained above, the impact force was estimated from each of the strain response data by means of the noncausal Wiener filtering technique. The results are shown in Fig.7. The results are unstable and unacceptable especially when they are estimated from $e_2(t)$ and $e_3(t)$. Figure 8 shows the result of the estimation utilizing both the noncausal Wiener filtering technique and the least squares method. From this estimate, it can be understood that the ball collided with the plate for three times by 4 ms on the time axis. However, the estimate is still unstable as a whole and unacceptable. Therefore, further development is necessary for obtaining a satisfactory estimate. An additional constraint condition such that the impact force cannot be negative might be effective for this objective.

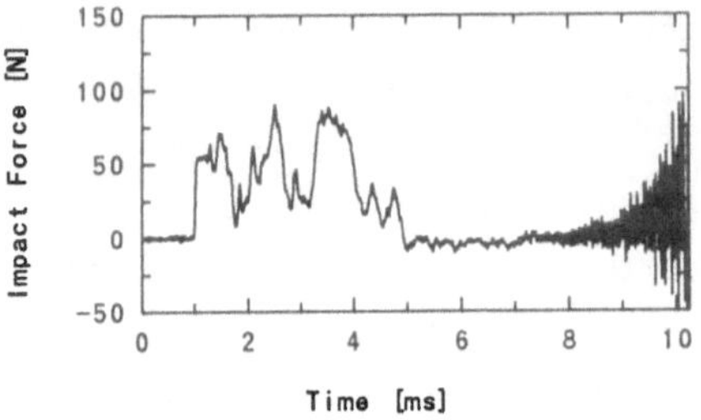

(a) Estimated from $e_1(t)$

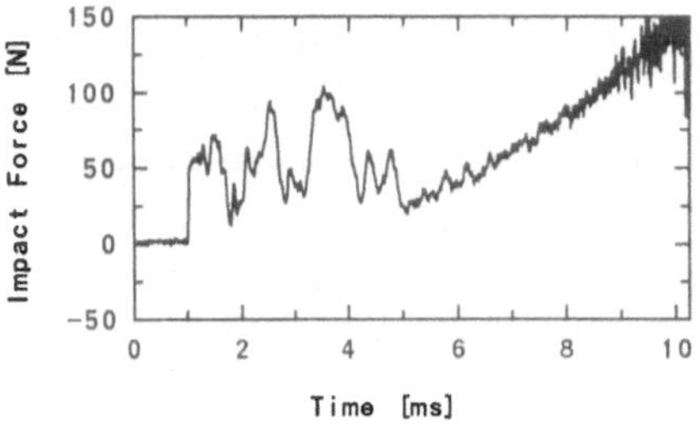

(b) Estimated from $e_2(t)$

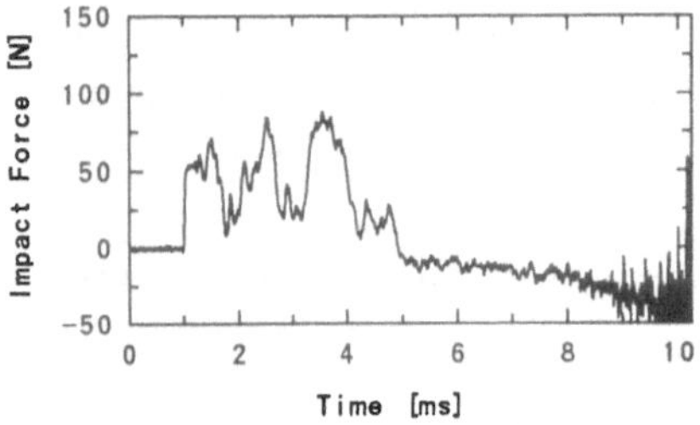

(c) Estimated from $e_3(t)$

Figure 5: Impact force of the rod estimated by applying the noncausal Wiener filtering technique.

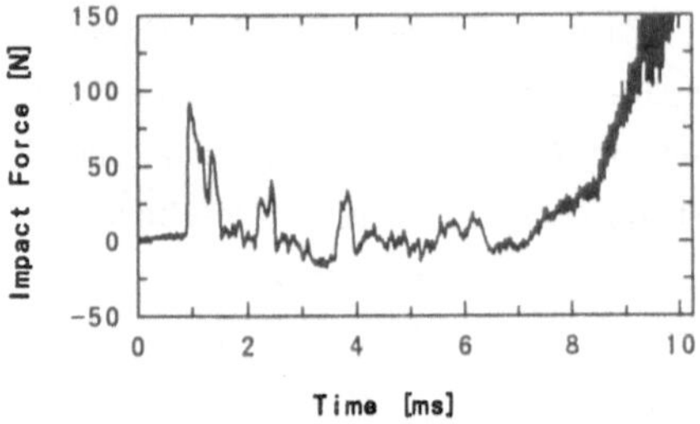

(a) Estimated from $e_1(t)$

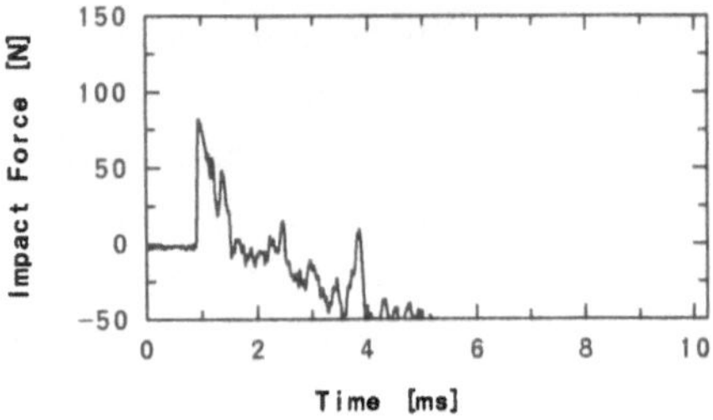

(b) Estimated from $e_2(t)$

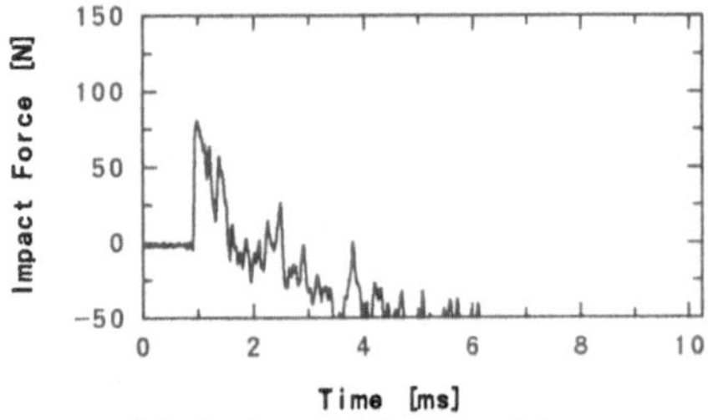

(c) Estimated from $e_3(t)$

Figure 7: Impact force of the ball estimated by applying the noncausal Wiener filtering technique.

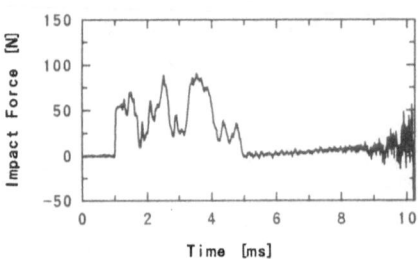

Figure 6: Impact force of the rod estimated by applying the noncausal Wiener filtering technique and the least squares method.

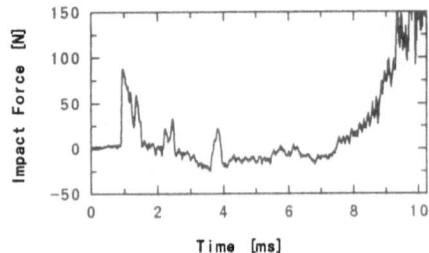

Figure 8: Impact force of the ball estimated by applying the noncausal Wiener filtering technique and the least squares method.

## 4. Conclusions

An inverse analysis technique of impact force has been studied in order to apply it to the measurement of the impact force acting on bodies of arbitrary shape. The noncausal Wiener filtering technique and the least squares method utilizing the SVD have been demonstrated to be effective in improving the estimation accuracy of the impact force by means of inverse analysis. However, it has been also understood that further developments are required for obtaining satisfactory estimates of the impact force.

The authors would like to thank Hidetoshi Okamoto, formerly undergraduate student in Tokyo Institute of Technology, for his helpful assistance in the experiments and data processing.

## References

1. Thornhill,R.J. and Smith,C.C.: Impact Force Prediction Using Measured Frequency Response Functions, *Trans. ASME, J. Dyn. Sys., Meas., Control*, **105** (1983) 227–231.

2. Whiston,G.S.: Remote Impact Analysis by Use of Propagated Acceleration Signals, I, *J. Sound Vib.*, **97** (1984) 35–51.

3. Jordan,R.W. and Whiston,G.S.: Remote Impact Analysis by Use of Propagated Acceleration Signals, II, *J. Sound Vib.*, **97** (1984) 53–63.

4. Doyle,J.F.: An Experimental Method for Determining the Dynamic Contact Law, *Exp. Mech.*, **24** (1984) 10–16.

5. Doyle,J.F.: Further Development in Determining the Dynamic Contact Law, *Exp. Mech.*, **24** (1984) 265–270.

6. Michaels,J.E. and Pao,Y.H.: The Inverse Source Problem for an Oblique Force on an Elastic Plate, *J. Acoust. Soc. Am.*, **77** (1985) 2005–2011.

7. Michaels,J.E. and Pao,Y.H.: Determination of Dynamic Forces from Wave Motion Measurements, *Trans. ASME, J. Appl. Mech.*, **53** (1986) 61–68.

8. Kishimoto,K., Kuroda,M., Aoki,S. and Sakata,M.: A Measuring System for Dynamic Stress Intensity Factors Using a FFT Analyzer, (in Japanese), *J. Soc. Mat. Sci., Japan*, **35** (1986) 850–853.

9. Doyle,J.F.: Determining the Contact Force during the Transverse Impact of Plates, *Exp. Mech.*, **27** (1987) 68–72.

10. Doyle,J.F.: Experimentally Determining the Contact Force during the Transverse Impact of an Orthotropic Plate, *J. Sound Vib.*, **118** (1987) 441–448.

11. Stevens,K.K.: Force Identification Problems—An Overview, *Proc. 1987 SEM Spring Conf. Exp. Mech.*, (1987) 838–844.

12. Nakao,T., Tanaka,C. and Takahashi,A.: Source Wave Analysis of Impact Force on Wood Based Panel Floor, (in Japanese), *J. Soc. Mat. Sci., Japan*, **37** (1988) 565–570.

13. Hojo,A., Chatani,A. and Uemura,F.: An Estimation of Impact Force by Deconvolution Integral, (in Japanese), *Trans. JSME*, A-**55** (1989) 477–482.

14. Hollandsworth,P.E. and Busby,H.R.: Impact Force Identification Using the General Inverse Technique, *Int. J. Impact Eng.*, **8** (1989) 315–322.

15. Chang,C. and Sun,C.T.: Determining Transverse Impact Force on a Composite Laminate by Signal Deconvolution, *Exp. Mech.*, **29** (1989) 414–419.

16. Rangel,R.V., Thornhill,R.J. and Smith,C.C.: Prediction of Impact Forces Using Hertzian Contact Theory and Measured Modal Structural Data, *Mech. Sys. Signal Processing*, **4** (1990) 287-294.

17. Bateman,V.I., Carne,T.G., Gregory,D.L., Attaway,S.W. and Yoshimura,H.R.: Force Reconstruction for Impact Tests, *Trans. ASME, J. Vib. Acoust.*, **113** (1991) 192–200.

18. Ödeen,S. and Lundberg,B.: Prediction of Impact Force by Impulse Response Method, *Int. J. Impact Eng.*, **11** (1991) 149–158.

19. Buttle,D.J. and Scruby,C.B.: Characterisation of Dust Impact Process at Low Velocity by Acoustic Emission, *Acoustic Emission: Current Practice and Future Directions, ASTM STP 1077*, (1991) 273–286.

20. Inoue,H., Shibuya,T., Koizumi,T. and Watanabe,R.: Measurement of Impact Force History by Deconvolution Method, *Proc. VI Int. Cong. Exp. Mech.*, (1988) 463–468.

21. Inoue,H., Watanabe,R., Shibuya,T. and Koizumi,T.: Measurement of Impact Force by the Deconvolution Method, *Trans. JSNDI*, **2** (1989) 63–73.

22. Inoue,H., Shibuya,T., Koizumi,T. and Fukuchi,J.: Measurement of Impact Force Applied to a Plate by the Deconvolution Method, *Trans. JSNDI*, **2** (1989) 74–83.

23. Inoue,H., Shibuya,T. and Koizumi,T.: Measurement of Impact Load in Charpy Test by Using Deconvolution Procedure, *Proc. 9th Int. Conf. Exp. Mech.*, (1990) 1989–1998.

24. Inoue,H., Ishida,H., Kishimoto,K. and Koizumi,T.: Measurement of Impact Load by Using an Inverse Analysis Technique, *JSME Int. J., Ser. I*, **34** (1991) 453-458.

25. Inoue,H., Kishimoto,K., Shibuya,T. and Koizumi,T.: Estimation of Impact Load by Inverse Analysis, (in Japanese), *Trans. JSME*, **A-57** (1991) 2727-2734.

26. Inoue,H., Kamibayashi,M., Kishimoto,K., Shibuya,T. and Koizumi,T.: Numerical Laplace Transformation and Inversion Using the Fast Fourier Transform, (in Japanese), *Trans. JSME*, **A-57** (1991) 2523–2528.

27. Baumeister,J.: *Stable Solution of Inverse Problems*, (1987), Friedr. Vieweg & Sohn.

28. Horn,R.A. and Johnson,C.R.: *Matrix Analysis*, (1985), Cambridge.

# Optimality of Internal Structure and Identification of Unknown Load

Yukio Tada[*] and Miho Kamada[**]

*  Department of Systems Engineering, Faculty of Engineering,
   Kobe University, Rokkodai, Nada, Kobe  657  Japan
** FA Systems Division, Computer Division,
   Electronics & Information Systems Division's Group,
   Nippon Steel Corporation, 5-10-1 Fuchinobe, Sagamihara,
   Kanagawa  229  Japan

Summary
This paper considers the relation between loads and configuration in a
structure optimized for a load.  First, applying the principle of the
minimum potential energy, an optimization problem for the distribution of
the members of a continuous truss structure is studied with respect to
density and direction.  Second, the change of optimal structure by loading
condition is considered.  Lastly, the possibility of identifying unknown
applied loads from the optimal shape is discussed.

## 1. Introduction

It is said that the configuration of animal's bone has an optimality
against the mechanical environment around it.  The environment, however, is
very complicated and its details have been unknown.  The relation between
the density of the bone and the strain energy density was studied by
earlier works.  The bone trabeculae have a particular pattern [1], and its
similarity to the Michell's structure was discussed.  On the other hand,
in the field of optimum structural design, the determination of the topology, that is, the allocation of members for the continuous truss structure,
which was composed of an infinite number of members, was considered aiming
the minimum weight [2].

In this paper, we consider an optimization problem in a general structure
which is a continuous truss structure by the use of Finite Element Method.
First, applying the principle of the minimum potential energy, an optimization problem for the distribution of the members of a continuous truss
structure is studied with respect to density and direction.  Second, the
change of optimal structure by loading condition is considered.  Lastly,

the possibility of identifying unknown applied loads from the optimal shape is discussed.

## 2. Continuous Truss Structure

The Finite Element Method for the continuous truss structure was proposed as "Structural Continua" by Ref.[2], in which the element was composed of two groups of parallel members intersecting each other by a certain angle and distributing continuously. In this paper, we consider an element which is one of continuous structures but whose two groups of members are intersecting each other perpendicularly as shown in Fig.1. Each element has the stiffness only for the tension and compression in the directions of those two groups. Therefore, the stiffness equation in the $\alpha$-$\beta$ local coordinate system of an element is written as

$$[k][w] = [f] \tag{1}$$

$$[k] = tAE[B]^T[D][B] \tag{2}$$

$$[B] = \frac{1}{2A} \begin{bmatrix} b_1 & 0 & b_2 & 0 & b_3 & 0 \\ 0 & c_1 & 0 & c_2 & 0 & c_3 \end{bmatrix} \quad (3), \qquad [D] = \begin{bmatrix} D_1 & 0 \\ 0 & D_2 \end{bmatrix} \tag{4}$$

where [w] and [f] are the nodal displacement vector and nodal force vector, and [k] is the stiffness matrix. E, A and t are the Young's modulus, area and thickness of the triangular element. [B] is the matrix which relates the strain vector $[e_\alpha, e_\beta]^T$ with the nodal displacement vector [w], where

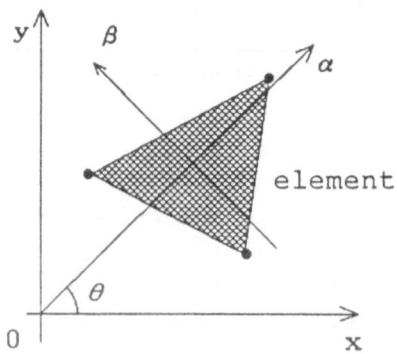

Fig.1 Element of continuous truss

$e_\alpha$ and $e_\beta$ are the tensile strains of $\alpha$-member and $\beta$-member, respectively. Therefore, $b_i$ and $c_i$ (i=1,2,3) are coefficients of linear interpolating functions as usual definition, that is, $b_i = \beta_j - \beta_k$ and $c_i = \alpha_k - \alpha_j$ (i,j,k=1,2,3: even permutation) in $\alpha$ and $\beta$ coordinate system of the element. Moreover, $D_1$ and $D_2$ are the densities of $\alpha$ and $\beta$ members in the element. Transforming Eq.(1) into the global coordinate system and assembling them, the equilibrium equation in the global system

$$[K][U] = [P] \tag{5}$$

can be obtained, where [U] and [P] are the vectors of nodal displacement and force in the global coordinate system, and [K] is the corresponding stiffness matrix.

## 3. Optimization Problem

### 3.1 Formulation of optimization problem

This paper formulates an optimization problem, which seeks the optimum directions and densities of members of a continuous truss under the condition of the volume constancy, according to the inverse variational principle with the potential energy [3]. That is, for a structure composed of n elements, the problem is written as

$$L = (1/2)[U]^T[K(\Theta, D)][U] - [U]^T[P] - g\{V(D)-C\}$$

$$\text{----> stationary } ([U], g, \Theta, D) . \tag{6}$$

The third and forth terms are incorporated as the constraint that the total volume of the continuous truss structure, V, is constant, that is,

$$V(D) = \sum_{i=1}^{n} t_i A_i (D_{1i}+D_{2i}) = C , \tag{7}$$

where C is a specified value and g is a Lagrange multiplier. $\Theta=(\theta_1, \theta_2, \ldots, \theta_n)^T$ and $D=(D_{11}, D_{21}, D_{12}, D_{22}, \ldots\ldots, D_{1n}, D_{2n})$ are sets of design variables. $\theta_i$ (i=1,...,n) is the angle between $\alpha$ axis in the i-th element and x axis of the global coordinate system, and $D_{1i}$ and $D_{2i}$ (i=1,...,n) are densities of $\alpha$ and $\beta$ members in the i-th element. The stationary conditions of Eq.(6) are the equilibrium condition [Eq.(5)], the constraint of the volume constancy [Eq.(7)] and the equations for the

optimality of members in directions and densities.    The last two of optimality conditions are written by

$$E_{ji} - g = 0 \qquad (j=1,2) \quad (i=1,\ldots,n) \tag{8}$$

$$Q_i = 0 \quad (0 < \theta_i < \pi/2) \qquad \text{or} \qquad \theta_i = 0 \qquad (i=1,\ldots,n) , \tag{9}$$

where

$$E_{ji} = (1/2)[U]^T(\partial[K]/\partial D_{ji})[U]/(\partial V/\partial D_{ji}) \qquad (j=1,2) \quad (i=1,\ldots,n) \tag{10}$$

$$Q_i = (1/2)[U]^T(\partial[K]/\partial\theta_i)[U] \qquad\qquad (i=1,\ldots,n) . \tag{11}$$

$E_{ji}$ corresponds to the strain energy density of the j-th member in the i-th
element,  and Eq.(8) requires that the strain energy density is uniform for
all members in the continuous truss structure.

### 3.2 Methods for optimization
The  optimization problem is solved  by using the energy-ratio  method  [3]
and Newton method.    First,  from the meaning of Eq.(8), a method which can
equalize  the  strain  energy densities of all members is  suited  for  the
optimization of the densities $D_{ji}$.  The stress is proportional to the nodal
displacement in an element  and  from the form of the stiffness matrix [Eq.
(2)]  the displacement is in inverse proportion to the  densities.    Therefore,  the stress, and then the strain energy density of each member can be
regarded to vary inversely as its density.    Then, for the density $D_{ji}^{(t)}$ at
the t-th iteration step, calculating $E_{ji}$ by Eqs.(5) and (10), the following
algorithm

$$D_{ji}^{*} = D_{ji}^{(t)} + w\{(E_{ji}-\overline{E})/\overline{E}\}D_{ji} \qquad (j=1,2) \quad (i=1,\ldots,n) \tag{12}$$

is used, where $\overline{E}$ is the mean value of all $E_{ji}$'s, that is,

$$\overline{E} = (\Sigma_{i=1}^{n} \Sigma_{j=1}^{2} E_{ji})/2n , \tag{13}$$

which is adopted as an estimated value of the Lagrange multiplier g,  and w
is  a  weighting factor.    As the new density $D_{ji}^{*}$ obtained above  may  not
satisfy the volume constraint, the second reformation

$$D_{ji}^{(t+1)} = D_{ji}^{*} \ C/V( \ D^{*}) \qquad (j=1,2) \ (i=1,\ldots,n) \qquad (14)$$

is performed based on the lineality of the volume in the density D [Eq. (7)]. These reformations are iterated until the condition

$$Max \ [(E_{ji}-\bar{E})/\bar{E}] < \varepsilon_{1} \qquad (15)$$

is satisfied, where $\varepsilon_1$ is a small positive value. This iterative algorithm is one of the energy-ratio method.

Next, the optimality equation (9) for the direction $\Theta$ is solved by the Newton method

$$\Theta^{(t'+1)} = \Theta^{(t')} - [J( \ \Theta^{(t')} )]^{-1} [Q_{1}( \ \Theta^{(t')} ), \ \ldots \ , \ Q_{n}( \ \Theta^{(t')} )]^{T} , (16)$$

where $[J( \ \Theta^{(t')} )]$ is the Hessian at $\Theta = \Theta^{(t')}$, whose elements are $J_{ij} = \partial Q_i( \ \Theta^{(t')} )/\partial\theta_j$. The condition for the convergence is

$$\varepsilon_{2} > \Sigma_{i=1}^{n} \ (Q_{i})^{2}/n , \qquad (17)$$

where $\varepsilon_2$ is a small positive value.

For the combined optimization of the densities and direction, the iteration algorithms are performed in turn by fixing other variables as parameters.

## 4. Example of Design

As an example, a continuous truss in the shape of a disk is optimized by the present method. The disk is subjected to compressive loads radially as shown in Fig.2 (a). From the symmetry of the shape and the loading condition, only a quarter part is analyzed, and the object domain is divided into eight elements as shown in Fig.2 (b) (Disk Model 1). Figure 3 (a) shows the distribution of all members in an initial assumption; that is, the densities of all members are uniform and they are directed horizontally and vertically. In the following figures, densities of members are represented as the densities in the number of parallel lines whose directions show those of members of the continuous truss in respective elements. Figure 3 (b) shows the continuous truss obtained by the present method and Table 1 also shows the result of the optimization. It is found from Table

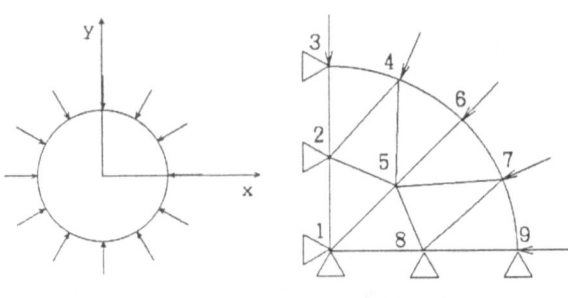

(a) disk        (b) Disk Model 1

Fig.2   Disk model

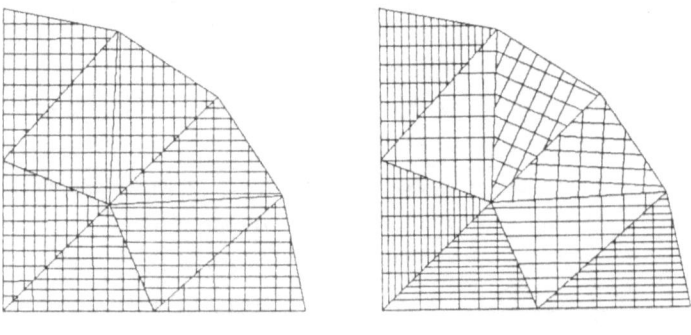

(a) initial body        (b) optimum body

Fig.3   Disk Model 1

Table 1   Chracteristic values

|  | initial body | optimum body |  |
|---|---|---|---|
| potential energy | $-8.75 \times 10^{-6}$ | $-7.76 \times 10^{-6}$ | Po.E. |
| mean value of strain energy density | $4.65 \times 10^{-11}$ | $4.05 \times 10^{-11}$ | M.S.E.D. |
| variance of strain energy density | $3.59 \times 10^{-22}$ | $0.00 \times 10^{-22}$ |  |
| sum of squares of $Q_i$'s | $1.17 \times 10^{-12}$ | $1.55 \times 10^{-13}$ |  |

1 that the distribution of the strain energy density becomes uniform and the sum of $Q_i$'s is decreased.

## 5. Possibility of Inverse Problem

### 5.1 Loading conditions and optimum structures

[Disk Model]

For two disk models, different loading conditions are considered. Figure 4

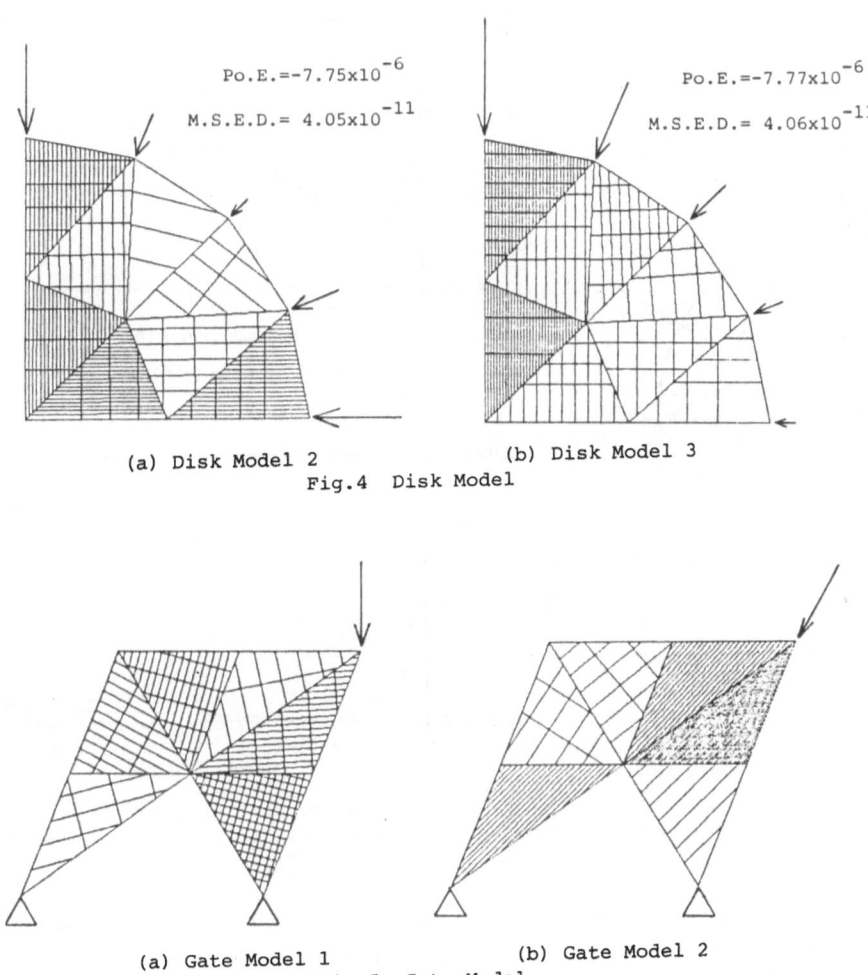

(a) Disk Model 2         (b) Disk Model 3

Fig.4   Disk Model

(a) Gate Model 1         (b) Gate Model 2

Fig.5   Gate Model

shows the optimum configuration of members of respective continuous trusses. It is found from these figures that the optimum didtributions of members are different from each other. The values associated with each figure are the potential energy (Po.E.) and the mean value of the strain energy density, (M.S.E.D.). Compared with Table 1, it is interesting that the potential energies and the mean values of the strain energy density have nearly same values in three cases irrespective of loading conditions.

[Gate Model]

For two continuous trusses with the same outward form as shown in Fig.5 (Gate Model 1 and 2), optimization problems are considerd under different

186

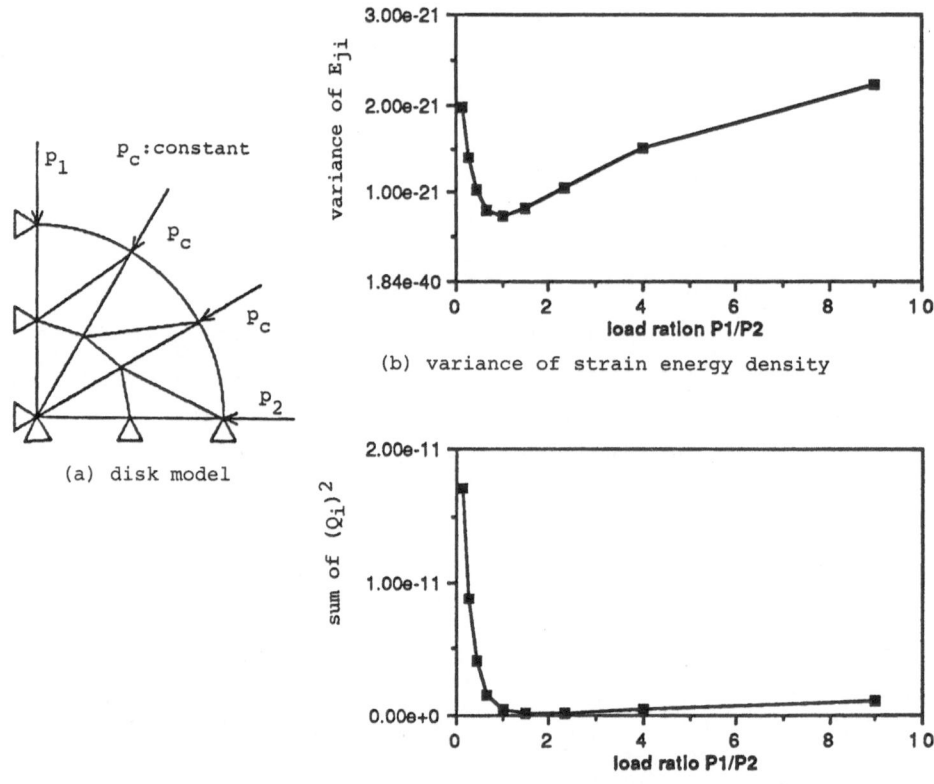

(a) disk model

(b) variance of strain energy density

(c) sum of squares of $Q_i$'s
Fig.6   Responses of Disc Model 4 for different ratio of two loads

loading conditions, that is, two concentrated loads acting from different directions. Figure 5 shows the optimum distributions of members in respective continuous trusses for their loads.

Thus, when different loading conditions are considered, the optimum distributions of directions and densities of members in a continuous truss are different depending on the loading conditions, which suggests the possibility of identifying unknown applied loads on the basis of the structural optimality.

5.2 Possibility of identifying unknown applied load
[Ratio in magnitudes of two loads]
First, a continuous truss as shown in Fig.6 (a) is optimized in the condi-

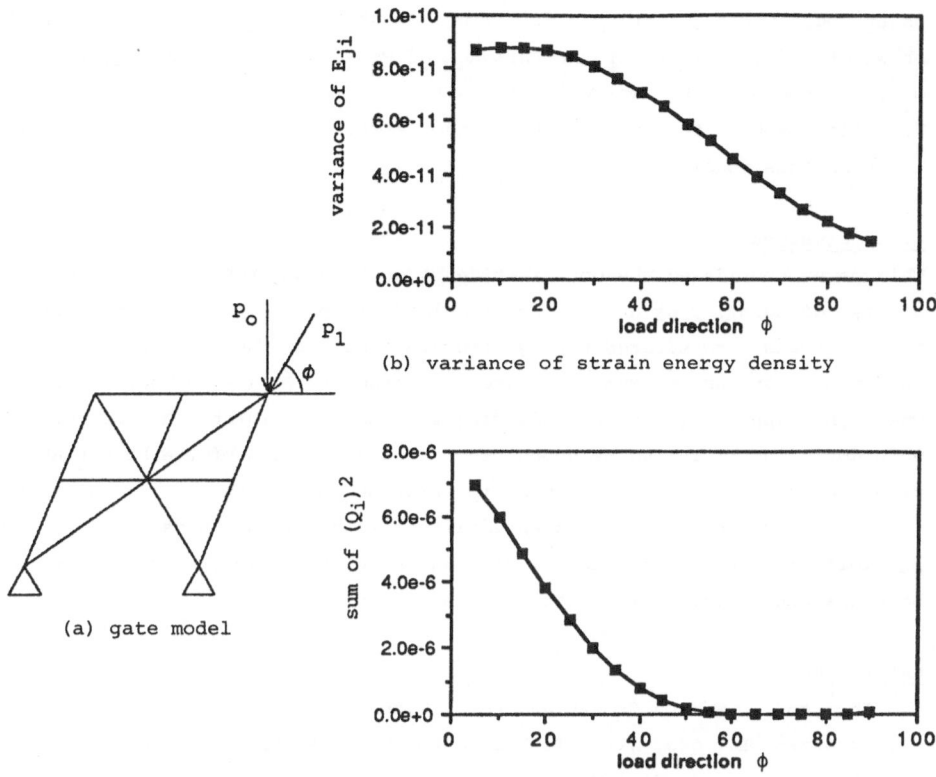

(b) variance of strain energy density

(a) gate model

(c) sum of squares of $Q_i$'s
Fig.7 Responses of Gate Model 1 for a load from various directions

tion that $p_1/p_2$=1 (Disk Model 4). Second, for the structure so optimized, structural responses are calculated by FEM for various cases of the ratio in the magnitudes of two loads, $p_1/p_2$. The considered responses are the variance of the quantities $E_{ji}$, which corresponds to the strain energy density as mentioned before, and the sum of squares of $Q_i$'s. Figure 6 shows the variations of these two values for the ratio of $p_1/p_2$. It is observed from the graphs that these values have their minimums at the true value of $p_1/p_2$, that is, at $p_1/p_2$=1.

[Direction of load]
Secondly, structural responses of Gate Model 1 which was optimized for the load from $\phi$=90° ($p_o$) in Sec.5.1 is considered when it is subjected to a concentrated load $p_1$ with the same magnitude as that of Sec.5.1 at the same

point but whose direction $\phi$ is varied as shown in Fig.7 (a). Figure 7 shows the variations of the variance of $E_{ji}$ and the sum of squares of $Q_i$'s. Though both two quantities attain to their minimums at the true direction $\phi=90°$, the sensitivity of the latter to the direction is not so large in this model.

## 6. Conclusions

This paper considered the optimization of continuous trusses and the possibility of identifying unknown applied loads from the optimality of structure. It was demonstrated through two examples that the variance of strain energy density and the sum of squares of parameter $Q_i$'s take their minimums when the applied loads coincide with the loads for which the structures were optimized. In the orthopedics, the distribution of the bone trabeculae are examined as the source of information about mechanical conditions [4]. If the bone has the adaptability for the acting loads, it may be possible to know the mechanical environment by measuring the structure of bone trabeculae [5] from the assumption of its optimality.

## References

1. Thompson, D.W.: On Growth and Form, Cambridge, (1942).

2. Yamazaki, K.; Oda, J.: A Technique to Obtain an Optimum Layout of Truss Structures by Finite Element Method (Minimum Weight Design of Structural Continua), Bulletin of the JSME, 24-193 (1981) 1109-1114.

3. Tada, Y.; Seguchi, Y.: Shape Determination of Structures Based on the Inverse Variational Principle/ The Finite Element Approach, Ch. 8 of New Directions in Optimum Structural Design (Ed. by Atrek, E. et al.), John Wiley & Sons, (1984) 197-209.

4. Shiba, R.; Nakabayashi, K.; Hirohata, K.: Analysis of Trabeculae of Femoral Head and Neck, Proc. of 1985 Annual Meeting of Jap. Soc. for Orthopaedic Biomech., 7 (1986) 59-66, (in Japanese).

5. Tada, Y.: Analysis and 3D Reconstruction of Bone Trabeculae by Computer Aided Technique, Proc. of 1991 Int'l Conf. on Industrial Electronics, Control and Instrumentation, Kobe, IEEE, 3 (1991) 1767-1772.

# Chapter 4
# Shape Determination and Optimization

# Shape Identification Using Acoustic Measurements: A Numerical Investigation Using BIE and Shape Differentiation

**Marc BONNET**

Laboratoire de Mécanique des Solides (CNRS, Ecole Polytechnique, Mines, Ponts et Chaussées),
Ecole Polytechnique, 91128 Palaiseau Cedex, FRANCE.

## 1 Preliminaries.

The purpose of this paper is to study numerically the problem of identifying a rigid 3D finite body $\Omega$ imbedded in an infinite acoustical medium $\Omega_e = \mathcal{R}^3 - \Omega$ (wave velocity $c$). The cavity is subjected to a known harmonic incident pressure $p^I(\mathbf{x}) \exp(-i\omega t)$, which satisfies Helmholtz' equation $(\Delta + k^2)p^I = 0$ inside $\Omega_e$ (with $k = \omega/c$. The scattered pressure field $p_\Gamma(\mathbf{x})$ induced by the presence of the obstacle (i.e. the solution of the *direct problem*) satisfies the so-called 'state equation':

$$\begin{cases} (\Delta + k^2)p = 0 & \text{in } \Omega_e \\ p_{,n} + p^I_{,n} = 0 & \text{on } \Gamma = \partial\Omega_e = \partial\Omega \quad ( \ (\cdot)_{,n} \equiv \frac{\partial}{\partial n}(\cdot)) \\ \text{(radiation condition)} \end{cases} \tag{1}$$

where the unit normal $\mathbf{n}$ is directed outside $\Omega_e$, i.e. is interior to $\Gamma$.

The shape of $\Omega_e$, ie the surface $\Gamma$, is unknown, and $p$ defined by (1) depends on $\Gamma$: $p = p_\Gamma$. Extra data is necessary if one is to solve the inverse problem, e.g.

$$p(\mathbf{x}) = \hat{p}(\mathbf{x}) \qquad \text{on } C \tag{2}$$

$C$ being a surface, or curve, exterior to $\Gamma$, on which the pressure field is measured. Thus $\Gamma$ is classically sought as

$$\min_\Gamma J(\Gamma) , \quad J(\Gamma) \equiv J(p_\Gamma, \Gamma) \tag{3}$$

where:

$$J(p, \Gamma) = \frac{1}{2} \int_C | p(\mathbf{y}) - \hat{p}(\mathbf{y}) |^2 \ dC_\mathbf{y} \tag{4}$$

The nonlinear minimization problem (3) is best solved (in terms of both computational efficiency and accuracy) using gradient methods, such as BFGS of conjugate gradient [11]. These algorithms need repeated computations of the derivative of $J(\Gamma)$ with respect to $\Gamma$ (or, in practice, the design parameters which define the current location of $\Gamma$). These derivatives may be computed using finite-difference methods. However, this is computationally expensive (because the evaluation of each partial derivative needs a complete solution of (1) on a perturbed geometry $\Gamma + \delta G$) and may be poor in terms of accuracy.

Hence an analytical differentiation of (4) is investigated instead. Its derivation relies on the *shape differentiation approach,* and two approaches are presented here: the direct differentiation approach (DDA) (section 3) and the adjoint problem approach (section 4).

The basic nature nature of the inverse problem under consideration (to search an unknown surface) suggests the use of boundary integral equations (BIE) and boundary elements for the numerical modelling of the direct problem (1). In the present problem, this indication is made even stronger by the infinite character of the domain $\Omega_e$.

Any solution $p(\mathbf{x})$ of Helmholtz' equation satisfies a well-known BIE derived from third Green's formula. In the present paper, we use its *regularized* version [1], [2], which reads:

$$\gamma p(\mathbf{x}) + \int_\Gamma \left\{ p(\mathbf{y})[G_{,n}(\mathbf{x},\mathbf{y}) - G^0_{,n}(\mathbf{x},\mathbf{y})] + [p(\mathbf{y}) - p(\mathbf{x})]G^0_{,n}(\mathbf{x},\mathbf{y}) - q(\mathbf{y})G(\mathbf{x},\mathbf{y}) \right\} dS_\mathbf{y} = 0 \quad (5)$$

In (5), $G(\mathbf{x},\mathbf{y}) = e^{ikr}/(4\pi r)$ and $G^0(\mathbf{x},\mathbf{y}) = 1/(4\pi r)$ are the dynamic and static Green functions, while $r = \parallel \mathbf{x} - \mathbf{y} \parallel$, $(\cdot)_{,n} \equiv n_s(\mathbf{y})(\cdot)_{,s} \equiv n_s(\mathbf{y})\partial(\cdot)/\partial y_s$, $q \equiv \partial p/\partial n$. The coefficient $\gamma$ in (5) depends only on the boundedness of the domain under consideration: $\gamma = 1$ (infinite medium), $1/2$ (half-space), 0 (bounded domain). For a given domain, eqn. (5) holds for interior *and* boundary points $\mathbf{x}$ using the *same* $\gamma$, which is therefore not to be mistaken with the conventional free-term coefficient. In the sequel, as infinite domains are considered, eqn. (5) will be used with $\gamma = 1$. The integrand in eqn. (5) is weakly singular, provided $p(\mathbf{x}) \in C^{0,\alpha}(\Gamma)$, thanks to the regularizing effect of the term $[p(\mathbf{y}) - p(\mathbf{x})]$, and contains no Cauchy principal value (CPV) integral. This point will prove crucial for the derivation of the *rate BIE* in section 3 below, as well as for all numerical computations. For $p(\mathbf{x})$ solution of (1), one puts $q(\mathbf{y}) = -p^I_{,n}(\mathbf{y})$ in (5).

## 2   The shape differentiation approach.

The shape differentiation approach deals with derivatives of functionals with respect to variable domains or boundaries (e.g. $J(\Gamma)$) involving fields which themselves depend on the geometry, notably the solutions of boundary-value problems like $p_\Gamma(\mathbf{y})$.

Let $\Gamma$ denote the current location of the unknown boundary during the minimization process, and consider a further (small) evolution of the surface $\Gamma$ defined by means of a normal 'velocity' field $\theta(\mathbf{y})$:

$$\mathbf{y} \in \Gamma \to \mathbf{y} + \theta(\mathbf{y})\mathbf{n}(\mathbf{y})\tau \quad \text{i.e. } \Gamma(\tau) = \Gamma + \theta\mathbf{n}\tau \qquad (\tau \geq 0, \tau \text{ 'small'}) \quad (6)$$

while the measurement area $C$ is kept fixed ($\theta(\mathbf{y}) = 0 \ \forall \mathbf{y} \in C$). Definition (6) is considered only for small values of $\tau$, consistently with the fact that we are only interested here by derivatives for $\tau = 0$, i.e. for the current $\Gamma$ (in the sequel, derivatives with respect to $\tau$ are always taken for $\tau = 0$).

The SDA defines several kinds of derivatives with respect to $\tau$ for fields $u(\mathbf{y},\tau)$. In a BIE approach of shape differentiation, it seems natural to use 'material' derivatives, i.e. to 'follow' the field $u$ while the field point $\mathbf{y}$ moves according to (6), in order to use only the information available on the boundary as it 'moves'. When, as here, the geometrical transformation is described by means of a *normal* velocity field $\theta$, the 'transformation derivative' $\overset{*}{u}$ [9] of $u = u(\mathbf{y},t)$ is introduced:

$$\overset{*}{u}(\mathbf{y},0) = u_{,\tau}(\mathbf{y} + \theta\mathbf{n}\tau,\tau) \mid_{\tau=0} = u_{,\tau}(\mathbf{y},0) + \theta(\mathbf{y})u_{,n}(\mathbf{y},0) \quad (7)$$

Various formulas are given in the literature (see e.g. [9], [7]) for the derivative of integrals with respect to variable volumes $\Omega$ or surfaces $\Gamma$, among which:

$$\frac{d}{d\tau} \int_\Omega a(\mathbf{y},\tau)\,dV_\mathbf{y} = \int_\Omega a_{,\tau}(\mathbf{y},\tau)\,dV_\mathbf{y} + \int_{\partial\Omega} a(\mathbf{y},\tau)\theta(\mathbf{y})\,dS_\mathbf{y} \quad (8)$$

$$\frac{d}{d\tau} \int_\Gamma a(\mathbf{y},\tau)\,dS_\mathbf{y} = \int_\Gamma \left\{ a_{,\tau}(\mathbf{y},\tau) + (a_{,n}(\mathbf{y},\tau) - 2K(\mathbf{y})a(\mathbf{y},\tau))\theta(\mathbf{y}) \right\} dS_\mathbf{y} \quad (9)$$

$$= \int_\Gamma \left\{ \overset{*}{a}(\mathbf{y},\tau) - 2K(\mathbf{y})a(\mathbf{y},\tau)\theta(\mathbf{y}) \right\} dS_\mathbf{y} \quad (10)$$

where $K(\mathbf{y})$ denotes the mean curvature at $\mathbf{y} \in \Gamma$. Equations (9), (10) hold only for a closed smooth surface, while in equation (8) $\theta$ refers to the unit normal $\mathbf{n}$ directed towards the exterior of $\Omega$. Generalization of above formulas to piecewise smooth surfaces is available [9] but will not be used here.

## 3 Rate BIE: the direct differentiation approach.

In this approach, (4) is differentiated with respect to $\tau$ directly, using eqn. (10). material derivative of integrals [9]. This yields:

$$\frac{d}{d\tau}J(\Gamma) = \int_C (p_\Gamma(\mathbf{y}) - \hat{p}(\mathbf{y}))\,\overset{*}{p}_\Gamma(\mathbf{y})\,dC_\mathbf{y} \tag{11}$$

in which the fact that $C$ is fixed has been taken into account.

The use of (11) implies, in turn, the calculation of $\overset{*}{p}_\Gamma$. This is achievable by deriving from the *state BIE* (5) a *rate BIE*, in which the fields $(\overset{*}{p}_\Gamma, \overset{*}{q}_\Gamma)$ appear. The rate BIE is indeed obtained upon application of formula (10) to (5). The validity of the differentiation of (5) using (10) relies upon the weakly singular character of the acoustic BIE (4) (as opposed to conventional CPV BIEs, for which extreme care must be taken when considering their differentiation with respect to a parameter).

As a result of this differentiation process, the *rate BIE* reads, in the present context:

$$\overset{*}{p}(\mathbf{x}) + \int_\Gamma \left\{ \overset{*}{p}(\mathbf{y})[G_{,n}(\mathbf{x},\mathbf{y}) - G^0_{,n}(\mathbf{x},\mathbf{y})] + [\overset{*}{p}(\mathbf{y}) - \overset{*}{p}(\mathbf{x})]G^0_{,n}(\mathbf{x},\mathbf{y}) - \overset{*}{q}(\mathbf{y})G(\mathbf{x},\mathbf{y}) \right\} dS_\mathbf{y} =$$

$$= \int_\Gamma q(\mathbf{y}) \left\{ [\theta(\mathbf{y})n_s(\mathbf{y}) - \theta(\mathbf{x})n_s(\mathbf{x})]G_{,s}(\mathbf{x},\mathbf{y}) - 2K(\mathbf{y})\theta(\mathbf{y})G(\mathbf{x},\mathbf{y}) \right\} dS_\mathbf{y}$$

$$- \int_\Gamma [\theta(\mathbf{y})n_r(\mathbf{y}) - \theta(\mathbf{x})n_r(\mathbf{x})] \left\{ D_{rs}p(\mathbf{y})G_{,s}(\mathbf{x},\mathbf{y}) - k^2 p(\mathbf{y})G(\mathbf{x},\mathbf{y}) \right\} dS_\mathbf{y} \tag{12}$$

where $D_{rs}$ is the tangential differential operator given by eqn. (30) of Appendix A. The derivation of (12) uses the integration by parts formula (35) and:

$$\frac{d}{d\tau}G(\mathbf{x},\mathbf{y}) = [\theta(\mathbf{y})n_s(\mathbf{y}) - \theta(\mathbf{x})n_s(\mathbf{x})]G_{,s}(\mathbf{x},\mathbf{y})$$

**Remark 1** The rate BIE as given by (12) is valid for a smooth surface, without edges or corners (but the present approach can be extended to piecewise smooth surfaces).

**Remark 2** All integrands in eqn. (12) are weakly singular, thanks to the regularizing effect of $[\overset{*}{p}(\mathbf{y}) - \overset{*}{p}(\mathbf{x})]$, $[\theta(\mathbf{y})n_s(\mathbf{y}) - \theta(\mathbf{x})n_s(\mathbf{x})]$ and $[p(\mathbf{y}) - p(\mathbf{x})]$.

**Remark 3** The validity of the derivation of (12) relies upon the validity of the differentiation of (5) using (10). In this respect, the weakly singular character of the acoustic BIE (5) is crucial. At any stage of the derivation, all integrals are at most weakly singular. In contrast, a very close attention should be exercised with respect to the handling of exclusion neighbourhoods and subsequent limit processes, were the same approach to be applied to conventional strongly singular BIEs.

As a consequence of (1), the expression of $\overset{*}{q}_\Gamma = -(p^I_{,n})^*$ on $\Gamma$ is given (see appendix B) by:

$$(p^I_{,n})^* = \left( 2Kp^I_{,n} - k^2 p^I \right) \theta - D_s(\theta D_s p^I) \tag{13}$$

and the right-hand side $\mathcal{F}(p,q)\theta$ in (12) becomes

$$\mathcal{F}(p,q)\theta = -\int_\Gamma p^I_{,n}(\mathbf{y})(\theta(\mathbf{y})n_s(\mathbf{y}) - \theta(\mathbf{x})n_s(\mathbf{x}))G_{,s}(\mathbf{x},\mathbf{y})\,dS_\mathbf{y}$$

$$+ \int_\Gamma \left\{ D_s(\theta D_s p^I)(\mathbf{y}) - (\theta(\mathbf{y})n_r(\mathbf{y}) - \theta(\mathbf{x})n_r(\mathbf{x}))D_{rs}p(\mathbf{y}) \right\} G_{,s}(\mathbf{x},\mathbf{y})\,dS_\mathbf{y} \tag{14}$$

The representation formula for interior values $p(\mathbf{x})$ may also be differentiated with respect to the variable domain. The result is:

$$\overset{*}{p}(\mathbf{x}) = -\int_\Gamma \left\{ [\theta(\mathbf{y})p^I_{,n}(\mathbf{y}) + \overset{*}{p}_\Gamma(\mathbf{y})]G_{,n}(\mathbf{x},\mathbf{y}) \right.$$

$$\left. + \theta(\mathbf{y})\left( k^2(p_\Gamma + p^I)(\mathbf{y})G(\mathbf{x},\mathbf{y}) - D_s(p_\Gamma + p^I)D_sG(\mathbf{x},\mathbf{y}) \right) \right\} dS_\mathbf{y} \tag{15}$$

Then, in order to evaluate the derivative (11), one has to

1. Solve the state BIE (5) for the unknown $p_\Gamma$.

2. Solve the rate BIE (14) for the unknown $\overset{*}{p}_\Gamma$. This must be done successively with each $\theta$ in turn (or rather, in practice, with a finite number of $\theta$s associated to design variables).

3. Compute the field $\overset{*}{p}_\Gamma$ on the measurement surface $C$ using representation formula (15).

4. Insert, for each $\theta$, $p_\Gamma$ and $\overset{*}{p}_\Gamma$ in $\frac{dJ(\Gamma)}{dt}$ (11).

Indeed, as soon as the state BIE (5) is solved, the right-hand side $\mathcal{F}(p,q)\theta$ of (12) is completely known. The successive solution of (14) for various $\theta$s may seem at first sight to be a formidable computational task. However, as the same integral operator governs the fields $p$ in (5) and $\overset{*}{p}$ in (12), the actual solution of (12) needs only, for each $\theta$, the building of its right-hand side $\mathcal{F}(p,q)\theta$ followed by a backsubstitution using the already factored matrix. In other words, each computation of the derivative needs the building and factorization of only one matrix.

The field $\overset{*}{p}_\Gamma$ appears clearly to be a linear form over $\theta$. Thus, with sufficiently regular $\Gamma$, $\theta$ and $p^I$, Riesz representation theorem may be used to state the existence of a kernel $\frac{\partial}{\partial\Gamma}p(\mathbf{x},\mathbf{y})$ (the derivative of $p$ with respect to $\Gamma$, such that:

$$\overset{*}{p}(\mathbf{x}) = \int_\Gamma \frac{\partial p}{\partial\Gamma}(\mathbf{x},\mathbf{y})\theta(\mathbf{y})\,dS_\mathbf{y} \tag{16}$$

hence the terminology of 'shape differentiation'. As a result, $dJ/d\tau$ itself is, from (11) and the previous remarks, a linear form over $\theta$, and the shape derivative kernel $\frac{\partial}{\partial\Gamma}J(\mathbf{y})$ may be defined as well.

The DDA has also been applied to Galerkin type BIEs in [8], for crack identification problems governed by Laplace's equation. As Galerkin-type BIEs are weakly singular (they are regularized by integrating by parts twice), the DDA approach does not raise difficulties.

## 4   The adjoint problem approach.

This is an alternative approach for deriving an analytical expression, which is known e.g. in the field of structural shape optimization [7] (see also [6] for thermal inverse problems). The problem (3) may be viewed as a constrained optimization problem, where $J(p,\Gamma)$ is to be minimized under the constraint $p = p_\Gamma$ (1). The latter can be expressed in weak form as:

$$\mathcal{A}(p,w;\Gamma) \equiv \int_{\Omega_\epsilon}(\nabla p \cdot \nabla w - k^2 pw)\,dV_\mathbf{y} + \int_\Gamma wp_{,n}^I\,dS_\mathbf{y} = 0 \quad \forall w \in \mathcal{V} \tag{17}$$

where $\mathcal{V} = \{w \in H^1(\Omega),\ w(\mathbf{y}) = 0\ (\mathbf{y} \in S)\}$. A lagrangian functional $\mathcal{L}$ is introduced:

$$\mathcal{L}(p,w,\Gamma) = J(p,\Gamma) + \mathcal{A}(p,w;\Gamma) \tag{18}$$

where $w$ is the Lagrange multiplier. Upon application of formulas (8), (9), taking into account identity (13) and using identity (34), the stationarity of $\mathcal{L}$ is expressed as

$$\frac{d}{d\tau}\mathcal{L} = \frac{\partial\mathcal{L}}{\partial p}p_{,\tau} + \frac{\partial\mathcal{L}}{\partial\Gamma}\theta \tag{19}$$

$$\frac{\partial\mathcal{L}}{\partial p}p_{,\tau} = \int_C p_{,\tau}(p - \hat{p})\,dC_\mathbf{y} + \int_{\Omega_\epsilon}(\nabla p_{,\tau} \cdot \nabla w - k^2 p_{,\tau}w)\,dV_\mathbf{y} \tag{20}$$

$$\frac{\partial\mathcal{L}}{\partial\Gamma}\theta = \int_\Gamma \theta\left[\nabla_S w \cdot \nabla_S(p + p^I) - k^2 w(p + p^I)\right]\,dS_\mathbf{y} \tag{21}$$

Now the choice of the Lagrange multiplier $w$ is restricted in such a way that $\frac{d}{d\tau}\mathcal{L} = 0$ for $\theta \equiv 0$, that is, we put:

$$\frac{\partial \mathcal{L}}{\partial p}p_{,\tau} = 0, \qquad \forall p_{,\tau} \in \mathcal{V} \tag{22}$$

This leads, in view of eqn. (20), to introduce the adjoint field $w_\Gamma$ as the solution of the following *adjoint problem*:

$$\begin{cases} (\Delta + k^2)w = -(p - \hat{p})\delta_C & \text{in } \Omega_e \\ w_{,n} = 0 & \text{on } \Gamma \\ \text{(radiation condition)} \end{cases} \tag{23}$$

The adjoint problem is equivalently formulated in terms of a BIE:

$$w(\mathbf{x}) + \int_\Gamma \left\{ w(\mathbf{y})n_s(\mathbf{y})[G_{,s}(\mathbf{x},\mathbf{y}) - G^0_{,s}(\mathbf{x},\mathbf{y})] + [w(\mathbf{y}) - w(\mathbf{x})]n_s(\mathbf{y})G^0_{,s}(\mathbf{x},\mathbf{y}) \right\} dS_{\mathbf{y}}$$

$$= \int_C [p(\mathbf{y}) - \hat{p}(\mathbf{y})]G(\mathbf{x},\mathbf{y}) \, dC_{\mathbf{y}} \tag{24}$$

Finally the *shape derivative* of $J$ is given by:

$$\frac{dJ}{d\tau} = \frac{d}{d\tau}\mathcal{L}(p_\Gamma, w_\Gamma, \Gamma) = \int_\Gamma \theta \left[ \nabla_S w_\Gamma \cdot \nabla_S (p_\Gamma + p^I) - k^2 w_\Gamma (p_\Gamma + p^I) \right] dS \tag{25}$$

Then, in order to evaluate the derivative $dJ/d\tau$, one has to

1. Solve the state BIE (5) for the unknown $p_\Gamma$.
2. Solve the *adjoint BIE* (24) associated to the adjoint problem (23) for the unknown $w_\Gamma$.
3. Insert $p_\Gamma$ and $w_\Gamma$ in $dJ/d\tau$ (25).

**Remark 1** The adjoint problem approach is not specifically BIE-oriented. Indeed, its establishment uses weak forms of boundary-value problems and is therefore equally well suited for FEM numerical treatments.

**Remark 2** The adjoint field $w$ does not depend on $\theta$. Therefore, the adjoint problem approach needs the solution of two distinct boundary-value problems.

**Remark 3** As for the DDA, the state and adjoint problems use the same integral operator. Therefore, each computation of the derivative needs the building and factorization of only one matrix.

## 5 Numerical implementation and examples.

Numerical tests for the solution of the inverse problem using shape differention have been conducted, for 3D situations. The basic numerical tool is the regularized collocation BIE (5), which is implemented in our BEM research code ASTRID. Isoparametric 8-noded curved surface elements were used throughout the present study, and $\Gamma$ is made of 24 elements, which amounts to 74 nodes. The incident wave $p^I(\mathbf{y})$ is here taken as a plane wave propagating along $Ox_3$ in the positive direction.

Here the unknown surface $\Gamma$ is searched as an ellipsoid. This choice reduces the inverse problem to the search of 9 design parameters $d_1, \ldots, d_9$ (coordinates $x_G, y_G, z_G$ of the center, Euler angles $\phi, \theta, \psi$ and principal axes $a, b, c$, in this order), it has been made in order to test the method on situations with moderate number of design parameters. In the next step (left for future work), mesh nodes coordinates will be taken as the design parameters. The following one-to-one mapping between $\Gamma$ and the unit sphere $\mathcal{S}$ is used:

$$\mathbf{Y} = (Y_1, Y_2, Y_3) \in \mathcal{S} \to \mathbf{y} = (y_1, y_2, y_3) \qquad \begin{cases} y_1 = x_G + r_{11}aY_1 + r_{12}bY_2 + r_{13}cY_3 \\ y_2 = y_G + r_{21}aY_1 + r_{22}bY_2 + r_{23}cY_3 \\ y_3 = y_G + r_{31}aY_1 + r_{32}bY_2 + r_{33}cY_3 \end{cases} \tag{26}$$

where $r_{ij} = r_{ij}(\phi, \theta, \psi)$ are the components of the rotation which thansforms the coordinate axes onto the principal axes of the ellipsoid (several $(\phi, \theta, \psi)$ triplets may define the same rotation). Hence, taking $\tau = d_i$, $\theta_i(\mathbf{y})$ is defined by considering an increment $\delta d_i$ added to $d_i$:

$$\delta \mathbf{y} = \frac{\partial \mathbf{y}}{\partial d_i} \delta d_i = \theta_i(\mathbf{y}) \delta d_i \tag{27}$$

To be more precise, the sphere $S$ is meshed using 24 elements (3 per octant) and the mesh of $\Gamma$ is defined by applying the mapping (26) between the *nodes* of $S$ and of $\Gamma$. Hence eqn. (27) defines the nodal values of $\theta$, the latter being is interpolated using the same shape functions as the geometry.

In all computations, the measurement surface $C$ was a sphere of radius 10 metres, centered at the origin, and the wave velocity was equal to unity. Two meshes of measurement points were used (respectively 24 elements, 74 points and 96 elements, 290 points). The elements (and the shape functions associated) were used for the computation of integrals like (4) or (25) over $C$. In some cases, measurement noise has been artificially introduced by multiplication of the data values by $1 + r$, $r$ being random numbers uniformly distributed in $[-\epsilon, \epsilon]$, with $\epsilon = 10^{-3}, 10^{-2}, 10^{-1}$).

Both conjugate gradient (CG) and BFGS variable metric algorithms have been applied to the minimization problem (3), using programs from [10]. All numerical computations have been done in double precision complex arithmetic, on HP-Apollo 400 type workstations.

Numerical results are presented below for four situations. Each of them is defined by the data of 'true' and initial values of $(d_1, \ldots, d_9)$, according to the table 1 below. In the last example (case 4), the 'true' obstacle is not an ellipsoid but a rectangular box of sides $2a$, $2b$, $2c$, in order to test the performance of the method on a case where the searched-for shape differs from the true one.

As the same ellipsoid can result from many combinations of Euler angles and permutations of principal axes, it is difficult to measure the accuracy of the identification of $\Gamma$ by means of a mere comparison of the identified parameters $d_i$ with those defining the 'true' $\Gamma$, used to compute the simulated data. Instead, the relative errors $e_V$, $e_A$, $e_I$ for the volume, boundary area and geometrical inertia tensor (with respect to the *fixed origin and axes* have been computed (the indicator $e_I$ being very sensitive to the orientation of $\Gamma$ in space), together with the ratio $J_n/J_0$, where $J_n = J(\Gamma_n)$, $\Gamma_n$ being the current $\Gamma$ after the $n$-th iteration of the minimization process. Our numerical values of $J_{final}/J_0$, $e_V$, $e_A$, $e_I$ obtained for the cases defined in table 1 are displayed in tables 2, 3, 4, 5.

**remark 1** Cases 1,2,3, where the 'true' cavity is also an ellipsoid, exhibit very good convergence and accuracy, especially for non-perturbed data.

**remark 2** The convergence and accuracy remain good for case 4, where the 'true' cavity is a rectangular box and exact convergence is hence impossible. The 'final' ellipsoid found by the algorithm has very similar volume and inertia tensor than the box (see table 5) and slightly different area. It even seems to be less sensitive to data noise.

**remark 3** At least in the range $\epsilon = 10^{-3}$ to $10^{-1}$], the error indicators $e_V$, $e_A$, $e_I$ vary linearly with $\epsilon$ in the results presented here, while $J_{final}/J_0$ vary quadratically. The numerical solution of the inverse problem hence behaves well with respect to measurement noise. This is probably a consequence of the strong assumption made on the unknown geometry, which is described using only 9 parameters.

**remark 4** Both CG and BFGS minimization methods have been tried, using library routines and without any optimization attempt. Neither of them seems to be significantly more accurate than the other: the final convergence and accuracy are similar. The CG method seems to exhibit slower convergence in the final stage.

# 6  Concluding comments.

A very strong assumption has been made here on the shape of Γ, and this is probably one of the reasons of the good convergence of the nonlinear optimization here. and of the fairly good behaviour with respect to data noise.

Hence, a crucial step in the numerical tests will be to relax this a priori information and to allow the search for more general surfaces, notably by using the node coordinates as the design variables. In this event, difficulties are expected:

- The general inverse problem is expected to be ill-posed (high sensitivity to data noise), and undesirable oscillations in the recovered shape of Γ are to be expected. Hence, in the formulations outlined in sections 4 and 3 above, the functional $J(\Gamma)$ may be replaced by $J(\Gamma) + \alpha\Omega(\Gamma)$, where $\alpha > 0$ and $\Omega(\Gamma)$ is a (positive) stabilizing functional (Tikhonov regularization [12]), in order to cater for the ill-posedness by adding qualitative a priori information. For instance, one may consider $\Omega(\Gamma) = \int_\Gamma (D_s n_s)^2 dS$, which is expected to damp oscillations. This achieves a trade-off between accuracy and stability with respect to data noise.

- Element distortion may occur as the minimization proceeds, and must be controlled.

The assumption made on Γ in the results presented here may indeed be viewed as a regularization, as some *selection* has been done in the *space of parameters* (i.e. of all possible Γ).

The work presented here Finally, let us mention that rate BIEs or the adjoint problem approach can be derived for elastodynamics as well; work is under progress.

# References

[1] Bonnet M. - Méthode des équations intégrales régularisées en élastodynamique tridimensionnelle. PhD thesis (Ecole Nationale des Ponts et Chaussées, Paris, France), Bulletin EDF/DER série C, n⁰ 1/2, 1987.

[2] Bonnet M. - Regularized Boundary Integral Equations for Three-dimensional Bounded or Unbounded Elastic Bodies Containing Curved Cracks of Arbitrary Shape Under Dynamic Loading. In "Boundary Element Techniques: Applications in Engineering (ed. C.A. Brebbia & N.G. Zamani), Computational Mechanics Publications (Southampton), 1989.

[3] Bonnet M., Bui H.D. - On Some Inverse Problems for Determining Volumic Defects by Electric Current Using B.I.E. Approaches : an Overview. 6th National Japonese Conf. on Boundary Elements Methods., Tokyo , December 1989.

[4] Bonnet M. - Shape differentiation of regularized BIE: application to 3-D crack analysis by the virtual crack extension approach. In "Boundary Elements in Mechanical and Electrical Engineering" (C.A. Brebbia and A. Chaudouet, eds.), Computational Mechanics Publications, Springer-Verlag, 1990.

[5] Bonnet M. and Bui H.D. - Regularization of the Displacement and Traction BIE for 3D Elastodynamics using indirect methods, To appear in: *Advances in Boundary Element Techniques*, J.H. Kane, G. Maier, N. Tosaka and S.N. Atluri, eds., Springer-Verlag, 1992.

[6] Bonnet M. and Bui H.D. - Identification of Heat Conduction Coefficient: Application to Nondestructive Testing. *IUTAM Symposium on Inverse Problems in Engineering Mechanics (Tokyo, 11-15 may 1992)*, H.D. Bui & M. Tanaka, eds., Springer-Verlag, 1992.

[7] Haug, Choi, Komkov - Design Sensitivity Analysis of Structural Systems, Academic Press, 1986.

[8] Nishimura N, Kobayashi S. - A boundary integral equation method for an inverse problem related to crack detection. Int. J. Num. Meth. Eng. **32**, 1371-1387, 1991.

198

[9] Petryk H., Mroz Z. - Time derivatives of integrals and functionals defined on varying volume and surface domains. Arch. Mech. vol. 38, n°5-6, pp 697-724, 1986.

[10] Press W.H., Flannery B.P., Teukolsky S.A., Vetterling W.T. - *Numerical recipes: the art of scientific computing.* Cambridge press, 1986.

[11] Stoer J., Bulirsch R. - *Introduction to numerical analysis.* Springer-Verlag, 1980.

[12] Tikhonov A.N., Arsenin V.Y. - Solutions to ill-posed problems. Winston-Wiley, New York, 1977.

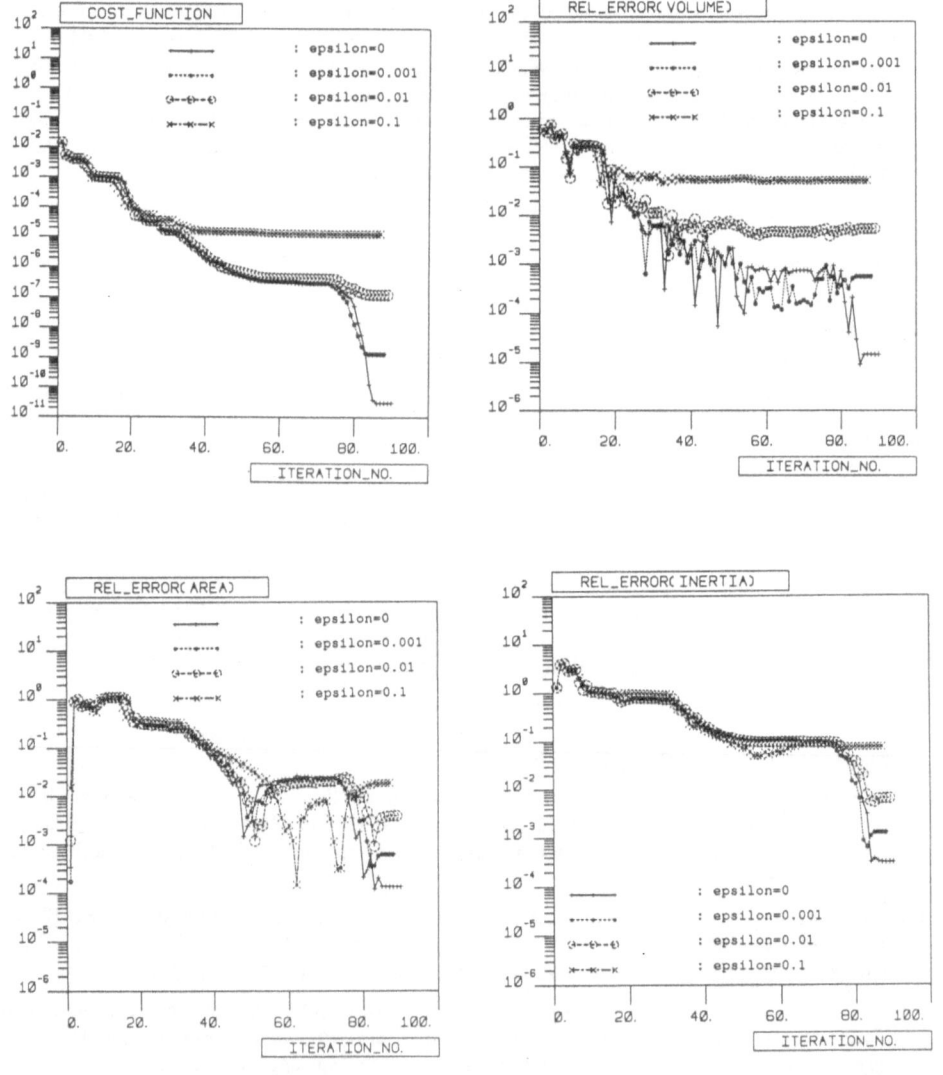

Figure 1: Convergence of $J_{final}/J_0$, $e_V$, $e_A$, $e_I$, for Case 2, with $\epsilon = 0.$, $10^{-3}$, $10^{-2}$, $10^{-1}$.

| | | $x_G$ | $y_G$ | $z_G$ | $\phi$ | $\theta$ | $\psi$ | $a$ | $b$ | $c$ |
|---|---|---|---|---|---|---|---|---|---|---|
| Case 1 | True | 0.0 | 0.0 | 0.0 | 0.0 | 0.0 | 0.0 | 1.0 | 2.0 | 1.0 |
| ($\omega/c = 1$) | Initial | 2.0 | $-1.0$ | 1.0 | 1.0 | 2.0 | 3.0 | 0.5 | 0.5 | 0.5 |
| Case 2 | True | 1.0 | 1.0 | 0.5 | 0.4 | 0.9 | 0.6 | 1.0 | 2.0 | 1.0 |
| ($\omega/c = .25$) | Initial | 0.0 | 0.0 | 0.0 | 0.0 | 0.0 | 0.0 | 5.0 | 5.0 | 5.0 |
| Case 3 | True | 1.0 | 1.0 | 0.5 | 0.4 | 0.9 | 0.6 | 1.0 | 2.0 | 1.0 |
| ($\omega/c = 1$) | Initial | 1.0 | 1.0 | 0.5 | 0.4 | 0.9 | 0.6 | 1.2 | 2.4 | 1.2 |
| Case 4 (box) | True | 1.0 | 0.0 | $-2.0$ | 0.4 | 0.9 | 0.6 | 1.0 | 3.0 | 1.0 |
| ($\omega/c = .5$) | Initial | 0.0 | 0.0 | 0.0 | 0.0 | 0.0 | 0.0 | 1.5 | 1.5 | 1.5 |

Table 1: Wavenumber values and 'true' and initial $(d_1, \ldots, d_9)$ used in the numerical calculations.

| Case 1 | $\epsilon = 0.$ | $\epsilon = 10^{-3}$ | $\epsilon = 10^{-2}$ | $\epsilon = 10^{-1}$ |
|---|---|---|---|---|
| $J_{final}/J_0$ | $3.04\,10^{-5}$ | $3.43\,10^{-6}$ | $1.35\,10^{-5}$ | $7.88\,10^{-4}$ |
| $e_V$ | $6.98\,10^{-4}$ | $1.28\,10^{-3}$ | $6.47\,10^{-3}$ | $5.80\,10^{-2}$ |
| $e_A$ | $5.44\,10^{-4}$ | $9.42\,10^{-4}$ | $4.59\,10^{-3}$ | $4.00\,10^{-2}$ |
| $e_I$ | $6.58\,10^{-3}$ | $6.78\,10^{-3}$ | $1.33\,10^{-2}$ | $1.08\,10^{-1}$ |

Table 2: Results for case 1 (convergence after about 40 CG iterations).

| Case 2 | $\epsilon = 0.$ | $\epsilon = 10^{-3}$ | $\epsilon = 10^{-2}$ | $\epsilon = 10^{-1}$ |
|---|---|---|---|---|
| $J_{final}/J_0$ | $1.57\,10^{-12}$ | $7.14\,10^{-11}$ | $6.57\,10^{-9}$ | $6.70\,10^{-7}$ |
| $e_V$ | $1.48\,10^{-5}$ | $5.65\,10^{-4}$ | $5.35\,10^{-3}$ | $5.31\,10^{-2}$ |
| $e_A$ | $1.34\,10^{-4}$ | $6.16\,10^{-4}$ | $3.87\,10^{-3}$ | $1.84\,10^{-2}$ |
| $e_I$ | $3.20\,10^{-4}$ | $1.29\,10^{-3}$ | $6.54\,10^{-4}$ | $7.73\,10^{-2}$ |

Table 3: Results for case 2 (convergence after about 90 BFGS iterations).

| Case 3 | $\epsilon = 0.$ | $\epsilon = 10^{-3}$ | $\epsilon = 10^{-2}$ | $\epsilon = 10^{-1}$ |
|---|---|---|---|---|
| $J_{final}/J_0$ | $4.78\,10^{-9}$ | $3.44\,10^{-7}$ | $3.45\,10^{-5}$ | $3.50\,10^{-3}$ |
| $e_V$ | $3.48\,10^{-6}$ | $5.47\,10^{-4}$ | $5.42\,10^{-3}$ | $5.43\,10^{-2}$ |
| $e_A$ | $7.61\,10^{-6}$ | $3.48\,10^{-4}$ | $3.41\,10^{-3}$ | $3.40\,10^{-2}$ |
| $e_I$ | $1.75\,10^{-5}$ | $7.82\,10^{-4}$ | $7.73\,10^{-3}$ | $7.82\,10^{-2}$ |

Table 4: Results for case 3 (convergence after about 25 CG iterations).

| Case 4 | $\epsilon = 0.$ | $\epsilon = 10^{-1}$ |
|---|---|---|
| $J_{final}/J_0$ | $2.79\,10^{-4}$ | $8.85\,10^{-4}$ |
| $e_V$ | $1.68\,10^{-2}$ | $3.66\,10^{-2}$ |
| $e_A$ | $1.51\,10^{-1}$ | $1.25\,10^{-1}$ |
| $e_I$ | $5.71\,10^{-2}$ | $2.37\,10^{-2}$ |

Table 5: Results for case 4 (convergence after 32 BFGS iterations).

# A  Tangential differential operators and integration by parts.

Let $S$ be a twice continuously differentiable *closed* $(C^2)$ surface, of unit normal $\mathbf{n}$ (open surfaces can be considered as well, see e.g. [5]). Consider a scalar field $u(\mathbf{y})$, $\mathbf{y} \in S$, which may be undefined outside $S$ (e.g. $u = n_i(\mathbf{y})$, $u = \theta(\mathbf{y})$). In this case, the cartesian derivatives $u_{,i}$ are generally meaningless, and one has to introduce tangential differential operators. The domain of definition of $u$ is extended in a neighbourhood $V$ of $S$ by introducing a continuation $\hat{u}$ of $u$ outside $S$ defined as: $\forall (\mathbf{y} \in V)$, $\hat{u}(\mathbf{y}) = u(P(\mathbf{y}))$, where $P(\mathbf{y})$ is the orthogonal projection of $\mathbf{y}$ onto $S$. Clearly the restriction of $\hat{u}$ to $S$ is equal to $u$. Moreover the normal derivative of $\hat{u}$ is equal to zero, i.e. the vector $\nabla\hat{u}$ is tangent to $S$; therefore it may be used to define the tangential gradient $\nabla_S u$ of the function $u$;

$$\nabla_S u = \nabla_S \hat{u} = \nabla\hat{u} \tag{28}$$

If $u$ is an arbitrary scalar function defined in $V$, one has, consistently with (28):

$$\nabla_S u = \nabla u - \mathbf{n}u_{,n} = \mathbf{e}_r \hat{D}_r u = \mathbf{e}_r(u_{,r} - n_r u_{,n}) \tag{29}$$

which defines the tangential partial derivatives $\hat{D}_r u$ (using the notation $(\cdot)_{,n} = \partial/\partial n(\cdot)$). In the following, the symbol (^) will be omitted, keeping in mind when necessary the extension. The operator

$$D_{rs}f = (n_r f_{,s} - n_s f_{,r}) \tag{30}$$

is also introduced. From (29), $D_{rs}f = n_r D_s f - n_s D_r f$: $D_{rs}f$ is a tangential differential operator.

An interesting consequence of (29) is the following identity for the Laplace operator:

$$\Delta u = u_{,nn} - 2Ku_{,n} + D_s D_s u \tag{31}$$
$$2K = -D_s n_s \tag{32}$$

The classical Stokes' identity for a vector field $\mathbf{U}$ defined over V reads:

$$\int_S \mathbf{n} \cdot rot(\mathbf{U})dS = 0 \tag{33}$$

Application of identity (33) to $\mathbf{U} = (\mathbf{n} \wedge \mathbf{e}_j)f$ and $\mathbf{U} = \mathbf{n} \wedge (\mathbf{e}_j \wedge \mathbf{n})f$ yields integration by parts identities associated to the tangential differential operators (29), (30):

$$\int_S (-n_r Kf + D_r f)\, dS_{\mathbf{y}} = 0 \tag{34}$$

$$\int_S D_{rs}u\, dS_{\mathbf{y}} = 0 \qquad \text{for any fixed pair } r, s, \quad r, s = 1, 2, 3 \tag{35}$$

Identity (35) is very interesting for BEM formulations: it allows integration by parts on surfaces using ordinary partial derivatives (i.e. without separation of tangential and normal derivatives), thanks to eqn. (30). It is a key tool for regularization techniques [6].

# B  Some auxiliary formulas for shape differentiation.

The following formulas hold (they can be found e.g. in [9]):

$$(u_{,s})^* = \overset{*}{u}_{,s} - u_{,n}\theta_{,s} - \theta D_r u D_s n_r \tag{36}$$
$$\overset{*}{\mathbf{n}} = -\nabla_S \theta = -D_r \theta \mathbf{e}_r \tag{37}$$

A combined application of (36) and (37) leads to:

$$(u_{,n})^* = \overset{*}{u}_{,n} - u_{,n}\theta_{,n} - D_s u D_s \theta \tag{38}$$

Moreover, if $u(\mathbf{y}, \tau) = u(\mathbf{y})$, then:

$$\overset{*}{u}(\mathbf{y}) = \theta(\mathbf{y})u_{,n}(\mathbf{y}) \tag{39}$$
$$(u_{,n})^* = \theta u_{,nn} - D_s u D_s \theta \tag{40}$$

In the particular case $u = p^I$, eqn. (40) holds and identity (31), combined with Helmholtz' equation, gives eqn. (13).

# Shape Determination of a Scattering Obstacle by Eigenfunction Expansion Embedding Integral Method

Takuo Fukui

Faculty of Engineering, Fukui University,
3-9-1 Bunkyo, Fukui 910, Japan

**Summary**

The inverse scattering problem for acoustic waves to recover the shape of a scatterer from the far-field pattern of the scattered field is concerned. A method for solving the problem by the embedding integral equation in eigenfunction expansion form is presented. The method is applied to sound soft obstacles, and recovered shapes are obtained numerically for some examples. The results show that the method recovers the obstacle shape well. The method of near field reproduction is also applied to hard obstacles.

## 1. Introduction

This paper is concerned with the inverse scattering problem for acoustic waves to find the shape of a scatterer from the knowledge of incident wave field and far-field pattern of scattered field. A combination of analytical and numerical techniques for solving the two-dimensional problem is proposed.

The embedding integral equation is conveniently used to formulate the solution method for shape determination problem[1] because it does not need a priori assumption of the boundary shape, which must be determined. Moreover, introducing an eigenfunction expansion form of the embedding integral equation,[2] the method is reduced to a Fourier series expression. The near field solution around the unknown scatterer is determined by the Fourier coefficients of the scattering amplitude.[3], [4]

In this paper, the near field solution representation is derived from the embedding integral equation in eigenfunction expansion form. The coefficients determined from the scattering amplitude data are modified by the regularization technique. The shape of soft scatterer is recovered by searching minimizing points for the near field wave amplitude. The similar scheme is proposed for hard scatterer. Numerical examples show the effectiveness of this method.

## 2. Shape Determination Problem

The inverse scattering problem is shown in Fig. 1. There is a scatterer whose boundary is $B$ in an infinite body. An incident wave $u^I$ impinges on the scatterer, and the scattering wave $u^S$ occurs.

On a large sphere (circle) $S$ of radius $a$ with its center inside of the scatterer, the scattering wave can be expressed by the scattering amplitude $f(\theta)$, which is a complex function of direction, i.e., the distance from the scatterer is eliminated. The problem is at first to reproduce the scattering wave solution near the scatterer from the scattering amplitude data and then to determine the scatterer's boundary shape $B$ on which the expected boundary condition is satisfied.

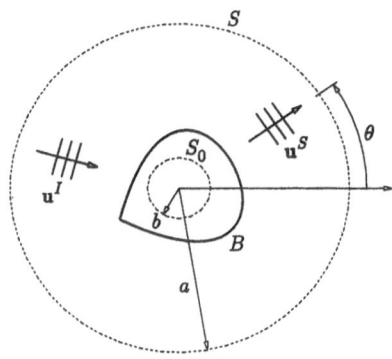

Fig. 1. Inverse scattering problem

## 3. Method of Inverse Analysis

### 3.1. Reproduction of near field solution

Two-dimensional acoustic problem is concerned in this paper. The governing equation is the Helmholtz equation

$$\nabla^2 u + k^2 u = 0 \tag{1}$$

where $\nabla^2$ is the Laplace operator and $k$ is the wave number. The total wave $u$ is the sum of the incident wave $u^I$ and scattering wave $u^S$, i.e., $u = u^I + u^S$. The scattering wave $u^S$ also satisfies the equation (1). Since the problem region is exterior, the scattering wave $u^S$ must satisfy the radiation condition

$$\frac{\partial u^S}{\partial r} - iku^S = o\left[\frac{1}{\sqrt{r}}\right] \qquad\qquad r \to \infty \tag{2}$$

where $r$ is the distance from the origin. The solution of (1) that satisfies the radiation condition (2) can be expressed by a simple layer potential due to a density $\sigma$ distributed on an auxiliary boundary $S_0$ that is provided inside of $B$ (see Fig. 1), i.e.,

$$u^S(\mathbf{x}) = \int_{S_0} \frac{i}{4} H_0^{(1)}(k|\mathbf{x}-\mathbf{y}|)\sigma(\mathbf{y})\,ds(\mathbf{y}) \tag{3}$$

where $H_n^{(1)}$ is the $n$-th order Hankel function of the first kind. Choosing the auxiliary boundary as a circle of radius $b$, the simple layer potential (3) can be expressed by an eigenfunction expansion form on the polar coordinate.[2] If the density $\sigma$ is expressed by Fourier series

$$\sigma(\mathbf{y}) = \frac{\tilde{A}_0}{2} + \sum_{m=1}^{\infty} \left[ \tilde{A}_m \cos m\theta_y + \tilde{B}_m \sin m\theta_y \right] \tag{4}$$

then the integral expression (3) becomes

$$u^S(\mathbf{x}) = \frac{i\pi}{2} \left[ \frac{A_0}{2} H_0^{(1)}(kr) + \sum_{n=1}^{\infty} \left[ A_n \cos n\theta + B_n \sin n\theta \right] H_n^{(1)}(kr) \right] \tag{5}$$

where the polar coordinates $(r,\theta)$ and $(b,\theta_y)$ represent the points $\mathbf{x}$ and $\mathbf{y}$, respectively. The coefficients $A_n = \tilde{A}_n b J_n(kb)$ and $B_n = \tilde{B}_n b J_n(kb)$ are introduced instead of $\tilde{A}_n$ and $\tilde{B}_n$ in (4). $J_n$ is the $n$-th order Bessel function.

From the radiation condition (2), the far field scattering wave is expressed by

$$u^S(\mathbf{x}) = \frac{e^{ikr}}{\sqrt{r}} \left[ f(\theta) + O\left(\frac{1}{r}\right) \right] \qquad r \to \infty \tag{6}$$

where $f(\theta)$ is the scattering amplitude. Comparing (6) and the far field form of (5), the scattering amplitude $f(\theta)$ is expressed as an eigenfunction expansion form

$$f(\theta) = i\sqrt{\frac{\pi}{2k}} e^{-i\frac{\pi}{4}} \left[ \frac{A_0}{2} + \sum_{n=1}^{\infty} e^{-i\frac{n\pi}{2}} \left[ A_n \cos n\theta + B_n \sin n\theta \right] \right] \tag{7}$$

Thus, the near field solution can be reproduced by determining the unknown coefficients of (5) as the Fourier coefficients of the scattering amplitude $f(\theta)$, $i.e.$,

$$A_n = \frac{C_n}{\pi} \int_0^{2\pi} f(\theta) \cos n\theta \, d\theta, \qquad B_n = \frac{C_n}{\pi} \int_0^{2\pi} f(\theta) \sin n\theta \, d\theta \tag{8}$$

where

$$C_n \equiv -i\sqrt{\frac{2k}{\pi}} e^{i\frac{(2n+1)\pi}{4}} \tag{9}$$

Although the near field expression (5) is derived from the embedding integral equation method, the same expression can be derived directly from scattering field consideration.[5] However, the embedding integral expression will be essential in regularizing the solution as shown in the following section.

## 3.2. Regularization

The coefficients (8) with (9) must converge faster than the usual Fourier coefficients because the Fourier coefficients of (5), $i.e.$,

$$\frac{i\pi}{2} H_n^{(1)}(kr) A_n \quad \text{and} \quad \frac{i\pi}{2} H_n^{(1)}(kr) B_n,$$

must converge (at least when $r$ is larger than a certain value). However, the coefficients numerically obtained from (8) may include round-off error, and then the scattering solution (5) may not converge. This fact corresponds to the irregularity of the integral equation (3).[1] To improve the convergence property of (8), a regularization scheme is introduced as follows.

Denote the right hand side term of (7) by $F(\theta;\mathbf{A})$ where $\mathbf{A} = \{A_0, A_1, B_1, A_2, B_2, \cdots \}$, then the Fourier coefficients of (8) are obtained by minimizing the functional

$$P(\mathbf{A}) = \| f(\theta) - F(\theta;\mathbf{A}) \|_{L^2}^2 \tag{10}$$

To improve the solution (8), instead of the functional (10), we introduce a new functional

$$P_\alpha(\mathbf{A}) = \| f(\theta) - F(\theta;\mathbf{A}) \|_{L^2}^2 + \alpha \| \sigma(\theta_y;\mathbf{A}) \|_{L^2}^2 \tag{11}$$

where $\alpha$ is a small parameter to control the improvement. Minimizing the functional (11), new Fourier coefficients are obtained. The result is given by the same expression (8), but $C_n$ is exchanged by new form

$$C_n = \frac{-i\sqrt{\dfrac{\pi}{2k}}\, e^{i\frac{(2n+1)\pi}{4}}}{\dfrac{\pi}{2k} + \dfrac{\alpha}{b^2 J_n^2(kb)}} \tag{12}$$

The functional (11) is the Tikhonov functional[1] in Fourier expansion form. The parameter $\alpha$ has the dimension of length. The limiting value of the coefficient $C_n$ in (12) as $\alpha \to 0$ is equal to (9). The radius of the auxiliary boundary $b$ appears in (12), but it is only a control parameter in this expression.

### 3.3. Determination of boundary shape

The shape of the scattering obstacle can be determined if the near field scattering solution (5) is obtained.

For a sound soft scatterer, the total wave is vanished on the boundary, i.e., $u = u^I + u^S = 0$. The boundary can be determined to search out the points where $|u|^2 = |u^I + u^S|^2$ is minimized. A simple scheme to do this is to search the radius $r$ for each fixed $\theta$ such that

$$\left. \frac{\partial |u|}{\partial r} \right|_\theta = 0 \tag{13}$$

This method is applicable only to convex scatterer whose shape is not so much distorted from a circle. Still it is useful to catch a brief image of the scatterer's shape.

For a hard scatterer, the normal derivative of the total wave is vanished on the boundary, *i.e.*, $\partial u/\partial n = 0$. In this case, there is difficulty in determining the boundary shape because the boundary condition includes the information about boundary shape itself. To avoid this difficulty, a necessary condition $\nabla(\mathrm{Re}\,u) = \nabla(\mathrm{Im}\,u)$ can be used. If we introduce a function

$$\phi = [\nabla(\mathrm{Re}\,u) \times \nabla(\mathrm{Im}\,u)] \cdot \boldsymbol{k}, \tag{14}$$

$\phi = 0$ must be satisfied on the boundary. Thus, we can use the similar way as (13) to search the boundary points.

## 4. Numerical Examples

The method is applied to some simple shapes of scatterers: circular, elliptic, and rectangular soft scatterers, and circular hard scatterer. For the hard scatterer, only the near field reproduction is shown. In the examples, 128 discrete point data of the scattering amplitude computed by the boundary element method are used.

### 4.1. Soft scatterer

### (1) Circular obstacle

If the boundary is a circle, the expression (5) gives the correct solution. Fig. 2 shows the reference BEM solution and the reproduced near field of wave amplitude $|u|$ around a circular obstacle of radius $a$. The wave number is $ka = 1$. The incident wave propagates from left to right along the horizontal axis. The terms up to $n = 7$ in the equation (5) are used, *i.e.*, the reproduction is made of 15 coefficients calculated by (8) and (9) numerically.

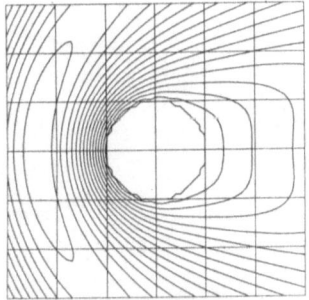 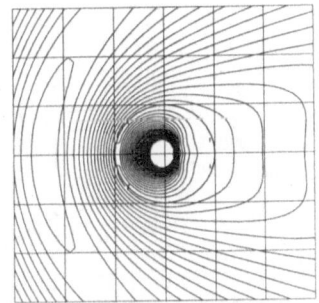

(a) Reference BEM solution      (b) Reproduced wave field: $n = 7$

Fig. 2.   Distribution of wave amplitude $|u|$ around a circular soft scatterer with radius $a$. Wave number is $ka = 1$.

The figures show that the reproduced wave field is close to the reference field and that the estimated boundary is correct. Note that there is a blank area inside of the scatterer in the figure of the reproduced field (b). In this blank area, the value of wave amplitude of the reproduced field is larger than a certain large value (2.0 in this figure), *i.e.*, the reproduced field (5) may diverge in the area. Even though the correct value of radius of which inside the expression (5) may diverge is not known a priori, this blank area suggests the approximate size of it.

### (2) Elliptic obstacle

An elliptic soft scatterer of the longer radius $1.5a$ and the shorter radius $a$ is examined. The axis of longer radius is inclined 60° from the horizontal axis that is the incident wave propagation direction. The reference BEM solution around the scatterer is shown in Fig. 3.

Fig. 4 shows the reproduced near field solution by (5) with the coefficient calculated by (8) and (9) of the original formulation. The figure shows the dependence of the reproduced field on the number of used terms. The larger the number of terms, the bigger the blank area in which the reproduced field may diverge.

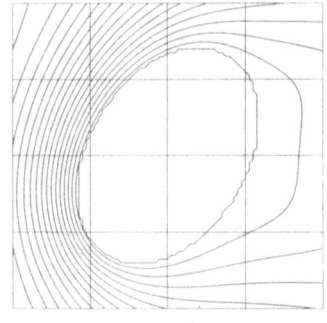

Fig. 3. Reference BEM solution of wave field $|u|$ around an elliptic soft scatterer: longer radius = $1.5a$, shorter radius = $a$, and wave number $ka = 1$.

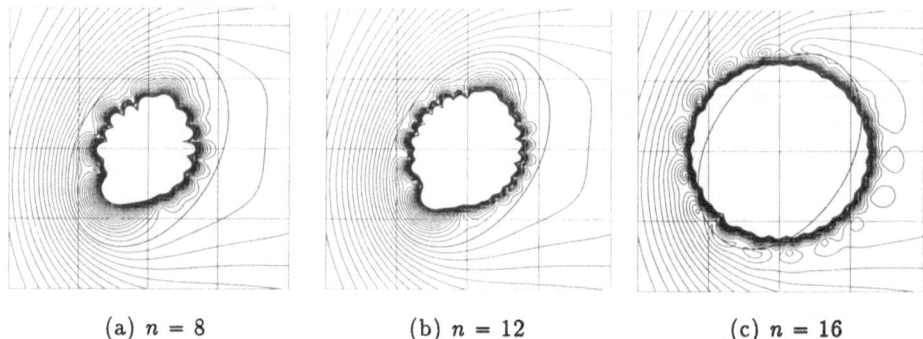

(a) $n = 8$      (b) $n = 12$      (c) $n = 16$

Fig. 4. Reproduced wave field $|u|$ around the elliptic scatterer (drawn in the figure). Wave number $ka = 1$.

Therefore, the near field solution (5) may not converge around the scatterer. This is caused by round-off error in numerical calculation of coefficient (8). Because the absolute value of the coefficient is very small for large number $n$, the magnitude of round-off error exceeds the actual value when $n$ is larger than a certain value.

Such poor results are improved by introducing the coefficient (12) instead of (9) as shown in Fig.5. The figure shows the same reproductions obtained by (12) with the parameter $\alpha = 10^{-14}a$ and the radius of auxiliary boundary $b = 0.8a$. By comparison with Fig.4, the effect of the modification in Fig.5 is clear.

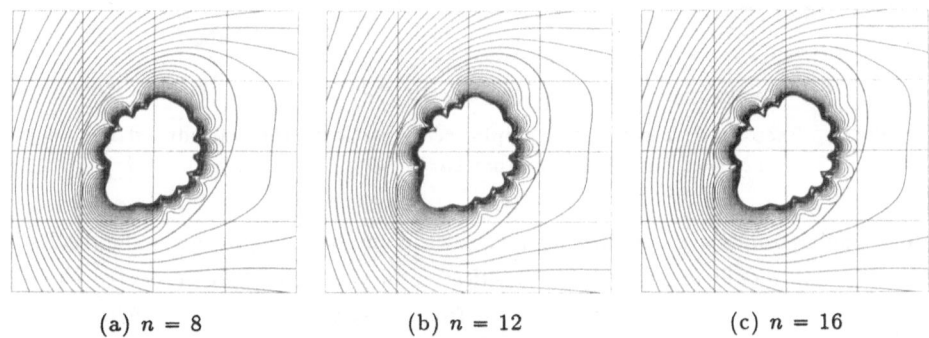

(a) $n = 8$ (b) $n = 12$ (c) $n = 16$

Fig.5. Reproduced wave field $|u|$ by the modified Fourier coefficients with $\alpha = 10^{-14}a$, $b = 0.8a$. Wave number $ka = 1$.

Fig.6 shows the effect of the magnitude of the parameter $\alpha$ on the reproduced field. The larger the parameter $\alpha$, the smaller the size of the blank area. However, for large parameter $\alpha$, the reproduced near field is distorted somehow.

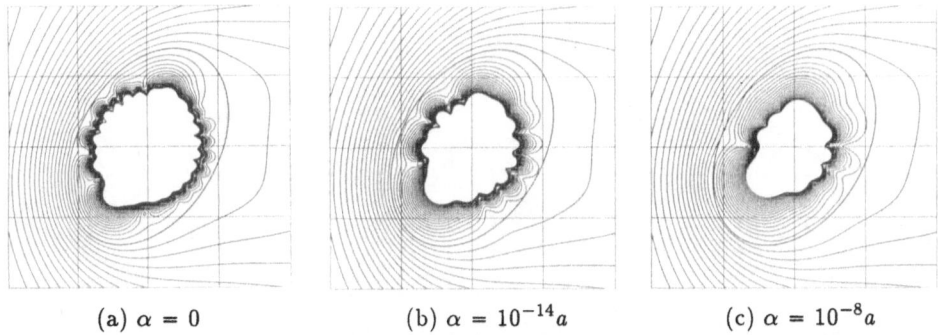

(a) $\alpha = 0$ (b) $\alpha = 10^{-14}a$ (c) $\alpha = 10^{-8}a$

Fig.6. Effect of the parameter $\alpha$ on the reproduced wave field $|u|$: $n = 12$, $b = 0.8a$. Wave number $ka = 1$.

Fig.7 shows the recovered shape of the scatterer boundary from the reproduced near field shown in Fig.6 determined by the condition (13). These figures recover the original shape well. Although the detail of the shape depends on boundary decision scheme, $\alpha = 0$ case is the best in these three figures. In other cases, the shape is rather smooth but distorted from the original shape.

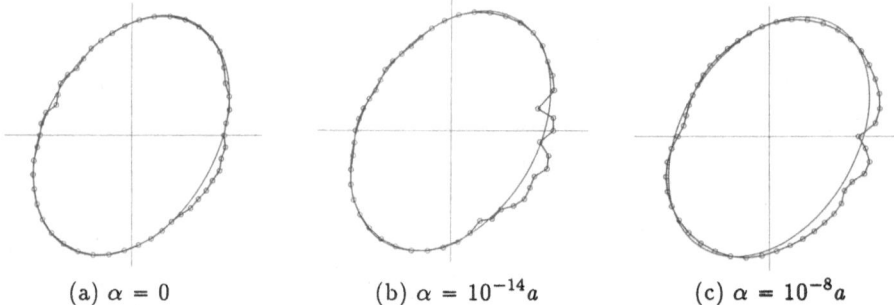

(a) $\alpha = 0$        (b) $\alpha = 10^{-14}a$        (c) $\alpha = 10^{-8}a$

Fig. 7.  Recovered shape of the elliptic scatterer from the reproduced field $|u|$: $n = 12$, $b = 0.8a$.  Wave number $ka = 1$.

The similar results are obtained for higher wave numbers.  Fig. 8 shows the results for $ka = 4$.  The modification (12) is still effective in this case.  The recovered shape is more distorted at back side, although the front side shape is recovered well.

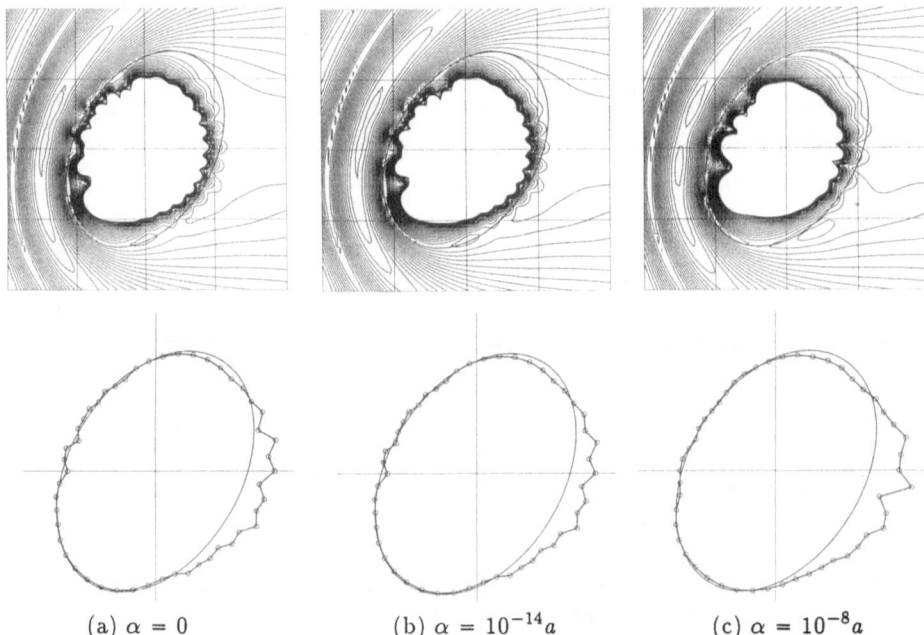

(a) $\alpha = 0$        (b) $\alpha = 10^{-14}a$        (c) $\alpha = 10^{-8}a$

Fig. 8.  Reproduced field $|u|$ (upper figure) and recovered shape (lower figure) of the elliptic scatterer for wave number $ka = 4$.  $n = 12$, $b = 0.8a$.

### (3) Rectangular obstacle

A rectangular soft scatterer of the longer side 2.5*a* and the shorter side 2*a* is examined here. The reference BEM solution is shown in Fig. 9. The longer axis is inclined 30° from the horizontal axis that is the propagating direction of the incident wave.

Fig. 10 shows the reproduced near field solution by (5) with the coefficients (8) and (9) of the original formulation in the same manner as Fig. 4. Because the shape is more complex in this case, the blank areas are larger than those of the elliptic case: even in $n = 12$ case the blank area is exceeding the actual boundary.

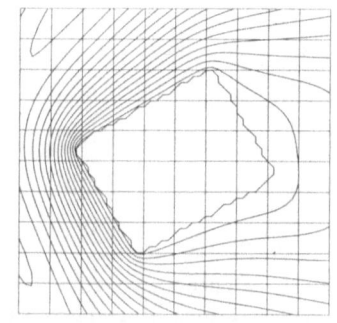

Fig. 9. Reference BEM solution of wave field $|u|$ around a rectangular soft scatterer: longer side = 2.5*a*, shorter side = 2*a*, and wave number $ka = 1$.

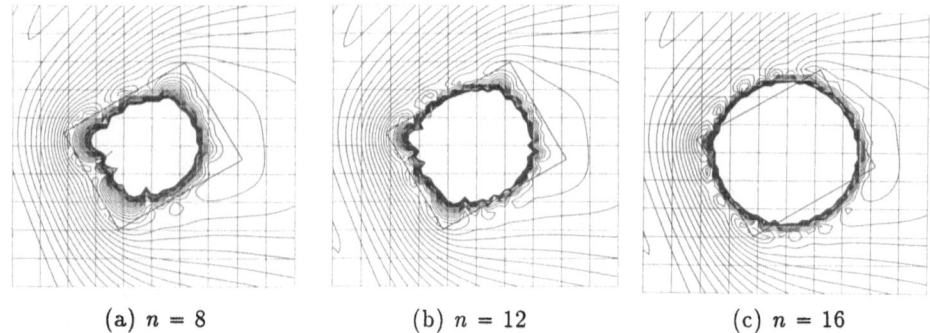

(a) $n = 8$      (b) $n = 12$      (c) $n = 16$

Fig. 10. Reproduced wave field $|u|$ around the rectangular scatterer (drawn in the figure). Wave number $ka = 1$.

Fig. 11 shows the effect of the modification on the reproduced near field. The modification is more effective than the elliptic obstacle case.

Fig. 12 shows the recovered shapes of variously inclined rectangular obstacles. The upper figures are obtained from the reproductions by (9), *i.e.*, $\alpha = 0$, and the lower figures are obtained by (12) with $\alpha = 10^{-14}a$ and $b = 0.8a$. Although the shape is distorted in detail, the outline of shape is good. The shape of the front side and the shorter side is better than that of the back side and the longer side. The modification (12) improves smoothness of the shape especially at the front side.

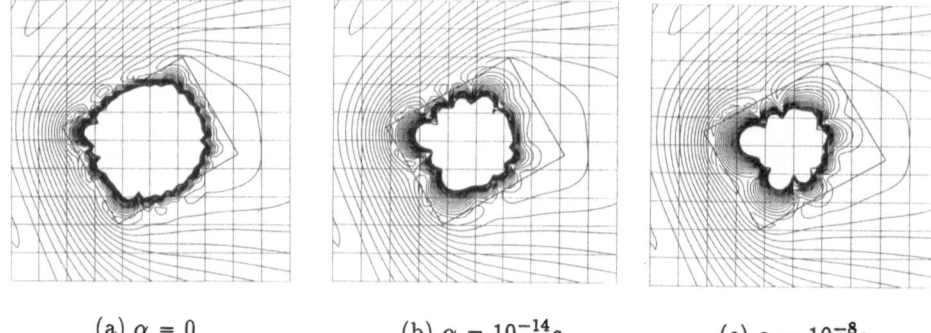

(a) $\alpha = 0$         (b) $\alpha = 10^{-14}a$         (c) $\alpha = 10^{-8}a$

Fig. 11.  Reproduced wave field $|u|$ by the modified Fourier coefficients with various $\alpha$: $n = 12$ and $b = 0.8a$.  Wave number $ka = 1$.

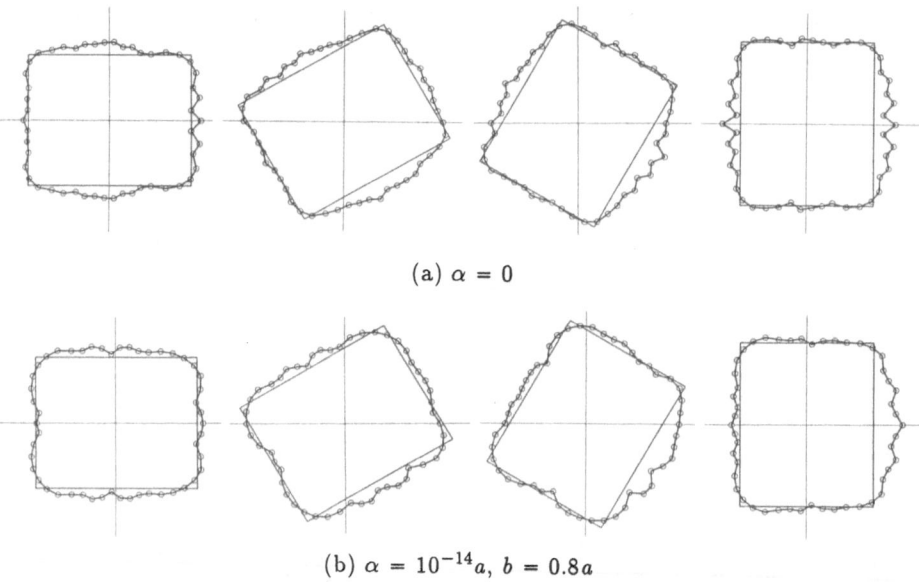

(a) $\alpha = 0$

(b) $\alpha = 10^{-14}a$, $b = 0.8a$

Fig. 12.  Estimated shape of the rectangular scatterers from the reproduced field $|u|$.  The longer axis is inclined $0°$, $30°$, $60°$, and $90°$ from the horizontal axis along which the incident wave propagates:  $n = 12$ and wave number $ka = 1$.

## 4.2.  Hard scatterer

Fig. 13 shows the reproduced near field solutions and their gradient direction field around a circular hard scatterer of radius $a$. The wave number is $ka = 1$. The direction of the incident wave is the same as the previous cases. The actual boundary shape is also drawn in the figures. It is shown in the lower figures that the directions of both the real and imaginary gradient fields on the boundary coincide with the tangential direction of the boundary. Therefore, the directions of the real and imaginary gradient fields are the same on the boundary. This fact suggests the validity of the boundary searching scheme proposed in the section **3.3**.

 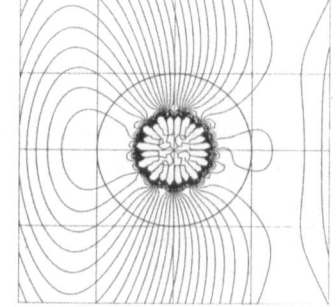

(a) Reproduced near field: real part (left) and imaginary part (right).

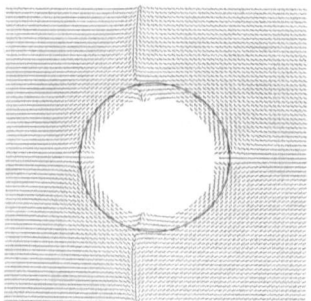 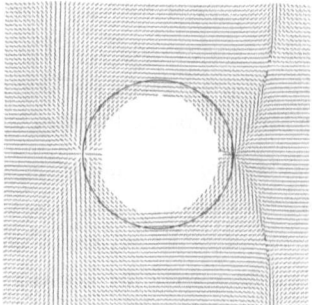

(b) Gradient direction field: real part (left) and imaginary part (right).

Fig. 13.   Reproduced wave field and the gradient direction field around a circular hard scatterer of radius $a$: $n = 12$ and wave number $ka = 1$.

Fig. 14 shows the distribution of the estimation function $\phi$ defined by (14). The contour lines show that the function value vanishes on the boundary, but that it also vanishes on other lines that are not on the boundary. Moreover, there are singular points (6 on the boundary in this case) at which two zero contour lines intersect each other. Thus, the proposed method of boundary determination may catch the

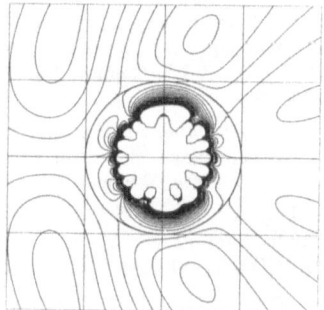

Fig. 14. Function $\phi$

actual boundary as well as other lines on which the gradient directions of the real and imaginary field coincide each other. To remove the false lines, the additional condition that the gradient direction of the wave field and the tangential direction of the boundary coincide must be used.

## 5. Conclusions

A simple numerical method for solving the two-dimensional acoustic inverse scattering problem to find the shape of a scatterer from the knowledge of incident wave field and far-field pattern of scattered field is proposed. The method is derived from the embedding integral equation in eigenfunction expansion form. The shape determination analysis is made of two steps. At first the near field solution is reproduced by determining the Fourier coefficients from the scattering amplitude data. Second the shape of scatterer is recovered by searching minimizing points of the estimation function derived from the near field.

The method was applied to some examples. The near field solution is reproduced well by the modified Fourier coefficients derived from the regularization technique. The method recovers sound soft scatterer's shape well. For hard scatterers, an estimation function to search the boundary point is proposed without numerical testing. To apply this estimation function, an additional condition is needed.

The advantage of this method is simplicity. It can be applied easily, and it needs only a little computing time. It can be applicable only to simple shape; however, it is workable to catch the outline of shape. To determine the detail of shape, elaborate iterative method would be needed. Even in the case, this method may give the first guess of the shape.

### References

(1) Kirsch, A.; Kress, R.: An optimization method in inverse acoustic scattering, *Boundary Elements IX*, eds. C.A. Brebbia, W.L. Wendland and G. Kuhn, Vol.3 Fluid Flow and Potential Applications, Springer-Verlag (1987) 3-18.

(2) Shaw, R.P.; Huang, S.C.: Element and eigenfunction expansion forms for an embedding integral equation approach to acoustics, *Advances in Boundary Elements*, eds. C.A. Brebbia and J.J. Connor, Vol. 2 Field and Fluid Flow Solutions, Springer-Verlag (1989) 319-339.

(3) Fukui, T.; Oono, S.: An application of the auxiliary boundary integral equation method to defect shape determination problems, *Proceedings of The Symposium on Non-Destructive Evaluation in Civil Engineering*, JSCE (1991) 57-60 (in Japanese).

(4) Fukui, T.; Oono, S.: Shape determination of an obstacle by embedding method, *Proceedings of the Eighth Japan National Symposium on Boundary Element Methods*, JASCOM (1991) 205-210 (in Japanese).

(5) Imbriale, W.A.; Mittra, R.: The two-dimensional inverse scattering problem, *IEEE Trans. Antennas and Propagations*, **AP-18** (1970) 633-642.

# Identification of Crack Shape Hidden in Solid by Means of Neural Network and Computational Mechanics

G. Yagawa*, S. Yoshimura*, Y. Mochizuki** and T. Oishi**

* Department of Nuclear Engineering, University of Tokyo
  7-3-1 Hongo, Bunkyo, Tokyo 113, Japan
** Graduate School of University of Tokyo
  7-3-1 Hongo, Bunkyo, Tokyo 113, Japan

Summary

This paper describes an application of the hierarchical neural network to the identification of a crack hidden in solid using the electric potential drop method. The present method consists of three subprocesses. First, a number of sample data of identification parameters vs. electric potential values are calculated by the finite element method. Second, the error-back-propagation neural network is trained using the sample data. Finally, the trained network is utilized for crack identification.

The present method is first applied to the identification of a two-dimensional crack in plate, and then applied to that of a semi-elliptical surface crack in plate. The accuracy and efficiency of the method are discussed in detail.

## Introduction

Nondestructive evaluation (NDE) techniques to determine sizes and locations of cracks and defects hidden in solid are of very importance to assure the structural integrity of operating plants and structures and to evaluate their residual life time. So far various NDE techniques using ultrasonic wave, X-ray, magnetic powder, eddy current and so on have been studied in view of accuracy, cost and applicability. Among various researches, much attention has been paid to the inverse analysis approaches based on the computational mechanics combined with kinds of optimization techniques. These approaches have the following advantages and disadvantages :

(a) These are good at quantitative evaluation.
(b) These are hardly applied to multiple local minima problems.
(c) These require a number of computational mechanics simulations in
    application processes of identification.

The item (b) is inevitable in conventional optimization methods.

Recently neural networks derived through modeling of human brain have attracted much attention [1, 2]. Among neural networks, the hierarchical networks have been applied to various mapping problems, i.e. qualitative mappings such as character recognition and quantitative mappings such as robotic control. These networks have been gradually applied to some structural mechanics problems including crack detection using ultrasonic data and eddy current data [3-12].

The attractive features of the hierarchical neural networks can be summarized as follows :

(a) One can automatically construct a nonlinear mapping from multiple input data to multiple output data in the network through a learning process of some or many learned patterns.
(b) The network has a feature of so-called "generalization", i.e. a kind of interpolation, such that the trained neural network estimates appropriate output data even for unlearned patterns.
(c) The trained network operates quickly in an application process. The CPU power required for the operation of the trained network may be equivalent to only that of a personal computer.

The present paper describes an inverse analysis approach based on the hierarchical neural network and the computational mechanics. Then its fundamental performances are clearly demonstrated through the identification of a crack hidden in plate using the electric potential drop method.

Network Architecture

Figure 1 shows a processing unit of the neural network which has multipe input and a single output data. The relation between the input and the output data is formulated as follows [1, 2] :

$$O_j = f(U_j) = 1 / \{1 + \exp(-2U_j / U_0)\} \qquad (1)$$

$$U_j = \sum_{i=1}^{l} W_{ji} \cdot I_i - \theta_j \qquad (2)$$

where $O_j$ is the output of the j-th unit, $U_j$ is the weighted input to the j-th unit, f is the activation function, i.e. the sigmoid function here, $U_0$ is the temperature constant, $W_{ji}$ is the connection weight between the i-th and the j-th units, $I_i$ is the input from the i-th to the j-th units, $\theta_j$ is the bias of the j-th unit, and l is the number of input data, respectively.

The hierarchical neural network consists of multiple layers, each of which posesses a number of processing units. The basic idea of training the network is as follows. At first the following error E is defined :

$$E = E_p = \sum_{k=1}^{n} \frac{1}{2}(T_{pk} - O_{pk})^2 \qquad (3)$$

where $E_p$ is the error for the p-th learned pattern, $T_{pk}$ is the teaching data corresponding to the k-th unit for the p-th learned pattern, $O_{pk}$ is the output from the k-th unit for the p-th learned pattern, and n is the number of learned patterns, respectively. In the training process, the connection weights $W_{ji}$ and the bias values $\theta_j$ are modified iteratively based on the steepest gradient method to minimize the above error. Through the training, the network attains the ability of outputting the similar data to the teaching data. This training algorithm is called the error back propagation [1, 2]. In the present study, the error back propagation algorithm with the moment method is employed to attain stable convergence in training processes.

It is theoretically proven that the hierarchical neural network can approximate any kinds of continuous mappings [13]. However, the network has some limitation in reality due to poor convergence in learning processes when the numbers of units and layers increase.

Procedure of Inverse Analyses
In the electric potential drop method, crack parameters such as size and location of a crack are determined from electric potential values measured at various points. This is one of inverse problems. On the other hand, if crack parameters are given, an electric potential distribution can be easily calculated using computational mechanics simulations. This is a direct problem. In the present study, we simply regard these direct and inverse problems to be nonlinear mapping problems from multiple data, say electric potential values, to multiple data, i.e. crack size and location. Thus, the following inverse analysis procedure using the hierarchical neural network is considered.

As shown in Fig. 2, this procedure consists of three subprocesses. At first , sample data of crack parameters, i.e. crack size, location and rotational angle, vs. electric potential values are calculated through computational mechanics simulations for many combinations of the crack parameters. Each piecewise relation between the crack parameters and the electric potential values is called here "learned pattern". Next an error-back-propagation neural network is trained using a number of learned patterns. Here, the crack parameters are given to the network as teaching data, while the electric potential values are given to it as input data. As a result, the trained network can identify crack parameters from measured electric potential values. This inverse analysis approach can be applied to any inverse problems by changing computational mechanics simulations [6, 12].

## Identification of 2D Crack

### Problem

Figure 3 shows a plate with a through-wall crack whose top and bottom edges are subjected to fixed potential values of +V and -V, respectively. By measuring electric potential values around the crack, the following four crack parameters are identified : X, Y (crack location), L (a half of crack length), θ (rotational angle).

### 6-point measurement case

At first, the four crack parameters are identified by using electric potential values measured at 6 points of A to F as shown in Fig. 3. Table 1 shows a range of each parameter and the number of its samples chosen for network training. For example, the parameter X is varied from 6 to 10, and 3 smaples, that is, 6, 8 and 10 are chosen from the range. As a result, 98 learned patterns are utilized to train the network. The neural network used is of three-layered type, and its input and output layers have six and four units, respectively.

Table 2 shows mean errors of estimation for learned patterns and unlearned patterns when the network with 30 units in its hidden layer is trained through 30,000 iterations. The definition of the mean error of estimation is as follows :

$$\text{Mean\_Error} = \frac{1}{r} \sum_{p=1}^{r} \left| O^p_{neuro} - O^p_{true} \right| \tag{4}$$

where $O^p{}_{neuro}$ is the output of the network for the p-th pattern, and $O^p{}_{true}$ is a true value for the p-th pattern, and r is the number of learned patterns or that of unlearned patterns. It should be noted here that input and output data of the neural network are normalized into a unit range of [0, 1]. Estimation errors for unlearned patterns are examined to check a feature of generalization of the network. Figure 4 shows some comparisons between actual crack shapes and estimated ones. Table 2 clearly shows that estimation errors in X, Y (crack location) and L (a half of crack length) are smaller than that in θ (rotational angle ). However, it can be confirmed from Fig. 4 that the trained network can estimate crack shapes with sufficient accuracy. It is in general considered that estimation errors for learned patterns are smaller than those for unlearned patterns. However, this is not true in the case of parameter X in Table 2. We cannot find any reasonable explanation for this result yet. Anyway, it should be noted that estimation accuracy is different parameter by parameter in the case of identification of multiple parameters.

Table 3 shows the mean errors of estimation for leanred patterns when four kinds of networks with 20, 30, 40 and 50 units in a hidden layer are trained through 30,000 iterations. When the number of units in a hidden layer increases from 20 to 30, estimation accuracy is improved for all the parameters.

When increasing from 30 to 40 or 50, this is not true. It should be also noted here that each parameter might have the different optimal number of units in a hidden layer.

## 40-point measurement case

The four crack parameters are identified from electric potential values measured at 40 points along the lines AB, BD, DC and CA. The number of samples and ranges of the four parameters employed for network training are the same as in Table 1. Units in a hidden layer are set to be either 20 or 30. Table 4 shows the mean errors of estimation for learned patterns after training the network through 4,000 iterations. For the purpose of comparison, the table also shows the estimation errors in the case of 6-point measurement. The table clearly shows that estimation accuracy is improved by increasing the number of electric potential values. To compare the efficiency of the two cases, we examine the following ratio :

Ratio of total numbers of renewals of W and $\theta$

$$= \frac{\text{Network scale used for 40-point measurement} \times \text{No. of iterations}}{\text{Network scale used for 6-point measuremet} \times \text{No. of iterations}} \quad (5)$$

$$= 0.582$$

where the number of units in a hidden layer for both cases is 30. The scale of network which is directly related to the CPU time consumed in one renewal of the network is defined here as follows :

Network scale = No.of connections between output and hidden layers
+ No. of connections between hidden and input layers
+ No. of output units
+ No. of hidden units     (6)

This result implies that the increase of electric potential values leads the increase of the network scale, but the number of training iterations reduces, and finally the training time reduces. This may be caused due to the improvement of the ill-poseness of the present inverse problem due to the increase of measurement points.

## Identification of 3D crack

Next we apply the present method to the identification of a semi-elliptical surface crack in plate using electric potential values measured at its backside surface as shown in Fig. 5. Electric potential values are measured at 11 x 11 (=121) grid points. For the purpose of the simple comparison between the previous 2D crack problem and the present 3D one, we fix the aspect ratio of the surface crack L/D to be two. Then, the four crack parameters of (X, Y, L, θ) are again identified. Learned patterns chosen are the same as in the previous 2D crack problem, i.e Table 1. The three-layered and the four-layered networks are utilized. The activation functions used in the output layer are the sigmoid function and the linear function. Table 5 shows the mean errors of estimation corresponding to the following three kinds of networks.

(a) Three-layered network with the sigmoid function in the output layer
(b) Three-layered network with the linear function in the output layer
(c) Four-layered network with the linear function in the output layer

The errors given in the table show the average error of the four parameters. All the networks are trained until the errors for unlearned patterns reach the minimum values. The final numbers of iterations for the three networks are 5,000, 5,000 and 2,000, respectively.

As for three-layered networks, the utilization of the linear activation function results in slight reduction of estimation error for unlearned patterns but in slight increase of that for learned patterns. Since in this problem crack parameters are output as an analog signal ranging from 0 to 1, the linear activation function seems suitable than the sigmoid activation function. However, Table 6 does not show such an effect.

Increase of hidden layers results in the improvement of estimation accuracy for both learned and unlearned patterns. However, the five-layered network cannot attain sufficient convergence of error of estimation for unlearned patterns.

Table 6 shows estimation errors for four parameters when using the four-layerd network with the linear activation function. Figure 6 shows some comparisons between actual crack shapes and estimated shapes. Comaperd with the results of 2D crack as shown in Table 4, estimation errors increase in the present 3D problem. Nevertheless, the accuracy of estimation is satisfactory in an engineering sense.

Concluding Remarks
The principal conclusions given in the present study can be summarized as follows :

(1) An inverse analysis method based on the hierarchical neural network and computational mechanics is proposed.

(2) This method is successfully applied to the identification of 2D and 3D cracks hidden in solid using the electric potential drop method.

(3) Its principal advantages are :

    (a) Easy application

    (b) Quick operation in an application process

## References

1.    Rumelhart, D. E. ; McClelland, J. L. ; the PDP Research Group : Parallel distributed processing (Explorations in the microstructure of cognition), MIT Press, Cambridge, MA, 1986.

2.    Rumelhart, D. E. ; Hinton, G. E. ; Williams, R. J. : Learning representations by back propagation errors, *Nature*, **323** (1986) 533.

3.    Baker, A. R. ; Windsor, C. G. : The classification of defects from ultrasonic data using neural networks (The Hopfield method), *NDT Int.*, **22** (1989) 97.

4.    Udpa, L. ; Udpa, S. S. : Eddy current defect characterization using neural networks, *Materials Evaluation*, **48** (1990) 342.

5.    Mann, J. M. ; Schmerr, L. W. ; Moulder, J. C. ; Kubovich, M. W. : Inversion of uniform field eddy current data using neural networks, Review of Progress in Quantitative Nondestructive Evaluation, Eds., D. O. Thompson and D. E. Chimenti, 9A (1990) 681, Plenum Press.

6.    Mochozuki, Y. ; Yagawa, G. ; Yoshimura, S. : Inverse analysis by means of the combination of multilayered neural network and computational mechanics ( Study on learning and estimating processes and its application to defect identification), *Trans. JSME*, **57**A (1991) 1922 (in Japanese).

7.    Hajela, P. ; Berke, L. : Neurobiological computational models in structural analysis and design, *Comp. & Struc.*, **41** (1991) 657.

8.    Thomsen, J. J. ; Lund, K. : Quality control of composite materials by neural network analysis of ultrasonic power spectra, *Materials Evaluation*, **49** (1991) 594.

9.    Pratt, D. ; Sansalone, M. : The use of a neural network for automating impact-echo signal interpretation, Review of Progress in Quantitative Nondestructive Evaluation, Eds., D. O. Thompson and D. E. Chimenti, 10A (1991) 667, Plenum Press.

10.   Kitahara, M. ; Achenbach, J. D. ; Guo, Q. C. ; Peterson, M. ; Ogi, T. ; Notake, M. : Depth determination of surface-breaking cracks by a neural network, Eds., D. O. Thompson and D. E. Chimenti, 10A (1991) 689, Plenum Press.

11.   Wu, X. ; Ghaboussi, J. ; Garrett, J. H. : Use of neural networks in detection of structural damage, *Comp. & Struc.*, **42** (1992) 649.

12.   Yoshimura, S. ; Yagawa, G. ; Toyonaga, K. ; Ohishi, T. ; Mochizuki, Y. : Structure identification by means of the combination of neural network and computational mechanics ( its application to one-dimensional beam), *Trans. JSME*, **58**C (1992), in Print ( in Japanese ).

13.   Funabashi, K. : On the approximate realization of continuous mappings by neural networks, *Neural Networks*, **2** (1989) 183.

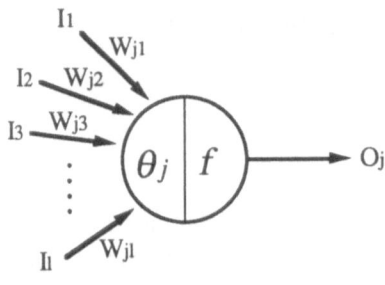

Fig. 1 Schematic view of unit
of neural network

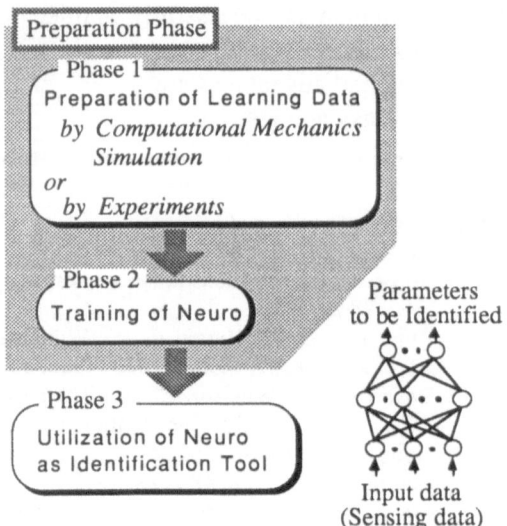

Fig. 2 Inverse analysis procedure based
on hierarchical neural network

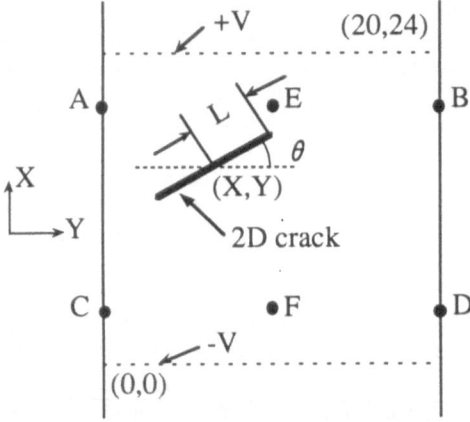

Fig. 3 Plate with a through-wall crack subjected
to prescribed electric potential conditions

Table 1 Number of samples and ranges of crack
parameters chosen for network training

| Parameter | Range | No. of Samples |
|-----------|-------------|----------------|
| X | 6.0 ~ 10.0 | 3 |
| Y | 12.0 ~ 16.0 | 3 |
| L | 2.0 ~ 5.0 | 3 |
| $\theta$(deg) | 0.0 ~ 45.0 | 4 |

Table 2  Mean errors of estimation (after 30,000 iterations)
(6-point measurement, 30 units in a hidden layer)

| | Mean Error of Estimation | |
|---|---|---|
| Parameter | Learned Patterns | Unlearned Patterns |
| X | 0.044 | 0.035 |
| Y | 0.017 | 0.028 |
| L | 0.010 | 0.040 |
| θ | 0.043 | 0.082 |

Table 3  Effects of the number of units in a hidden layer to estimation accuracy
(after 30,000 iterations) (2D crack)

| | Mean Error of Estimation for Learned Patterns | | | |
|---|---|---|---|---|
| Parameter | Units of Hidden Layer | | | |
| | 20 | 30 | 40 | 50 |
| X | 0.083 | 0.044 | 0.026 | 0.033 |
| Y | 0.038 | 0.017 | 0.013 | 0.015 |
| L | 0.018 | 0.010 | 0.007 | 0.007 |
| θ | 0.116 | 0.043 | 0.060 | 0.071 |

Table 4  Comparison of estimation accuracy between 6-point measurement
and 40-point measurement (2D crack)

| | Mean Error of Estimation for Learned Patterns | | |
|---|---|---|---|
| | 40-point Measurement (Iterations=4,000) | | 6-point Measurement (Iterations=30,000) |
| Parameter | Units of Hidden Layer | | Units of Hidden Layer |
| | 20 | 30 | 30 |
| X | 0.019 | 0.014 | 0.044 |
| Y | 0.013 | 0.010 | 0.017 |
| L | 0.004 | 0.009 | 0.010 |
| θ | 0.018 | 0.015 | 0.043 |

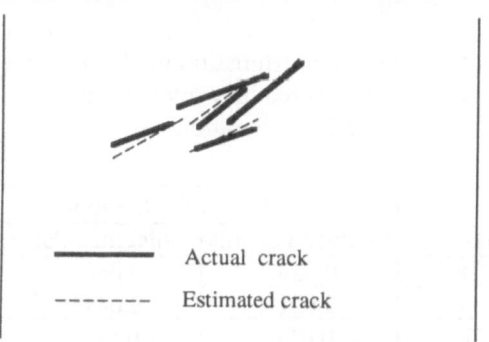

———————— Actual crack

- - - - - - - Estimated crack

Fig. 4  Comparison between actual and estimated cracks (2D crack)

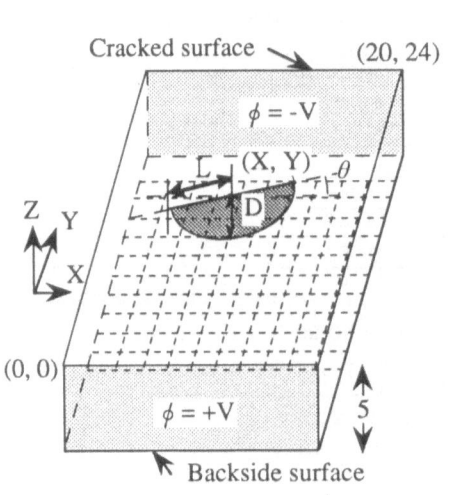

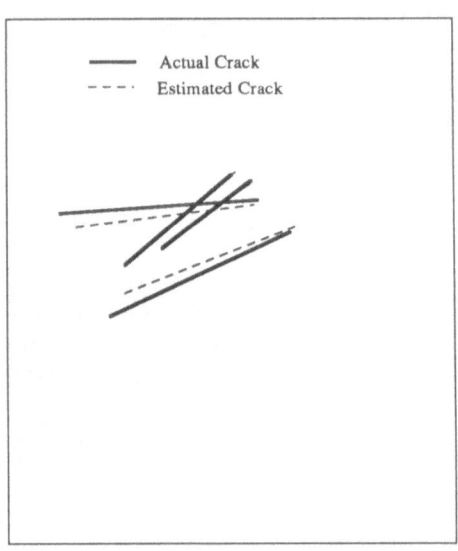

Fig. 5  Plate with a semi-elliptical surface    Fig. 6  Comparison between actual
crack subjected to prescribed electric            and estimated cracks (3D crack)
potential conditions

Table 5  Comparison of estimation accuracy among
three kinds of networks in 3D crack identification

| Network | Function of Output Layer | No. of Learning | Mean Error of Estimation | |
|---|---|---|---|---|
| | | | Learned Patterns | Unlearned Patterns |
| 3-Layered | sigmoid | 5,000 | 0.019 | 0.076 |
| 3-Layered | linear | 5,000 | 0.049 | 0.054 |
| 4-Layered | linear | 2,000 | 0.024 | 0.051 |

Table 6  Mean errors of estimation in 3D crack identification
when four-layered network is utilized
(after 2,000 iterations)

| Parameter | Mean Error of Estimation | |
|---|---|---|
| | Learned Patterns | Unlearned Patterns |
| X | 0.033 | 0.036 |
| Y | 0.021 | 0.081 |
| L | 0.011 | 0.046 |
| $\theta$ | 0.031 | 0.039 |

# Approximate Recovery and Redesign
# of a Curved Surface Based on a Flattened Form

**T. SHIMADA\*,Y. Tada\*\***

\*   University of Marketing & Distribution Sciences
    3-1 Gakuen-nishi-machi, Nishi-ku, Kobe 651-21, Japan
\*\* Department of Systems Engineering, Kobe University
    Rokkodai 1-chome, Nada-ku, Kobe 657, Japan

Summary
In designing for curved shapes, the reconstruction algorithm based on two
dimensional information has a prominent role, being especially in
dispensable to Computer Aided Engineering.    When designing a curved
surface, we need to consider the following process.    First, a curved
surface is transformed into the flattened    form.    Second, the    curved
surface is recovered from the flattened form.    Third, a new curved
surface has to be made corresponding to the modified flattened shape.
The second and third processes are both inverse problems for the    first
process.       We have already proposed a method to develop any curved
surface    into a flattened form using the finite element method.we have
already proposed methods for developing any curved    surfaces using the
finite element method.    In this paper,    we deal with recovery of 3D
curved surface based on the corresponding flattened forms using pseudo
plane stress problem in FEM.

## Recovery of Curved Surface

First, trying to design a product composed of curved surfaces,    we assume
a trial curved surface and the objective is approximately transformed onto a
plane[1] .    Next a scheme by which the    curved surfaces are    recovered
from    corresponding modified flattened forms is needed. Generally speaking,
a curved surface in 3D which is corresponding    to a flattened shape on 2D
can not be determined uniquely.    Then, we need to constraint the curved
surface by prescribing normal directions of respective elements composing
the curved surface.    However, in redesigning of a curved surface, as a
curved surface corresponding to a modified flattened    form is assumed to
have a shape resembling to a basic trial one,    the normal directions    of
elements composing the modified curved    surface can be regarded as    almost
equal to those of original ones.

Then, we show a method through which a curved surface can be recovered from the shapes of the objective flattened form and normal directions of elements in the following.

We decompose the objective flattened shape into the finite number (n-1) of adjacent regions (i.e. region 1, region 2, region 3,····, region n-1) and divide each region into m-1 rectangular elements. Let each node be denoted by the pair (i,j) indicating the volume and row, respectively, and consider that the whole region consists of triangular elements $e_\Omega$ formed by these nodes $P_{i,j}$ (i=1 ～ m,j=1 ～ n) as in the illustrative example shown in Fig.1. If this flattened form has an elastic characteristic, we can constitute out-of-plane elements using two sides in the plane and new one side outside the plane in horizontal and vertical directions. Then, in order to determine normal directions of in-plane elements, we adopt a reference curved surface corresponding to the mesh of flattened forms on the plane. Let denote out-of-plane in horizontal (Fig.2(1)) and vertical directions (Fig.2(2)) be $f_R$ and angles corresponding to the elements be $\theta_R$. Those elements $f_R$ and angles $\theta_R$ are additional information to recover the curved surface from a flattened form. Then, for the flattened shape, new triangular elements are named $f_\Omega$ which is composed of an angle $\theta_R$ and sides $L_{i,j-1,j}$ and $L_{i,j,j+1}$ in horizontal direction and $\theta_R$, $L_{i-1,i,j}$ and $L_{i,j+1,j}$ in vertical direction. On the while, assuming an initial trial curved surface having following elements in and out-of the plane. the notation $e_C$ is considered to be the in-plane triangles constructed by the nodal coordinates $C_{i,j}$ (i=1 ～ m,j=1 ～ n) of mesh into which the curved surface corresponding to the flattened form is divided(Fig.3). Moreover, the notation $f_C$ is considered to be the out-of triangular elements in horizontal (Fig.4(1)) and vertical directions (Fig.4(2)) corresponding to the triangular elements of reference curved surface mentioned above. Assuming that an initial trial shape is composed of a group of elements mentioned above, our algorithms are as follows.

(1) Assuming a reference flattened form and reference curved surface.

(2) Assuming an initial trial curved surface.

(3) Each triangle $e_C$ on the initial trial curved surface is compared with the corresponding one $e_\Omega$ on the reference flattened form, and the differences between the triangles on the curved surface and ones on the plane are qualified as transformation matrices.

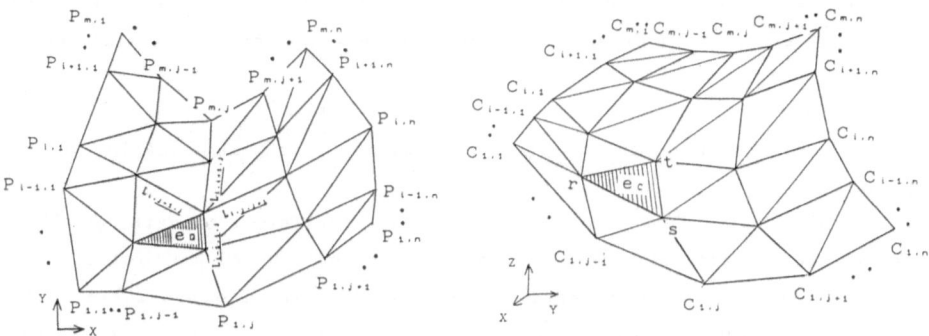

Fig.1 Reference flattened shape

Fig.3 Triangles on the trial curved surface

(1)

(1)

(2)

(2)

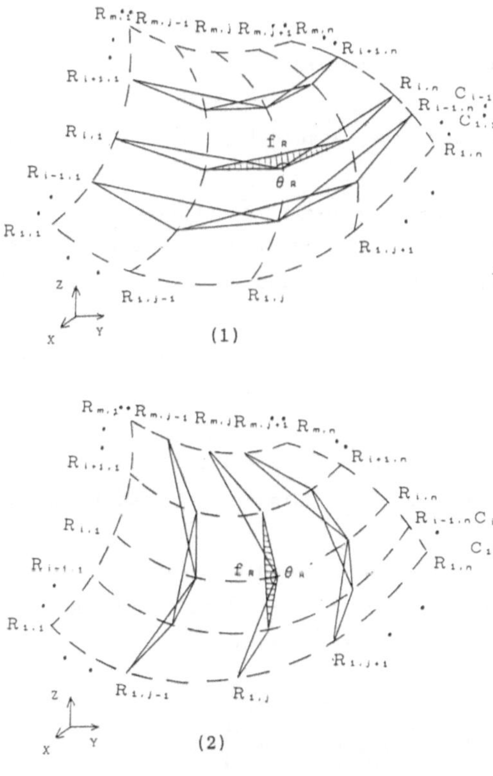

Fig.2 Additional information to prescribe normal directions
(1) Outplane triangles to prescribe horizontal directions
(2) Outplane triangles to prescribe vertical directions

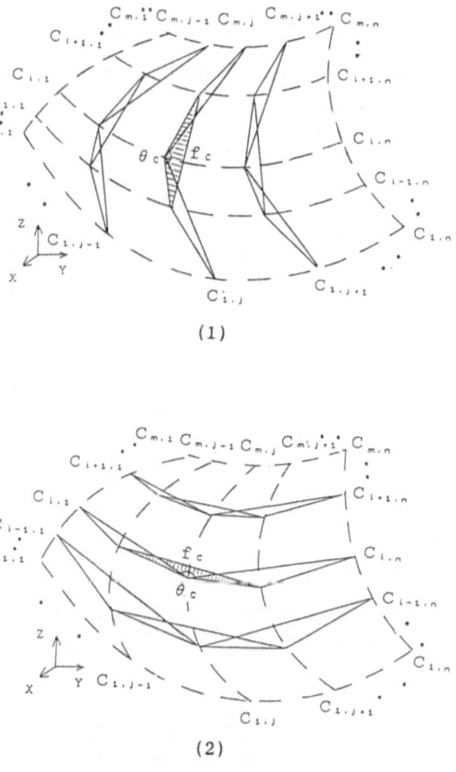

Fig.4 Outplane triangles of the trial curved surface
(1) Outplane triangles to prescribe horizontal directions
(2) Outplane triangles to prescribe vertical directions

(4) Assuming new triangular elements $f\omega$ composed of angles $\theta_R$ and corresponding sides to prescribe normal directions of an initial trial curved surface.

(5) The differences between the triangles $f\sigma$ from on the initial curved surface and ones $f\omega$ from the plane are qualified as transformation matrices.

(6) Using Singular Value Decomposition, the transformation matrices are divided into operator with functions of the deformation and the rotation.

(7) Formulating as a pseudo-plane-stress problem in 3D space where the notation $E\sigma$ is the Young's modules and the quantity $\nu_\sigma$ is Poisson's ratio of in-plane elements and $E_\tau$ and $\nu_\tau$ are those for out-of-plane elements.

(8) Deriving equivalent forces by the use of nodal coordinates on an initial trial curved surface.

(9) Obtaining stiffness matrices of an initial trial curved surface

(10) Fixing rotational components of nodal coordinates X,Y,Z as constraint conditions.

(11) Global displacements are derived by assembling each equilibrium equation.

(12) Adopting the total strain energy as the convergence criterion

(13) If the criterion of the convergence is attained, the trial curved surface is regarded as the recovered one corresponding to the flattened form.

(14) Otherwise, the improvement processes are successively performed and returned (3).

First, we can qualify the differences between the triangles $e\sigma$ on the initial curved surface and ones $e\omega$ on the plane as transformation matrices.
In Fig.3, that is, considering the global nodal coordinates denoted by the notation $(X,Y,Z)$, unit vectors $X\sigma$ $(=(\lambda_{XX}, \lambda_{YY}, \lambda_{ZZ})^T)$ in the side $\underline{rs}$ direction of the triangular element $e\sigma$ are expressed as

$$X_\sigma = \begin{pmatrix} \lambda_{XX} \\ \lambda_{YY} \\ \lambda_{ZZ} \end{pmatrix} = \frac{1}{L_{sr}} \begin{pmatrix} X_{sr} \\ Y_{sr} \\ Z_{sr} \end{pmatrix} = \frac{1}{L_{sr}} \begin{pmatrix} X_s - X_r \\ Y_s - Y_r \\ Z_s - Z_r \end{pmatrix} \quad (1)$$

where

$$L_{sr} = \sqrt{(X_{sr})^2 + (Y_{sr})^2 + (Z_{sr})^2} \tag{2}$$

And unit vectors $Z_c$ in normal direction on the triangular elements are obtained as outer products by unit vectors of two triangular sides.

$$Z_c = \begin{pmatrix} \lambda_{zx} \\ \lambda_{zy} \\ \lambda_{zz} \end{pmatrix} = \frac{1}{2L_n} \begin{pmatrix} Y_{sr}Z_{tr} - Z_{sr}Y_{tr} \\ Z_{sr}X_{tr} - X_{sr}Z_{tr} \\ X_{sr}Y_{tr} - Y_{sr}X_{tr} \end{pmatrix} \tag{3}$$

where

$$L_n = \sqrt{(\lambda_{zx})^2 + (\lambda_{zy})^2 + (\lambda_{zz})^2} \tag{4}$$

And unit vectors $Y_c$ in normal directions to both vectors $X_c$ and $Z_c$ are expressed as

$$Y_c = \begin{pmatrix} \lambda_{yx} \\ \lambda_{yy} \\ \lambda_{yz} \end{pmatrix} = \begin{pmatrix} \lambda_{zy}\lambda_{xz} - \lambda_{zz}\lambda_{xy} \\ \lambda_{zz}\lambda_{xx} - \lambda_{zx}\lambda_{xz} \\ \lambda_{zx}\lambda_{xy} - \lambda_{zy}\lambda_{xx} \end{pmatrix} \tag{5}$$

Then, the transformation from coordinates $(X_c, Y_c)$ on the triangular elements $e_c$ represented by the local system to coordinates $(X_a, Y_a)$ on the ones $e_a$ by global system can be given by following matrices

$$\begin{pmatrix} X_a \\ Y_a \end{pmatrix} = A \begin{pmatrix} X_c \\ Y_c \end{pmatrix} . \tag{6}$$

Hence, using Singular Value Decomposition, the matrix A can be divided into matrices $R(\beta)$, $R(\phi)$ and P as

$$A = R(\phi) \cdot P \cdot R(\beta) \tag{7}$$

where

$$R = \begin{pmatrix} \cos\beta & , -\sin\beta \\ \sin\beta & , \cos\beta \end{pmatrix} \tag{8}$$

$R(\beta)$ and $R(\phi)$ are rotational matrices with rotating angles of $\beta$ and $\phi$, respectively. P is a diagonal matrix, the components of which are square roots of eigenvalues of $(A^T \cdot A)$. Similarly, the transformation from $f_c$ to $f_a$ are expressed by similar equations as Eqs. from (6) to (8).

If a virtual material has these anisotropies, we can consider a thermal stress problem. In the plane stress problem, the notation [B] is a matrix connecting the displacement with the strain. [k] is a stiffness matrix and [D] is a matrix of the elastic coefficient which combines the stress with the strain. {u} is a displacement vector and $\{\varepsilon_t\}$ is a vector of the thermal strain. The equilibrium equation is written as

$$[k] \cdot \{u\} = \{f\} + \{g\} ,\qquad(9)$$

where

$$[k] = \int [B]^T \cdot [D] \cdot [B]\ dv\qquad(10)$$

$$\{g\} = \int [B]^T \cdot [D] \cdot \{\varepsilon_t\} dv\qquad(11)$$

and {f} is a vector of nodal force. The global nodal force vector is assembled from {f} and {g}. The former is zero in this case and {g} is an equivalent nodal force. Assuming that the principal direction p of the thermal strain $\{\varepsilon_t'\}$ meets the reference axis x at $\beta$ degrees, the thermal strain can be expressed as the product

$$\{\varepsilon_t\} = [[T]^T]^{-1} \cdot \{\varepsilon_t'\} ,\qquad(12)$$

where

$$[T] = \begin{pmatrix} \cos2\beta & , \sin2\beta & , -2\sin\beta\cos\beta \\ \sin2\beta & , \cos2\beta & , 2\sin\beta\cos\beta \\ \sin\beta\cos\beta & , -\sin\beta\cos\beta & , \cos2\beta - \sin2\beta \end{pmatrix}\qquad(13)$$

and

$$\{\varepsilon_t'\} = \begin{pmatrix} \alpha_p \\ \alpha_q \\ 0 \end{pmatrix}\qquad(14)$$

, $\alpha_p$ and $\alpha_q$ are anisotropy coefficients of linear expansion. Substitutions of Eqs. (14) into Eq. (12) makes it

$$\{\varepsilon_t\} = [[T]^T]^{-1} \cdot \begin{pmatrix} d_p-1 \\ d_q-1 \\ 0 \end{pmatrix}\qquad(15)$$

, dp and dq are expansion ratios of anitropic materials.
Then, the equivalent nodal force {g} can be calculated by the thermal strain $\{\varepsilon_t\}$ as

$$\{g\} = \int [B]^T \cdot [D] \cdot [[T]^T]^{-1} \begin{pmatrix} d_p-1 \\ d_q-1 \\ 0 \end{pmatrix} dv. \qquad (16)$$

Finally, substituting Eq.(16) into Eq.(9), the equilibrium equation can be expressed as follows;

$$[k] \cdot \{u\} = \{f\} + \int [B]^T \cdot [D] \cdot [[T]^T]^{-1} \begin{pmatrix} d_p-1 \\ d_q-1 \\ 0 \end{pmatrix} dv \qquad (17)$$

By solving the pseudo plane stress problem for a virtual object composed of in-plane elements and out-of-plane elements, the global displacement vectors $\{U_{i,j}\}$ can be obtained as follows. Consider the notation E to be the Young's modules and the quantity $\nu$ to be the Poisson's ratio. Let $\varepsilon_x$, $\varepsilon_y$ and $\gamma_{xy}$ be strains. The area of a triangular element is S. The strain energy due to the displacement is expressed as

$$J = \sum_{e=1}^{N} (\varepsilon_x^2 + \varepsilon_y^2 + 2\nu \varepsilon_x \varepsilon_y + \gamma_{xy}^2 (1-\nu) \cdot [E/\{2(1-\nu^2)\}]) S. \qquad (18)$$

By introducing a parameter $\alpha$ which is an appropriate coefficient with respect to the convergence of the shape, revised nodal coordinates are obtained by adding the displacements multiplied by a parameter $\alpha$ to the previous nodal coordinates. Accordingly, the revised shape on a plane becomes a new trial shape in the next step of our algorithm.
The criterion of the convergence is represented as

$$J \leqq \varepsilon_o , \qquad (19)$$

where $\varepsilon_o$ is a small positive number given by the comparison of the accuracy and the cost concerning the calculation.
If the convergence is attained, the trial curved surface is regarded as the recovered one corresponding to the objective. Otherwise, the following process is successively performed for the improvement of the trial shape.

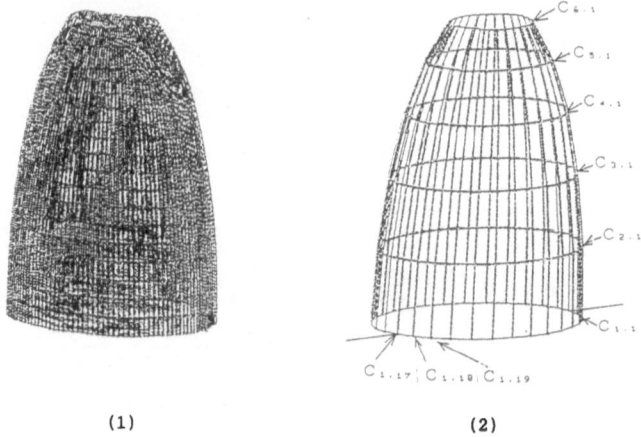

(1)                                    (2)

Fig.5 Truncated half ellipsoid
(1) Curved surface
(2) Reference curved surface
    to prescribe normal directions of elements

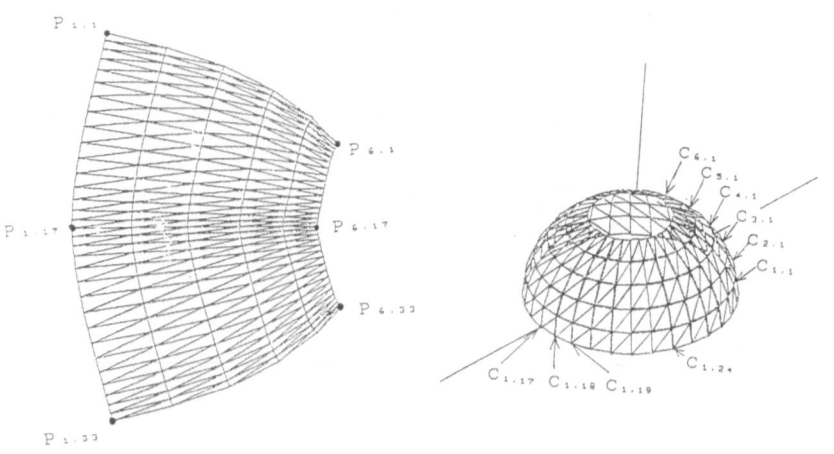

Fig.6 Reference flattened form          Fig.7 Initial trial curved surface

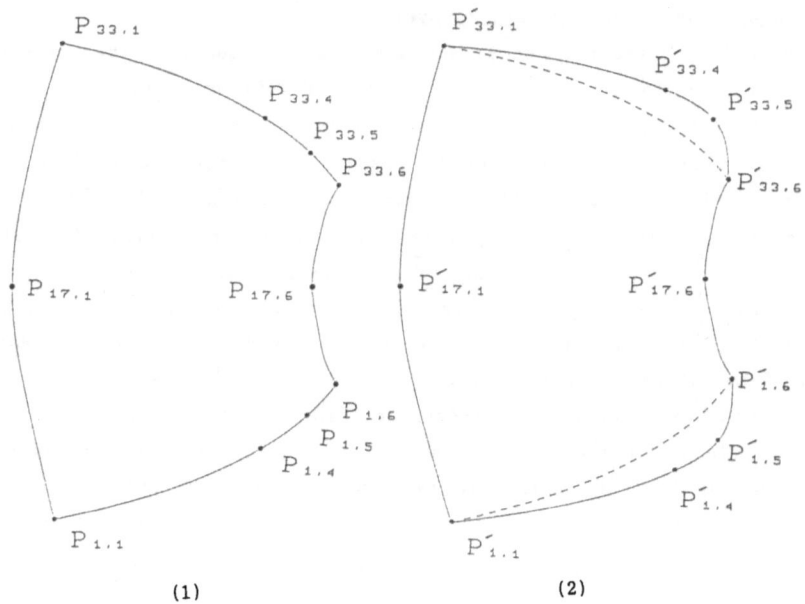

$P_{33,1}$
$P_{33,4}$
$P_{33,5}$
$P_{33,6}$
$P_{17,1}$
$P_{17,6}$
$P_{1,6}$
$P_{1,5}$
$P_{1,4}$
$P_{1,1}$

(1)

$P'_{33,1}$
$P'_{33,4}$
$P'_{33,5}$
$P'_{33,6}$
$P'_{17,1}$
$P'_{17,6}$
$P'_{1,6}$
$P'_{1,5}$
$P'_{1,4}$
$P'_{1,1}$

(2)

Fig.8 External shape modification

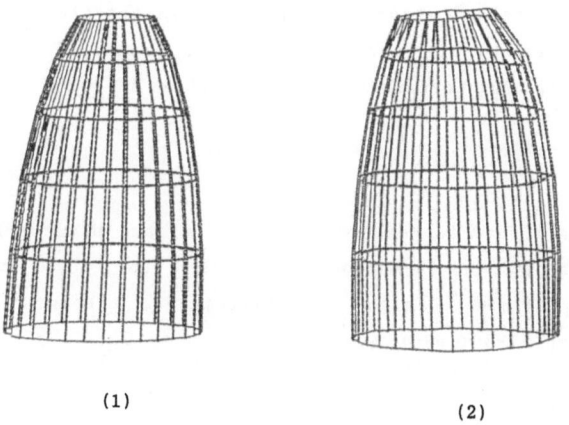

(1)                          (2)

Fig.9 Recovered curved surface

## Example of  the Curved Surface Recovery

The objective curved surface is shown in Fig.5.   Consider the corresponding flattened form to be the reference shape (Fig.6) on a plane.   Assume the reference curved surface to  prescribe normal directions of elements.   Let an initial trial curved surface be the partial cubic one (Fig.7).   A new curved can be made by flattened shape modification.   one and additional information.   The developing process of a truncated ellipsoid is shown in Fig.8.   $P_{33, 4}'$ ,  $P_{33, 5}'$   and  $P_{1, 4}'$ ,  $P_{1, 5}'$   is offset 10mm in outer direction.    Real line in Fig.8(2) is shown modified external shape. Similarly, the modified external shape is divided into mesh.   Moreover, after formulating as pseudo plane stress  problem, the trial shape is obtained as the recovery curved surface (Fig.9(2)).    Comparison Fig.9(1) and (2) yields following results.   This   indicates that reconstruction algorithm based on the principle is   presented to be able to recover the 3D curved shape from the  flattened one and additional information.

## Conclusion

In this paper, we dealt with recovery of 3D curved surface using  pseudo plane stress problem in FEM from the  relation between elements of the flattened   form   on   a   plane   and   ones   on   a   curved   surface.    The reconstruction algorithm was proposed, in which the objective curved surface was   modeled   by   a   group   of   finite   elements   and   which   qualified   the difference of geometrical shapes as total strain energies.   The proposed method   was   applied   to   the   curved   surface   of   truncated   ellipsoid   and   is shown   to   be   applicable   because   of   its   ability   of   broad   application   and extension.

## References

1.Shimada,T.;Tada,Y.:Development of a Curved Surface Using a Finite Element Method   in   Brebbia,C   and   Hernandez,   S   (eds)   Proc.   1st   Int.   Conf. Computer-Aided   Optimum   Design   of   Structures,   Recent   Advances Springer-Verlag,pp.23-30(1989).

# Optimum Design of a Sound-Insulating Wall by the Boundary Element Method

M. Tanaka*, T. Matsumoto* and M. Shirotori**

*   Faculty of Engineering, Shinshu University
    500 Wakasato, Nagano 380, Japan
** Graduate School of Shinshu University
    500 Wakasato, Nagano 380, Japan

## Summary

This paper is concerned with optimpum design of a sound-insulating wall by using the boundary element method in the semi-infinite two-dimensional space, which can be formulated as an inverse problem. The influence of the ground, which is extended infinitely, is taken into consideration by introducing the Green's function generated by superposition of the two free-space fundamental solutions. It is assumed that a pure sound is radiated from a point source to the 2-D semi-infinite acoustic field, and it should be reduced at some evaluation points by attaching a control board to the given insulating wall. The acoustic field is calculated by the boundary element method, and the optimal set of parameters describing the location, length and inclination angle of the control borad are sought by the standard optimization procedure. Computational results on such optimum design reveal the potential usefulness of the proposed procedure.

## Introduction

Reduction of noise is one of the most important subjects in engineering. Recently, in addition to passive reduction in which noise energy is absorbed by means of muffler or other absorbing materials, active reduction has been attracting much attention, in which noise is reduced by interference with other noise[1-8]. The authors have so far formulated such active noise reduction or control problems into the acoustic inverse problem where an optimal set of parameters should be sought, and carried out numerical simulation for active control of duct noise[9]. Attempt was also made for its application to optimum design of a sound-insulating wall[10-12]. The idea of this noise reduction is to produce a different sound by attaching a control board to the main insulating wall. Through numerical simulation, the effectiveness of the noise reduction method was investigated in detail for a simpler case where the insulating wall was located in the two-dimensional free space. In practical use, however, the effect of the ground can not be neglected, and in general two equal sound-insulating walls are equipped on both sides of a railway or highway. In this paper, all these effects mentioned above and also a more complicated shape of the control borad are taken into consideration.

In the boundary element method, the acoustic field with a symmetric plane can be expressed by using the Green's function which the free-space fundamental solution is superposed by the fundamental solution with a source at the mirror point. Since the boundary conditions on the symmetric plane are exactly satisfied in this way, no boundary element discretization is required if such a Green's function is used in the formulation. The effect of the ground can be taken into account in this manner. The optimal set of parameters describing the shape, location and inclination angle of the control bord are sought such that a sum of specific power of sound at some evaluation points is minimized. Computational results for several example problems are shown, whereby the potential usefulness of the proposed method is demonstrated.

## BEM for Acoustic Direct Problems

In this paper, the so-called regularized boundary integral equation formulation is used for boundary element analysis of the acoustic problem under consideration. Since no singular integrals to be evaluated in the sense of Cauchy principal value appear in this formulation, all the integrals can be evaluated by the standard Gaussian quadrature.

Under the assumption of steady-state vibration with a small amplitude, the acoustic problem is governed by the Helmholtz differential equation. As has been well known, an equivalent integral form of the Helmholtz equation can be expressed as

$$p(x)+\int_S q^*(x,y)p(x)dS(x) = -j\omega\rho\int_S p^*(x,y)v(x)dS(x)+Fp^*(x_s,y) \tag{1}$$

where $S$ denotes the boundary of the acoustic field $V$, $j = \sqrt{-1}$, and $w$ and $r$ respectively the angular frequency and the mass density of the medium. Moreover, $p$ is the sound pressure, $v$ the particle velocity outward normal to the boudary and $F$ the intensity of point source located at point $x_s$ in the domain $V$. The asterisked function $p^*(x,y)$ is the fundamental solution of the Helmholtz euqation, and $q^*(x,y)$ its normal derivative at point $x$ on the boundary. These functions can be given, e.g. for two-dimensional problems, as follows:

$$p^*(x,y) = -\frac{j}{4}H_0^{(2)}(kr) , \quad q^*(x,y) = \frac{jk}{4}H_1^{(2)}(kr)\frac{\partial r}{\partial n} \tag{2}$$

where $k$ is the wave number, $r$ the distance between two points $x$ and $y$. The functions $H_0$ and $H_1$ are the zeroth and the first order Hankel function of second kind, respectively.

Now, we introduce the fundamental solution of the Laplace equation denoted by $P^*(x,y)$ and define the corresponding flux $Q^*(x,y)$ as $Q^*(x,y)= \partial P^*(x,y)/\partial n$. Then, we can finally derive from Eg. (1) the following integral equation expression :

$$\left[ \int_S \left\{ q^*(x,y)-Q^*(x,y)\right\}dS(x) \right]p(y)+\int_S q^*(x,y)\left\{p(x)-p(y)\right\}dS(x)$$

$$= -j\omega\rho \int_S p^*(x,y)v(x)dS(x)+Fp^*(x_s,y) \tag{3}$$

It is noted here that the above integral euqation always holds both in the domain and on the boundary if the sound pressure $p$ satisfies the Hölder continuity[13]. In this study, the reguralized boundary integral equation (3) is employed, and discretized by means of the isoparametric quadratic boundary element.

Once the sound pressure $p$ and the particle velocity $v$ on the boundary are obtained by boundary element analysis, the sound pressure $p$ and other quantities in the domain $V$ can be computed using the following integral equation:

$$p(y) = -\int_S \left\{ q^*(x,y)-Q^*(x,y)\right\}dS(x)\cdot p(x_0) - \int_S q^*(x,y)\left\{p(x)-p(x_0)\right\}dS(x)$$

$$-j\omega\rho \int_S p^*(x,y)v(x)dS(x)+p(x_0) + Fp^*(x_s,y) \quad , \quad y\in V \tag{4}$$

where $p(x_0)$ is the sound pressure at point $x_0$ on the boundary, nearest to the source point $y$. The particle velocity component $v_i$ can be computed using the following expression:

$$v_i(y) = \frac{-1}{j\omega\rho}\frac{\partial p(y)}{\partial y_i} = \frac{1}{j\omega\rho}\int_S \left( \frac{\partial q^*(x,y)}{\partial y_i} - \frac{\partial Q^*(x,y)}{\partial y_i} \right)dS(x)\cdot p(x_0) +$$

$$\frac{1}{j\omega\rho}\int_S \frac{\partial q^*(x,y)}{\partial y_i}\left\{p(x)-p(x_0)\right\}dS(x) + \int_S \frac{\partial p^*(x,y)}{\partial y_i}v(x)dS(x) - \frac{F}{j\omega\rho}\frac{\partial p^*(x_s,y)}{\partial y_i} \tag{5}$$

It is interesting to note that derivatives of the fundamental solutions $p^*$ and $q^*$ in two-dimensional problems can be expressed as

$$\frac{\partial p^*(x,y)}{\partial y_i} = -\frac{jk}{4} H_1^{(2)}(kr) \frac{\partial r}{\partial x_i} \tag{6}$$

$$\frac{\partial q^*(x,y)}{\partial y_i} = -\frac{jk}{4}\left[ kH_0^{(2)}(kr) \frac{\partial r}{\partial n} \frac{\partial r}{\partial x_i} + \frac{1}{r} H_1^{(2)}(kr)\left( n_i - 2\frac{\partial r}{\partial n} \frac{\partial r}{\partial x_i} \right) \right] \tag{7}$$

$$\frac{\partial Q^*(x,y)}{\partial y_i} = -\frac{1}{2\pi r^2}\left( 2\frac{\partial r}{\partial n} \frac{\partial r}{\partial x_i} - n_i \right) \tag{8}$$

Now, we will briefly show in the following how to treat the symmetry conditions. Symmetry can be taken into account in such away that the free-space fundamental solution is superposed by the other fundamental solution with a source at the mirror point[14]. In this manner, we can consider two kinds of the symmetry condition, that is, the complete reflection (v=0) and the complete absorbtion (p=0). Since the Green's function thus obtained by superposition satisfies the boundary conditions on the symmetry plane, no boundary element discretization is required, which is the most important advantage of the present approach. It is noted that superposition is made in the following way:

(i)  For the symmetry condition of particle velocity $v=0$ (complete deflection),

$$p^* = p^*(x,y) + p^*(x,y') \tag{9}$$

(ii) For the symmetry condition of sound pressure $p=0$ (complete absorbtion),

$$p^* = p^*(x,y) - p^*(x,y') \tag{10}$$

where $y'$ denote the mirror point of $y$ with respect to the symmetry plane.

If we discretize integral equations (3), (4) and (5) by the boundary element method,we can obtain their discretized versions as follows:

$$[H]\{p\}_S = [G]\{v\}_S + \{f\}_V \tag{11}$$

$$\{p\}_V = -[A]\{p\}_S + [B]\{v\}_S + \{f\}_V \tag{12}$$

$$\{v_i\}_V = -[C_i]\{p\}_S + [D_i]\{v\}_S + \{f,_i\}_V \tag{13}$$

where the matrices $[H]$, $[G]$, $[A]$, $[B]$, $[C_i]$ and $[D_i]$ can be computed from the fundamental solutions. In the above equations, subscripts $S$ and $V$ denote the quantities in domain and on the boundary. Applying the given boundary conditions to equation (11), we can determine all the unknowns on the boundary. Then we may use these results to equations (12) and (13) to calculate the sound pressure and the particle velocity components.

## Optimum Design of Sound-Insulating Wall

Now, we shall apply an analysis system of acoustic inverse problems, which combines the boundary element method with the standard optimization procedure, to optimization of the sound-insulating wall under consideration.

Optimum design of the sound-insulating wall is schematically shown in Fig.1. As the cost function for optimization we choose a square sum of absolute value at evaluation points. The non-dimensional cost function $R$ can be expressed as

$$R = \frac{1}{N} \sum_{i=1}^{N} \left( \frac{|p_i|}{|p_{0i}|} \right)^2 \tag{14}$$

where $N$ is the number of evaluation points, and $p_i$ and $p_{oi}$ are the sound pressure at point $i$ with and without the control board, respectively. In this study, we employ the steepest descent method, one of the standard optimization procedures [15,16]. Iterative computation is stopped if the following inequality is satisfied:

$$\left| R^k - R^{k-1} \right| < \varepsilon \tag{15}$$

where $\varepsilon$ is a given tolerance limit and $k$ is the iteration number.

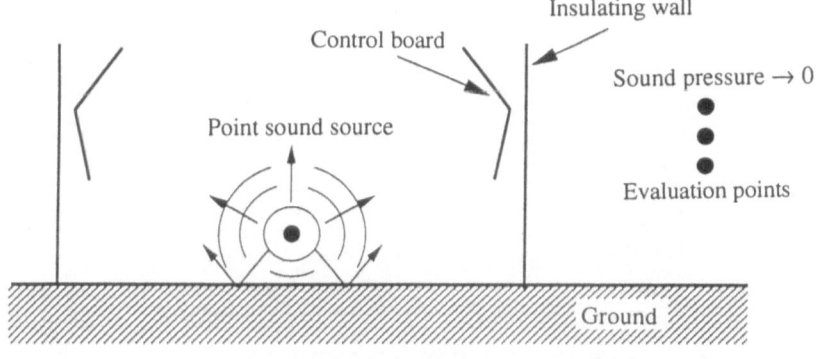

Fig.1 Schematic illustration of noise reduction
by sound insulating walls

## Computational Results and Discussion

We now show some of the numerical results obtained for a typical example problem. It is assumed that a point source is located on the symmetrical plane of a road covered on each side by an equal insulating wall with a control boad as shown in Figs.1 and 2. We consider three evaluation points as shown in Fig. 2. Because of symmetry, only a half space is discretized by the boundary element method as shown in Fig.3. No discretization is required for the symmetrical plane IDE because the fundamental solution taking account of this fact is used for analysis. In order to take into account different boundary conditions, only the road surface EF is divided into boundary elements, while the other parts of the ground are not discretized. The method of sub-regions is applied to the solution of this acoustic problem so that the whole space into two sub-regions as shown in Fig.3. The given sound-insulating wall and the control board are treated as completely thin, i.e. of no thickness. Two sides of the wall and the control board are equally divided into 19 and 10 boundary elements, respectively. The road surface EF is divided into 16 equal elements. In the computer program developed, an impedance boundary condition can be taken into consideration, but we assume in this study the zero impeadance, i.e. $z=v/p=0$. The interface boundary CD or AG is divided into at most 14 elements. During iterative computations, re-meshing is done appropriately to maintain numerical accuracy. The material constants and the computational parameters are assumed as sound velocity $c_0$ =340 m/s, mass density $\rho$ =1.2 kg/m$^3$ and $\varepsilon$ =10$^{-5}$. Computation is carried out for the case where a pure sound with frequency 500Hz is radiated from the point source.

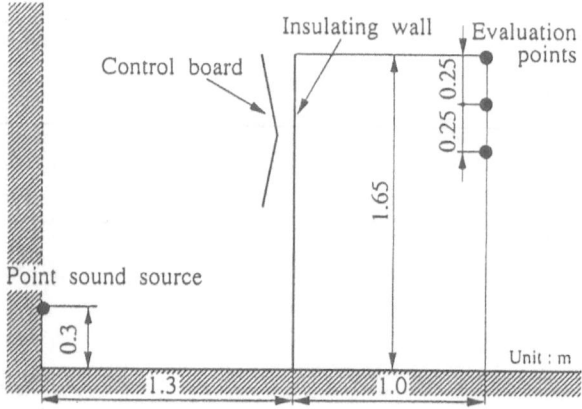

Fig.2 Simulation model of sound insulating wall with control board

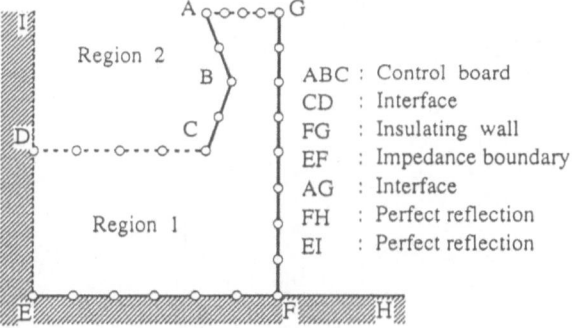

Fig.3 Boundary element discretization

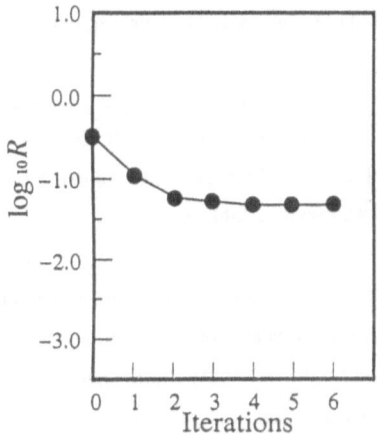

Fig.4 Convergence property of nondimensional cost function

Table 1 Parameter values at initial and optimal states

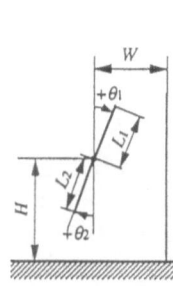

| | Initial state | Optimal state |
|---|---|---|
| $W$ [m] | 0.1 | 0.202 |
| $H$ [m] | 1.225 | 1.216 |
| $L_1$ [m] | 0.425 | 0.471 |
| $\theta_1$ [rad] | 0.0 | 0.012 |
| $L_2$ [m] | 0.425 | 0.481 |
| $\theta_2$ [rad] | 0.0 | 0.003 |

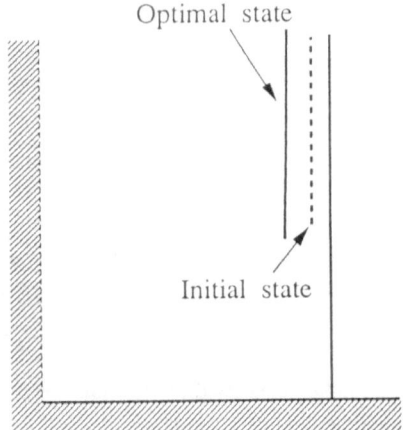

Fig.5 Initial and optimal states of control board

Figure 4 shows a typical convergence property of iterative computations for optimization with respect to the nondimensional cost function $R$. In this case, after six ietrations the optimal set of parameters are found. The optimal and initial states of the control board are illustrated in Fig.5, while the detailed results are summarized in Table 1.

In Figs. 6 and 7 are shown the contours of sound pressure level for the acoustic fields without and with the control board. If the control board is attached to the main insulating wall under the optimal shape, location and inclination angle estimated, noise can be reduced in amount of about 10 dB, and a wider range near the evaluation points is made calm.

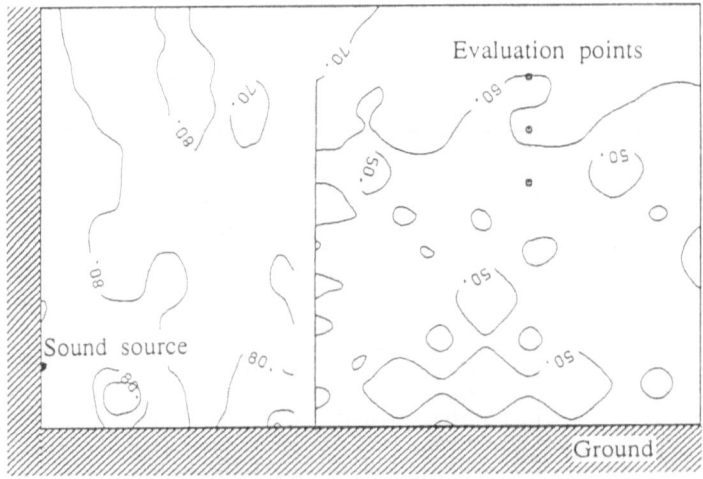

Fig.6 Contours of sound pressure level (no control board)

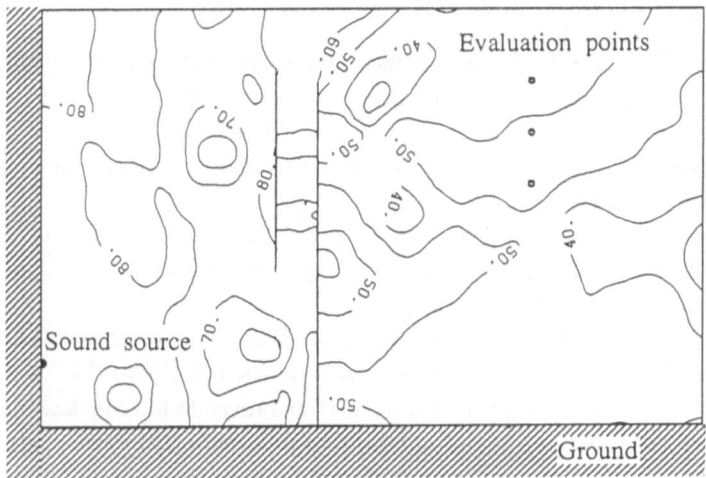

Fig.7 Contours of sound pressure level at optimal noise reduction

## Conclusing Remarks

The real ground should be modelled as an impeadance boundary condition which shows absorbtion of noise power to some extent. In the present computation, the ground was modelled as the rigid plane of complete reflection. However, it can be seen that the essential feature of the sound-insulating wall was captured in the present computer simulation, because interference between two sounds with diffrent phases produced by the control board plays the most important role for noise reduction in this way.

It might be concluded that the present inverse analysis system can be successfully applied to a wide variety of optimal design problems in acoustics. Such computer simulation is no doubt important and effective not only for optimum design of acoustic fields, but also finding an innovative way for active control or active reduction of noise.

As the future research work in this direction, it might be recommended to investigate the feasibilities of optimal design under more complicated boundary conditions. In addition, computer simulation is needed for real noise consisting of different frequency components. It is interesting to note that using the present acoustic inverse analysis system we can easily study more sophisticated models of such real noise.

242

## References

1. Hamada, H. and Imai.: Recent studies for active sound and vibration control - Quick survey of Intergantional Symposium on Active Control of Sound and Vibration -, (in Japanese), *J. Acoust. Soc. Japan,* **47** (1991) 685-693.

2. Stevens, J.C. and Ahuja, K.K.: Recent advances in active noise control, *AIAA J.,* **29** (1991) 1085-1067.

3. Cuneface, K.A. and Koopmann, G.H.: A boundary element approach to optimization of active noise control sources on three-dimensional structures, *ASME J. Vibration and Acoustics,* **113** (1991) 387-394.

4. Molo, C.G. and Bernhard, R.J.: Numerical evaluation of the performance of active noise control systems, *ASME J. Vibration and Acoustics,* **112** (1990) 230-236.

5. Salikuddin, M., Tanna, H.K., Burrin, R.H. and Khan, M.M.: Application of active noise control to model propeller noise, *J. Sound and Vibration,* **137** (1990) 9-41.

6. Bullmore, A.J., Nelson, P.A. and Elliott, S.J.: Theoretical studies of the active control of propeller-induced cabin noise, *J.Sound and Vibration,* **140** (1990) 191-217.

7. Kido, K., Kanai, H. and Abe, M.: Active reduction of noise by additional noise source and its limit, *ASME J. Vibration, Stress, and Rliability in Design,* **111** (1989) 480-485.

8. Molo, C.G. and Bernhard, R.J.: Generalized method of predicting optimal performance of active noise controllers, *AIAA J.,* **27** (1989) 1473-1478.

9. Tanaka, M., Yamada, Y. and Shirotori, M.: Computer simulation of active noise control by the boundary element method, *Boundary Elements XII,* Vol.2, ed. by M. Tanaka, C.A. Brebbia and T. Honma, CP Publications and Springer-Verlag, (1990) 147-158.

10. Tanaka, M., Matsumoto, T. and Shirotori, M.: Boundary element method simulation for control of acoustic fields by an insulating wall, (in Japanese), *Proc. 8th Japan Natl. Symp. on BEM,* JASCOME, (1991) 211-216.

11. Tanaka, M., Yamada, Y. and Shirotori, M.: Computer simulation by the boundary element method for optimal control of acoustic fields using a control board, *Abstracts of IABEM-91 Symp. in Kyoto/Japan,* (1991) 153-155.

12. Tanaka, M., Matsumoto, T. and Shirotori, M.: BEM simulation for acoustic field control by sound insulating wall, (in Japanese), *Proc. 13th Computational Electromagnetics Symp. in Japan,* Japan Soc. Simulation Technology, (1992) 283-288.

13. Tanaka, M. and Matsumoto, T.: Application of accurate boundary element methods to sensitivity analyses, (in Japanese), *Computational Mechanics [III],* ed. by N. Tosaka and G. Yagawa, Yokendo/Tokyo, (1992) 1-18.

14. Walker, S.: Fundamental solutions, *Progress in Boundary Element Methods,* Vol.2, ed. by C.A. Brebbia, Pentech Press, (1981) 13-44.

15. Fox. R.L.: *Optimization Methods for Engineering Design,* Addison-Wesley Publishing Co., (1971) 38-116.

16. Papalambros, P.Y. and Wilde, D.J.: *Principles of Optimal Dsign,* Cambridge Univ. Press (1988).

# Optimum Design of Cooling Lines of Injection Mold Based on Boundary Element Design Sensitivity Analysis

T. Matsumoto*, M. Tanaka* and H. Hirata**

* Faculty of Engineering, Shinshu University
  500 Wakasato, Nagano 380, Japan
** Graduate School of Shinshu University
  500 Wakasato, Nagano 380, Japan

**Summary**
In this paper, a boundary integral design sensitivity formulation for steady-state heat conduction problems is presented and applied to the optimum design of the arrangement of the cooling lines in injection mold. The cost function for the present optimization problem is approximated with the first-order Taylor series expansion and its first-order design sensitivities with respect to the design variables are calculated from the boundary element design sensitivity analysis. The sensitivity boundary integral equation which relates the boundary temperatures and heat fluxes with their sensitivity coefficients are derived by differentiating directly a regularized boundary integral equation for the steady-state heat conduction problems with respect to design variables. Some two-dimensional numerical examples illustrate the effectiveness of the present approach.

**Introduction**

In recent years, the boundary element method has been effectively applied to shape optimization of continuum structures in various engineering problems. Shape optimization is one of the fields in which the boundary element method is effectively applied. One of the reasons for this is that the boundary element method can treat a problem by discretizing only the boundary in most of the linear problems. In addition, the boundary element method can usually give us more accurate boundary solutions than the methods based on the domain discretization. Therefore, the boundary element method is suitable for such an optimization problem in which the objective function and the constraints consist of the structural responses only on the boundary or at small number of points in the domain.

In this paper, we present a boundary integral design sensitivity formulation for steady-state heat conduction problems and applies it to the design of injection molds. It is important to obtain the optimal arrangement of cooling lines in a mold to maintain a uniform temperature distribution on the cavity surface for high quality production. The pioneering work has been done by Barone and Caulk [1] in this field. In the iterative process of optimization, the accurate and efficient computation of the sensitivities of the system response is important in order to obtain the optimal state successfully and economically. The boundary integral design sensitivity formulation based on the implicit differentiation of the boundary integral equation with respect to a design parameter has been successful in obtaining accurately and efficiently the design sensitivities in elastostatic problems [2,3] and acoustic problems [4]. In this method, the boundary integral equation or its discretized version is differentiated implicitly with respect to a design parameter such as a shape design variable and derived is a boundary integral equation or its discretized version relating the sensitivity coefficients to the boundary responses.

In the formulation, every parameter which affects the system responses such as the boundary condition itself, shape change, thermal conductivity, heat transfer coefficient and ambient temperature can be adopted as a design variable. The boundary integral equation which can treat the convective boundary condition is first regularized up to weakly singular order. Next, it is differentiated with respect to a design variable and a boundary integral equation relating the sensitivities of the boundary temperature and heat flux is derived. The sensitivity boundary integral equation can be solved by discretizing it and applying all the boundary temperatures and heat fluxes obtained by the standard BEM analysis and the boundary conditions for the temperature and heat flux sensitivity.

In the design process of cooling lines, the objective function is defined from the nodal temperatures over the cavity surface of the mold. The sensitivity coefficients of the boundary nodal temperatures obtained by the boundary element design sensitivity analysis is used for the calculation of the first order design sensitivities of the linearized objective function.

## Optimum Design of Cooling Lines of Injection Mold

Injection molding is an approach to manufacture products by injecting the molten polymer into the cavity of the mold through the runners and cooling it until it is solidified. The heat of the polymer is taken out by the coolant running through the cooling lines in the mold. In the standard CAE systems for injection molding, the steady-state temperature field of the mold is analyzed in order to estimate the cycle-averaged temperature distribution of the cavity surface. The purpose of such analyses is the reduction of the cycle time of molding and the manufacturing cost of molds by simulating the effect of the cooling lines arrangement.

However, the desired cycle-averaged temperature over the cavity surface or the cycle-averaged heat flux from the polymer to the cavity surface can be estimated from the desired cycle time of molding. Hence, in the present study, we consider an optimum design of the arrangement of cooling lines to make the temperature distribution over the cavity as uniform as possible.

In the present case, the cost function to be minimized can be written as follows:

$$f(\mathbf{b}) = \frac{1}{N}\sum_{i=1}^{N}\left(\frac{\theta^{i}(\mathbf{b})}{T}-1\right)^{2} \tag{1}$$

where $\theta^{i}(i=1,2,...,N)$ are the temperatures at the reference points over the cavity, $\mathbf{b}$ is the design variable vector and $T$ is the target temperature. In the present study, we consider a two-dimensional model for simplicity. Then, one can choose as the design variables the positions of the centers of the cooling lines, their radii, the temperature of the coolant and the heat transfer coefficient between the mold and the coolant.

Now we consider the following first-order Taylor expansion of $f$ around the specific values $\mathbf{b}^{*}$ of the design variables:

$$f(\mathbf{b}^{*}+\delta\mathbf{b}) \approx f(\mathbf{b}^{*}) + \nabla f(\mathbf{b}^{*})\cdot\delta\mathbf{b} \tag{2}$$

The vector $\nabla f(\mathbf{b}^{*})$ is the direction in which the value of $f$ increases. If we write the unit vector in this direction as $\mathbf{d}$, then,

$$\mathbf{d} = \nabla f(\mathbf{b}^{*}) / \left|\nabla f(\mathbf{b}^{*})\right| \tag{3}$$

Therefore,

$$\frac{\partial f}{\partial d} = \nabla f(\mathbf{b}^{*})\cdot\mathbf{d} = \left|\nabla f(\mathbf{b}^{*})\right| \tag{4}$$

We now take the small change $\delta\mathbf{b}$ in the same direction as $\mathbf{d}$. Then,

$$\delta\mathbf{b} = (\delta d)\mathbf{d} \tag{5}$$

In this case, we can write Eq.(2) as follows:

$$f(\mathbf{b}^{*}+\delta\mathbf{b}) \approx f(\mathbf{b}^{*}) + \frac{\partial f}{\partial d}\delta d \tag{6}$$

Denoting a small change in $f$ by $\delta f$,

$$\delta f = \frac{\partial f}{\partial d}\delta d \tag{7}$$

From Eq.(4) and (7), we obtain

$$\delta d = \delta f \Big/ \frac{\partial f}{\partial d} = \frac{\delta f}{\left|\nabla f(\mathbf{b}^{*})\right|} \tag{8}$$

Substituting Eqs.(3) and (8) into Eq.(5), we finally obtain the small change of the design vector in the direction in which $f$ increases/decreases by giving a positive/negative small change $\delta f$ as follows:

$$\delta b = \frac{(\delta f)\nabla f(\mathbf{b}^*)}{|\nabla f(\mathbf{b}^*)|^2} \tag{9}$$

Since $f$ is defined as positive, we can search the minimum point of $f$ by using Eq.(9) with the appropriate value of $\delta f$ satisfying $0 < -\delta f \le f(\mathbf{b}^*)$.

**Boundary Integral Design Sensitivity Formulation for Steady-State Heat Conduction**

The gradients of $f$ with respect to the design variables become as follows:

$$\nabla f(\mathbf{b}) = \frac{2}{NT}\sum_{i=1}^{N} \nabla\theta^i(\mathbf{b})\left(\frac{\theta^i(\mathbf{b})}{T} - 1\right) \tag{10}$$

Therefore, in order to calculate $\nabla f(\mathbf{b}^*)$ in the iterative process of searching the minimum point of $f$, we must calculate the temperature distribution over the cavity surface and its design sensitivity coefficients with respect to the design variables.

The integral representation used in the boundary element analysis of steady-state heat conduction problems for an isotropic body is

$$\theta(y) + \int_{\Gamma} q^*(x,y)\theta(x)d\Gamma(x) = \int_{\Gamma} \theta^*(x,y)q(x)d\Gamma(x), \qquad y\in\Omega \tag{11}$$

where $\Omega$ is the domain of the body, $\Gamma$ is its boundary, $\theta$ is the temperature and $q$ is the heat flux. $\theta^*$ is the fundamental solution of the Laplace equation and satisfies the following differential equation:

$$\theta^*_{,ii} = \frac{1}{\lambda}\delta(x-y) \tag{12}$$

where $\lambda$ is the thermal conductivity.

$q^*$ is the heat flux related to $\theta^*$ and defined as

$$q^* = -\lambda\frac{\partial\theta^*}{\partial n} \tag{13}$$

For two-dimensional problems, $\theta^*$ and $q^*$ are given as follows:

$$\theta^* = \frac{-1}{2\pi\lambda} \ln \frac{1}{r}, \qquad q^* = \frac{-1}{2\pi r} \frac{\partial r}{\partial n} \qquad (14)$$

where $r = |x - y|$.

Equation (11) can be rearranged into the following form:

$$\int_\Gamma q^*\{\theta - \theta(y)\}d\Gamma = \int_\Gamma \theta^* q \, d\Gamma \qquad (15)$$

Equation (15) is a regularized form [5] of (11) and can be applied to a point $y$ on the boundary. In heat conduction problems, the treatment of the convective boundary condition is important. The convective boundary condition is expressed in the form

$$q = h \, (\theta - \theta_0) \qquad\qquad \text{on } \Gamma_C \qquad (16)$$

where $h$ is the heat transfer coefficient and $\theta_0$ the ambient temperature. Denoting the rest part of the boundary by $\Gamma_R$ and applying equation (16) to (15), we obtain

$$\int_\Gamma q^*\{\theta - \theta(y)\}d\Gamma = \int_{\Gamma_R} \theta^* q \, d\Gamma + \int_{\Gamma_C} \theta^* h(\theta - \theta_0) d\Gamma \qquad (17)$$

In Eq.(17), only the boundary temperature is the unknown function on $\Gamma_C$.

Discretization of Eq.(17) results in the following set of linear algebraic equations:

$$[H]\{\theta\} = [G]\{q\} \qquad (18)$$

Notice that in Eq.(18) the convective boundary condition has already been applied.

Let us assume that all the field functions in Eq.(17) depend on design variables, such as the prescribed boundary temperature and heat-flux distribution, the heat transfer coefficient, the ambient temperature, the thermal conductivity and the shape design parameters. Let us also assume that the system responses are $C^0$ continuous with respect to the design variables.

Differentiating both sides of Eq.(17) when $y \in \Omega$ results in

$$
\begin{aligned}
\int_{\Gamma} q^* \{\dot{\theta} - \dot{\theta}(y)\} d\Gamma = & \int_{\Gamma_R} \theta^* \dot{q} d\Gamma + \int_{\Gamma_C} \theta^* h(\dot{\theta} - \dot{\theta_0}) d\Gamma \\
& - \int_{\Gamma} \dot{q}^* \{\theta - \theta(y)\} d\Gamma - \int_{\Gamma} q^* \{\theta - \theta(y)\} d\dot{\Gamma} \\
& + \int_{\Gamma_R} \dot{\theta}^* q d\Gamma + \int_{\Gamma_R} \theta^* q d\dot{\Gamma} \\
& + \int_{\Gamma_C} (\dot{\theta}^* h + \theta^* \dot{h})(\theta - \theta_0) d\Gamma + \int_{\Gamma_C} \theta^* h(\theta - \theta_0) d\dot{\Gamma}
\end{aligned}
\tag{19}
$$

where a over-scribed dot ( ˙ ) denotes the differentiation with respect to a design variable.

Equation (19) is applicable continuously from $y \in \Omega$ to $y \in \Gamma$ and all the boundary integrals involved in Eq.(19) are at most weakly singular when $y$ is located on $\Gamma$.

If we choose the design variable as a shape parameter, the derivatives with respect to the design variable become as follows:

$$
\dot{\theta}^* = \theta_{,i}^* (\dot{x_i} - \dot{y_i})
\tag{20}
$$

$$
\dot{q}^* = -\lambda \theta_{,ij}^* n_i (\dot{x_i} - \dot{y_i}) - \lambda \theta_{,i}^* \dot{n_i}
\tag{21}
$$

$$
d\dot{\Gamma} = \{(\dot{x_i})_{,i} - (\dot{x_i})_{,j} n_i n_j\} d\Gamma
\tag{22}
$$

$$
\dot{n_i} = (\dot{x_k})_{,l} n_k n_l n_i - (\dot{x_k})_{,i} n_k
\tag{23}
$$

Discretizing Eq.(19) results in the following set of algebraic equations:

$$
[H]\{\dot{\theta}\} = [H]\{\dot{q}\} - [C]\{\theta\} + [D]\{q\}
\tag{24}
$$

In order to calculate the design sensitivity coefficients of the boundary temperature and heat flux by using Eq.(24), we must calculate the unknown boundary temperatures and heat fluxes by applying the boundary conditions to Eq.(18). Notice here that the design sensitivities of the temperatures or the heat fluxes are also known on the boundary where the values of the temperatures or the heat fluxes are prescribed as the boundary conditions.

## Numerical Examples

First we check the accuracy of the present design sensitivity analysis approach. In Figure 1 is shown a thick cylinder with a uniform temperature distribution along the inner wall and a uniform convective boundary condition along the outer wall. A quarter region of the section of the cylinder is divided into 22 quadratic elements as shown in Figure 2. In this example, the design variables are chosen as the radius of the inner wall $a$, thermal conductivity $\lambda$, ambient temperature $\theta_0$ and heat transfer coefficient $h$. The theoretical solution of the temperature in the radial direction is given as follows:

$$\theta = \frac{h\,(\theta_0 - \theta_a)\,\ln(r/a)}{h\,\ln(b/a) + \lambda/b} + \theta_a \tag{25}$$

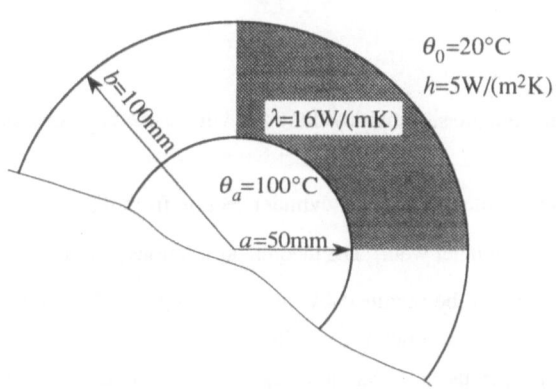

Fig.1 Section of cylinder with uniform temperature distribution over inner wall.

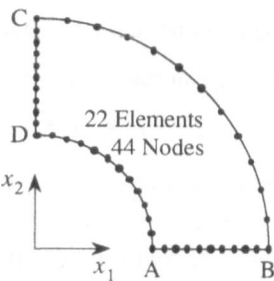

Fig.2 Boundary element discretization for quarter region of cylinder section.

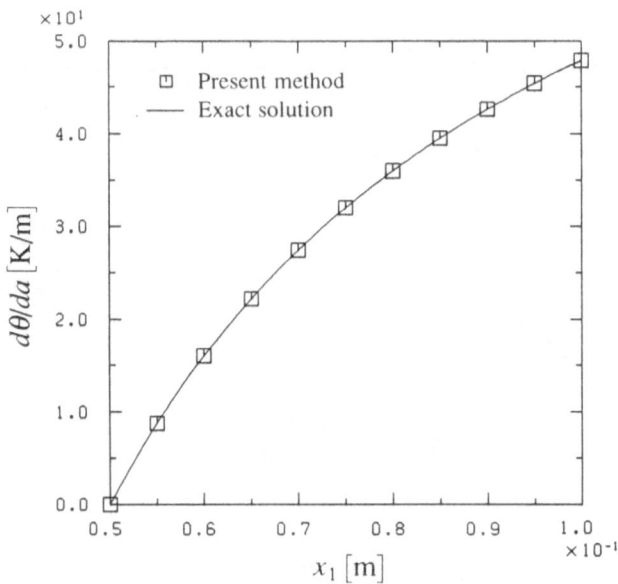

Fig.3 Temperature sensitivities along AB with respect to radius of inner wall.

where $r$ is the distance of a point in the cylinder section from the center of the cylinder and $\theta_a$ the temperature over the inner wall. The theoretical solutions for the design sensitivity coefficients of temperature can be obtained by differentiating Eq.(25) with respect to the design variables. In Figures 3 is shown the results for the temperature sensitivities along AB in Figure 2. The present results show very good agreement with the theoretical solutions.

Next we show an example of the actual design of cooling lines. In Fig.4 is shown the half section of a stationary infection mold. In this example the positions of two cooling lines (holes in Fig.2) and their radii are determined so that the temperature over the cavity surface becomes $60\pm0.8°C$. The range of the cost function in this case becomes as follows:

$$0 \leq f \leq 1.78\times10^{-4} \tag{26}$$

The initial values of the coordinates of the centers and the radii of two holes are set as follows:

$$\left.\begin{array}{lll} x_1 = 0.1\text{m}, & y_1 = 0.025\text{m}, & r_1 = 0.004\text{m} \\ x_2 = 0.05\text{m}, & y_2 = 0.025\text{m}, & r_2 = 0.004\text{m} \end{array}\right\} \tag{27}$$

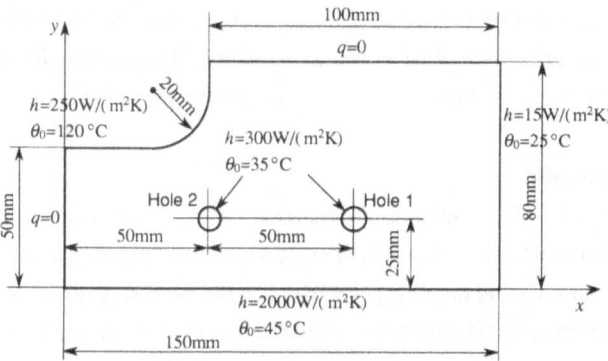

Fig.4 Half section of stationary injection mold

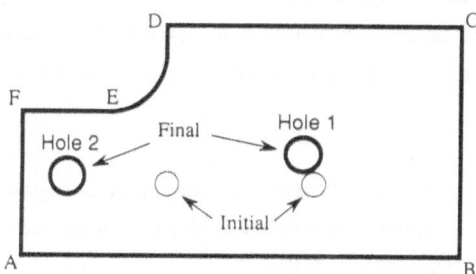

Fig.5 Designed arrangement of cooling lines

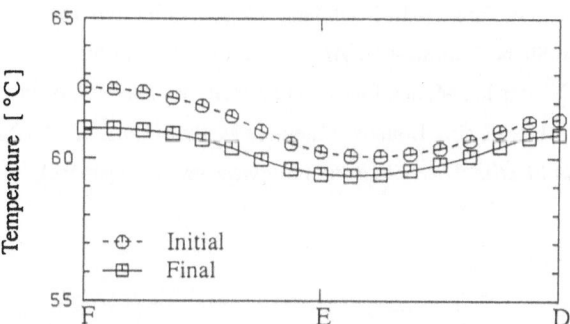

Fig.6 Temperature distributions over cavity surface

In Figs.5 and 6 are shown the final positions of two holes obtained and the temperature distributions in the initial and final states, respectively. The solution for the final state was obtained by the three times searches in the design space from the initial state.

## Concluding Remarks

A boundary integral design sensitivity formulation has been presented for steady-state heat conduction problems and applied to the optimum design of cooling lines in injection molds. The design sensitivity formulation in this paper is based on the direct differentiation of the boundary integral equation with respect to the design variables. The present approach can give us accurate design sensitivity coefficients and is efficient in the sense that the re-analyses for the model with the infinitesimally changed design variables are not needed for calculating the design sensitivity coefficients of the boundary temperatures. Some numerical test examples have shown that the present approach is accurate and effective. Such design problems as those for injection molds do not always have the state satisfying the design requirements. Therefore, it is important to know the possibility of the design at a smaller cost. The boundary element design sensitivity analysis may also be effective in judging the possibility of the required design.

## References

1. Barone, M.R. and Caulk, D.A.: Special Boundary Integral Equations for Approximate Solution of Laplace's Equation in Two-Dimensional Regions with Circular Holes, *Q. J. Mech. Appl. Math.*, **3**4(1981), 265-268 .

2. Kane, J.H.: Shape Optimization Utilizing a Boundary Element Formulation, *BETECH 86: Proc. 2nd Boundary Element Technology Conf.*, Computational Mechanics Publications, (1986), 781-803.

3. Barone, M.R. and Yang, R.J.: Boundary Integral Equations for Recovery of Design Sensitivities in Shape Optimization, *AIAA J.*, **2** (1988), 589-594.

4. Matsumoto, T., Tanaka, M. and Egawa, H.: Sensitivity Analysis of Frequency Response in Acoustic Problem Using Boundary Element Method, *Proc. JSST 12th Symposium on Calculations in Electrical and Electronics Engineering*, (in Japanese), (1991), 433-437.

# Efficient Shape Optimization Technique Based on Boundary Element Method

K. Yamazaki* and J. Sakamoto*

*   *Department of Mechanical Systems Engineering, Kanazawa University,*
    *2-40-20 Kodatsuno, Kanazawa, 920 Japan*

## Summary

An efficient approximation method to determine the optimum shape of minimum weight subjected to stress and displacement constraints is suggested by using the boundary element method. The objective function of weight is approximated to an expansion of a second-order Taylor series and the stress and displacement constraints to expansions of first-order Taylor series, based on the boundary element sensitivity analysis at current design point. Then the approximated subproblem is solved by a linear complementary pivot method. Design variable reduction techniques of isoparametric interpolation and trigonometric series interpolation for the boundary shape are also adopted for reducing the degree of freedom of the design problems.

## Introduction

Shape optimization is an important topic in the structural design research area. There are some papers[1-6] in the literature that treat the shape optimization based on the boundary element method(BEM). These works are motivated by advantage that the BEM can analyze the boundary values more precisely than the finite element analysis and the mesh generation and remeshing are relatively straightforward and inexpensive. On the other hand, the system matrices are unsymmetric and full dense. Therefore, for the efficient implementation of the shape optimization in the practical use of large scale structures, an effective and exact sensitivity analysis and a good approximation method are required to reduce the iteration of the mathematical programming procedure and of the structural analysis. Considerable effort has been devoted to developing efficient design sensitivity analysis techniques[2,4,7-11]. These techniques are classified into the direct differentiation method of the discretized integral equation[7-10] and the adjoint variable method[1,2,4,11]. The former method has been shown to be an effective strategy for the formulation of the design sensitivity analysis, since the method allows for the multiple reuse of the LU factorization of the system matrix composed for the displacement analysis. Therefore, the direct differentiation method is adopted here for the efficient sensitivity analysis.

In this paper, an efficient approximation method to determine the optimum shape of minimum weight subjected to stress and displacement constraints is suggested by using the boundary element method. Design variable reduction techniques of isoparametric interpolation and trigonometric series interpolation for the boundary shape are adopted for reducing the degree of freedom of the design problems. The objective function of weight is approximated to an expansion of a second-order Taylor series and the stress and displacement constraints to expansions of first-order Taylor series, based on the boundary element sensitivity analysis at current design point. Then the approximated subproblem is solved by a linear complementary pivot method[12]. In the design variable reduction of isoparametric interpolation, an adaptive mesh refinement technique is also adopted to keep the accuracy of the structural analysis. The efficiency of the approximation method suggested here is confirmed by applying to the minimum weight design problems to determine hole shapes in a plate under biaxial loadings.

## Efficient Shape Optimization Using BEM
### Approximation method of minimum weight design problems

Minimum weight design subject to stress constraints is one of the most important problems in the shape design problems for the practical use. Consider the minimum weight design of a two-dimensional body occupying a region $\Omega$ with a boundary $\Gamma$. When the shape design parameters which describe the boundary shape are taken as the design variables,

$$b = (b_1, b_2, \cdots, b_{Nd})^T \quad (N_d : \text{Number of design variables})$$ (1)

the minimum weight design problem is stated mathematically as minimizing the weight $f$

$$f = \frac{\rho}{2} \int_\Gamma r_0 \frac{\partial r_0}{\partial n} d\Gamma \rightarrow \min$$ (2)

subject to the displacement and stress constraints

$$g_j(b) = u_j/u_{aj} - 1 \leq 0, \quad j = 1, \cdots, M_u \quad (M_u : \text{Number of displacement constraints})$$ (3)

$$g_k(b) = \sigma_k/\sigma_{ak} - 1 \leq 0, \quad k = 1, \cdots, M_\sigma \quad (M_\sigma : \text{Number of stress constraints})$$ (4)

and the side constraints

$$b_i^L \leq b_i \leq b_i^U, \quad i = 1, \cdots, N_d$$ (5)

where $\rho$ and $n$ denote the specific gravity of the material and an unit normal vector to the boundary. $u_{aj}$ and $\sigma_{ak}$ are the allowable upper limits of the displacement $u_j$ and the stress $\sigma_k$ at the observation point $j$ and $k$, and $b_i{}^L$ and $b_i{}^U$ are the lower and upper bounds of the design variable $b_i$. When we denote an arbitrary point on the boundary as $(x_1, x_2)$, $r_0$ in Eq.(2) indicates a distance from an origin $(x_1{}^0, x_2{}^0)$ to the boundary point given as

$$r_0^2 = (x_\alpha - x_\alpha^0)(x_\alpha - x_\alpha^0)$$ (6)

in which the repeated index must be summed up in the usual tensor symbolism manner, and Greek indices refer to the Cartesian coordinate directions as $\alpha = 1, 2$.

For solving the minimum weight design problems defined above, the mathematical programming method will be used which requires the evaluation of the objective function, the displacement and stress constraint function values, and their sensitivities. It is easy and inexpensive to evaluate the objective function value for the design variable change, however, the structural analysis and its sensitivity analysis have to be done to evaluate the displacement and stress constraints for each design variable change. Then, it is not rare that iterations of the structural and sensitivity analyses more than several hundred times are required until obtaining the optimum shape finally in the large scale optimization problems. Therefore, if we can form an approximated problem for which the optimum solution will be searched directly in the approximated design space, it may be expected that the number of the structural analysis and the sensitivity analysis based on the boundary element method until getting the final solution will be reduced.

For this purpose, a direct approximation to the actual shape optimization problem is constructed by estimating the objective function using a second-order Taylor series at the current design variable $b_0$ as

$$f(b) \cong \frac{1}{2}(b - b_0)^T C (b - b_0) + (\nabla f)^T (b - b_0) + f(b_0)$$ (7)

and the displacement and stress constraints using a first-order Taylor series as

$$g_j(b) \cong (\nabla g_j)^T (b - b_0) + g_j(b_0) \leq 0$$ (8)

where

$$\nabla f(b) = (\frac{\partial f}{\partial b_1}, \frac{\partial f}{\partial b_2}, \cdots, \frac{\partial f}{\partial b_{Nd}})^T \quad \nabla g_j(b) = (\frac{\partial g_j}{\partial b_1}, \frac{\partial g_j}{\partial b_2}, \cdots, \frac{\partial g_j}{\partial b_{Nd}})^T$$

$$C = \begin{bmatrix} \dfrac{\partial^2 f}{\partial b_1^2} & \cdots & \dfrac{\partial^2 f}{\partial b_1 \partial b_{Nd}} \\ \cdots & \cdots & \cdots \\ \dfrac{\partial^2 f}{\partial b_{Nd} \partial b_1} & \cdots & \dfrac{\partial^2 f}{\partial b_{Nd}^2} \end{bmatrix}$$

The evaluation of first derivative of the constraints requires the displacement and stress sensitivity analysis based on BEM.

The approximated subproblem is solved by a nonlinear programming optimization algorithm with appropriate move limits. The move limits are employed to insure that a new design point remains in the vicinity of the current point $b_0$ around which the Taylor series was expanded. The move limits are typically specified by a limit factor $\delta$ to determine the upper and lower bounds as

$$(1-\delta)b_0 \leq b \leq (1+\delta)b_0, \quad 0 < \delta < 1 \quad (i = 1, \cdots, N_d)$$ (9)

The move limits of Eq.(9) are applied as side constraints in stead of equation (5), if they are more restrictive than the minimum and maximum gage constraints.

### Design sensitivities of weight and static response

An efficient and exact sensitivity analysis technique is very important to develop an efficient shape optimization procedure. When the boundary $\Gamma$ is discretized into boundary elements, the coordinate $x_\alpha$ at any point of parametric coordinate $\xi$ in an element is interpolated by the shape function $N_J(\xi)$ for node $J$ as

$$x_\alpha = N_J x_{\alpha J}$$ (10)

in which $x_{\alpha J}$ is a coordinate $x_\alpha$ at the nodal point $J$ in the element, and the upper-case index refers to the nodes in the element. Then, Eq.(2) is discretized as

$$f = \frac{\rho}{2} \sum_e \int_{-1}^{1} r_0 \frac{\partial r_0}{\partial n} |J| \, d\xi = \frac{\rho}{2} \sum_e \int_{-1}^{1} (x_\alpha - x_\alpha^0) n_\alpha |J| \, d\xi$$ (11)

by considering the following relationship

$$\frac{\partial r_0}{\partial n} = \frac{(x_\alpha - x_\alpha^0)n_\alpha}{r_0} \tag{12}$$

where $\Sigma_e$ means the summation with respect to every boundary element. When the derivative with respect to the parametric coordinate $\xi$ is expressed as $\partial(\ )/\partial\xi = (\ )_{,\xi}$, the Jacobian determinant and direction cosine $n_\alpha$ in the discretized form are given as

$$|J|^2 = x_{\alpha,\xi}x_{\alpha,\xi} = N_{I,\xi}N_{J,\xi}x_{\alpha I}\,x_{\alpha J}$$
$$n_1 = x_{2,\xi}/|J| = N_{I,\xi}x_{2I}/|J|, \quad n_2 = -x_{1,\xi}/|J| = -N_{I,\xi}x_{1I}/|J| \tag{13}$$

Then, differentiating Eq.(11) directly the first derivative of the objective function with respect to the design parameter $b_i$ is derived as

$$\frac{\partial f}{\partial b_i} = \frac{\rho}{2}\sum_e \int_{-1}^1 [\{\frac{\partial x_\alpha}{\partial b_i}n_\alpha + (x_\alpha - x_\alpha^0)\frac{\partial n_\alpha}{\partial b_i}\}\,|J| + (x_\alpha - x_\alpha^0)n_\alpha\frac{\partial |J|}{\partial b_i}]d\xi \tag{14}$$

where the derivatives of the Jacobian determinant and the direction cosines are given

$$\frac{\partial |J|}{\partial b_i} = \frac{1}{|J|}N_{I,\xi}N_{J,\xi}x_{\alpha I}\frac{\partial x_{\alpha J}}{\partial b_i}$$
$$\frac{\partial n_1}{\partial b_i} = \frac{n_2}{|J|}(n_1\frac{\partial x_{1,\xi}}{\partial b_i} + n_2\frac{\partial x_{2,\xi}}{\partial b_i}) , \quad \frac{\partial n_2}{\partial b_i} = -\frac{n_1}{|J|}(n_1\frac{\partial x_{1,\xi}}{\partial b_i} + n_2\frac{\partial x_{2,\xi}}{\partial b_i}) \tag{15}$$

Furthermore, the second derivative of the objective function is also given by the direct differentiation of Eq.(14).

$$\frac{\partial^2 f}{\partial b_i \partial b_j} = \frac{\rho}{2}\sum_e \int_{-1}^1 [\{\frac{\partial^2 x_\alpha}{\partial b_i \partial b_j}n_\alpha + \frac{\partial x_\alpha}{\partial b_i}\frac{\partial n_\alpha}{\partial b_j} + \frac{\partial x_\alpha}{\partial b_j}\frac{\partial n_\alpha}{\partial b_i} + (x_\alpha - x_\alpha^0)\frac{\partial^2 n_\alpha}{\partial b_i \partial b_j}\}\,|J|$$
$$+ \{\frac{\partial x_\alpha}{\partial b_i}n_\alpha + (x_\alpha - x_\alpha^0)\frac{\partial n_\alpha}{\partial b_i}\}\frac{\partial |J|}{\partial b_j} + \{\frac{\partial x_\alpha}{\partial b_j}n_\alpha + (x_\alpha - x_\alpha^0)\frac{\partial n_\alpha}{\partial b_j}\}\frac{\partial |J|}{\partial b_i} + (x_\alpha - x_\alpha^0)n_\alpha\frac{\partial^2 |J|}{\partial b_i \partial b_j}]d\xi \tag{16}$$

On the other hand, the displacement sensitivity is given by solving a system of the boundary element sensitivity equation. When the displacement $u_\alpha$ and the traction $p_\alpha$ at any point in an element are interpolated by the same shape function introduced in Eq.(10) as

$$u_\alpha = N_J u_{\alpha J}^e , \quad p_\alpha = N_J p_{\alpha J}^e \tag{17}$$

where $u_{\alpha J}^e$ and $p_{\alpha J}^e$ denote the displacement and traction at node $J$ in the element $e$. Then, the discretized boundary integral equation for a source point $y_\alpha$ on the smooth boundary is described as

$$\frac{1}{2}u_\kappa(y) + \sum_e \int_{-1}^1 p_{\kappa\alpha}^* N_I u_{\alpha J}^e |J|d\xi = \sum_e \int_{-1}^1 u_{\kappa\alpha}^* N_I p_{\alpha J}^e |J|d\xi \qquad (\kappa=1,2) \tag{18}$$

where $u_{\kappa\alpha}^*$ and $p_{\kappa\alpha}^*$ denote the Kelvin's fundamental solution of the displacement and traction given as

$$u^*_{\kappa\alpha} = \frac{1}{8\pi(1-\nu)\mu}\{(3-4\nu)\delta_{\kappa\alpha}\ln\frac{1}{r} + r_{,\kappa}r_{,\alpha}\}$$

(19)

$$p^*_{\kappa\alpha} = \sigma^\kappa_{\alpha\beta}n_\beta = -\frac{1}{4\pi(1-\nu)r}[\{(1-2\nu)\delta_{\kappa\alpha}+2r_{,\kappa}r_{,\alpha}\}r_{,n} + (1-2\nu)(r_{,\alpha}n_\kappa - r_{,\kappa}n_\alpha)]$$

(20)

$$\sigma^\kappa_{\alpha\beta} = -\frac{1}{4\pi(1-\nu)r}\{(1-2\nu)(\delta_{\kappa\alpha}r_{,\beta}+\delta_{\kappa\beta}r_{,\alpha}-\delta_{\alpha\beta}r_{,\kappa})+2r_{,\alpha}r_{,\beta}r_{,\kappa}\}$$

(21)

in which $\mu$ and $\nu$ are the shear modulus and Poisson's ratio. $\delta_{\alpha\beta}$ represents Kroneker's delta, and

$$r^2 = (x_\alpha-y_\alpha)(x_\alpha-y_\alpha) = N_I N_J (x_{\alpha I} - y_{\alpha I})(x_{\alpha J} - y_{\alpha J}) , \quad r_{,\alpha}=(x_\alpha-y_\alpha)/r , \quad r_{,n}=r_{,\alpha}n_\alpha$$

(22)

Differentiating the boundary integral equation (18), the direct differentiation method leads to following discretized boundary integral sensitivity equation to be solved for $\dot{u}^e_{\alpha I}=\partial u^e_{\alpha I}/\partial b_i$ and $\dot{p}^e_{\alpha I}=\partial p^e_{\alpha I}/\partial b_i$.

$$\frac{1}{2}\dot{u}_\kappa(y)+\sum_e \int_{-1}^1 p^*_{\kappa\alpha}N_I \dot{u}^e_{\alpha I}|J|d\xi+\sum_e \int_{-1}^1 (\dot{p}^*_{\kappa\alpha}|J|+p^*_{\kappa\alpha}|\dot{J}|)N_I u^e_{\alpha I}\,d\xi$$

$$= \sum_e \int_{-1}^1 u^*_{\kappa\alpha}N_I \dot{p}^e_{\alpha I}|J|d\xi+\sum_e \int_{-1}^1 (\dot{u}^*_{\kappa\alpha}|J|+u^*_{\kappa\alpha}|\dot{J}|)N_I p^e_{\alpha I}\,d\xi$$

$(\kappa=1,2)$

(23)

where $(\dot{\ })$ represents a derivative of corresponding component with respect to the design variable. The derivatives of the fundamental solution appeared are calculated from

$$\dot{u}^*_{\kappa\alpha} = u^*_{\kappa\alpha,\beta}(\frac{\partial x_\beta}{\partial b_i} - \frac{\partial y_\beta}{\partial b_i}) = \frac{1}{8\pi(1-\nu)\mu}\{-(3-4\nu)\delta_{\kappa\alpha}\frac{r_{,\beta}}{r} +r_{,\kappa\beta}r_{,\alpha}+r_{,\kappa}r_{,\alpha\beta}\}(\frac{\partial x_\beta}{\partial b_i} - \frac{\partial y_\beta}{\partial b_i})$$

(24)

$$\dot{p}^*_{\kappa\alpha} = \sigma^\kappa_{\alpha\beta,\gamma}(\frac{\partial x_\beta}{\partial b_i} - \frac{\partial y_\beta}{\partial b_i})n_\beta+\sigma^\kappa_{\alpha\beta}\dot{n}_\beta$$

(25)

$$\sigma^\kappa_{\alpha\beta,\gamma} = -\frac{1}{4\pi(1-\nu)r^2}[(1-2\nu)\{\delta_{\kappa\alpha}(\delta_{\beta\gamma}-2r_{,\beta}r_{,\gamma})+\delta_{\kappa\beta}(\delta_{\alpha\gamma}-2r_{,\alpha}r_{,\gamma}) -\delta_{\alpha\beta}(\delta_{\kappa\gamma}-2r_{,\kappa}r_{,\gamma})\}$$

$$+2(\delta_{\alpha\gamma}r_{,\beta}r_{,\kappa}+\delta_{\beta\gamma}r_{,\alpha}r_{,\kappa}+\delta_{\kappa\gamma}r_{,\alpha}r_{,\beta}- 4r_{,\alpha}r_{,\beta}r_{,\kappa}r_{,\gamma}\}]$$

(26)

A square system of algebraic equations for the sensitivities of the surface displacement and traction is obtained by locating the source point $y_\alpha$ of Eq.(23) at every node on the boundary.

$$[H]\{\dot{U}\}+[\dot{H}]\{U\} = [G]\{\dot{P}\}+[\dot{G}]\{P\}$$

(27)

where $[H]$ and $[G]$ denote the usual coefficient matrices of the system of boundary integral equations (18) which correspond to the first and second terms of left hand side and the first term of right hand side of Eq.(23). $\{U\}$ and $\{P\}$ are column vectors of displacement and traction components of all nodes. If the coefficient matrix of Eq.(18) after processing the boundary conditions has been decomposed like $LU$ and the solution of $\{U\}$ and $\{P\}$ have been obtained, we can solve the above equation for the displacement and traction sensitivities by forming the only the terms of $[\dot{H}]\{U\}$ and $[\dot{G}]\{P\}$. For evaluating the derivatives of $[H]$ and $[G]$, it is expected that not all elements are integrated numerically, because some design parameter perturbation may not change all parts of the boundary shape. This concept can be used to improve the computational efficiency of the sensitivity analysis.

On the other hand, the third term of left hand side of boundary integral sensitivity equation (23) includes the singularity of $O(1/r)$ when $x \rightarrow y$, however, we can apply the simple rigid body displacement technique for evaluating Cauchy principal values of $[\dot{H}]$ indirectly for the usual continuous design parameters.

The efficiency of the direct differentiation method is compared with the semi-analytic method in which the derivatives of the coefficient matrices are evaluated by the finite difference. The sensitivity of stress components is also calculated by solving the derivative form of boundary displacement and surface traction.

Reduction technique of design variables
To achieve the effective implementation of shape optimization, it is also important to reduce the number of the design variables as little as possible. An isoparametric interpolation and trigonometric series interpolation techniques for representing the arbitrary boundary shape are adopted here, where the perturbation of the boundary shape is controlled by a few design parameters at some key points.

In the isoparametric interpolation technique, a design boundary, the shape of which should be determined, is divided into several sub-design boundaries as shown in Fig.1, and the boundary shape of each sub-boundary is interpolated by the shape function $N_K(\zeta_J)$ of key point $K$ as

$$x_\alpha^J = x_\alpha^0 + N_K(\zeta_J)x_{\alpha K} \tag{28}$$

where the coordinates $x_\alpha^0$ and $x_\alpha^J$ are the coordinates $x_\alpha$ at a proper origin and any point $\zeta_J$ of parametric coordinate on the boundary, and $x_{\alpha K}$ denotes the coordinate $x_\alpha$ at key point $K$ which is positioned by a shape vector $b_K$ in an interpolated sub-boundary. The magnitude of the shape vector $b_K$ at the key points is taken directly as the design variables. This interpolation technique can represent any complicated design boundary shape with sequence of piecewise and smooth curves and can reduce the total number of design variables. However, the iterative application of the technique causes undesirable boundary meshes after the shape modification, therefore an adaptive mesh refinement technique which repositions the boundary nodes according to the stress gradient along the boundary is also adopted to keep the accuracy of the boundary element based structural analysis[5].

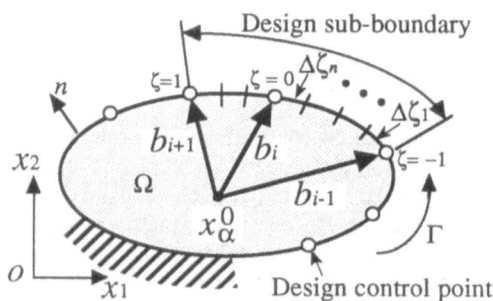

Fig.1 Shape design model and isoparametric interpolation.

In the trigonometric interpolation technique, a design boundary shape is interpolated by

$$x_\alpha^J = \sum_n A_{\alpha n}\cos(n\theta_J) + \sum_m B_{\alpha m}\sin(m\theta_J) \tag{29}$$

where $A_{\alpha n}$ and $B_{\alpha m}$ are the coefficients of trigonometric terms and are taken as the design variables directly. $\theta_J$ denotes an angle parameter of node $J$ on the design boundary. It is convenient and enough to remain the first several terms to represent the arbitrary hole and corner shapes.

## Numerical Examples

To show the effectiveness of the shape optimization procedure suggested here, the techniques are applied to the sensitivity analysis and the shape optimization of a hole in a plate subject to the stress constraints under biaxial tension loading. A quadratic isoparametric boundary element of three nodes is adopted and the plane strain condition is assumed. The material properties are assumed to be $\mu=79.2$ GPa and $\nu=0.3$. The complementary pivot method[12] is also adopted as a practical sequential quadratic programming method for optimization.

A square plate of width $2W$ with an elliptic hole subject to biaxial uniform tension is considered as shown in Fig.2. The tension loadings of $\sigma_0$ and $2\sigma_0$ are applied in the $x_1$- and $x_2$-directions. Because of symmetry, boundaries of a quarter region of the model are discretized into 26 quadratic elements and 21 nodes covers the hole boundary. The hole shape is interpolated by the trigonometric interpolating function as

$$x_1 = a_1\cos\theta, \quad x_2 = a_2\sin\theta \tag{30}$$

where $\theta$ denotes an angle between a rotating radius and the major axis $a_1$ and controls the node position on the hole boundary. The major axes $a_1$ and $a_2$ are selected as the design variables.

Table 1 Stresses and their sensitivities around elliptic hole in a plate.

| $a_1/a_2$ | | 1.0 | 2.0 | 4.0 |
|---|---|---|---|---|
| Stress | BEM | 5.104 | 9.169 | 17.28 |
| $(\sigma_\theta)_{\theta=0^\circ}/\sigma_0$ | Theory* | 5.000 | 9.000 | 17.00 |
| Stress | BEM | 0.9496 | -0.0548 | -0.5478 |
| $(\sigma_\theta)_{\theta=90^\circ}/\sigma_0$ | Theory* | 1.0000 | 0.0000 | -0.5000 |
| Sensitivity | BEM | 4.294 | 6.341 | 10.49 |
| $(\partial\sigma_\theta/\partial a_1)_{\theta=0^\circ} R/\sigma_0$ | SAM | 4.222 | 6.223 | 10.29 |
| | Theory* | 4.000 | 6.000 | 10.00 |
| Sensitivity | BEM | -2.148 | -0.8263 | -0.3674 |
| $(\partial\sigma_\theta/\partial a_1)_{\theta=90^\circ} R/\sigma_0$ | SAM | -2.150 | -0.8214 | -0.3610 |
| | Theory* | -2.000 | -0.7500 | -0.3125 |
| Sensitivity | BEM | -4.084 | -12.17 | -40.55 |
| $(\partial\sigma_\theta/\partial a_2)_{\theta=0^\circ} R/\sigma_0$ | SAM | -3.983 | -11.91 | -39.73 |
| | Theory* | -4.000 | -12.00 | -40.00 |
| Sensitivity | BEM | 2.044 | 1.486 | 1.228 |
| $(\partial\sigma_\theta/\partial a_2)_{\theta=90^\circ} R/\sigma_0$ | SAM | 2.027 | 1.475 | 1.223 |
| | Theory* | 2.000 | 1.500 | 1.250 |

\* : Theoretical values for infinite plate with an elliptic hole.
BEM : Implicit differentiation method.
SAM : Semi-analytical method.

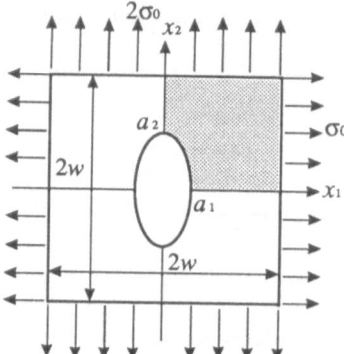

Fig.2 Design model of a plate with an elliptic hole under biaxial tension.

Table 2 The comparison of design variables, objective function and stress constraints.

|  | Initial | Optimal |
|---|---|---|
| $a_1/W$ | 0.2000 | 0.3480 |
| $a_2/W$ | 0.2000 | 0.6652 |
| $f/W^2$ | 0.9382 | 0.6694 |
| $g_1$ | -0.8050 | $1.802 \times 10^{-5}$ |
| $g_2$ | -0.2204 | $-4.902 \times 10^{-6}$ |
| $g_3$ | 0.3575 | -0.1185 |

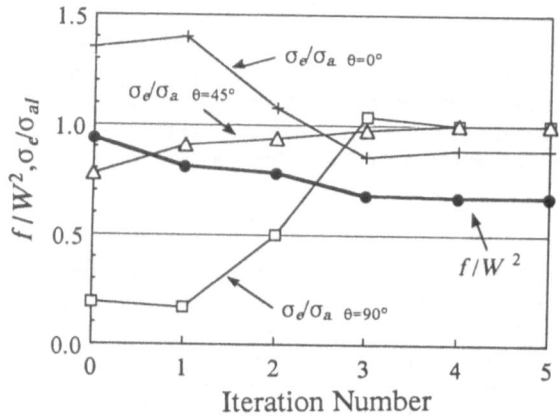

Fig.3 Variations of objective function and stress constraints.

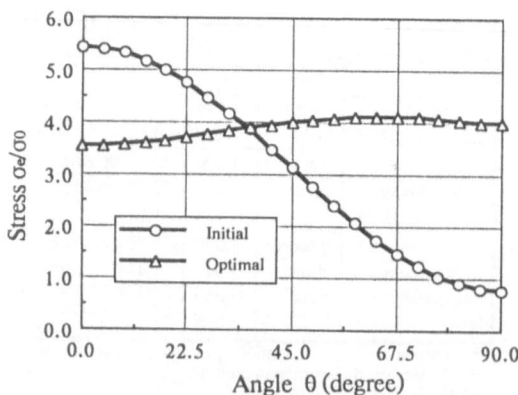

Fig.4 Von Mises tress distributions of initial and optimum shapes.

The stress sensitivity analysis based on the direct differentiation method is performed for the Von Mises stress on the elliptic hole boundary, when the aspect ratios of major axes are $a_1/a_2 = 1, 2, 4$, and $a_1/W = 1/10$. The stress and its sensitivity values on the major axes are tabulated in Table 1, in which $R = (a_1+a_2)/2$, in comparison with the values obtained by the semi-analytic method using the central difference and the perturbation $\Delta a_1/a_1 = \Delta a_2/a_2 = 0.01$. Both values of the stress sensitivities for each aspect ratio show good agreement. The theoretical values of infinite plate with an elliptic hole are also tabulated in the table for reference. It takes 2613 ms to calculate the displacement sensitivity around the hole by the direct differentiation method, while 3144 ms by the semi-analytic method with a computer FACOM M760-20. These results show the efficiency of the direct differentiation method and this tendency becomes more clear when the sensitivity analysis with much more design variables is carried out.

The minimum weight design of elliptic hole shape subject to the stress constraints is considered. The Von Mises stress values $\sigma_e$ on the elliptic hole boundary are restricted less than the allowable stress $\sigma_a = 4\sigma_0$, and the stress constraints are imposed at $\theta = 0°$, 45°, 90° in practice as

$$g_1 = \sigma_{e\,\theta=90°}/\sigma_a - 1 \le 0, \quad g_2 = \sigma_{e\,\theta=45°}/\sigma_a - 1 \le 0, \quad g_3 = \sigma_{e\,\theta=0°}/\sigma_a - 1 \le 0 \tag{31}$$

Figure 3 shows the optimization process of the objective function and the stress constraint values from an initial shape of $a_1/W = a_2/W = 0.2$. The optimum solution of the design variables, objective function and the stress constraints obtained after five iteration are tabulated in Table 2 in comparison with the initial values. The stress constraints at $\theta = 45°$ and 90° are active in the optimum shape. The stress distributions of the initial and optimum shapes around the elliptic hole are shown in Fig.4. From these results, it is confirmed that the approximation technique and the interpolation methods can achieve efficient convergence to the optimum solution.

## Concluding Remarks

For an efficient procedure of two-dimensional shape optimization, an approximation technique is suggested to decrease the iteration steps to get the optimum solution and interpolation methods to reduce the number of design variables is adopted. The objective function of weight is approximated to an expansion of a second-order Taylor series and the stress and displacement constraints to expansions of first-order Taylor series, based on the BE sensitivity analysis formulated by the direct differentiation method. The numerical results show the approximation technique can attain the optimum solution within several iteration steps.

## References
1. Mota Soares, C.A.; Rodrigues, H.C.; Choi, K.K.; Shape optimal structural design using boundary elements and minimum compliance techniques, *Trans. ASME, J. Mech. Trans. Automat. Des.*, 106(1984), 518-523.
2. Choi, J.H.; Kwak, B.M.; Boundary integral equation method for shape optimization of elastic structures, *Int. J. Num. Methods Eng.*, 26(1988), 1579-1595.
3. Sandgren, E.; Wu, S.J.; Shape optimization using the boundary element method with substructuring, *Int. J. Num. Methods Eng.*, 26(1988), 1913-1924.
4. Meric, R.A.; Shape design sensitivity analysis for non-linear anisotropic heat conducting solids and shape optimization by the BEM, *Int. J. Num. Methods Eng.*, 26(1988), 109-120.
5. Hajela, P.; Jih, J.; Adaptive grid refinement in a BEM-based optimal shape synthesis, *Int. J. Solids Struct.*, 26(1990), 29-41.

262

6. Yang, R.J.; Component shape optimization using BEM, Comp. Struct., 37(1990), 561-568.
7. Barone, M.R.; Yang, R.J.; Boundary integral equations for recovery of design sensitivities in shape optimization, *AIAA J.*, 26(1988), 589-594.
8. Kane, J.H.; Saigal, S.; Design sensitivity analysis of solid using BEM, *Trans. ASCE, J. Eng. Mech.*, 11(1988), 1703-1722.
9. Choi, J.H.; Choi, K.K.; Direct differentiation method for shape design sensitivity analysis using boundary integral formulation, *Comp. Struct.*, 34(1990), 499-508.
10. Saigal, S.; Kane, J.H.; Design sensitivity analysis of boundary element substructures, *AIAA J.*, 28(1990), 1277-1284.
11. Dems, K.; Mroz, Z.; Variational approach by means of adjoint systems to structural optimization and sensitivity analysis II.Structure shape variation, *Int. J. Solids Struct.*, 20(1984), 527-552.
12. Bazaraa, M.S.; Shetty, C.M.; *Nonlinear Programming, Theory and Algorithms*, Jhon Wiley (1973), 437-453.

# Chapter 5
# Material Property Characterization

# Inversion of Surface Acoustic Wave Data to Determine the Elastic Constants of Nitride Films

J. D. Achenbach and J. O. Kim

Center for Quality Engineering and Failure Prevention
Northwestern University, Evanston, Illinois 60208 U.S.A.

Summary
The elastic constants of single-crystal transition-metal nitride films were determined from the inversion of surface acoustic wave (SAW) dispersion data obtained by line-focus acoustic microscopy. Measurements were carried out on the (001) plane of cubic crystal TiN and NbN films grown on cubic crystal MgO substrates. Dispersion curves of SAWs propagating along the symmetry axes, namely the [100] and [110] directions, were obtained from the wave velocities for various film thicknesses and wave frequencies. Using a modified simplex algorithm, an inversion of the SAW dispersion data yields the elastic constants of cubic symmetry, namely $c_{11}$, $c_{12}$, and $c_{44}$. The inversion procedure essentially consists of seeking a least-square fit of the experimental data by the dispersion relation. The Rayleigh wave velocity calculated from the determined elastic constants agrees well with results measured by Brillouin scattering spectroscopy.

## 1. Introduction

Thin film materials are becoming widely used, for example, to improve the hardness of a surface. The elastic constants of the film on a substrate are, however, difficult to measure. By a technique which was recently discussed[1] the elastic constants of single-crystal films can be obtained by the use of a line-focus acoustic microscope, by measuring the velocities of surface acoustic waves (SAWs) propagating over the thin-film/substrate specimen. This technique requires an efficient inverse method to obtain the elastic constants from the measured SAW data. Such a technique is presented in this paper. A similar method has been applied to the inversion of Lamb wave data by Karim et al.[2], to the inversion of plate wave dispersion data by Mal et al.[3], and to the inversion of acoustic material signature data of layered solids by Kundu[4]. The method is based on a systematic function minimization procedure, known as the simplex method, introduced by Nelder and Mead[5].

Acoustic microscopy has proven to be a useful technique for obtaining SAW dispersion data of a thin film deposited on an elastic substrate[1]. The technique is based on the measurement of the V(z) curve, also often referred to as the acoustic material signature[6]. The V(z) curve is a record of the transducer voltage output $V$ with the variation of the distance $z$ between the acoustic lens and the specimen. The SAW velocity can be obtained from the periodic variation of the V(z) curve. Weglein[7] measured SAW velocities of various thin films using a point-focus acoustic microscope. The line-focus acoustic microscope allows, however, the measurement of SAW velocity in a single prescribed direction and hence it can be used to investigate the anisotropic nature of elastic materials[8]. The instrument has been used to measure the anisotropic dependence of SAW velocities on the propagation direction on single crystals. The measurement of anisotropic SAW dispersion data for single-crystal thin films grown on single-crystal solids has been carried out by Kim and Achenbach[1].

In the present paper, acoustic microscopy measurements and the inverse method have been combined to determine the elastic constants of cubic crystal TiN and NbN films grown on cubic crystal MgO substrates. These transition-metal nitride films are commonly used as hard, protective coatings for softer surfaces. They are also the components of a TiN/NbN[9] superlattice film, which has been shown to exhibit much higher hardness than homogeneous films. The elastic constants obtained in this paper have been used to calculate the effective elastic constants of the superlattice material[10]. The directional measurements by line-focus acoustic microscopy have produced the phase velocities as functions of the angle of propagation. Dispersion curves of SAWs propagating along the symmetry axes, namely the [100] and [110] directions, have been obtained by measuring the wave velocities for various film thicknesses and frequencies. By seeking a least-square fit of the dispersion relation using a modified simplex algorithm, an inversion of the SAW dispersion data has yielded the elastic stiffness constants of cubic symmetry, $c_{11}$, $c_{12}$, and $c_{44}$, for TiN and NbN. The Rayleigh surface wave velocity calculated from the determined elastic constants and the known mass density has been compared with a measurement by Brillouin scattering spectroscopy (BSS) reported elsewhere[11].

## 2. Inversion of SAW Data

### 2.1 Calculation of SAW Velocity

The dependence of the surface wave velocities on the film thickness and the frequency, i.e. the effect of dispersion, has been studied theoretically by Farnell and Adler[12]. Here we follow the approach of Ref. [12] with appropriate modifications to account for the specific anisotropies of the materials being considered here. In this paper both the film and the substrate are assumed to display cubic symmetry. The crystalline axes of the film, which are parallel and normal to the free surface, are taken to coincide with those of the substrate.

Consider a surface wave propagating along a symmetry axis, either the [100] or [110] direction, on the (001) plane of a system of cubic crystals. The elastic constants and mass density of the substrate are $\widehat{c}_{11}$, $\widehat{c}_{12}$, $\widehat{c}_{44}$, and $\widehat{\rho}$ and those of the film are $c_{11}$, $c_{12}$, $c_{44}$, and $\rho$. The displacement solution in the perfectly elastic, homogeneous, but anisotropic media is assumed to have components of the form

$$u_i = U_i \exp[i\, k\, l_3\, x_3]\, \exp[i\, k\, (x_1 - v\, t)] , \tag{1}$$

where $U_i$ is the amplitude, $k$ is the wavenumber, $x_1$ is the propagation direction in the free surface, $x_3$ is the axis normal to the free surface, and $l_3$ is the direction cosine.

Substitution of Eq. (1) into the displacement equations of motion gives the following homogeneous set of equations for the substrate

$$\left[\hat{e}_{jk} - \delta_{jk}\,\hat{\rho}\,v^2\right]\{\hat{U}_j\} = \{0\}, \qquad (j, k = 1, 2) \tag{2}$$

where $\hat{e}_{11} = \hat{c}_x + \hat{l}_3^{\,2}\,\hat{c}_{44}$, $\hat{e}_{22} = \hat{c}_{44} + \hat{l}_3^{\,2}\,\hat{c}_{11}$, and $\hat{e}_{12} = \hat{e}_{21} = \hat{l}_3\,(\hat{c}_{12} + \hat{c}_{44})$

The corresponding equations for the film are

$$\left[e_{jk} - \delta_{jk}\,\rho\,v^2\right]\{U_j\} = \{0\}, \qquad (j, k = 1, 2) \tag{3}$$

where $e_{11} = c_x + l_3^{\,2}\,c_{44}$, $e_{22} = c_{44} + l_3^{\,2}\,c_{11}$, and $e_{12} = e_{21} = l_3\,(c_{12} + c_{44})$.

For propagation along a crystalline axis, e.g., the [100] direction, $\hat{c}_x = \hat{c}_{11}$ and $c_x = c_{11}$, and for propagation along the direction of $45°$ from a crystalline direction, e.g., the [110] direction, $\hat{c}_x = (\hat{c}_{11} + \hat{c}_{12})/2 + \hat{c}_{44}$ and $c_x = (c_{11} + c_{12})/2 + c_{44}$.

From the traction-free boundary condition at the free surface of the film of thickness $h$ and from the continuity of traction and displacement at the interface between the film and the substrate, the following equation is obtained.

$$F(\overline{h}, v, c_{11}, c_{12}, c_{44}) \equiv |a_{ij}| = 0, \qquad (i, j = 1,2,...,6) \tag{4}$$

where $\overline{h}\ (= h/\lambda_s)$ is the normalized film thickness and $\lambda_s\ (= \sqrt{\hat{c}_{44}/\rho}\ /f)$ is the wavelength of transverse waves in the substrate at the frequency $f$. For $j = 1$ and $2$ the elements of the determinant in Eq. (4) are

$$a_{1j} = \hat{U}_1^{(m)}, \quad a_{2j} = \hat{U}_2^{(m)}, \quad a_{3j} = \hat{c}_{12}\,\hat{U}_1^{(m)} + \hat{l}_3^{(m)}\,\hat{c}_{11}\,\hat{U}_2^{(m)},$$
$$a_{4j} = \hat{c}_{44}\left[\hat{l}_3^{(m)}\,\hat{U}_1^{(m)} + \hat{U}_2^{(m)}\right], \quad a_{5j} = \hat{a}_{6j} = 0$$

while for $j = 3, 4, 5,$ and $6$ the elements are

$$a_{1j} = -U_1^{(n)}, \quad a_{2j} = -U_2^{(n)}, \quad a_{3j} = -c_{12}\,U_1^{(n)} - l_3^{(n)}\,c_{11}\,U_2^{(n)},$$
$$a_{4j} = -c_{44}\left[l_3^{(n)}\,U_1^{(n)} + U_2^{(n)}\right], \quad a_{5j} = \left[c_{12}\,U_1^{(n)} + l_3^{(n)}\,c_{11}\,U_2^{(n)}\right]\exp(-i\,l_3^{(n)}k\,h),$$
$$a_{6j} = \left[l_3^{(n)}\,U_1^{(n)} + U_2^{(n)}\right]\exp(-i\,l_3^{(n)}k\,h).$$

Here $\hat{U}_i^{(m)}$ and $U_i^{(n)}$ are wave amplitudes defined by Eq. (1). The superscripts $m\ (= j)$ and $n$ $(= j - 2)$ denote the m-th and n-th solutions of the eigenvalue problems defined by Eqs. (2) and (3), respectively. When the elastic constants $c_{11}, c_{12},$ and $c_{44}$ are known, Eq. (4) is the dispersion equation which yields the phase velocity $v$ as a function of the normalized thickness $\overline{h}$. On the other hand, when $v$ is known as a function of $\overline{h}$, Eq. (4) can be used to determine the elastic constants $c_{11}, c_{12},$ and $c_{44}$.

For known elastic constants the computation of the SAW velocity proceeds as follows. Initially a value of the SAW velocity $v$ is guessed and substituted into Eqs. (2) and (3). The eigenvalue problems defined by Eqs. (2) and (3) are solved for $(\hat{l}_3^{(m)}, \hat{U}^{(m)})$ and $(l_3^{(n)}, U^{(n)})$, respectively. Next the boundary-condition determinant given by Eq. (4) is evaluated. The phase velocity $v$ is searched iteratively until this determinant becomes close to zero within a given accuracy. This method of calculation yields the SAW dispersion curve for the thin-film coated crystals.

## 2.2 Modified Simplex Method

When the elastic constants are unknown, they can be determined from the measured SAW velocities by solving the dispersion equation given by Eq. (4). First this equation is rewritten as

$$F_{Aq} \equiv F(\overline{h}_q, v_q^A, c_{11}, c_{12}, c_{44}) \quad \text{and} \quad F_{Bq} \equiv F(\overline{h}_q, v_q^B, c_{11}, c_{12}, c_{44}), \quad (5a,b)$$

where $(\overline{h}_q, v_q^A)$ and $(\overline{h}_q, v_q^B)$ denote sets of data (q = 1, 2, $\cdots$ , $N$) for the normalized film thickness $\overline{h}$ and the measured value of the SAW velocity $v$ for the [100] and [110] directions, respectively. The inversion procedure determines a set of constants $(c_{11}, c_{12}, c_{44})$ that corresponds to the minimum of the following function defined by $N$ data points for each direction

$$y = \sum_{q=1}^{N} \left( w_{Aq} F_{Aq}^* F_{Aq} + w_{Bq} F_{Bq}^* F_{Bq} \right). \quad (6)$$

Here $F_{Aq}^*$ and $F_{Bq}^*$ are the complex conjugates of $F_{Aq}$ and $F_{Bq}$ and $w_{Aq}$ and $w_{Bq}$ are weighting coefficients assigned for each set of data. The weighting coefficients are initially assumed to be unity. They are adjusted such that for a calculation of Eq. (4) with an initial solution of the elastic constants and for a unit change of the velocity the change of the terms in Eq. (6) is the same for each set of data. A systematic function minimization procedure, known as the simplex method, has been introduced by Nelder and Mead[5]. In this paper the method is modified to determine the elastic constants from the experimental SAW data.

The procedure is summarized in Fig. 1. Consider the minimization of function $y$ of $n$ variables ($n$ is 3 for a cubic crystal). Select initially $(n+1)$ sets of $n$ variables, $P_0, P_1, \cdots, P_n$. Here each $P_j$ ( j = 0, 1, $\cdots$ , $n$) defines a point in $n$-dimensional space, and the $n+1$ points define the current simplex. The function value corresponding to $P_j$ is denoted by $y_j$. The subscripts j is assigned so that the $y_j$'s are arranged in increasing order of magnitude, i.e. $y_0 < y_1 < \cdots < y_{n-1} < y_n$. A midpoint M between $P_0$ and $P_1$ is calculated for later use. In order to eliminate the worst estimate $y_n$, $P_n$ is replaced by a new point according to four kinds of operations: reflection, expansion, contraction, and shrinkage.

*Reflection*: The reflection of $P_n$ is denoted by R and its coordinates are defined by the relation

$$R - M = M - P_n. \quad (7)$$

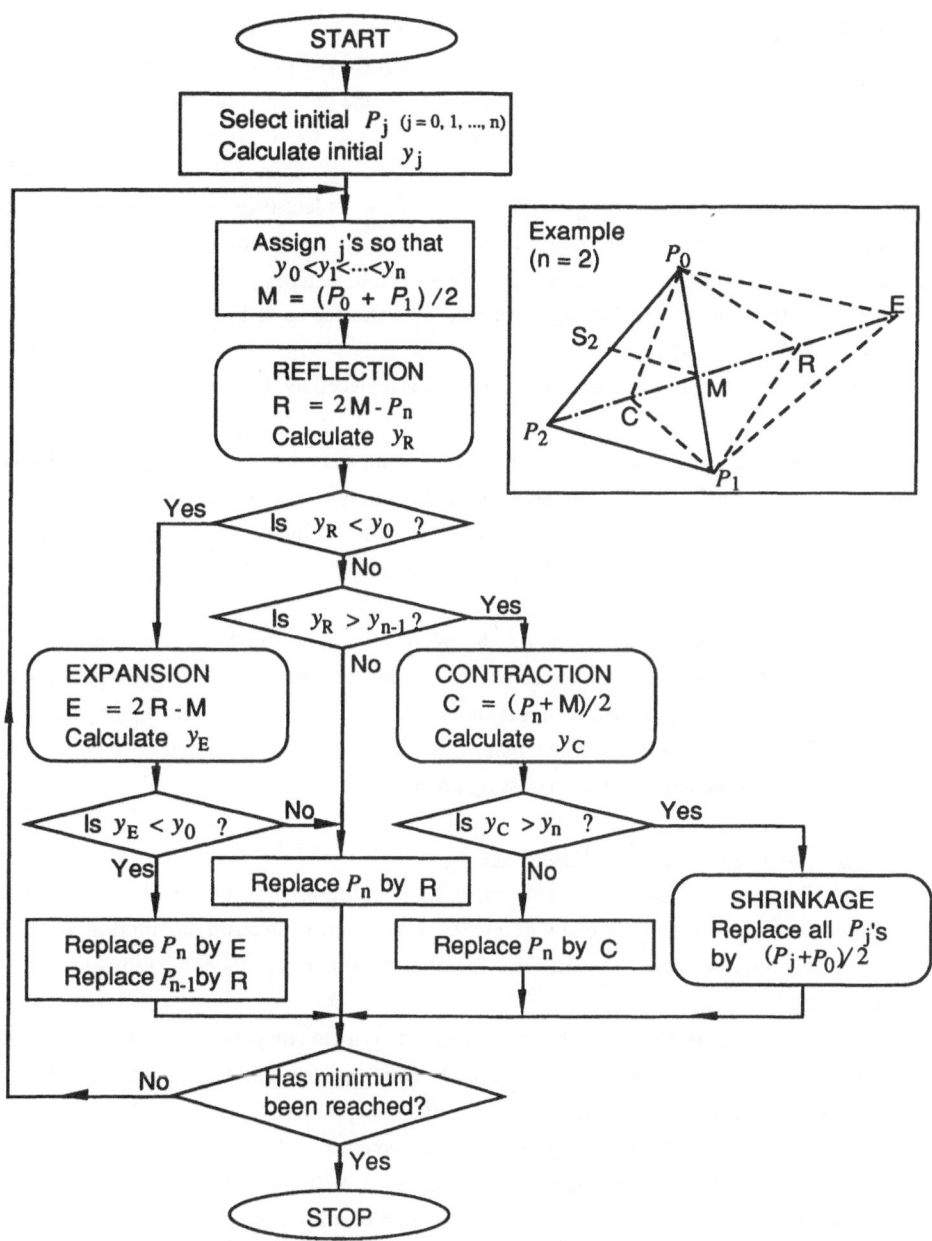

Fig. 1: Flow chart for the algorithm of a modified simplex method to determine a set $P$ minimizing a function $y$. $P_j$ ($j = 0, 1, ..., n$) is a set of $n$ unknown variables and $y_j = y(P_j)$. M is a midpoint of $P_0$ and $P_1$.

Thus R is a point on the line joining $P_n$ and M, on the far side of M from $P_n$. The corresponding value of the function obtained from Eq. (6) is $y_R$. If $y_R$ lies between $y_0$ and $y_{n-1}$, then $P_n$ is replaced by R. With the new simplex, the process is restarted.

*Expansion*: If $y_R$ is less than $y_0$, then R is expanded to E by the relation

$$E - R = R - M.$$  (8)

The function value corresponding to E is $y_E$. If $y_E$ is less than $y_0$, $P_n$ and $P_{n-1}$ are replaced by E and R and the process is restarted. If $y_E$ is not less than $y_0$, then the expansion has failed and only $P_n$ is replaced by R before restarting.

*Contraction*: If $y_R$ is greater than $y_{n-1}$, then $P_n$ is contracted to C by the relation

$$C - M = (P_n - M)/2.$$  (9)

The function value corresponding to C is $y_C$. $P_n$ is replaced by C, unless $y_C$ is greater than $y_n$.

*Shrinkage*: If all three operations - reflection, expansion, and contraction - fail, then all $P_j$'s are replaced by $(P_j + P_0)/2$ and the process is restarted with new sets of $P_j$'s.

The overall procedure is repeated until the function value reaches a minimum within a certain accuracy. The elastic constants $c_{11}$, $c_{12}$, and $c_{44}$ corresponding to the minimum are the solution to the inverse problem.

In Section 4 numerical results for the SAW velocity calculated by the use of elastic constants obtained by the method discussed in this section have been compared with experimental data. The numerical results do not include the effect of water loading on the wave velocity. This effect is assumed to be small as compared to the experimental error ($< \pm 0.2\%$) of the measurement system and the processing procedure.

## 3. Experimental Apparatus and Procedures

The thin-film/substrate specimens were made by the ultra-high vacuum reactive sputtering technique described in detail elsewhere[13]. Low-energy electron diffraction (LEED) measurements have been used to verify that the films are single crystals of the same orientation as the substrates.

A line-focus acoustic microscope consists of four main components: the acoustic probe, the measurement system for transmitting and receiving tone-burst electrical signals, the mechanical systems for alignments and movements to record V(z) curves, and a computer for controlling the system and processing the recorded waveforms. Figure 2 shows schematically the configuration of the acoustic probe and the specimen. A ZnO-film transducer generates and detects longitudinal acoustic waves at the flat surface of a Z-cut sapphire rod. The acoustic beam is focused by an acoustic lens with a cylindrical concave surface at the other end of the rod. The cylindrical concave surface has a radius of 1.0 mm and an aperture half-angle of $60^\circ$. The operating frequency is around 225 MHz and the focal length of the lens is 1.15 mm. For efficient transmission of acoustic waves through the lens-couplant interface, a chalcogenide glass film with a quarter wavelength thickness was deposited on the cylindrical concave surface. A specimen which is placed on a mechanical stage, can be translated in the vertical direction and rotated around the axis of the rod.

An r.f. tone-burst, of width 0.5 μsec and repetition rate 20 kHz, excites the transducer to generate pulsed plane acoustic waves. The plane waves propagate along the buffer rod, are refracted at the concave surface of the lens and are focused through the couplant into the specimen. Two kinds of rays return to the transducer, which now acts as a receiver. One kind is the specular reflection from normal incidence on the specimen, and the other kind is the radiation of the surface waves excited on the specimen by critical angle incidence. The voltage output of the transducer displays the interference of these two kinds of rays. Due to the vertical translation of the stage, the voltage output of the transducer is recorded to form a V(z) curve. Figure 3 shows a V(z) curve recorded for a measurement along the [100] direction on the (001) plane of 1.1 μm NbN film grown on an MgO substrate.

The surface wave velocity $v$ is a function of the interval $\Delta z$ of the periodic dips of the V(z) curve according to the following relation[6]:

$$ v = v_w \left[ 1 - \left( 1 - v_w/2f\Delta z \right)^2 \right]^{-1/2} = \left( v_w \cdot f \cdot \Delta z \right)^{1/2} \left( 1 - v_w/4f\Delta z \right)^{-1/2} , \tag{10} $$

where $v_w$ is the wave velocity in water, viz. $v_w = 1490$ m/s, and $f$ is the wave frequency. Equation (10) implies that the surface wave velocity $v$ is approximately proportional to the square root of the interval $\Delta z$ for $v_w/(4\,f\,\Delta z) \ll 1$. The periodic dip interval $\Delta z$ and thus the surface wave velocity of the specimen are obtained by processing the V(z) data. The processing procedure consists of three main steps: subtraction of the geometric effect of the acoustic lens

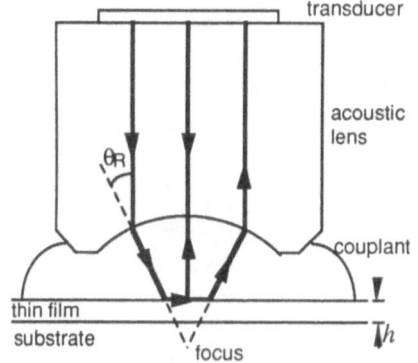

Fig. 2: The configuration of the acoustic probe and the thin-film/substrate specimen.

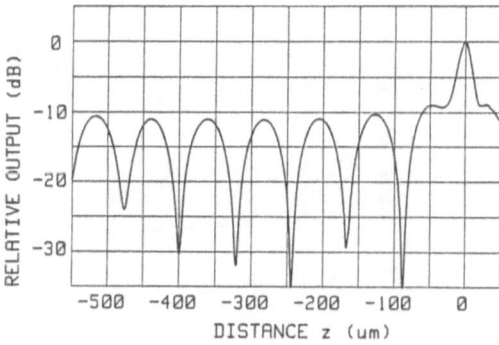

Fig. 3: V(z) curve measured at 225 MHz for propagation along the [100] direction of 1.1 μm NbN film grown on an MgO substrate.

272

from the V(z) data, a digital low-pass filtering, and a fast Fourier transform waveform analysis as described in detail by Kushibiki and Chubachi[8].

The experimental results reported in this paper were obtained with a Honda AMS-5000 ultrasonic measurement system, with a line-focus acoustic lens provided by Tohoku University.

## 4. Results and Discusstion

The anisotropic dependence of the SAW velocity on the propagation direction for the configuration of a cubic crystal thin film grown on the (001) plane of a cubic crystal substrate is investigated first. SAW velocities were measured using the line-focus acoustic microscope by the procedure described in Section 3, along wave propagation directions varying incrementally from [100] to [010]. Figure 4 shows the anisotropic dependence of the measured SAW velocities on the propagation direction for the cases of TiN and NbN films grown on the (001) plane of MgO substrates. The angle represents the direction relative to the [100] crystalline axis. The symbols in Fig. 4 show the results for a 1.1 μm NbN film at the wave frequencies 195, 225 and 255 MHz and for 1.2 and 2.8 μm TiN films and bare MgO at 225 MHz.

The symmetric anisotropy curves for TiN and NbN films in Fig. 4 support the LEED observation that the symmetry axes of the films coincide with those of the substrate. It is noted from Fig. 4 that the curves have somewhat different shapes for different film thicknesses and

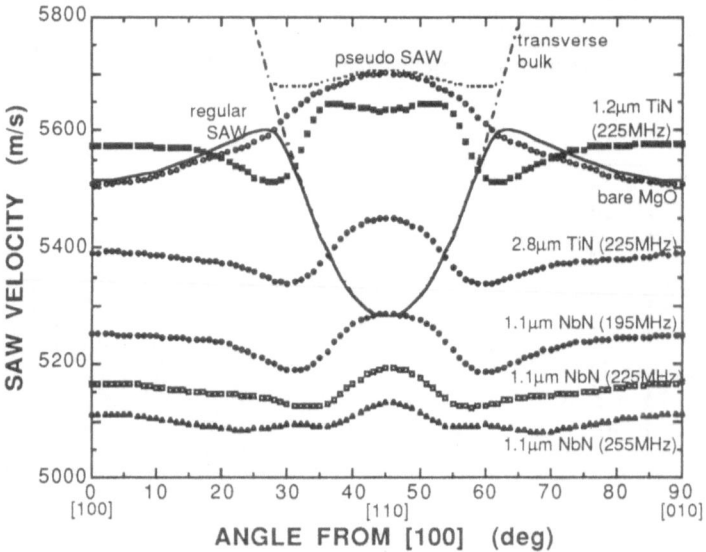

Fig. 4: Anisotropic dependence of the SAW velocity measured on the (001) plane of TiN and NbN films grown on MgO substrates. Lines show the calculated results for bare MgO.

wave frequencies. For specified material properties the film thickness and the wave frequency determine the phase velocities of the surface waves. The film thickness and the frequency are combined to define a normalized parameter and the variations of the velocities with thickness and frequency for specified direction are discussed.

In Fig. 4 the lines show the theoretical results for a bare MgO substrate. The solid and dashed lines represent the regular and pseudo SAWs, respectively, and the dash-dot line represents the transverse bulk wave, as reported elsewhere[1]. It is noted from Fig. 4 that the measured velocities near the [100] and [010] propagation directions are the phase velocities of the regular SAW. The measured velocities near the [110] direction are the phase velocities of the pseudo SAW that are observed in anisotropic crystals and that have phase velocities higher than those of the transverse bulk waves propagating in the same directions. The reason has been explained in detail elsewhere[1]. Near $30°$ from [100], the measured velocity is neither for the regular SAW nor for the pseudo SAW but for a superposition of these waves, because near this angle both regular and pseudo SAWs contribute to the phase shift of the tone-burst signal output. The measurements along the [100] and [110] directions can most easily be compared with theoretical results.

In Fig. 5, the SAW velocities measured at various frequencies for TiN films of four thicknesses grown on the (001) plane of MgO substrates are displayed for various values of the normalized thickness $h/\lambda_s$. Here $h$ is the film thickness and $\lambda_s$ is the wavelength of transverse waves in the substrate. Since the thicknesses of the films used in the experiment are much smaller than the SAW wavelength, the surface wave penetrates through the film into the substrate. In Fig. 5, the solid squares are the experimental results for the [100] direction while the open squares are for the [110] direction. The measurements were carried out for 0.1 mm at 225 MHz, 0.44 μm film at 225 and 255 MHz, and for 1.2 and 2.8 μm films at 195, 225, and 255 MHz, to yield data for nine data points. An additional data point was obtained for $h/\lambda_s = 0$, where the phase velocity equals the one for a bare substrate.

The elastic constants of TiN determined from the inversion of SAW data discuss :d in Section 2.2 are $c_{11} = 625$, $c_{12} = 165$, and $c_{44} = 163$ GPa. The lines in Fig. 5 are the dispersion curves calculated for these elastic constants and for mass density $5.39 \times 10^3$ kg/m$^3$[14]. The Rayleigh surface wave velocities along the [100] and [110] directions on the (001) plane of a TiN half-space were calculated from the elastic constants and mass density mentioned above as 5274 m/s and 5207 m/s respectively, according to a procedure similar to the one described in Section 2.1. These predicted velocities are close to the value, 5370 m/s, measured by BSS[11]. The wave direction was not stated in Ref. [11] but it is assumed to be between maximum (i.e. [100]) and minimum (i.e. [110]) directions.

In Fig. 6 the SAW velocities measured at various frequencies for NbN films of two thicknesses grown on the (001) plane of MgO substrates are displayed for various values of the normalized thickness $h/\lambda_s$. In Fig. 6, the solid squres are the experimental results for the [100]

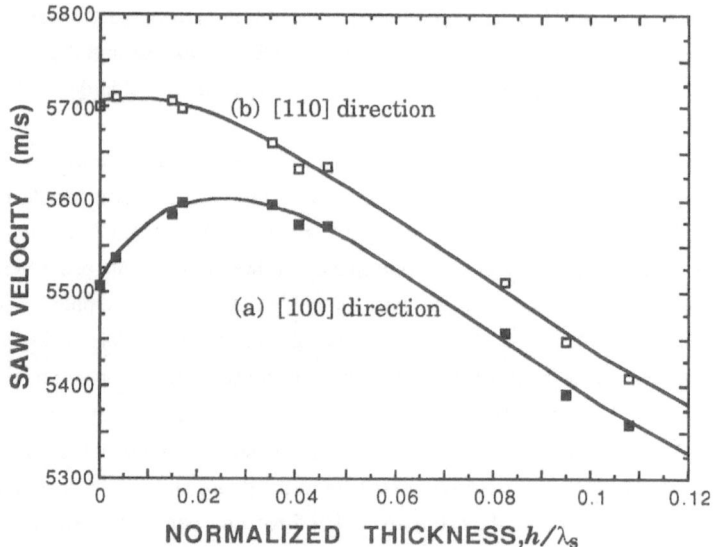

Fig. 5: SAW dispersion curves: (a) for the [100] direction and (b) for the [110] direction of TiN film grown on the (001) plane of MgO. The squares and lines show the experimental and theoretical results, respectively.

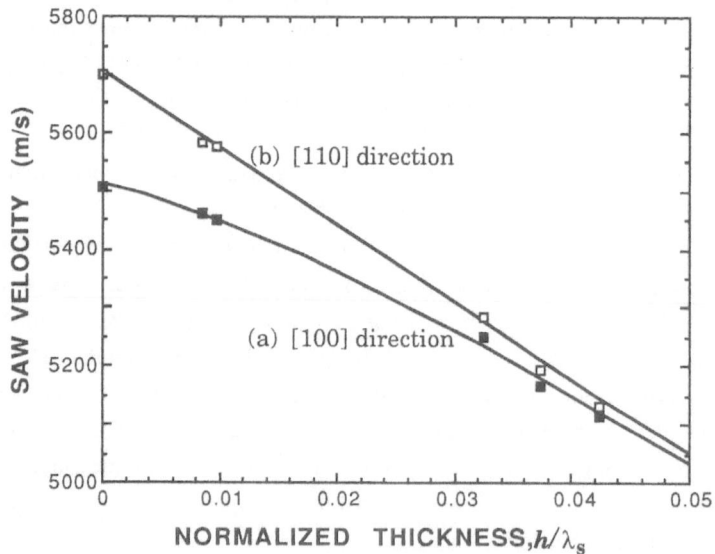

Fig. 6: SAW dispersion curves: (a) for the [100] direction and (b) for the [110] direction of NbN films grown on the (001) plane of MgO. The squares and lines show the experimental and theoretical results, respectively.

direction and the open squares are for the [110] direction. The measurements were carried out for a 0.25 μm film at 225 and 255 MHz and a 1.1 μm film at 195, 225, and 255 MHz, to yield data for five data points. An additional data point was obtained for $h/\lambda_s=0$.

The elastic stiffness constants of NbN determined from the inversion of SAW data discussed in Section 2.2 are $c_{11} = 556$, $c_{12} = 152$, and $c_{44} = 125$ GPa. The lines in Fig. 6 are the dispersion curves calculated from the determined elastic constants and mass density $8.43{\times}10^3$ kg/m$^3$[14]. The Rayleigh wave velocities along the [100] and [110] directions on the (001) plane of an NbN half-space calculated from the elastic constants and mass density mentioned above are 3718 m/s and 3688 m/s, respectively. These predicted velocities are close to the value, 3700 m/s, measured by BSS[11].

## 5. Conclusion

The elastic constants of single-crystal transition-metal nitride films grown on MgO substrates have been determined from SAW dispersion data obtained by the use of a line-focus acoustic microscope. Measurements were carried out at various frequencies, 195, 225, and 255 MHz, for TiN and NbN films of various thicknesses grown on the (001) plane of cubic crystal MgO substrates. The phase velocities measured as functions of the angle of propagation display the expected anisotropic nature of cubic crystals. The dispersion curves of SAWs propagating along the symmetry axes, namely the [100] and [110] directions, were obtained by measuring the wave velocities for various film thicknesses and frequencies. Using a modified simplex algorithm, an inversion of the SAW dispersion data has yielded the elastic constants of TiN and NbN single crystals. The Rayleigh surface wave velocities calculated from the determined elastic constants agree with values measured by Brillouin scattering spectroscopy. The results of this paper show that the combination of SAW measurements by line-focus acoustic microscopy and the proposed inverse method provides an effective means of determining the elastic constants of an anisotropic thin film on an anisotropic substrate.

## Acknowledgment

The authors are pleased to acknowledge the help of Prof. S.A. Barnett, Mr. P.B. Mirkarimi, and Ms. M. Shinn for providing us with the thin film specimens. The measurements were carried out with a line-focus acoustic lens provided by Tohoku University through the courtesy of Professors N. Chubachi and J. Kushibiki. Helpful discussions with Dr. J. Kushibiki are also gratefully acknowledged. This work was carried out in the course of research sponsored by the Office of Naval Research under Contract N00014-89-J-1362.

## References
1. Kim, J. O.; Achenbach, J. D.: Line-focus acoustic microscopy to measure anisotropic acoustic properties of thin films, *Thin Solid Films*, to appear (1992).

2. Karim, M. R.; Mal, A. K.; Bar-Cohen, Y.: Inversion of leaky Lamb wave data by simplex algorithm, *J. Acoust. Soc. Am.*, **88** (1990) 482-491.
3. Mal, A. K.; Gorman, M. R.; Prosser, W. H.: Material characterization of composite laminates using low-frequency plate wave dispersion data, *Review of Progress in Quantitative Nondestr. Eval.*, **11** (1992) to appear.
4. Kundu, T.: Inversion of acoustic material signature of layered solids, *J. Acoust. Soc. Am.*, **91** (1992) 591-600.
5. Nelder, J. A.; Mead, R.: A simplex method for function minimization, *Computer J.*, **7** (1965) 308-313.
6. Weglein, R. D.: A model for predicting acoustic material signatures, *Appl. Phys. Lett.*, **34** (1979) 179-181.
7. Weglein, R. D.: Acoustic micro-metrology, *IEEE Trans. Sonics Ultrason.*, **32** (1985) 225-234.
8. Kushibiki, J.; Chubachi, N.: Material characterization by line-focus-beam acoustic microscope, *IEEE Trans. Sonics Ultrason.*, **32** (1985) 189-212.
9. Shinn, M.; Hultman, L.; Barnett, S. A.: Growth, structure, and microhard-ness of epitaxial TiN/NbN superlattices, *J. Mater. Res.*, **7** (1992) 901-911.
10. Kim, J. O.; Achenbach, J. D.; Shinn, M.; Barnett, S. A.: Effective elastic constants and acoustic properties of single-crystal TiN/NbN superlattices, *J. Mater. Res.*, August (1992).
11. Mirkarimi, P. B.; Shinn, M.; Barnett, S. A.; Kumar, S.; Grimsditch, M.: Elastic properties of TiN/(V$_x$Nb$_{1-x}$)N superlattices measured by Brillouin scattering, *J. Appl. Phys.*, May (1992).
12. Farnell, G. W.; Adler, E. L.: Elastic wave propagation in thin layers, in *Physical Acoustics*, vol. 9, edited by W. P. Mason and R. N. Thurston (Academic Press, New York, 1972), Chap. 2.
13. Mirkarimi, P. B.; Shinn, M.; Barnett, S. A.: An ultrahigh vacuum, magnetron sputtering system for the growth and analysis of nitride superlattices, *J. Vac. Sci. Technol.*, **A 10** (1992) 75-81.
14. Holleck, H.: Material selection for hard coatings, *J. Vac. Sci. Technol.*, **A 4** (1986) 2661-2669.

# Inversion of Source, Material, and Defect Characteristics Using Guided Waves

S. K. Datta* and T. H. Ju**


*Department of Mechanical Engineering
University of Connecticut
191 Auditorium Road, Storrs, CT 06269-3139, USA

**Department of Mechanical Engineering
University of Colorado
Boulder, CO 80309-0427, USA

Summary
In this paper we have presented a review of the current literature on the use of guided elastic waves to characterize source excitation, material properties, and embedded horizontal cracks in a laminated composite plate. Results are presented for dispersion of guided waves and dynamic response in the presence of a delamination in a laminated plate.

Introduction

Propagation of guided Rayleigh-Lamb waves in a plate is of interest in seismology, electrical devices, ultrasonic material characterization, and ultrasonic nondestructive evaluation of defects. There have been numerous investigations of this problem since the early works of Rayleigh and Lamb, and reviews of early literature can be found in [1-3]. In recent years attention has been focused on the surface response of a plate due to buried or surface sources in the context of acoustic emission. For references the reader is referred to [4-6].

Dispersion of guided elastic waves in a laminated composite plate and inversion of anisotropic elastic properties of the plate have also received considerable attention in recent years. Extensive references can be found in [6-9]. All the experimental studies that deal with inversion of elastic properties have considered fluid-coupled plates.

In this paper we present a combined theoretical and experimental study of guided waves in a free-free plate. First we consider the inversion of the time dependence of a vertical line source modeling the impact of an elastic ball on an elastic plate. The model studied is a simulation of the experimental set-up shown in Fig. 1. The theoretical study deals with the surface response of a plate, which can in general be anisotropic and laminated, due to a vertical transient line force acting on the surface of the plate. It is assumed that the plate is a cross-ply laminated plate, in which each lamina is transversely isotropic with the symmetry axis lying in the plane of the lamina. This is to model a [0°/90°] lay-up of continuous uniaxial fiber reinforced laminae. The special case of an isotropic plate can be derived by taking the properties of the laminae to be the same and isotropic.

Dispersion of guided waves in the cross-ply laminated plate has been modelled next. Again we confine our attention to the two-dimensional case when the propagation is in the direction of the fibers in the 0° ply. The experimental results were obtained by using variable-angle-wedge contact transducers operated in a pitch-catch mode. Comparison of these with the model results leads to the determination of the elastic constants of a ply.

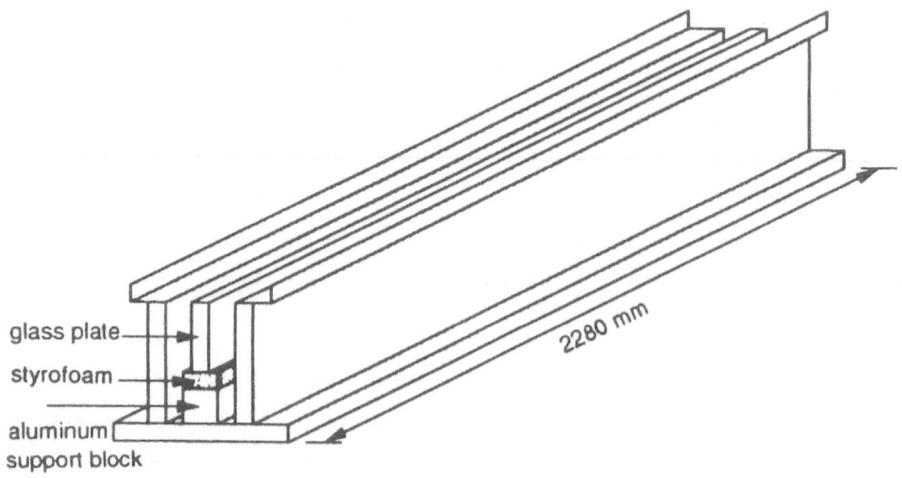

Fig. 1   The glass specimen is a 25.4 mm by 5.6 mm by 2.28 m plate

Finally we model the scattering of guided waves by a delamination between the first two plies ($0°$ and $90°$) in the cross-ply plate. It is shown that observation of surface displacements in time and frequency domains can be used to determine the length and depth of delamination.

Formulation and Solution

I. Green's displacement function for the plate

Figure 2 shows the geometry of the cross-ply plate with a delamination between the first two plies. Each ply is assumed to be homogeneous and transversely isotropic. A transversely isotropic material has five independent elastic stiffnesses, namely $C_{11}$, $C_{33}$, $C_{12}$, $C_{13}$, and $C_{55}$. All deformations are assumed to be small so that the problem can be analyzed using linear elastodynamic theories. The layers are assumed to be perfectly bounded to one another, so that the displacements and stress vectors are continuous across the interfaces. We will first compute the Green's functions, $G_{ki}(\mathbf{x}, \mathbf{x}'; t)$, which are defined as the displacements in the $i^{th}$ direction at a field point $\mathbf{x}$ due to a unit impulsive load in the $k^{th}$ direction acting at $\mathbf{x}'$. As shown in Fig. 2, a uniform unit line source along the y axis acts on the surface $z = 0$.

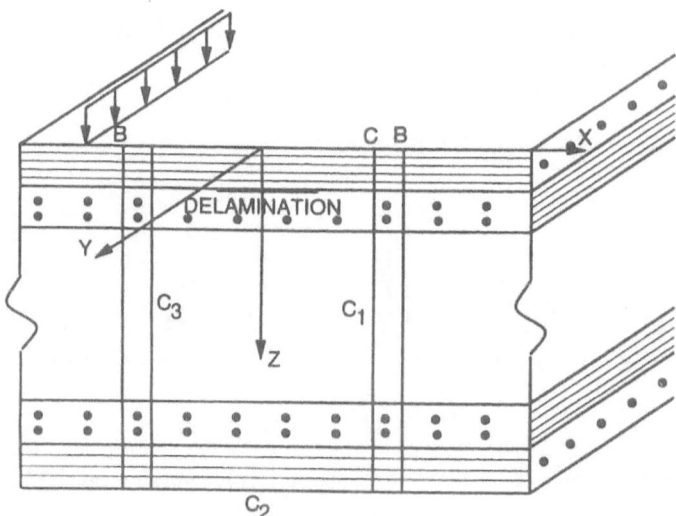

Fig. 2    Geometry of the $[0°/90°]_{2s}$ plate with delamination and the applied vertical line source

The equations of motion in each ply are,

$$C_{11}\widehat{u}_{xx} + (C_{13} + C_{55})\widehat{w}_{xz} + C_{55}\widehat{u}_{zz} = \rho\widehat{u}_{tt} \qquad (1.a)$$

$$C_{33}\widehat{w}_{zz} + (C_{13} + C_{55})\widehat{u}_{xz} + C_{55}\widehat{w}_{xx} = \rho\widehat{w}_{tt} \qquad (1.b)$$

where it is assumed that the wave propagation is in the x-z plane. $\widehat{u}(t, x, z)$ and $\widehat{w}(t, x, z)$ are displacements in the x and z directions, respectively. The subscripts denote the derivatives of these displacements with respect to space variables (x and z) and time (t). After taking Fourier transforms of these equations with respect to time and space variable x, such as

$$\phi(\omega, k, z) = \int_{-\infty}^{\infty}\int_{-\infty}^{\infty} \widehat{\phi}(t, x, z)e^{i(\omega t - kx)}\, dt\, dx$$

where $\widehat{\phi}$ is an abitrary function, equations (1) becomes

$$C_{55}u_{zz} + (C_{13} + C_{55})(ik)w_z + (\rho\omega^2 - C_{11}k^2)u = 0 \qquad (2.a)$$

$$C_{33}w_{zz} + (C_{13} + C_{55})(ik)u_z + (\rho\omega^2 - C_{55}k^2)w = 0 \qquad (2.b)$$

The solution of equation (2) when there is an impulsive loading at $(x', z')$ can be derived using the method outlined in [10]. For details the reader is refered to [10]. It can be shown that the displacement-stress vector, $[u, w, \sigma_{zz}, \sigma_{xz}]^T$, at any level z in the $m^{th}$ lamina can be written as,

$$Q(m)E(z, m)A(m) = T(m) , \qquad m = 1, 2, ...N. \qquad (3)$$

where

$$Q(m) = \begin{bmatrix} 1 & 1 & 1 & 1 \\ r_1 & r_2 & -r_1 & -r_2 \\ P_1 & P_2 & P_1 & P_2 \\ Q_1 & Q_2 & -Q_1 & -Q_2 \end{bmatrix}$$

$$E(z, m) = diag(e^{is_1(z - z_{m-1})}, e^{is_2(z - z_{m-1})}, e^{is_1(z_m - z)}, e^{is_2(z_m - z)})$$

$$T(m) = \{u \quad w \quad \sigma_{zz} \quad \sigma_{xz}\}$$

The $4 \times 1$ column matrix, $A(m) = \{A_1, A_2, A_3, A_4\}$, contains the arbitrary constants that have to be found from the boundary conditions and for the given line force and its point of application. Note that $m^{th}$ layer is bounded by the planes $z = z_{m-1}$ and $z = z_m$. The quantities $r_1, r_2, P_1, P_2, Q_1, Q_2$, and $s_1, s_2$ have been defined in [10].

Using the conditions of continuity of the displacement-stress vector at the interfaces between adjacent laminae, the appropriate jump conditions for the stress components $\sigma_{zz}$ and $\sigma_{xz}$ at the point where the force is applied, and the boundary conditions at the top and bottom surfaces of the plate, the 4N constants can be evaluated. Here N is the number of laminae. Once these are known then the displacement-stress vector can be calculated by Eq. (3). After taking the inverse Fourier transforms with respect to x and t the Green's displacement tensor, $G_{ki}(x, z; x', z'; t)$ for the plate is obtained.

## II. Dispersion of guided waves

The impulsive response of the plate has some characteristic features that are determined by the dispersive guided waves that can be supported by the plate. These dispersive waves are influenced greatly by the anisotropic properties of the plate and the interface conditions. In recent years several studies on the determination of lamina and interface properties by using guided wave dispersion has been reported [7-10]. In all these studies except [10] fluid-coupled plates have been considered. In this paper we present a summary of our work on guided waves in a free-free plate.

As in the previous section we consider the solution (3) for a lamina and using the interface continuity conditions and the stress-free conditions at the top and bottom surfaces of the plate we obtain the dispersion equation for guided waves in the plate. For details the reader is refered to [11,12]. In these papers an alternative numerical scheme based on a stiffness method has been presented also to obtain the frequency-wavenumber dispersion relation. This latter method has the advantage that a large number of layers can be considered.

## III. Scattering by a delamination defect

Figure 2 shows the geometry of the laminated plate with the delamination defect. As shown, the defect lies inside a fictitious contour C, which is composed of $C_1$, $C_2$ and $C_3$. We define the interior region $R_I$ to be bounded outside by the imaginary boundary B, which is composed of two verticle lines parallel to $C_1$ and $C_3$. The interior region $R_I$ contains the delamination defect and is divided into finite elements.

Now the total field in the region, $R_E$, exterior to C can be written as a boundary integral,

$$u_i(x', z') = \oint_C (u_j \Sigma_{ijk} - G_{ij}\sigma_{jk})n_k dC + u_i^{(f)}(x', z').\qquad (4)$$

where $G_{ij}(x, z; x', z')$ is the Green's function discussed in the foregoing, $\Sigma_{ijk}(x, z; x', z')$

is the stress derived from $G_{ij}$, and $u_i^{(f)}$ is the incident field in the absence of the delamination. Equation (4) can be used to write the displacements on B in terms of u and $\sigma$ on C. Combining this with the finite element representation of the interior field it is possible to solve for the displcements at the nodes on B and those at the nodes in $R_I$. This has been discussed in details in [13]. Once these displacements are known, then the stresses $\sigma_{ij}$ in the interior of B can be calculated. Equation (4) can then be used to find u at any point outside B.

## Numerical Results and Discussion

To use the guided waves to determine the frequency characteristics of the source of excitation we used a ball drop set-up as shown in Fig. 3 to excite the waves in the glass specimen shown in Fig. 1. The sensor to measure the surface strain in the y-direction is a PVDF sensor, which was used as shown in Fig. 3. Because the dominant frequency of the excited wave was less than 0.25 MHz, the wavelengths are much larger than the width of the plate. So the problem is two dimensional (plane stress).

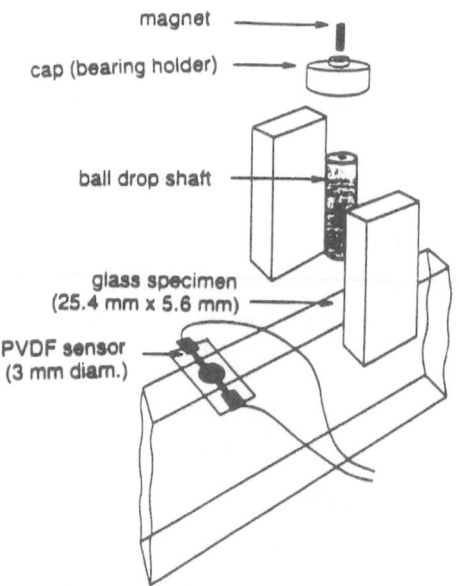

Fig. 3   Ball drop source setup and PVDF sensor

In the experiment the location of ball drop is fixed. The receivers are placed at distances of 5H and 8H from the source (ball drop). Here H is he height of the plate. The sensor measures the strain, $e_{yy}$, at these locations. Using an FFT (fast Fourier transform) the frequency spectrum is found to be $R_1(f)$ at (5H,0), which is given by S(f). D(f), where S(f) is the frequency spectrum of the source and $D_1(f)$ is the Fourier transform of the strain derived from the Green's function for the glass plate. This latter is calculated by the general technique discussed in the foregoing. The S(f) is known from the expression, $R_1(f)/D_1(f)$. Figure 4 shows the time domain response at 5H as measured by the PVDF sensor. The frequency spectrum of the source function is shown in Fig. 5.

The source function derived from the measured $e_{yy}$ at 5H is then used to predict the strain $e_{yy}$ at a distance 8H. This is shown in Fig. 6 and Fig. 7 shows the measured results. It is seen that these agree very well, thus demonstrating the accuracy of the inverted source function.

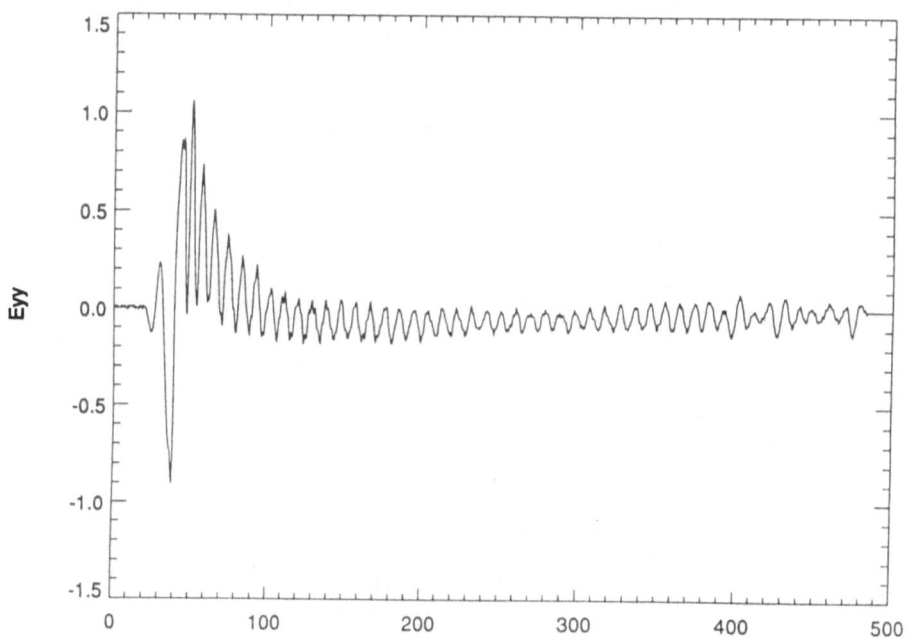

Fig. 4    Measured $e_{yy}$ in time domain at 5H

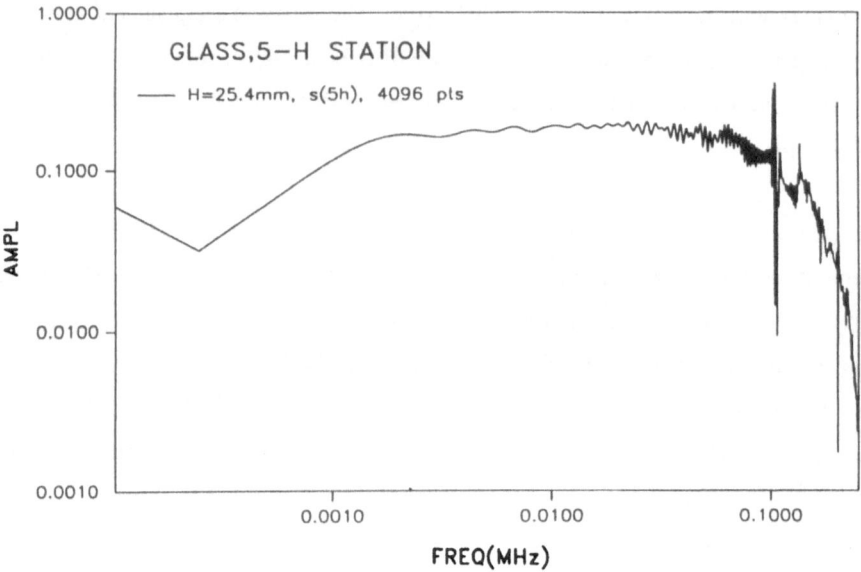

Fig. 5   Source spectrum extracted from the measured $e_{yy}$ at 5H

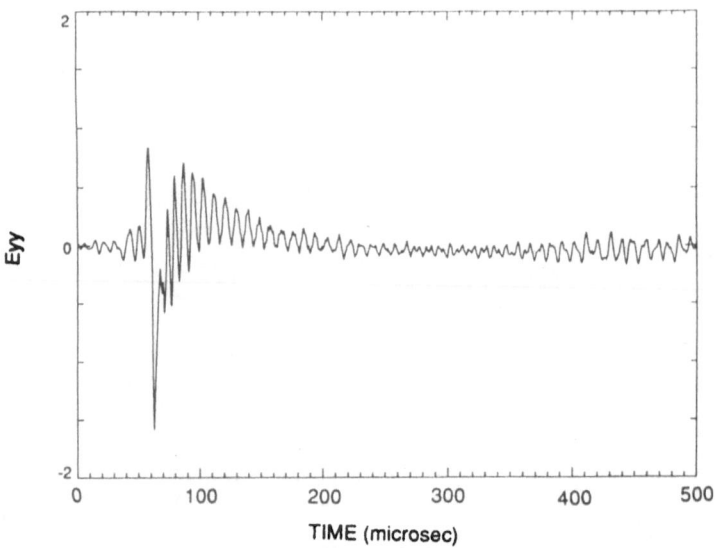

Fig. 6   Simulated $e_{yy}$ in time domain at 8H from the source

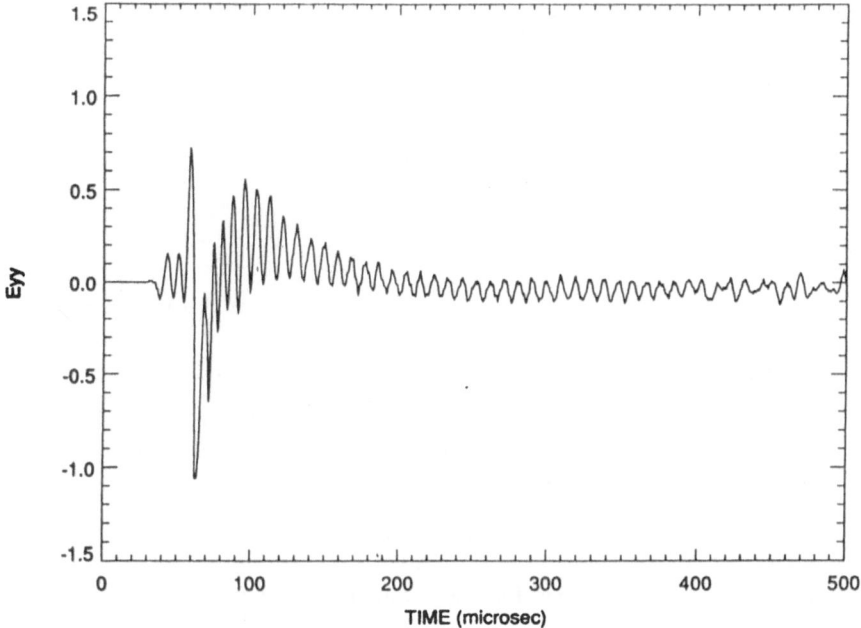

Fig. 7 Measured $e_{yy}$ in time domain at 8H from the source

Use of guided wave dispersion in a laminated plate provides an excellent means to determine the elastic stiffnesses of the laminae. To illustrate this we consider a 8-layer cross-ply $[0°/90°]_{2s}$ graphite-epoxy plate. The experimental set-up used variable-angle-wedge contact transducers operated in pitch-catch mode. This is shown in Fig. 8. The phase velocity is evaluated from the phase shift of a continuous sine wave between two positions along the plate. The transmitting transducer is a 1 MHz broad band which was excited by a continuous wave. The angle of incidence (20 - 50° from normal) and frequency (250 kHz -1.2 MHz) are fixed to select a particular guided wave mode. At the receiving transducer amplitude peaks are observed when a mode is excited. Figure 9 shows the experimentally measured velocities at different frequencies. The properties of lamina were varied in the numerical computations to obtain a good fit between the theoretical and experimental results. Dotted lines in Fig. 9 are the theoretically generated best-fit dispersion curves. Corresponding elastic stiffnesses of a lamina are given in Table 1. Note that the results in Fig. 9 are for propagation in the fiber direction in the topmost ply. This allows determination of $C_{11}$, $C_{33}$, $C_{13}$ and $C_{55}$ and $C_{44}$. It was

**Experimental Apparatus**

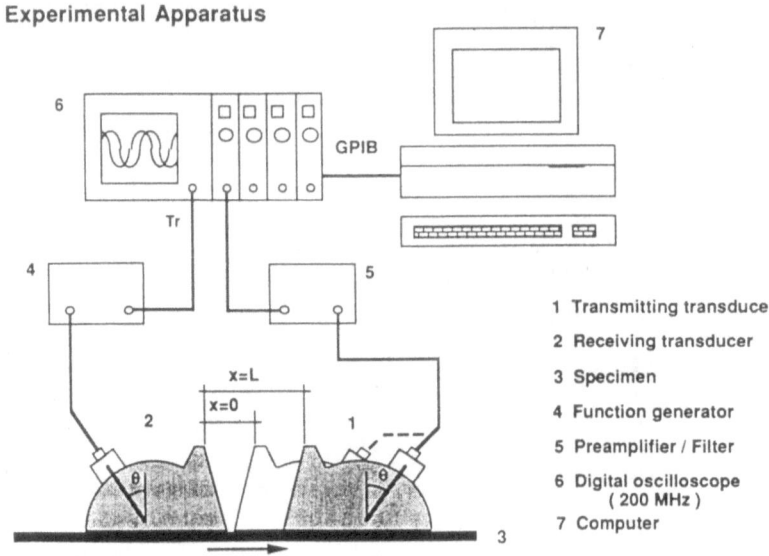

1 Transmitting transducer
2 Receiving transducer
3 Specimen
4 Function generator
5 Preamplifier / Filter
6 Digital oscilloscope
  ( 200 MHz )
7 Computer

Fig. 8   Experimental arrangement for the measurement of phase velocity

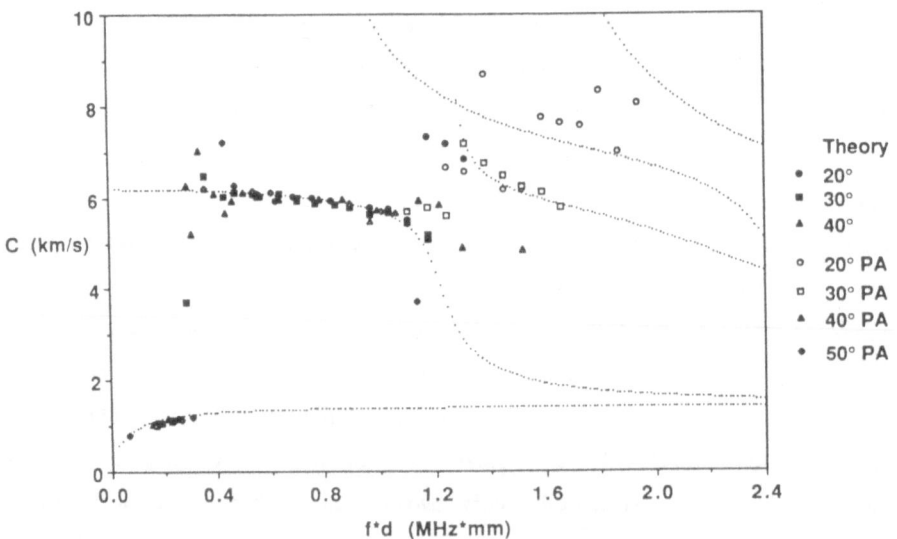

Fig. 9   Comparison of the experimental results with the theoretically generated dispersion curves

Table 1.   Inversion of elastic constants of a lamina in a graphite-epoxy plate

Material parameters

| Nominal Properties | | Equivalent Stiffness | | Matched Stiffness | | Derived Properties | |
|---|---|---|---|---|---|---|---|
| $\rho$(g/cc) | 1.6 | | | | | $\rho$(g/cc) | 1.48 |
| $E_L$(GPa) | 140 | $C_{11}$(GPa) | 142 | $C_{11}$(GPa) | 106.5 | $E_L$(GPa) | 104 |
| $E_T$(GPa) | 10 | $C_{22}=C_{33}$ | 12 | $C_{22}=C_{33}$ | 9.6 | $E_T$(GPa) | 8 |
| $\nu_{LT}$ | 0.25 | $C_{12}=C_{13}$ | 4.26 | $C_{12}=C_{13}$ | 4.26 | $\nu_{LT}$ | 0.31 |
| $\nu_{TL}$ | 0.018 | $C_{23}$(GPa) | 4.8 | $C_{23}$(GPa) | 3.0 | $\nu_{TL}$ | 0.024 |
| $\mu_L$(GPa) | 4.0 | $C_{44}$(GPa) | 3.6 | $C_{44}$(GPa) | 3.3 | $\mu_L$(GPa) | 3.2 |
| $\mu_T$(GPa) | 3.6 | $C_{55}=C_{66}$ | 4.0 | $C_{55}=C_{66}$ | 3.2 | $\mu_T$(GPa) | 3.3 |

assumed in our modeling that $C_{13}$ was the same as the assumed nominal property. It is seen that this is an efficient way to find a set of elastic stiffnesses of individual laminae in a laminated plate. Once the individual lamina properties are known then it is possible to study the scattering of elastic waves in a laminated plate. For illustrative purposes we present some model results for scattering of an impact wave by a delamination defect in an 8-layered cross-ply graphite-epoxy plate. Details of the computational procedure can be found in [13]. The properties of the 0° ply are: $C_{11}$=160.7 GPa, $C_{33}$=13.96 GPa=$C_{22}$, $C_{55}$=7.07 GPa=$C_{66}$, $C_{44}$=3.50 GPa, $C_{13}$=6.44 GPa, $\rho$=1.8$gcm^{-3}$. The thickness of the plate is 5.08 mm and the delamination is between the top 0° ply and the adjacent 90° ply. Its length is 6.4 mm (see Fig. 2). The line impact load is applied at 5.68 mm horizontal distance away from the left tip of the delamination. The time dependence of the impact is,

$$f(t) = sin[2\pi(t - T)/\tau], \quad T < t < T + \tau$$
$$= 0, \qquad\qquad\qquad t < T, t > T + \tau$$

(5)

For the numerical computations it was assumed that T=5$\mu s$ and $\tau$=2$\mu s$. Figure 10 shows the spectra of the vertical surface displacements at equi-spaced point on the surface above the delamination. The sharp peaks at the normalized frequency of 1.2 is close to the fundamental resonance frequency of the thin plate above the delamination with simply supported edges. It may be noted that this resonance frequency is given by

$$\omega = \left(\frac{\pi}{l}\right)^2 h \sqrt{\frac{C_{11} - \frac{C_{13}^2}{C_{33}}}{12\rho}} \qquad (6)$$

Here $l$ is the length of the plate (delamination) and $h$ is its thickness (depth of the delamination). Thus for the properties considered in this example the normalized frequency is 1.13, which is pretty close to 1.2. thus the frequency of the first sharp peak in Fig. 10 gives a relation between $h$ and $l$. To find another relation it is necessary to observe the time of travel of trapped Rayleigh wave as shown in Fig. 11. The time of travel of this wave from the left tip to the right is about 3.8 $\mu s$ and at the velocity of this wave at the central frequency of 0.5 MHz gives close to the actual delamination length. thus it is concluded that the measurement of surface displacement in the frequency and time domains can be used to inversely estimate a delamination location and its length.

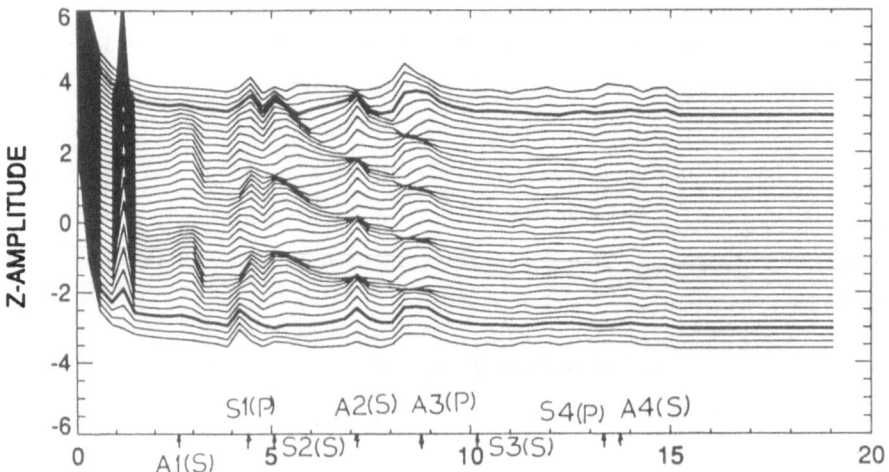

Fig. 10   Spectra of the vertical displacement when the load is an impulse

## SURFACE RESPONSE IN TIME-DOMAIN

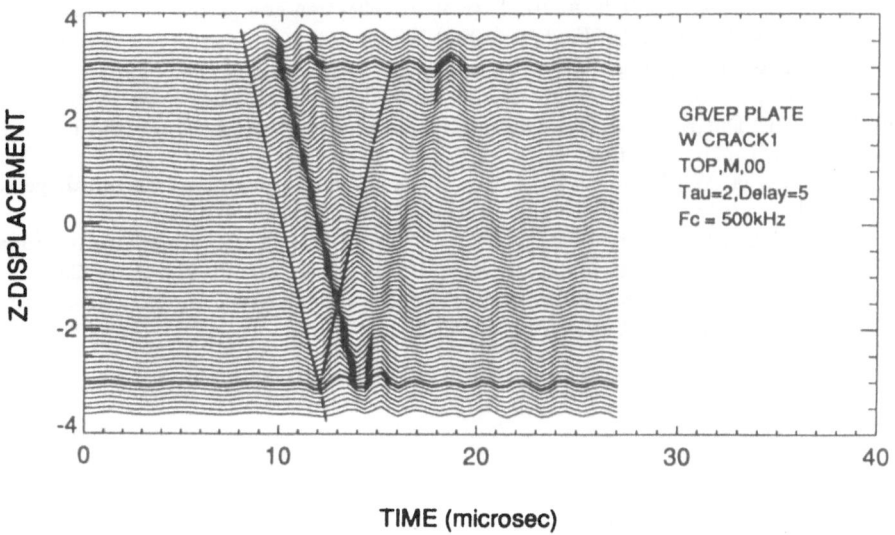

GR/EP PLATE
W CRACK1
TOP,M,00
Tau=2,Delay=5
Fc = 500kHz

TIME (microsec)

Fig. 11  Time domain vertical surface displacement when the source time function is given by Eq. (5)

Acknowledgement
The work reported here was supported in part by a grant from Office of Naval Research (grant # N00014-92-J-1346; Scientific Officer: Dr. Y. D. S. Rajapakse). Financial support provided to the second author (T. H. Ju) by University of Connecticut is gratefully acknowledged.

References
1. Mindlin, R. D.: Waves and vibrations in isotropic, elastic plates, in Structural Mechanics (J. N. Goodier and N. J. Hoff, eds.), Pergamon Press, New York (1960) 199—232.
2. Miklowitz, J.: The Theory of Elastic Waves and Waveguides, North-Holland, Amsterdam (1978) 179—230.
3. Auld, B. A.: Acoustic Fields and Waves in Solids, Vol.II, R. E. Krieger Publishing Co., Malabar, Florida (1990) 63—220.
4. Ceranoglu, A. N.; Pao, Y. H.: Propagation of elastic pulse and acoustic emission in a plate. Part 1, 2 and 3, J. appl. Mech., 48 (1981) 125—147.
5. Paffenholz, J.; Fox, J. W.; Gu, X.; Jewett, G. S.; Datta, S. K.; Spetzler, H. A.: Experimental and theoretical study of Rayleigh-Lamb waves in a plate containing a surface-breaking crack, Res. Nondestr. Eval., 1 (1990) 197—217.
6. Datta, S. K.; Achenbach, J. D.; Rajapakse, Y. S., eds.: Elastic Waves and Ultrasonic Nondestructive evaluation, North-Holland, Amsterdam (1990).

7. Mal, A. K.; Ting, T. C. T., eds.: Wave Propagation in Structrual Composites, AMD-Vol. 90, The American Society of Mechanical Engineers, New York (1988).
8. Chimenti, D. E.; Nayfeh, A. H.: Ultrasonic reflection and guided waves in fluid-coupled laminates, *J. Nondestr. Eval.*, **9** (1990) 51—69.
9. Mal, A. K.; Rajapakse, Y. D. S., eds.: Impact response and Elastodynamics of Composites, AMD-Vol. 116, The American Society of Mechanical Engineers, New York (1990).
10. Ju, T. H.; Datta, S. K.: Dynamics of a laminated composite plate with interface layers, submitted for publication.
11. Karunasena, W.; Shah, A. H.; Datta, S. K.: Wave propagation in a multilayered laminated cross-ply composite plate, *J. Appl. Mech.*, **58** (1991) 1028—1032.
12. Karunasena, W.; Bratton, R. L.; Datta, S. K.; Shah, A. H.: Elastic wave propagation in laminated composite plates, *J. Eng. Matls. Tech.*, **113** (1991) 413—420.
13. Ju, T. H.; Datta, S. K.: Pulse propagation in a laminated composite plate and nondestructive evaluation, *Comp. Eng.*, **2**(1992) 55—66.

# Inverse Problems Associated with Nondestructive Evaluation of Plastic Damages in Solids

T. Mura[1], T. Koya[2], S. C. Hsieh[3], Z. A. Moschovidis[4] and Z. Gao[5]

### Extended Abstract

It will be shown that solving inversely Green's boundary integral equations leads to a non-destructive evaluation of damage in terms of non-elastic strains (plastic strains, Somigliano's dislocations). Elasticity problems with partially overprescribed boundary conditions are also inverse problems, one of which is the pull-out test of fiber in a composite material. A numerical scheme is proposed to avoid the difficulty of integrations due to high singularity of the integrands.

A special idea is proposed for prediction of the damage domain. Due to the page limitation, only descriptions of the numerical examples are given by omitting mathematics and resulting figures and pictures.

### Green's Formula

Plastic strains $\varepsilon_{kl}^p$ are accumulated in a subdomain $\Omega$ of the given body $D$ (see Figure. 1) after a series of unknown loadings. The subdomain $\Omega$ is called the damage domain, and its shape and location are generally unknown.

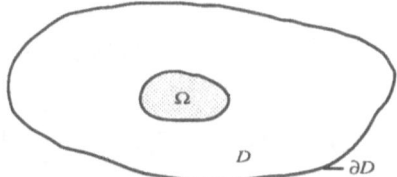

Figure 1. The domain occupied by the material is denoted by $D$. Plastic strains are accumulated in $\Omega$, a subdomain of $D$.

Due to the symmetry of the elastic moduli $C_{ijkl}$, the Betti-Maxwell reciprocal relation holds.

\* This research was supported by U. S. Army Research Office and partially by Amoco Co.

[1] Department of Civil Engineering, Northwestern University, Evanston, IL, U.S.A.

[2] Department of Civil Engineering, Northwestern University, Evanston, IL, U.S.A.

[3] Department of Civil Engineering, Northwestern University, Evanston, IL, U.S.A.

[4] Amoco Production Co., Tulsa, OK, U.S.A.

[5] Department of Mechanical and Aeronautical Engineering, Clarkson University, Potsdam, NY, U.S.A.

$$C_{ijkl}G_{km,l}(\mathbf{x}-\mathbf{x}')\big(u_{i,j}(\mathbf{x})-\varepsilon_{ij}^{p}(\mathbf{x})\big) = C_{ijkl}\big(u_{k,l}(\mathbf{x})-\varepsilon_{kl}^{p}(\mathbf{x})\big)G_{im,j}(\mathbf{x}-\mathbf{x}'), \qquad (1)$$

where $G_{km}(\mathbf{x}-\mathbf{x}')$ is the Green's function for an infinite elastic medium and satisfies the equation of equilibrium for a point force with the unit magnitude.

When (1) is integrated with respect to $\mathbf{x}$, Green's formula is obtained

$$\int_{\Omega} C_{ijkl}G_{km,l}(\mathbf{x}-\mathbf{x}')\varepsilon_{ij}^{p}(\mathbf{x})dx = \int_{\partial D} C_{ijkl}G_{km,l}(\mathbf{x}-\mathbf{x}')u_{i}(\mathbf{x})n_{j}dS(\mathbf{x}) + \beta u_{m}(\mathbf{x}'), \qquad (2)$$

where $\beta = 1$ for $\mathbf{x}' \in D$ and $\beta = 1/2$ for $\mathbf{x}' \in \partial D$.

When $\mathbf{x}'$ is considered on $\partial D$, the equation (2) is the Fredholm integral equation of the first kind for unknown $\varepsilon_{ij}^{p}$ under the given $u_{i}(\mathbf{x}')$ on $\partial D$. Since plastic strains are incompressible, the condition $\varepsilon_{kk}^{p} = 0$ is imposed to (2).

When $D$ is a half-space, $G_{km}(\mathbf{x}-\mathbf{x}')$ is taken as Green's function founded by Mindlin [2] for the half-space. The half-space Green's function has properties as follows. Denoting the boundary of the half-space by $x_{3} = 0$,

$$
\begin{aligned}
&C_{ijkl}G_{km,lj}(\mathbf{x},\mathbf{x}') + \delta_{im}\delta(\mathbf{x},\mathbf{x}') = 0, \quad \text{for } x_{3} \geq 0,\\
&C_{ijkl}G_{km,l}(\mathbf{x},\mathbf{x}')n_{j} = \delta_{im}\delta_{S}(\mathbf{x},\mathbf{x}'), \quad \text{on } x_{3} = 0,\\
&\int_{D} f(\mathbf{x})\delta(\mathbf{x},\mathbf{x}')dx = \beta f(\mathbf{x}'),\\
&\int_{\partial D} f(\mathbf{x})\delta_{S}(\mathbf{x},\mathbf{x}')dS(\mathbf{x}) = f(\mathbf{x}'), \quad \text{on } x_{3} = 0.
\end{aligned}
\qquad (3)
$$

Then, (2) becomes

$$\int_{\Omega} C_{ijkl}G_{km,l}(\mathbf{x},\mathbf{x}')\varepsilon_{ij}^{p}(\mathbf{x})dx = \frac{3}{2}u_{m}(\mathbf{x}'). \qquad (4)$$

Differentiating (4) with respect to $x_{n}'$ leads to

$$\int_{\Omega} C_{ijkl}\frac{\partial}{\partial x_{n}'}G_{km,l}(\mathbf{x},\mathbf{x}')\varepsilon_{ij}^{p}(\mathbf{x})dx = \frac{3}{2}u_{m,n}(\mathbf{x}'). \qquad (5)$$

When $\mathbf{x}'$ is considered on $\partial D$, (5) is the Fredholm integral equation of the first kind for unknown $\varepsilon_{ij}^{p}$ under given surface tilts $u_{m,n}(\mathbf{x}')$ on $\partial D$.

### The Regularization Method

We write equation (2) or (5) in the form

$$\mathbf{KV} = \mathbf{U}. \qquad (6)$$

Tikhonov (1963) considered the following variational method.

*Minimize* $\|\mathbf{V}\|^{2}$ *subjected to the constraint condition* $\|\mathbf{KV} - \mathbf{U}\|^{2} = 0$, \qquad (7) where $\|\ \|^2$ is the square of the $L_2$ norm, i.e., the inner product of the function with itself on the domain it is defined.

$$\|V(x)\|^2 = \int_\Omega [V(x)]^T V(x) dx, \qquad (8)$$

and

$$\left\|\int_\Omega K(x,x')V(x)dx - U(x')\right\|^2 = \int_{\partial D}\left\{\left[\int_\Omega K(x,x')V(x)dx - U(x')\right]^T\right.$$
$$\left.\bullet\left[\int_\Omega K(x,x')V(x)dx - U(x')\right]\right\}dx', \qquad (9)$$

where the superscript "$T$" indicates the transpose of the matrix.

For an engineering problem, our experiments always contain error. $U(x')$ is only an approximation of the exact displacements $U^\circ(x')$ such that

$$\left\|U(x') - U^\circ(x')\right\|^2 \le \varepsilon,$$

where $\varepsilon$ is a small, positive number. Therefore, the problem is changed into

$$\begin{cases} \text{Minimize} & \|V(x)\|^2 \\ \text{subject to} & \left\|\int_\Omega K(x,x')V(x)dx - U(x')\right\|^2 = \varepsilon \end{cases} \qquad (10)$$

Koya [4] avoided the integration difficulty in (9) by writing (10) as follows.
  *Minimize*

$$\sum_{i=1}^M \int_{\Omega_i} [V(x)]^T V(x)dx, \qquad (11)$$

*with constraint conditions*

$$\left[\sum_{i=1}^M \int_{\Omega_i} K\left(x,x'_{(j)}\right)V(x)dx - U\left(x'_{(j)}\right)\right]^T\left[\sum_{i=1}^M \int_{\Omega_i} K\left(x,x'_{(j)}\right)V(x)dx - U\left(x'_{(j)}\right)\right] = \varepsilon. \qquad (12)$$

The integration in (11) and (12) are carried out by assuming $V$ is constant in each subdomain $\Omega_i$.

### Overprescribed Boundary Value Problems

Suppose there is a body whose domain is denoted by $D$ and whose boundary is denoted by $\partial D = S_1 + S_2$. The overprescribed boundary value problem is to find the elastic fields of displacement and stress when both displacement and traction boundary values $u_i^\circ$ and $t_i^\circ$ are given on a part of the boundary named $S_1$, Green's formula becomes

$$-\int_{S_2} G_{im}(\mathbf{x}-\mathbf{x}')t_i(\mathbf{x})dS(\mathbf{x})+\int_{S_2} C_{ijkl}G_{km,l}(\mathbf{x}-\mathbf{x}')n_j u_i(\mathbf{x})dS(\mathbf{x})$$

$$=\int_{S_1} G_{im}(\mathbf{x}-\mathbf{x}')t_i^\circ(\mathbf{x})dS(\mathbf{x})-\int_{S_1} C_{ijkl}G_{km,l}(\mathbf{x}-\mathbf{x}')n_j u_i^\circ(\mathbf{x})dS(\mathbf{x})+\beta u_m(\mathbf{x}). \quad (13)$$

This is an integral equation to find $u_i$ and $t_i$ on $S_2$ when $u_i$ and $t_i$ on $S_1$ are given as $u_i^\circ$ and $t_i^\circ$. This is an inverse problem the same as equation (9).

### Statement of the Problem for Shape Determination of an Inhomogeneity

An elastic block $ABCD$ is subjected to the compression force $P$ at the boundaries $AD$ and $BC$. The displacements $u_i^\circ$ at the boundaries $AB$, $BC$, $CD$ and $AD$ are known. This block $ABCD$ contains an inhomogeneity $A'B'C'D'$ which has Young's modulus $E^*$ and Poisson's ratio $v^*$. While the matrix $ABCD - A'B'C'D'$ has Young's modulus $E$ and Poisson's ratio $v$ as shown in Figure 2. Find the size and location of the inhomogeneity.

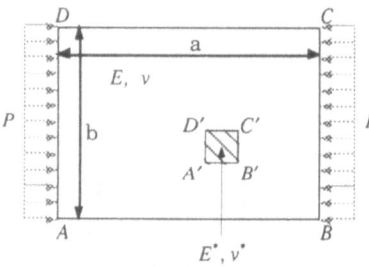

Figure 2.

### The Scanning Method

Consider a block as shown in Figure 3 where the inhomogeneity is assumed to be $A'BCD'$. We calculate the displacements of the block along edges $AB$, $BC$, $CD$ and $AD$. It can be done by using Green's formula.

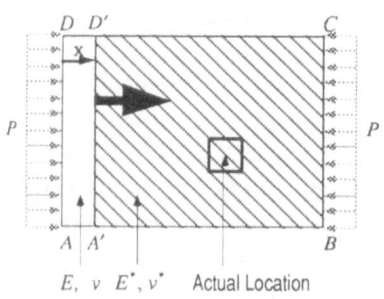

Figure 3.

The Betti-Maxwell reciprocal relation,

$$C_{ijkl}G_{km,l}(\mathbf{x},\mathbf{x}')u_{i,j}(\mathbf{x}) = C_{ijkl}G_{im,j}(\mathbf{x},\mathbf{x}')u_{k,l}(\mathbf{x}), \tag{14}$$

where $G_{ij}(\mathbf{x},\mathbf{x}')$ is the fundamental solution in the elastic body with elastic moduli $C_{ijkl}$.

Both sides of (14) are integrated in the domain $AA'D'D$ with respect to $\mathbf{x}$. The integration by parts reveals

$$\int_{\Gamma_1+\Gamma_2} C_{ijkl}G_{km,l}(\mathbf{x},\mathbf{x}')u_i(\mathbf{x})n_j d\Gamma + \beta u_m(\mathbf{x}') = \int_{\Gamma_1+\Gamma_2} G_{im}(\mathbf{x},\mathbf{x}')t_i(\mathbf{x})d\Gamma, \tag{15}$$

where $\Gamma_1$ is the edge $A'D'$, and $\Gamma_2$ is the edges $AA', D'D$, and $AD$. The values of $\beta$ are $\frac{1}{2}$ on $\Gamma_1$ and $\Gamma_2$ except at the corners and $\frac{1}{4}$ at the corners.

Another Betti-Maxwell reciprocal relation is considered in domain $A'BCD'$,

$$C_{ijkl}^* G_{km,l}^*(\mathbf{x},\mathbf{x}')u_{i,j}^*(\mathbf{x}) = C_{ijkl}^* G_{im,j}^*(\mathbf{x},\mathbf{x}')u_{k,l}^*(\mathbf{x}), \tag{16}$$

where $G_{ij}^*(\mathbf{x},\mathbf{x}')$ is the fundamental solution of the elastic body with the elastic moduli $C_{ijkl}^*$. The integration by parts with respect to $\mathbf{x}$ reveals

$$\int_{\Gamma_1+\Gamma_3} C_{ijkl}^* G_{km,l}^*(\mathbf{x},\mathbf{x}')u_i^*(\mathbf{x})n_j d\Gamma + \beta u_m^*(\mathbf{x}') = \int_{\Gamma_1+\Gamma_3} G_{im}^*(\mathbf{x},\mathbf{x}')t_i^*(\mathbf{x})d\Gamma, \tag{17}$$

where $\Gamma_1$ is the edge $A'D'$, and $\Gamma_3$ is the edges $A'B, BC$, and $CD'$.

Equations (15) and (17) are solved simultaneously for the displacements after having been imposed the following boundary conditions:

$$\begin{aligned} t_1 = -P, \quad t_2 = 0, \quad &\text{along } AD \text{ and } BC, \\ t_1 = 0, \quad t_2 = 0, \quad &\text{along } AB \text{ and } CD. \end{aligned} \tag{18}$$

and

$$\begin{aligned} u_i^* &= u_i, \\ t_i^* &= -t_i, \end{aligned} \tag{19}$$

along $\Gamma_1$ combine (15) and (17).

These computed displacements $u_i$ and $u_i^*$ are compared with the known displacements $u_i^\circ$ by the following formula:

$$\Delta = \int_{\Gamma_2+\Gamma_3} \left[\bar{u}_i(\mathbf{x}) - u_i^\circ(\mathbf{x})\right]^2 d\Gamma, \tag{20}$$

where $\bar{u}_i$ are $u_i$ when $\mathbf{x}$ is on $\Gamma_2$ and $u_i^*$ when $\mathbf{x}$ is on $\Gamma_3$. The $L_2$ norm $\Delta$ must be a function of the distance between the assumed $A'D'$ and $AD$ in Figure 3. Its minimum value is achieved when the assumed $A'D'$ aligns with the actual one as shown in Figure 4.

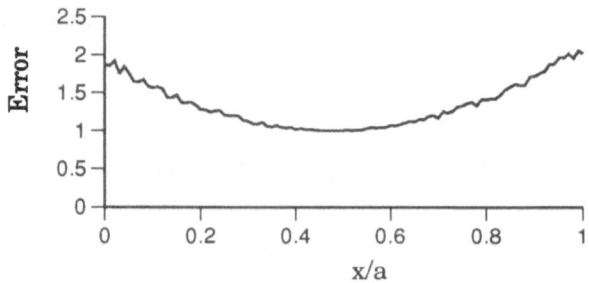

Figure 4.

The similar idea can be applied in order to determine the actual $C'D'$. Consider the block shown in Figure 5. The boundary integral equation similar to (15) is considered. The deviation $\Delta$ from the known displacements must be a function of the distance between the assumed $C'D'$ and $CD$ in Figure 5. The remaining boundaries $A'B'$ and $B'C'$ can be determined by the similar idea.

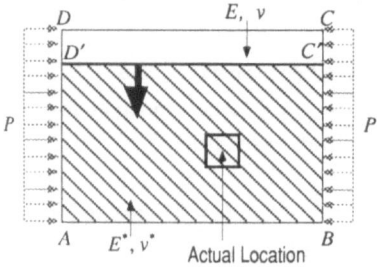

Figure 5.

## Examples

The dotted curve in Figure 6 represents the magnitude of damage in terms of Burgers vector of dislocations caused on the interface of inhomogeneity $\Omega$ (Figure 7) by repeated heating. The solid curve in Figure 6 is an estimation by the present inverse problem.

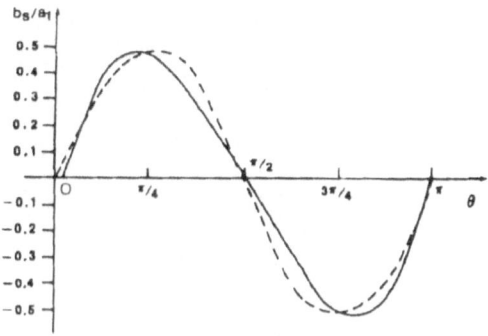

Figure 6. Displacement jump $b_s$ on the interface. Dashed line is the exact solution. Solid line is the numerical result.

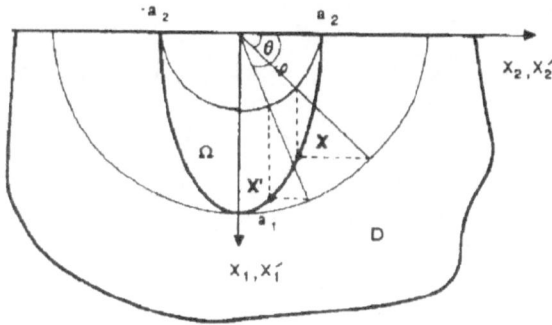

Figure 7. Half space $x_1 \geq 0$ is denoted by $D$. $\Omega$ is an inhomogeneity.

Figure 8 represents a damage zone $\Omega$ where plastic strain components are given. The solid curves in Figure 9 are residual stress components caused by this damage. The dotted curves in Figure 9 are those predicted by the present inverse method.

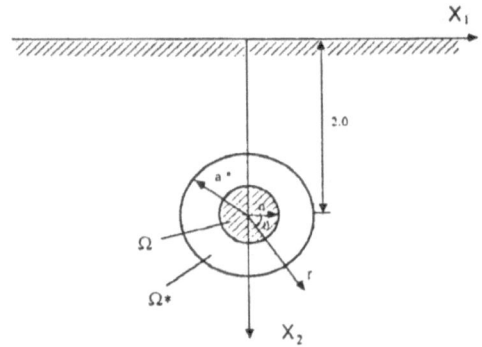

Figure 8. $\Omega$ is the circular domain in a half space $x_2 \geq 0$. $\Omega^*$ covers $\Omega$.

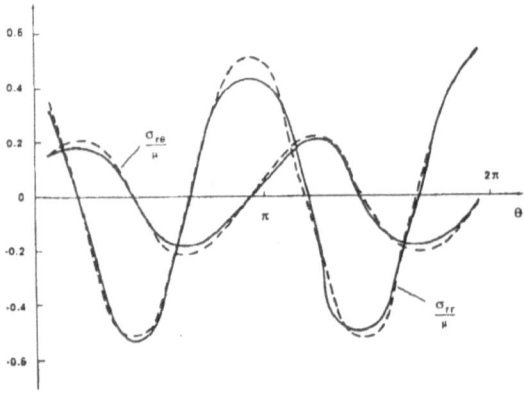

Figure 9. The residual stresses $\sigma_{rr}$ and $\sigma_{r\theta}$ at $r/a^* = 1.5$ caused by the given plastic strains are identical to those by the computed $\varepsilon_{ij}^p$ in $\Omega^*$. The solid lines are the computed results and the dashed lines are induced by the given plastic strains.

Figure 10 represents the given surface tilts caused by equivalent inclusion $\Omega$ defined by $-1 \leq x_2 \leq 1$ and $1 \leq x_1 \leq 2$ with equivalent eigenstrain $\gamma_{12}^* = 1$. Figure 11 represents the choice of the meshes which covers the real domain gives a result very close to the real solution. Therefore, the geometry, position and orientation of the cavity are recovered successfully.

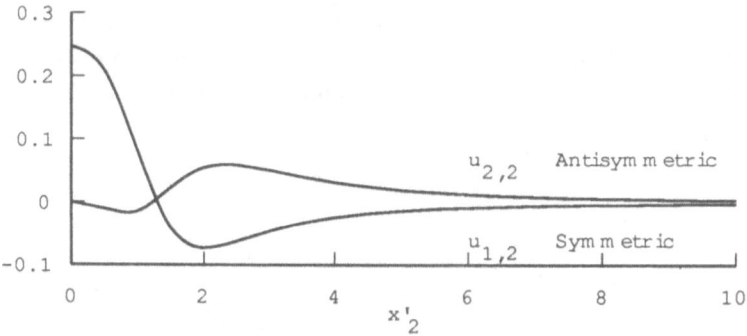

Figure 10. The given surface tilts.

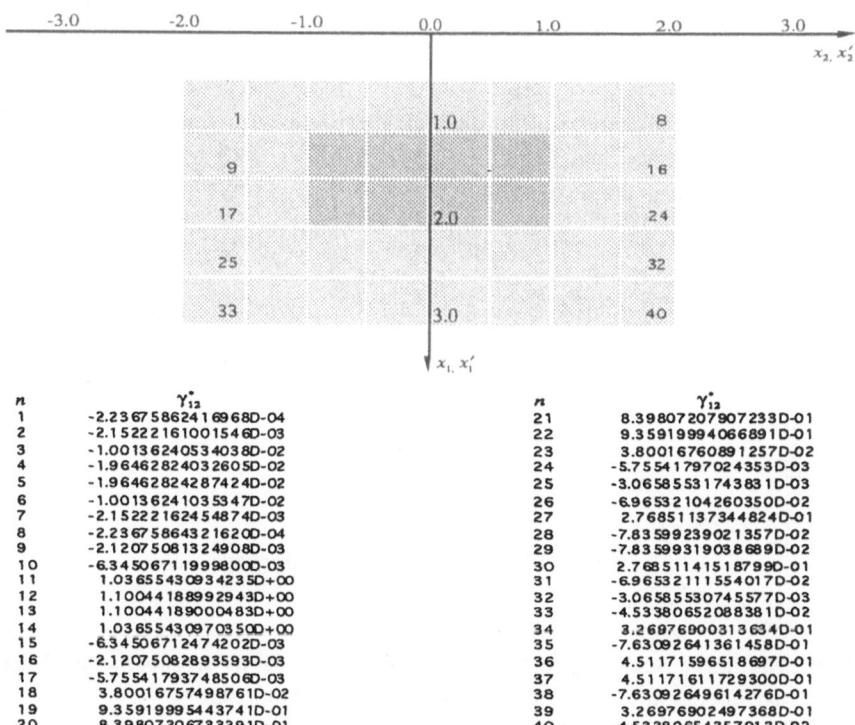

| $n$ | $\gamma_{12}^{*}$ | $n$ | $\gamma_{12}^{*}$ |
|---|---|---|---|
| 1 | -2.2367586241 6968D-04 | 21 | 8.39807207907233D-01 |
| 2 | -2.152221610015 46D-03 | 22 | 9.35919994066891D-01 |
| 3 | -1.0013624053 40 38D-02 | 23 | 3.80016760891257D-02 |
| 4 | -1.9646282403 2605D-02 | 24 | -5.75541797024353D-03 |
| 5 | -1.96462824287424D-02 | 25 | -3.0658553174383 1D-03 |
| 6 | -1.00136241035347D-02 | 26 | -6.96532104260350D-02 |
| 7 | -2.152221624548 74D-03 | 27 | 2.7685113734 4824D-01 |
| 8 | -2.2367586432 1620D-04 | 28 | -7.8359923902 1357D-02 |
| 9 | -2.12075081324908D-03 | 29 | -7.8359931903 8689D-02 |
| 10 | -6.345067119998 00D-03 | 30 | 2.768 51141518799D-01 |
| 11 | 1.0365543093423 5D+00 | 31 | -6.96532111554017D-02 |
| 12 | 1.10044188992943D+00 | 32 | -3.0658553074 5577D-03 |
| 13 | 1.10044189000483D+00 | 33 | -4.533806520 88381D-02 |
| 14 | 1.03 65543097 03 500D+00 | 34 | 3.2697690031363 4D-01 |
| 15 | -6.345067124742 02D-03 | 35 | -7.63092641361458D-01 |
| 16 | -2.12075082893593D-03 | 36 | 4.5117159651 8697D-01 |
| 17 | -5.7554179374850 6D-03 | 37 | 4.5117161172 9300D-01 |
| 18 | 3.80016757498761D-02 | 38 | -7.63092649614276D-01 |
| 19 | 9.3591999544374 1D-01 | 39 | 3.2697690249 7368D-01 |
| 20 | 8.39807206733291D-01 | 40 | -4.5338065435701 2D-02 |

Figure 11. Guess domain and computed equivalent eigenstrains.

Consider a statically equilibrated linear elastic plate as shown in Figure 12. The Cauchy data , displacements and tractions, is prescribed along the nodes #65 to #80, the remaining boundaries are free. The constant boundary element method is employed to discretize the original integral equation into the Nystrom form. Figure 13 shows the regularized solutions and the mixed boundary value solutions. They are in very good agreement.

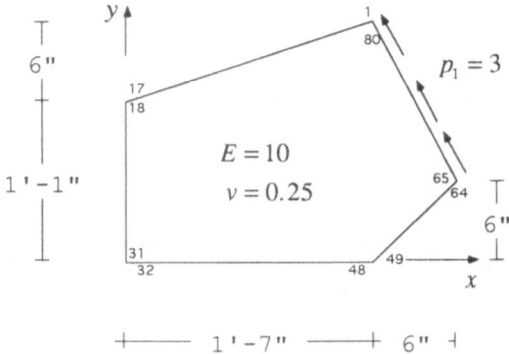

Figure 12. Statically equilibrated elastic plate with Cauchy data prescribed along the nodes #65 to #80. The other boundaries are free.

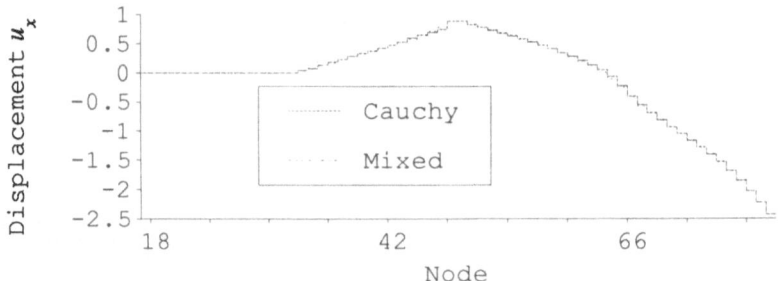

Figure 13. Displacement in $x$-direction.

## References

[1]  M. Kitahara, *Boundary integral equation methods in eigenvalue problems of elastodynamics and thin plates.* Elsevier, Amsterdam, 1985

[2]  R. D. Mindlin, *Force at a point in the interior of a semi-infinite solid,* Physics **7**, 195-202 (1936)

[3]  A. N. Tikhonov, *The solution of ill-posed problems,* Doklady Akad. Nauk SSSR **151**, 3 (1963)

[4]  T. Koya, *Some inverse problems in elasticity and their solution methods,* Ph.D. dissertation, Northwestern University, 1991

# Identification Aspects of Inhomogeneous Materials

M.A.N. Hendriks[*] and C.W.J. Oomens[†]

[*]TNO Building and Construction Research, P.O. Box 49, 2600 AA Delft, The Netherlands.
[†] Eindhoven University of Technology, Faculty of Mechanical Engineering, P.O. Box 513, 5600 MB Eindhoven, The Netherlands.

**Summary**

This paper presents an inverse method to determine parameters in constitutive equations. The method is especially suitable to study the mechanical behavior of inhomogeneous materials. The method is based on: numerical analysis, strain distribution measurement and system identification. By means of simulations it will be shown that for a solid with a varying fiber direction it is possible to estimate stiffness parameters a well as the local fiber directions from one single test.

**Introduction**

Material properties in plant and animal tissues can vary with the anatomical site. Also technical materials may have inhomogeneous properties, *e.g.* reinforced composites with short fiber like particles, processed by a molding operation. These composites may be described in terms of effective mechanical behavior, *i.e.* composites are considered on a scale, several times the dimensions of the constituent materials. On this scale, a certain smoothness of the material properties is assumed. In the present paper the concept of inhomogeneity refers to a larger scale and may, for instance, be caused by different orientations of the alignment of the fibers in the material. Ideally, the inhomogeneity of the material meets the mechanical demands of the object.

Mathematical modeling of inhomogeneous materials, *e.g.* by means of a finite element model, does not lead to fundamental problems. Experimental determination of inhomogeneous properties, however, is an arduous task. A possibility to measure some of the inhomogeneous properties by means of common mechanical tests, such as uniaxial strain tests and biaxial tests, is to extract samples at different positions in the material. A disadvantage of this approach is the disruption of the structure by cutting fibers in the manufacturing of the samples. Particularly for inhomogeneous materials, an inverse approach offers better possibilities than the common traditional testing. Despite the increasing interest for inverse methods in the realm of continuum mechanics, the identification of inhomogeneous material behavior has hardly attracted any attention. An exception is an example presented by Nappi[1] of a geotechnical problem.

The approach used in this paper is based on the combination of three elements:[2] (i) the use of digital image analysis for the measurement of non-homogeneous strain distributions on multi-axially loaded objects with arbitrary geometry , (ii) finite element modeling and (iii) application of systems identification. The third element comprises the comparison between experimental data and the outcomes of the finite element model, followed by the determination of the material parameters.

Recent publications describe the testing of this identification approach in practice, by means of experiments on an orthotropic elastic membrane.[2][3] Now the applicability of the method will be demonstrated by means of numerical simulations for inhomogeneous materials. These are carried out by computing a displacement distribution with a given constitutive model and known parameters. Subsequently, these displacements are used as fictitious 'measured data' for parameter estimation. In this way the influence of observation noise and model errors can be determined.

**Identification method**

In this section an outline of the identification method used is described. The reader can find further details in Hendriks[2] and Norton.[4] The method is based on the sequential minimum variance approach. The observational data are assumed to consist of a set of columns with data $\{y_k\}$, $k = 1,...,N$. The observational data of the 'experiments' described in this paper are displacement components of an inhomogeneous displacement field and are collected in a single column $y_1$. The displacements are considered to be a nonlinear function of a set of material parameters:

$$y_1 = h_1(x) + v_1 \tag{1}$$

where $x$ is a column with unknown material parameters, $h_1$ is a finite element model for the measured displacements, and $v_1$ is a column of observation errors.

The basic estimation problem is the use of the observed displacements $y_1$ to estimate parameter column $x$. The estimator can be specified from the model (Eq. 1), an uncertainty model for $v_1$ and a priori knowledge of $x$. The optimal parameter column minimizes the following quadratic expression:

$$S_1 = (y_1 - h_1(x))^T R_1^{-1} (y_1 - h_1(x)) + (\hat{x}_0 - x)^T (P_0 + Q_1)^{-1} (\hat{x}_0 - x) \tag{2}$$

where $\hat{x}_0$ is an initial guess for the parameter column $x$. In weighted least squares estimation the matrices $R_1$ and $P_0$ are chosen on the basis of engineering judgement. Matrix $Q_1$ is a nonnegative symmetric matrix. It is obvious that the introduction of $Q_1$ makes it possible to put less weight to the *a priori* estimate $\hat{x}_0$ (and more weight to the displacements $y_1$). The least squares estimate does not make any use of the statistics of the observation errors. In many applications, it is not uncommon for the mean and variance of the observation error to be known. Minimum variance estimates utilize this extra information, which results in specific choices for $R_1$ and $P_0$. In minimum variance estimation $R_1$ represents the covariance matrix of the observation error $v_1$. Matrix $P_0$ represents the covariance matrix of the estimation error in $\hat{x}_0$. Generally: the larger $P_0$, the smaller the influence of $\hat{x}_0$.

Solving the nonlinear inverse problem, defined by Eqs. (1) and (2), leads to an iterative scheme, which results in an estimation $\hat{x}_1$ for $x$ and in a covariance matrix of the estimation error $P_1$:

$$\hat{x}^{(i+1)} = \hat{x}^{(i)} + K^{(i+1)} (y - h(\hat{x}^{(i)})) \tag{3}$$

$$K^{(i+1)} = (P^{(i)} + Q) H^{(i+1)^T} (R + H^{(i+1)} (P^{(i)} + Q) H^{(i+1)^T})^{-1} \tag{4}$$

$$P^{(i+1)} = (I - K^{(i+1)}H^{(i+1)}) (P^{(i)} + Q) (I - K^{(i+1)}H^{(i+1)})^T + K^{(i+1)}RK^{(i+1)^T} \tag{5}$$

where the superscripts refer to the iteration number and where the subscripts are temporarily dropped. In each iteration $n + 1$ finite element calculations are executed, where $n$ is the number of parameters. The $n$ calculations are carried out to determine a matrix $H_1^{(i)}$ numerically, as a linearization of $h_1$ with respect to the most recent estimation $\hat{x}_1^{(i-1)}$.

The sequential property of the estimator is clear when a column $y_2$ with new observational data would become available. This can be data from another load case or from another point in time. These data can be used together with the initial conditions $\hat{x}_1$ and $P_1$ resulting in an improved estimation $\hat{x}_2$ and $P_2$. In the examples of this paper it will be shown that the data of a single load case contained in $y_1$ is sufficient for the characterization of the material behavior. The above estimator is implemented as an extra module in the finite element code DIANA.[5]

## Two approaches

Ideally, the material properties of an inhomogeneous material are determined with respect to each point of the material. In practice, however, regions surrounding a point are considered. Approaches for the identification of inhomogeneous materials can be distinguished by the size of these regions and by the inhomogeneity assumed in each region. In the example, described in the next section, two approaches are distinguished.

In the first approach a model of the entire loaded object is confronted with the 'experimental' data. The inhomogeneous properties are modeled with help of a continuous function over the finite element model. This function will be identified together with the stiffness parameters.

The influence of the model errors, depends on the suitability of this function to describe the true inhomogeneity.

In the second approach only a part of the loaded object is modeled and confronted with the 'measured' displacements of that region. The properties of this region are assumed to be homogeneous, which leads to model errors, depending on the size of the region and the level of inhomogeneity. The second approach has a very important consequence from finite element viewpoint. The boundaries of the region are no longer the actual boundaries of the loaded object. Practically, now only prescribed displacements can be used as boundary conditions and no forces. It is obvious that, with such a model, no stiffness parameters can be determined. The next section, however, will show that it is still possible to estimate the ratios between the different stiffness parameters.

**Example: curvilinear orthotropy**

Curvilinear orthotropy is the term used to describe a material, in which the orientation of the orthotropic symmetry coordinate system is different from point to point.[6]

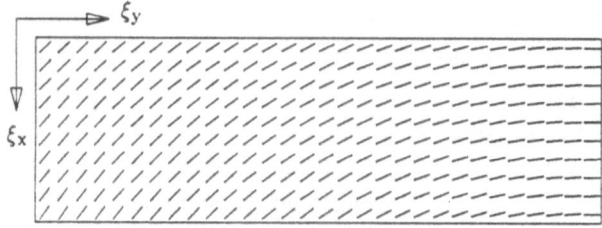

Fig. 1 Sample shape and orientation of local planes of symmetry.

Figure 1 shows a flat membrane (dimensions: $1 \times 3 \times 0.02$) with curvilinear orthotropic behavior. An orthotropic material has three mutually perpendicular planes of symmetry with respect to each point of the material. In the present example, it is assumed that one plane of symmetry coincides with the plane of the sample. The normal of one of the other planes of symmetry is indicated in the figure with a short line. These lines may be interpreted as the orientation of fiber like particles in a reinforced composite. The axes shown in figure 1 are tangent to concentric circles, where $(\xi_x = 3.0, \xi_y = 3.0)$ denotes the center. This type of circumferential orthotropy is typical for wood, where one axis is tangent to the growth rings.

In each point of the sample the stiffness properties with respect to the local symmetry axes are the same. The material parameters are chosen arbitrarily: $E_1 = 1.0$, $E_2 = 0.2$, $v_{12} = 0.3$, $G_{12} = 0.2$, where $E_1$ is the stiffness in material 1-direction as indicated in figure 1, $E_2$ is the stiffness in perpendicular direction and $v_{12}$ and $G_{12}$ denote the Poisson's ratio and shear modulus respectively.

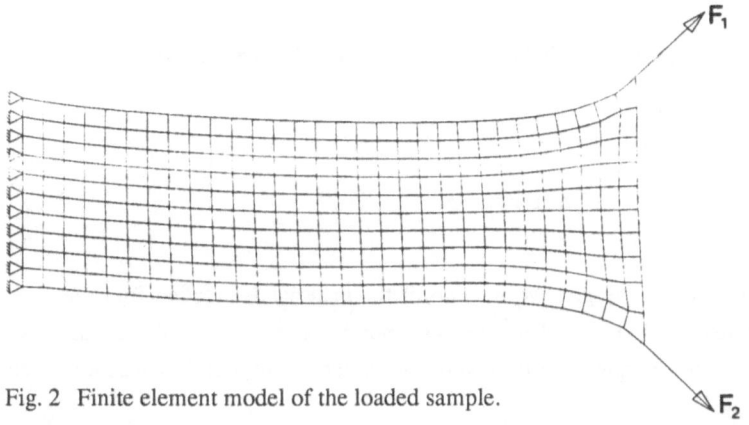

Fig. 2  Finite element model of the loaded sample.

Figure 2 shows the finite element model of the sample, used for the artificial generation of the displacement data. The model consists of 4-noded plane stress elements. The membrane is symmetrically loaded with two equal forces working in the plane of the sample. It will be clear that the deformation is not symmetrical, which is caused by the varying fiber direction.

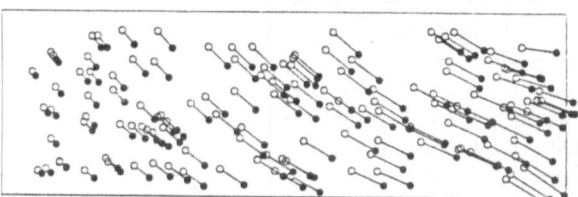

Fig. 3  Measured displacements for approach 1.

Two sets of measured displacements will be distinguished. The first set consists of the displacement components of 128 material points, as shown in figure 3. The initial positions of these points are a realization of a 2-dimensional uniform random distribution. The second set of measured displacements is demonstrated in figure 4.

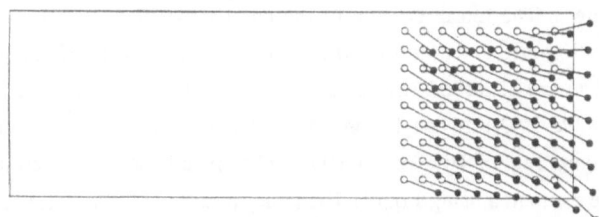

Fig. 4  Measured displacements for approach 2.

*Approach 1*

For the fiber direction two models will be distinguished. Model 1 is given by:

$$\alpha(\xi) = \begin{cases} -arctan\left(\dfrac{\xi_x - c_1}{\xi_y - c_2}\right) & \text{for } \xi_y \neq c_2 \\ \dfrac{1}{2}\pi & \text{for } \xi_y = c_2 \end{cases} \tag{6}$$

In this equation $\alpha$ denotes the positive rotation of the material 1-direction from the model $\xi_x$-axis. This rotation is a function of the position coordinates $\xi_x$ and $\xi_y$. The parameters $c_1$ and $c_2$ in Eq. (6) can be interpreted as the coordinates of the centroid of the concentric circles. Model 2 is given by:

$$\alpha(\xi) = b_0 + b_1\xi_x + b_2\xi_y \tag{7}$$

This bilinear function with unknown parameters $b_0$, $b_1$ and $b_2$ is used to investigate the influence of model errors. Clearly this function cannot pinpoint the actual inhomogeneity, as shown in figure 1, with any set of parameters $b_i$.

| Para- | Exact | Initial | Estimations | | |
|-------|-------|---------|-------------|---------------|--------------|
| meter | value | guess | No noise | $\sigma = 0.001$ | $\sigma = 0.01$ |
| $E_1$ | 1.000 | 0.666 | 1.000 | 0.993 | 0.931 |
| $E_2$ | 0.200 | 0.133 | 0.200 | 0.200 | 0.198 |
| $v_{12}$ | 0.300 | 0.200 | 0.300 | 0.301 | 0.305 |
| $G_{12}$ | 0.200 | 0.133 | 0.200 | 0.201 | 0.211 |
| $c_1$ | 3.000 | 2.000 | 2.998 | 3.004 | 3.055 |
| $c_2$ | 3.000 | 2.000 | 2.999 | 3.010 | 3.106 |

Table 1: Estimation results after 10 iterations with the tangential function (6).

Table 1 shows the estimation results with Eq. (6) Here 6 parameters are estimated using the experimental data shown in figure 3. The fourth column of the table shows the estimates of these parameters without observation errors. The fifth and sixth column show the estimation results when the displacement data are disturbed with a zero mean normal distribution. The average displacement of the sample is 0.1 It can be observed that identification approach works well, even with a noise signal rate of 10% ($\sigma = 0.01$) Similar results are presented in table 2, but now 7 parameters are estimated using Eq. (7) The table shows that in this case the obvious model errors scarcely effect the estimations for $E_1$, $E_2$, $v_{12}$ and $G_{12}$. Some discussion on the exact values is worthwhile in this case The model errors make the use of the

| Para-meter | Exact value | Initial guess | Estimations | | |
|---|---|---|---|---|---|
| | | | No noise | $\sigma = 0.001$ | $\sigma = 0.01$ |
| $E_1$ | 1.000 | 0.666 | 1.031 | 1.022 | 0.944 |
| $E_2$ | 0.200 | 0.133 | 0.199 | 0.199 | 0.200 |
| $v_{12}$ | 0.300 | 0.200 | 0.294 | 0.296 | 0.304 |
| $G_{12}$ | 0.200 | 0.133 | 0.193 | 0.194 | 0.209 |
| $b_0$ | *does* | -0.784 | -0.755 | -0.753 | -0.735 |
| $b_1$ | *not* | 0.262 | 0.200 | 0.199 | 0.193 |
| $b_2$ | *apply* | -0.262 | -0.274 | -0.273 | -0.263 |

Table 2: Estimation results after 10 iterations with the bilinear function (7).

term "exact" misleading. The exact values of column 2 are no longer necessarily the optimal parameters, in the sense that they minimize expression (2). Exact values for the parameters $b_i$ can not be given. Nevertheless, a comparison of the fiber directions calculated with Eq. (6) and Eq. (7) show hardly any difference[2]. This is visualized in figure 5 where the bilinear inhomogeneity is drawn based on the estimated $b_i$ parameters, whereas figure 1 shows the actual circumferential orthotropy. Evidently, in this case, the estimation results are neither very sensitive to this type of model errors, nor are they sensitive for the combination of model errors and random observation errors.

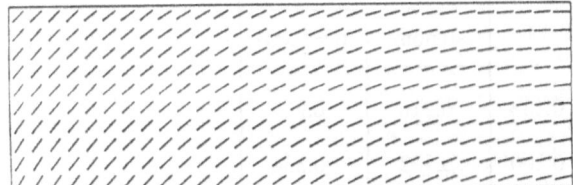

Fig. 5  Estimated inhomogeneity using a bilinear function.

*Approach 2*

Now the finite element model is based on the measured displacements, shown in figure 4. The figure shows that the material points are positioned in a square. For this square a finite element model is derived (figure 6). The prescribed displacements of the four edges are derived from the displacements of the outer material points. The displacements of the inner material points are considered as measured data. It is assumed that the material properties, the material orientation included, are homogeneous over the sample part.

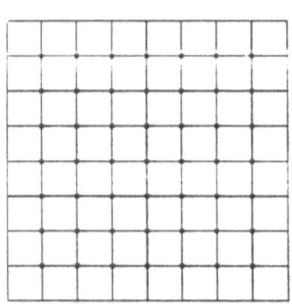 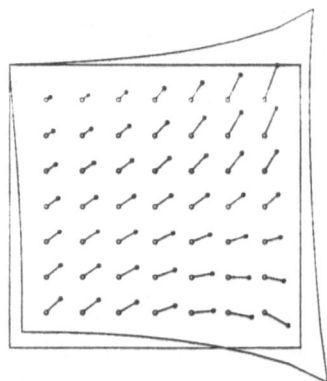

Fig. 6 Finite element model of a square part of the sample (left panel); Kinematic boundary conditions and measured data (right panel).

It is obvious that stiffness parameters can not be identified with the model presented here. In the present simulation we will investigate whether or not the combination of model and measured data does contain information about the ratios between the stiffness parameters. In the present example the following dimensionless parameters will be identified:

$$x^T = (\frac{E_2}{E_1}, v_{12}, \frac{G_{12}}{E_1}, cotan(\alpha)) \qquad (8)$$

where $\alpha$ denotes the positive rotation of the material axes system.

| Para-meter | Exact value | Initial guess | Estimations | |
|---|---|---|---|---|
| | | | No noise | $\sigma = 0.001$ |
| $\frac{E_2}{E_1}$ | 0.200 | 0.133 | 0.216 | 0.221 |
| $v_{12}$ | 0.300 | 0.200 | 0.322 | 0.330 |
| $\frac{G_{12}}{E_1}$ | 0.200 | 0.133 | 0.202 | 0.220 |
| $cotan(\alpha)$ | [-0.444,-0.052] | -0.100 | -0.093 | -0.096 |

Table 3: Estimation results after 10 iterations using a homogeneous model for a part of the sample.

Table 3 shows the estimation results for the dimensionless parameters. The true values are given in the second column. For the cotangential value of the rotation of the material axes, a range is given representing the true occurring values. In the case of perfect observations, it can be observed that there is a good agreement between the estimation results and the true parameters, although the comparison is less favorable as in the first approach. However, also here a discussion on the specification of "true" or "exact" values is in place. In the model it is assumed that the sample part has homogeneous properties. This model error makes the term "exact" misleading. The biased parameters of the fourth column may give better results in the homogeneous model than the true parameters[2].

Returning to the results of table 3, it can be observed that the results for the cases with disturbed data is less favorable. If the standard deviation of the noise is 1% of the average displacement of the sample ($\sigma = 0.001$), the estimation results differ slightly from the results in the perfect observation case. However, if the standard deviation increases to 10% the identification fails after two iterations, since the thermodynamical constraints on the stiffness parameters are violated. Apparently, this approach is more sensitive to measuring errors than the first approach. A possible explanation is that in the second approach measurement errors on the displacements enter as model errors via the specification of the kinematic boundary conditions. Hence the originally random observation errors cause systematic errors in the model.

**Concluding remarks**

For the identification of inhomogeneous materials a mixed numerical-experimental approach is favorable. Via two approaches, but using the same identification idea, it is shown that a nondestructive characterization is possible.

The advantages of the first approach, where the entire sample is identified, are:

• The procedure leads to a complete quantification of the entire sample.

• The identification is not sensitive to observation errors.

The advantages of the second approach are:

• The *a priori* specification of a function representing the inhomogeneity can be omitted.

• The finite element models are smaller.

• In general the models contain less parameters.

• The method meets to practical problems of determining the exact geometry and boundary conditions.

Note that the two approaches are in fact two extreme cases of a whole range of possible approaches. The boundary conditions may be partly kinematic and partly dynamic. In addition, also in the second approach an inhomogeneous model for the sample part can be considered. Experimental investigations have to learn whether these kind of model errors will disturb the identification process.

**Acknowledgement**

The results have been obtained in a research project under supervision of J.D. Janssen and J.J. Kok of the Eindhoven University of Technology. Their advice is greatfully acknowledged. The finite element calculations and the parameter estimation have been carried out using the DIANA finite element package of TNO Building and Construction Research.

**References**

1. Nappi, A.: Structural identification of nonlinear systems subjected to quasistatic loading, in "Application of system identification in engineering", ed. H.G. Natke, Springer Verlag, Berlin, New York and Tokyo, (1988).

2. Hendriks, M.A.N.: Identification of the mechanical behavior of solid materials, Ph.D.-thesis, Eindhoven university of technology, (1991).

3. Hendriks, M.A.N.; Oomens, C.W.J.; Jans, H.W.J.; Janssen, J.D.; Kok, J.J.: A numerical experimental approach for the mechanical characterization of composites, in "proceedings of the 9$^{th}$ international conference on experimental mechanics", ed. V. Askegaard, Aaby Tryk, Copenhagen, (1990) 552-561.

4. Norton, J.P.: An introduction to identification, Academic Press, New York and London, (1986).

5. de Borst, R.; Roddeman, D.G.: Computational mechanics: recent developments in DIANA, *Heron*, **36** , special issue no. 2, (1991).

6. Cowin, S.C,; Mehrabadi, M.M.: Identification of the elastic symmetry of bone and other materials, *J. Biomechanics*, **22** , (1989) 503-515.

# A Parameter Identification Procedure as a Dual Boundary Control Concept for Wave-Propagation Problem

Y. Ichikawa*, T. Kyoya*, T. Ohkami** and Xu Wu*

\* Dept. of Geotechnical and Environmental Engineering,
  Nagoya University, Nagoya 464-01, Japan
\*\* Dept. of Architecture and Civil Engineering,
  Shinshu University, Nagano 380, Japan

Summary
In geotechnical engineering, the material identification is an essential technology not only for the geological survey but also for the observational construction procedure. We have presented a generalized material parameter identification theory which is based on a boundary control concept using duality of displacement and force, and is applied to the linear elastic, viscoelastic and damage mechanics problems (Ichikawa and Ohkami 1992, Ohkami and Ichikawa 1992, Ohkami 1991). The finite element discretization and the Newton's iteration scheme are used for solving the nonlinear problem. We here apply the same dual boundary control concept for the wave propagation problem, and identify the material parameters in a frequency domain.

Identification of linear elastic material parameters
Let us first review the identification procedure of linear elastic material parameters as a dual boundary control concept (Ichikawa and Ohkami 1992).

The linear elastic problem is written as followings:

(Governing equation)

$$\nabla \cdot \boldsymbol{\sigma} = \mathbf{o} \qquad \text{in } \Omega \subset \boldsymbol{R}^n (n = 1, 2, 3) \tag{1}$$

(Boundary conditions)

$$\boldsymbol{u} = \hat{\boldsymbol{u}} \qquad \text{on } \partial\Omega_u \tag{2}$$

$$\boldsymbol{\sigma n} = \hat{\boldsymbol{t}} \qquad \text{on } \partial\Omega_t \tag{3}$$

(Constitutive equation: Hooke's law)

$$\boldsymbol{\sigma} = \boldsymbol{D}\boldsymbol{\varepsilon} \tag{4}$$

where $\boldsymbol{\sigma}$ is the stress tensor, $\boldsymbol{u}$ the displacement vector, $\boldsymbol{\varepsilon} = [(\nabla\boldsymbol{u}) + (\nabla\boldsymbol{u})^T]/2$ the strain tensor, $\boldsymbol{n}$ the outward unit normal vector on the boundary $\partial\Omega$, and $\boldsymbol{D}$ the Hookean tensor which involves Young's modulus $E$ and Poisson's ratio $\nu$ for linear isotropic materials.

Material parameters involved in the Hookean tensor $\boldsymbol{D}$ are identified by measuring a displacement $\bar{\boldsymbol{u}}$ on (a part of) $\partial\Omega_t$ where the traction $\hat{\boldsymbol{t}}$ is given, or conversely by measuring a traction $\bar{t}$ on (a part of) $\partial\Omega_u$ where the displacement $\hat{u}$ is given. The conditions are written by

(Observational or control boundary conditions)

$$u = \bar{u} \qquad \text{on } \partial\Omega_{t'} \subset \partial\Omega_t \tag{5}$$

$$\sigma n = \bar{t} \qquad \text{on } \partial\Omega_{u'} \subset \partial\Omega_u \tag{6}$$

The virtual work equation corresponding to (1), (2) and (3) is written as followings:

$$\int_\Omega \sigma : \delta\varepsilon dv - \int_{\partial\Omega_t} \hat{t} \cdot \delta u dS = 0. \tag{7}$$

The finite element method is applied to the virtual work equation (7), and we get the discretized equation which involves the unknown material parameters $P$ (for the isotropic linear elastic body with $m$-different regions, $P = \{E_i, \nu_i\}; (i = 1, 2, \ldots, m)$:

$$K(P)U = F \tag{8}$$

$$K(P) = \int_\Omega B^T D(P) B dv, \qquad F = \int_{\partial\Omega_t} N^T \hat{t} dS$$

where $N$ is the matrix of shape functions, and $B$ the strain-displacement matrix.

The observational boundary conditions (5) and (6) are discretized as

$$S_u U = \bar{U} \tag{9}$$

$$S_t F = \int_{\partial\Omega_u} N^T \bar{t} dS = \bar{F} \tag{10}$$

where $S_u$ and $S_t$ are diagonal matrices for choosing observational boundary nodes of displacement and traction, respectively. That is, if $i$-th node is the displacement observational boundary, and if $j$-th node is the traction one, we have

$$
S_u = \begin{bmatrix} 0 & & & & \\ & \ddots & & \mathbf{o} & \\ & & 1 & & \\ & \mathbf{o} & & \ddots & \\ & & & & 0 \end{bmatrix} \cdots i\text{-th row}; \qquad
S_t = \begin{bmatrix} 0 & & & & \\ & \ddots & & \mathbf{o} & \\ & & 1 & & \\ & \mathbf{o} & & \ddots & \\ & & & & 0 \end{bmatrix} \cdots j\text{-th row}
$$

$$\qquad\qquad \vdots \qquad\qquad\qquad\qquad\qquad\qquad\qquad \vdots$$
$$\qquad\quad i\text{-th column} \qquad\qquad\qquad\qquad\qquad j\text{-th column}$$

Now the Newton's iteration scheme is applied for (8), (9) and (10). That is, at the $k$-th iteration step, we get

$$-dF^k + K(P^k)dU^k + \left(\frac{\partial K}{\partial P}\right)^k dP^k U^k = [F - KU]^k$$

$$S_u dU^k = [\bar{U} - S_u U]^k = \bar{U} - S_u U^k \tag{11}$$

$$S_t dF^k = [\bar{F} - S_t F]^k = \bar{F} - S_t F^k$$

where

$$dU^k = U^{k+1} - U^k, \qquad d\boldsymbol{F}^k = \boldsymbol{F}^{k+1} - \boldsymbol{F}^k, \qquad d\boldsymbol{P}^k = \boldsymbol{P}^{k+1} - \boldsymbol{P}^k$$

The equations (11) are rewritten as

$$\boldsymbol{G}d\boldsymbol{x} = \boldsymbol{R} \tag{12}$$

where

$$
\boldsymbol{G} = \begin{bmatrix} \boldsymbol{K}(\boldsymbol{P}^k) & -\boldsymbol{I} & (\frac{\partial \boldsymbol{K}}{\partial \boldsymbol{P}}\boldsymbol{U})^k \\ \boldsymbol{S}_u & \boldsymbol{0} & \boldsymbol{0} \\ \boldsymbol{0} & \boldsymbol{S}_t & \boldsymbol{0} \end{bmatrix}, \qquad d\boldsymbol{x} = \begin{bmatrix} d\boldsymbol{U}^k \\ d\boldsymbol{F}^k \\ d\boldsymbol{P}^k \end{bmatrix}, \qquad \boldsymbol{R} = \begin{bmatrix} \boldsymbol{F}^k - \boldsymbol{K}(\boldsymbol{P}^k)\boldsymbol{U}^k \\ \bar{\boldsymbol{U}} - \boldsymbol{S}_u \boldsymbol{U}^k \\ \bar{\boldsymbol{F}} - \boldsymbol{S}_t \boldsymbol{F}^k \end{bmatrix}
$$

Since the number of observed data may exceed the unknowns, (12) is an overdetermined system. Thus, we introduce a least square method by the error function such that

$$\mathcal{E} = \frac{1}{2}(\boldsymbol{G}d\boldsymbol{x} - \boldsymbol{R}) \cdot (\boldsymbol{G}d\boldsymbol{x} - \boldsymbol{R}) \tag{13}$$

and get its normal equation

$$\boldsymbol{G}^T \boldsymbol{G}d\boldsymbol{x} = \boldsymbol{G}^T \boldsymbol{R} \tag{14}$$

$$
\boldsymbol{G}^T \boldsymbol{G} = \begin{bmatrix} \boldsymbol{K}^2 + \boldsymbol{S}_u & -\boldsymbol{K} & \boldsymbol{K}(\frac{\partial \boldsymbol{K}}{\partial \boldsymbol{P}}\boldsymbol{U}) \\ -\boldsymbol{K} & \boldsymbol{I} + \boldsymbol{S}_t & -\frac{\partial \boldsymbol{K}}{\partial \boldsymbol{P}}\boldsymbol{U} \\ (\frac{\partial \boldsymbol{K}}{\partial \boldsymbol{P}}\boldsymbol{U})^T \boldsymbol{K} & -(\frac{\partial \boldsymbol{K}}{\partial \boldsymbol{P}}\boldsymbol{U})^T & (\frac{\partial \boldsymbol{K}}{\partial \boldsymbol{P}}\boldsymbol{U})^T \frac{\partial \boldsymbol{K}}{\partial \boldsymbol{P}}\boldsymbol{U} \end{bmatrix}^k
$$

$$
\boldsymbol{G}^T \boldsymbol{R} = \begin{bmatrix} \boldsymbol{K}(\boldsymbol{F} - \boldsymbol{K}\boldsymbol{U}) + \bar{\boldsymbol{U}} - \boldsymbol{S}_u \boldsymbol{U} \\ -\boldsymbol{F} + \boldsymbol{K}\boldsymbol{U} + \bar{\boldsymbol{F}} - \boldsymbol{S}_t \boldsymbol{F} \\ (\frac{\partial \boldsymbol{K}}{\partial \boldsymbol{P}}\boldsymbol{U})^T(\boldsymbol{F} - \boldsymbol{K}\boldsymbol{U}) \end{bmatrix}^k
$$

Iteratively solving (14), we can determine the material parameters $\boldsymbol{P}$.

Identification procedure for the dynamic problem
We apply the above mentioned identification procedure to the dynamic problem defined in the frequency domain.

The equation of motion is given by

$$\rho\frac{d^2\boldsymbol{u}}{dt^2} - \nabla \cdot \boldsymbol{\sigma} = \boldsymbol{0} \qquad \text{in } \Omega \subset \boldsymbol{R}^n(n = 1, 2, 3) \tag{15}$$

and for simplicity we only introduce the displacement observational boundary such that

$$\boldsymbol{u} = \bar{\boldsymbol{u}} \qquad \text{on } \partial\Omega_{t'} \subset \partial\Omega_t \tag{16}$$

The virtual work equation corresponding to (15) is

$$\int_\Omega \rho \ddot{\boldsymbol{u}} \cdot \delta \boldsymbol{u} \, dv + \int_\Omega \boldsymbol{\sigma} : \delta \boldsymbol{\varepsilon} \, dv - \int_{\partial\Omega} \hat{\boldsymbol{t}} \cdot \delta \boldsymbol{u} \, ds = \boldsymbol{o}. \tag{17}$$

The finite element approximation is applied to this, and we get

$$\boldsymbol{M}\ddot{\boldsymbol{U}} + \boldsymbol{C}\dot{\boldsymbol{U}} + \boldsymbol{K}\boldsymbol{U} = \boldsymbol{F} \tag{18}$$

where

$$\boldsymbol{M} = \int_\Omega \rho \boldsymbol{N}^T \boldsymbol{N} \, dv$$

is the mass matrix, and $\boldsymbol{C}$ is the damping matrix which consists of a kind of Rayleigh damping term such that $\alpha \boldsymbol{M}$ and a viscous boundary term which represents a semi-infinite domain (Lysmer and Kuhlemeyer 1969).

We input an acceleration wave $\ddot{\boldsymbol{U}}_b$ at some (basement) boundary nodes, so that (18) can be rewritten as

$$\begin{bmatrix} \boldsymbol{M}_1 & \boldsymbol{M}_2 \\ \boldsymbol{M}_3 & \boldsymbol{M}_4 \end{bmatrix} \begin{bmatrix} \ddot{\boldsymbol{U}} \\ \ddot{\boldsymbol{U}}_b \end{bmatrix} + \begin{bmatrix} \boldsymbol{C}_1 & \boldsymbol{C}_2 \\ \boldsymbol{C}_3 & \boldsymbol{C}_4 \end{bmatrix} \begin{bmatrix} \dot{\boldsymbol{U}} \\ \dot{\boldsymbol{U}}_b \end{bmatrix} + \begin{bmatrix} \boldsymbol{K}_1 & \boldsymbol{K}_2 \\ \boldsymbol{K}_3 & \boldsymbol{K}_4 \end{bmatrix} \begin{bmatrix} \boldsymbol{U} \\ \boldsymbol{U}_b \end{bmatrix} = \begin{bmatrix} \boldsymbol{F} \\ \boldsymbol{F}_b \end{bmatrix}$$

The first equation is

$$\boldsymbol{M}_1\ddot{\boldsymbol{U}} + \boldsymbol{M}_2\ddot{\boldsymbol{U}}_b + \boldsymbol{C}_1\dot{\boldsymbol{U}} + \boldsymbol{C}_2\dot{\boldsymbol{U}}_b + \boldsymbol{K}_1\boldsymbol{U} + \boldsymbol{K}_2\boldsymbol{U}_b = \boldsymbol{F}$$

We now assume that no traction force is applied to this system ($\boldsymbol{F} = \boldsymbol{o}$). Applying Fourier transformation yields

$$(-\omega^2 \boldsymbol{M}_1 + i\omega \boldsymbol{C}_1 + \boldsymbol{K}_1)\tilde{\boldsymbol{U}} = -(-\omega^2 \boldsymbol{M}_2 + i\omega \boldsymbol{C}_2 + \boldsymbol{K}_2)\tilde{\boldsymbol{U}}_b \tag{19}$$

where $\tilde{\boldsymbol{U}}$ and $\tilde{\boldsymbol{U}}_b$ are the Fourier coefficients of $\boldsymbol{U}$ and $\boldsymbol{U}_b$, respectively. Let us separate $\tilde{\boldsymbol{U}}$ and $\tilde{\boldsymbol{U}}_b$ into real and imaginary parts:

$$\tilde{\boldsymbol{U}} = \boldsymbol{U}^r + i\boldsymbol{U}^i, \qquad \tilde{\boldsymbol{U}}_b = \boldsymbol{U}_b^r + i\boldsymbol{U}_b^i$$

Then, (19) can be written as

real part: $$(-\omega^2 \boldsymbol{M}_1 + \boldsymbol{K}_1)\boldsymbol{U}^r - \omega \boldsymbol{C}_1 \boldsymbol{U}^i = -(-\omega^2 \boldsymbol{M}_2 + \boldsymbol{K}_2)\boldsymbol{U}_b^r + \omega \boldsymbol{C}_2 \boldsymbol{U}_b^i$$

imaginary part: $$\omega \boldsymbol{C}_1 \boldsymbol{U}^r + (-\omega^2 \boldsymbol{M}_1 + \boldsymbol{K}_1)\boldsymbol{U}^r = -\omega \boldsymbol{C}_2 \boldsymbol{U}_b^r - (-\omega^2 \boldsymbol{M}_2 + \boldsymbol{K}_2)\boldsymbol{U}_b^i$$

We now eliminate $\boldsymbol{U}^i$ term. Since the equation includes unknown material parameters $\boldsymbol{P} = \{K_i, G_i, h_i\}$ where $K_i$ and $G_i$ are the bulk and shear moduli, respectively, and $h_i$ the damping factor of $i$-th region, we get

$$\boldsymbol{A}(\boldsymbol{P})\boldsymbol{U} = \boldsymbol{B}(\boldsymbol{P}) \tag{20}$$

where

$$\boldsymbol{A}(\boldsymbol{P}) = \{(-\omega^2 \boldsymbol{M}_1 + \boldsymbol{K}_1) + \omega^2 \boldsymbol{C}_1(-\omega^2 \boldsymbol{M}_1 + \boldsymbol{K}_1)^{-1}\boldsymbol{C}_1\}\boldsymbol{U}^r$$

$$\boldsymbol{B}(\boldsymbol{P}) = -\{\omega^2 \boldsymbol{C}_1(-\omega^2 \boldsymbol{M}_1 + \boldsymbol{K}_1)^{-1}\boldsymbol{C}_2 + (-\omega^2 \boldsymbol{M}_2 + \boldsymbol{K}_2)\}\boldsymbol{U}_b^r$$

$$+ \{-\omega \boldsymbol{C}_1(-\omega^2 \boldsymbol{M}_1 + \boldsymbol{K}_1)^{-1}(-\omega^2 \boldsymbol{M}_2 + \boldsymbol{K}_2) + \omega \boldsymbol{C}_2\}\boldsymbol{U}_b^i.$$

The displacement observational boundary condition (16) is discretized as

$$S_u U = \bar{U} \tag{21}$$

which is same as (9).

Let us apply the similar procedure to (11) to (14), then we obtain the following iteration scheme:

$$G^T G dx = G^T R \tag{22}$$

where

$$G = \begin{bmatrix} A(P^k) & (\frac{\partial A}{\partial P} U)^k - (\frac{\partial B}{\partial P})^k \\ S_u & 0 \end{bmatrix}, \quad dx = \begin{bmatrix} dU^k \\ dP^k \end{bmatrix}, \quad R = \begin{bmatrix} B(P^k) - A(P^k)U^k \\ \bar{U} - S_u U^k \end{bmatrix}$$

and

$$\frac{\partial A}{\partial P} = \{\frac{\partial A}{\partial K}, \frac{\partial A}{\partial G}, \frac{\partial A}{\partial h}\}$$

$$\frac{\partial A}{\partial K} = \frac{\partial K_1}{\partial K} + \omega^2 C_1 \frac{\partial(-\omega^2 M_1 + K_1)^{-1}}{\partial K} C_1$$

$$\frac{\partial A}{\partial G} = \frac{\partial K_1}{\partial K} + \omega^2 C_1 \frac{\partial(-\omega^2 M_1 + K_1)^{-1}}{\partial G} C_1$$

$$\frac{\partial A}{\partial h} = \omega^2 \frac{\partial C_1}{\partial h}(-\omega^2 M_1 + K_1)^{-1} C_1 + \omega^2 C_1(-\omega^2 M_1 + K_1)^{-1}\frac{\partial C_1}{\partial h}$$

$$\frac{\partial B}{\partial P} = \{\frac{\partial B}{\partial K}, \frac{\partial B}{\partial G}, \frac{\partial B}{\partial h}\}$$

$$\frac{\partial B}{\partial K} = -\{\omega^2 C_1 \frac{\partial(-\omega^2 M_1 + K_1)^{-1}}{\partial K} C_2 + \frac{\partial K_2}{\partial K}\}U_b^r$$

$$+\{-\omega^2 C_1 \frac{\partial(-\omega^2 M_1 + K_1)^{-1}}{\partial K}(-\omega^2 M_2 + K_2) - \omega C_1(-\omega^2 M_1 + K_1)^{-1}\frac{\partial K_2}{\partial K}\}U_b^i$$

$$\frac{\partial B}{\partial G} = -\{\omega^2 C_1 \frac{\partial(-\omega^2 M_1 + K_1)^{-1}}{\partial G} C_2 + \frac{\partial K_2}{\partial G}\}U_b^r$$

$$+\{-\omega^2 C_1 \frac{\partial(-\omega^2 M_1 + K_1)^{-1}}{\partial G}(-\omega^2 M_2 + K_2) - \omega C_1(-\omega^2 M_1 + K_1)^{-1}\frac{\partial K_2}{\partial K}\}U_b^i$$

$$\frac{\partial B}{\partial h} = -\{\omega^2 \frac{\partial C_1}{\partial h}(\omega^2 M_1 + K_1)^{-1} C_2\}U_b^r$$

$$+\{-\omega \frac{\partial C_1}{\partial h}(-\omega^2 M_1 + K_1)^{-1}(-\omega^2 M_2 + K_2)\}U_b^i.$$

## Numerical example

We now check the validity of this method by a simple example. The finite element model is shown in Figure 1. Firstly, we calculate the acceleration at points ● by the usual finite elements using the central time difference scheme. The material properties which are used for this ordinary FEM and which will be determined by the identification procedure are

316

$$K = 10,000/3 \quad (tf/m^2) \qquad E = 5,000 \quad (tf/m^2)$$
$$G = 2,000 \quad (tf/m^2) \qquad \nu = 0.25$$
$$\rho = 1.0 \quad (tfs^2/m^4)$$
$$\alpha = 8$$

The damping factor $\alpha = 8$ corresponds to 20 % damping of the first mode. Note that in the ordinary analysis, we set $\Delta t = 0.005$(sec).

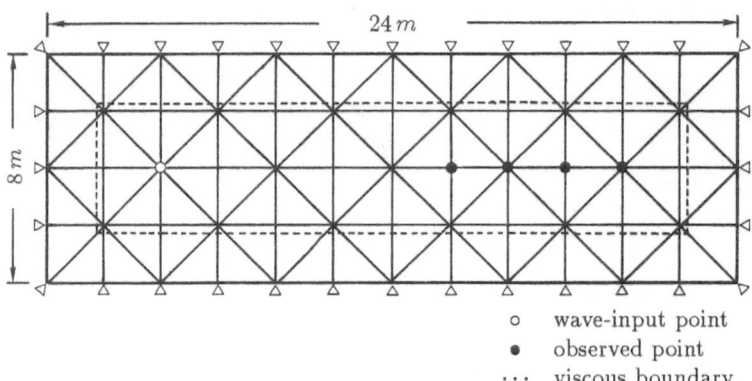

$\circ$    wave-input point
$\bullet$    observed point
$\cdots$    viscous boundary
$\triangle$    fixed boundary

Figure 1. FE model.

The results of identification are shown in Table 1, 2 and 3 in which initial values for iteration are changed.

Table 1.

| | Young's modulus $E(tf/m^2)$ | Poisson's ratio $\nu$ | damping factor $\alpha$ |
|---|---|---|---|
| initial | 4800 | 0.2 | 10 |
| true | 5000 | 0.25 | 8 |
| converged | 4996.15 | 0.2496 | 7.976 |
| error(%) | 0.08 | 0.16 | 0.3 |

Table 2.

| | Young's modulus $E(tf/m^2)$ | Poisson's ratio $\nu$ | damping factor $\alpha$ |
|---|---|---|---|
| initial | 5500 | 0.4 | 1.0 |
| true | 5000 | 0.25 | 8 |
| converged | 4995 | 0.249 | 7.773 |
| error(%) | 0.5 | 0.028 | 2.8 |

Table 3.

| | Young's modulus $E(tf/m^2)$ | Poisson's ratio $\nu$ | damping factor $\alpha$ |
|---|---|---|---|
| initial | 3000 | 0.1 | 20 |
| true | 5000 | 0.25 | 8 |
| converged | 4940 | 0.249 | 7.558 |
| error(%) | 1.2 | 0.4 | 5.5 |

Acknowledgment
We wish to express thanks to Mr. A. Takeichi who helps us in computation.

References

1. Ichikawa, Y., and Ohkami, T.(1992): "A Parameter Identification Procedure as a Dual Boundary Control Problem for Linear Elastic Materials," *Soils and Foundations*, to be appeared in June.

2. Lysmer, J., and Kuhlemeyer, G. (1969): "Finite Dynamic Model for Infinite Media," *EM4, ASCE.*, pp.859-877.

3. Ohkami, T., and Ichikawa, Y. (1992): "A Parameter Identification Procedure as a Dual Boundary Control Problem for Viscoelastic Materials," *Int. J. Num. Anal. Meth. Geomech.*, presented.

4. Ohkami, T. (1991): "A Generalized Back Analysis Method and Its Applications to Rock Mechanics," Dr. Eng. Thesis, Nagoya University.

# Identification of Dynamic In Situ Soil Properties at Existing Structure Site

KUNIHITO MATSUI, MANABU MATSUSHIMA AND SUSUMU OHTAKI

Professor, Tokyo Denki University, Saitama, Japan.
Director, Tokyo Electric Power Service Co. Ltd., Tokyo, Japan.
Graduate Student, Tokyo Denki University, Saitama, Japan

## Summary

This paper aims to develop a reliable procedure to estimate structural parameters of soil-foundation system from measured data by using a system identification technique.

Prior to identify the soil-foundation system, a set of laboratory tests are conducted by harmonically exciting a simple structural model and their structural parameters are identified to confirm an effectiveness of the procedure used.

A field observation has been made on a circuit breaker-footing system for which input earthquake and acceleration responses are measured. To make an identification possible, it has been observed that the maximum responses have to be greater than 3 gal and that a duration of relatively large input energy has to be used instead of using entire history of records. A procedure to find the duration is presented.

## Introduction

In recent years, considerable attention has been directed towards a seismic safety of structure as well as its integrity. Dynamic responses of existing civil and architectural structures have been measured so as to investigate their dynamic characteristics directly relevant to earthquake-resistant design.

In common approach in using these data has been to compare the observed response data with the response of a synthesized linear model subjected to observed base excitation. A number of researchers have studied structural identification problems for which some takes a time domain formulation and the other takes a frequency domain approach. Majority of researches are confined to a numerical simulation. Relatively small number of studies have been made by using a set of observed data.

Success of the identification depends on various factors: 1) number of sensors, 2) their locations, 3) their accuracy or a noise in observed data, 4) a discrepancy between actual structure and its synthesized linear model and [5] identification procedure.

A sweeping frequency test has been performed on a simple structural model in a laboratory prior to the field test on a circuit breaker-footing system in order to confirm whether the identification procedure is workable and also to detect a possible difficulty which may be encountered. The identification procedure used in this study is Gauss-Newton method. Between May 1990 and May 1991, seventeen earthquake records and the corresponding acceleration responses are obtained. Lateral and rotational stiffnesses and their corresponding damping coefficients are identified from these records.

## Identification procedure

When an excitation force acts on a structure, a behavior of structure can be described by the following equation of motion,

$$M\ddot{u} + C\dot{u} + Ku = Q(t) \tag{1}$$
$$u(t_0) = a, \ \dot{u}(t_0) = b$$

where $M, C$ and $K$ are mass, damping and stiffness matrices, $\ddot{u}$, $\dot{u}$ and $u$ are acceleration, velocity and displacement response vectors, and $Q(t)$ is a excitation force vector. Let an observed acceleration response at i be $\ddot{v}_i$ and its corresponding analytical acceleration be $\ddot{u}_i$, then one can write as,

$$\ddot{v}_i = \ddot{u}_i + \varepsilon_i, \ i \in A \tag{2}$$

in which $\varepsilon_i$ is a noise in observed record, and A is a set of observation points. If unknown parameters such as damping coefficient and stiffnesses are expressed by $X = (x_1, x_2, ..., x_M)^T$, then $\ddot{u}_i$ depends on $X$. $X$ can be determined so as to minimize the following evaluation function,

$$J = \frac{1}{2} \int_{t_0}^{t_1} \sum_{i \in A} w_i \{\ddot{v}_i - \ddot{u}_i(X)\}^2 dt \tag{3}$$

in which $t_0 - t_1$ indicates a duration of records used for identification. Weight function $w_i$ in Eq. 3 is assumed to be

$$w_i = \frac{1}{S_i}, \quad S_i = \int_{t_0}^{t_1} \ddot{v}_i^2 \, dt \tag{4}$$

Since Eq. 3 is a nonlinear least square problem, a minimization process requires an iterative computation. By applying Gauss-Newton method, $X$ can be updated by $\delta X$ calculated from

$$\sum_{j=1}^{M} \left\{ \int_{t_0}^{t_1} \left( \sum_{i \in A} w_i \frac{\partial \ddot{u}_i}{\partial x_k} \frac{\partial \ddot{u}_i}{\partial x_j} \right) dt \right\} \delta x_j$$

$$= - \int_{t_0}^{t_1} \sum_{i \in A} w_i \ (\ddot{v}_i - \ddot{u}_i) \frac{\partial \ddot{u}_i}{\partial x_k} dt \tag{5}$$
$$(k = 1 \cdots M)$$

Formulation of Eq. 5 necessitates a sensitivity of $\ddot{u}_i$ with respect to $x_k$, which can be evaluated from

$$M\frac{\partial \ddot{u}}{\partial x_k} + C\frac{\partial \dot{u}}{\partial x_k} + K\frac{\partial u}{\partial x_k}$$

$$= \frac{\partial Q}{\partial x_k} - \frac{\partial M}{\partial x_k}\ddot{u} - \frac{\partial C}{\partial x_k}\dot{u} - \frac{\partial K}{\partial x_k}u \qquad (6)$$

$$(k=1\cdots M)$$

Eq. 6 can be solved by the same algorithm used for Eq. 1. Fig. 1 illustrates a flow diagram for the identification procedure in which a move limit is placed on $\delta x_j$

## Laboratory test

A simple model of Fig. 2 is placed on a shake table and a frequency sweep test is conducted from 1 to 30 Hz with an increment of 1 Hz( 0.2Hz in a neighborhood of resonant frequency). Both input and response accelerations are recorded. Prior to the sweep test, static loading test is made and the stiffness was found to be 17.79kgf/cm. From its geometry the mass is obtained as 0.02371kgf.s2/cm. Hence undamped natural frequency is computed as 4.31Hz. Also from damped free oscillation test and resonant curve, damping coefficient is estimated as 0.00290kgf.s/cm and 0.00859kgf.s/cm respectively. In order to verify an effectiveness of Gauss-Newton method, a damping coefficient and stiffness are estimated from the recorded data and the results are presented in Fig.3. The figure shows that the stiffness estimate remains constant when external frequency is less than the resonant frequency and that it starts increasing after the resonant frequency. It is due to the reason that 1 degree of freedom system can not manifest a higher mode behavior which appears on the experimental model when the model is excited at the frequency higher than its natural frequency. If we use the same set of data to estimate its natural frequency,the frequency remains constant regardless of the external

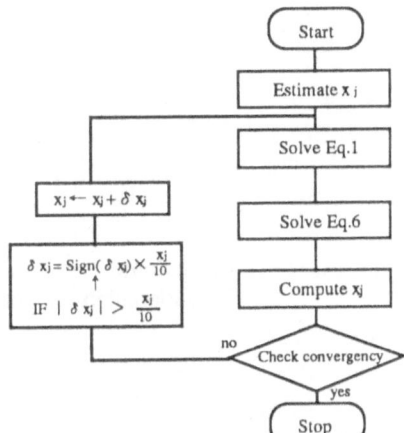

Fig.1 Flow diagram for identification procedure

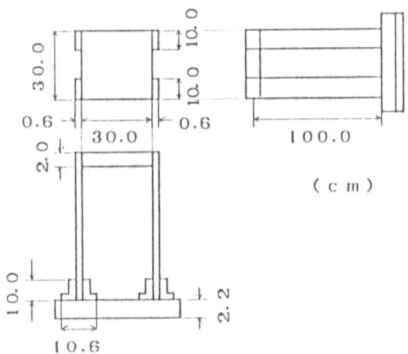

Fig.2 One-degree-of-freedom model

frequency. Fig. 4 illustrates the estimated modal parameters. Damping coefficient in Fig.3 and damping constant in Fig.4 show that they tend to vary in a wide range. To compare the agreement of observed and computed responses, SN ratio is computed from the following equation.Diagram of SN ratio-external frequency is drawn in Fig.5. As long as the diagram is concerned, the modal parameter identification is superior to structural parameter identification.

$$\text{SN Ratio}(\%) = \frac{\displaystyle\int_{t_0}^{t_1} \varepsilon(t)^2\, dt}{\displaystyle\int_{t_0}^{t_1} \text{Acc.}(t)^2\, dt} \times 100 \tag{7}$$

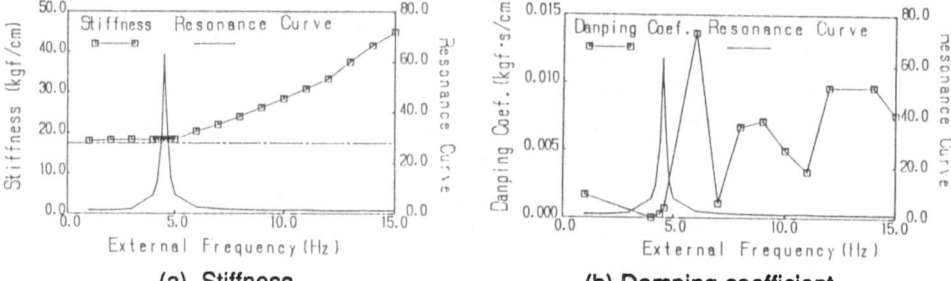

(a) Stiffness  (b) Damping coefficient
Fig.3 Identified results for one-degree-of-freedom model

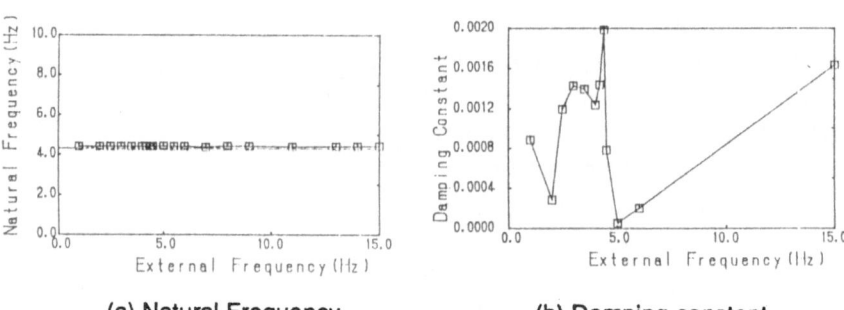

(a) Natural Frequency  (b) Damping constant
Fig.4 Identified mode parameters

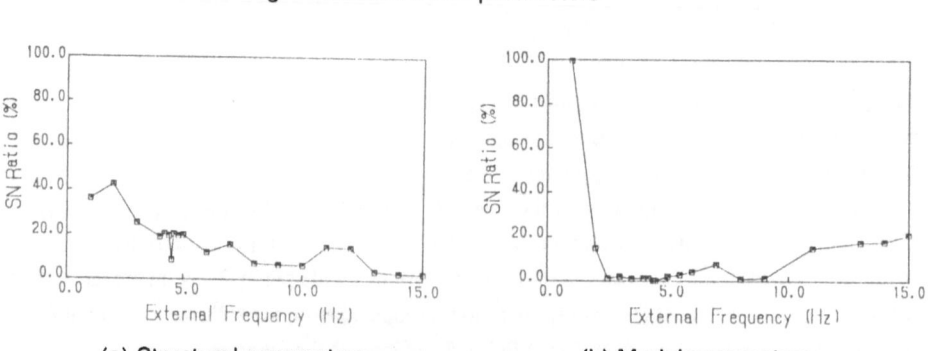

(a) Structural parameters  (b) Modal parameters
Fig.5 SN ratio

## Circuit Breaker-Foundation System

Existing circuit-breaker mounted on a foundation is modeled as Fig.6. Their structural characteristics are determined based on the seismic design code and given in Table 1. The dampers and stiffnesses for the ground are estimated from a shear wave velocity measured in advance.

## Observations of Earthquake

Earthquake wave and acceleration responses of the structure in Fig.6 are recorded at every occurrence of earthquake during May 1990 and May 1991. The maximum accelerations of all recorded data are tabulated in Table 2                        Iden

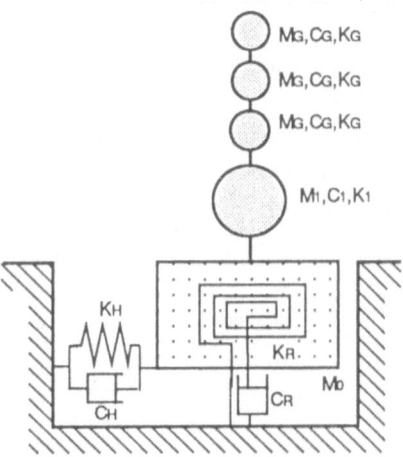

Fig.6 Circuit breaker-fiundation system

Table 1 Discription of circuit breaker-foundation model

| | | |
|---|---|---|
| Footing Size(m) | width | 2.35 |
| | length | 4.40 |
| | heigh | 0.90 |
| Mass of footing | $M_0$ (tf · s$^2$/m) | 1.83 |
| Super structure | mass | $M_G$ (tf · s$^2$/m) | 0.79 |
| | | $M_1$ (tf · s$^2$/m) | 0.46 |
| | stiffness | $K_G$ (tf/m) | 2000 |
| | | $K_1$ (tf/m) | 46000 |
| | damping | $C_G$ (tf · s/m) | 2.1 |
| | | $C_1$ (tf · s/m) | 8.7 |
| Soil stiffness | $K_H$ (tf/m) | 32584 |
| | $K_R$ (tf · m/rad) | 37665 |
| Soil damping | $C_H$ (tf · s/m) | 178.0 |
| | $C_R$ (tf · s · m/rad) | 95.0 |

Table.2 Observed earthquake

| Observed date 1990~ 1991 | Max. ground acc. | Footing | |
|---|---|---|---|
| | | Rocking acc. | Lateral acc. |
| 5//3 | 8.73 | 0.017 | 8.60 |
| 5/7 | 3.81 | 0.007 | 4.29 |
| 5/14 | 2.94 | 0.012 | 2.81 |
| 5/19 | 1.20 | 0.005 | 1.99 |
| 5/29 | 1.48 | 0.004 | 2.10 |
| 6/5 | 11.91 | 0.031 | 10.69 |
| 6/15 | 2.16 | 0.007 | 2.34 |
| 6/20 | 2.55 | 0.010 | 2.34 |
| 7/4 | 8.79 | 0.021 | 8.25 |
| 10/6 | 5.06 | 0.013 | 5.21 |
| 12/16 | 11.42 | 0.017 | 11.90 |
| 2/11 | 3.45 | 0.011 | 3.45 |
| 3/1 | 5.84 | 0.018 | 6.39 |
| 3/15 | 3.71 | 0.016 | 4.36 |
| 4/25 | 3.05 | 0.006 | 3.08 |
| 5/3 | 1.67 | 0.004 | 2.26 |
| 5/18 | 3.38 | 0.001 | 2.77 |

## Itification of structural parameters

Identified parameters in this study are $C_H$, $C_R$, $K_H$ and $K_R$. Since observed data is always contaminated with some degree of noise. When input and output accelerations are small, noise effect could become significant. In order to reduce the effect of noise on the estimates, it is preferable to use a duration of relatively large input energy. The duration is determined from the following dimensionless energy,

$$E_p(\%) = \frac{E_0}{E_t} \times 100 = \frac{E_1}{E_t} \times 100 \tag{8}$$

where

$$E_t = \int_0^{t_e} \ddot{z}_0^2 dt, \quad E_0 = \int_0^{t_0} \ddot{z}_0^2 dt, \quad E_1 = \int_{t_1}^{t_e} \ddot{z}_0^2 dt$$

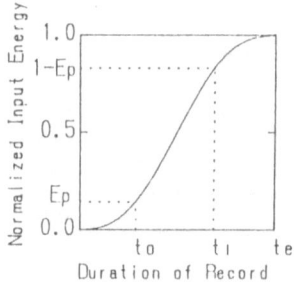

Fig.7 Conceptual diagram of dimensionless energy

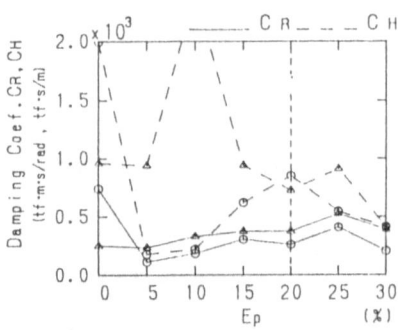

(a) Damping coefficient

The conceptual diagram for the dimensionless energy is prescribed in Fig.7. By choosing a different value of $E_p$, the parameters are estimated and the results are prescribed in Fig. 8. From the figure one can expect that a stable convergence when $E_p$ is greater than 15%. However $E_p$ exceeds 30%, a convergence tends to become difficult. This will be due to the reason that initial conditions for Eq.1 are assumed to be zero. Hence $E_p = 20\%$ is adopted for the duration. Fig.9 shows a trend for convergence by changing the values for initial estimates. Fig.10 presents the comparison in Fourier spectrum of observed and computed responses. Identification was made by using all the records listed in Table 2, but it was not necessarily always successful. A success and failure

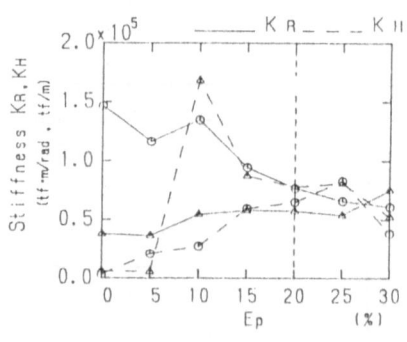

(b) Stiffness

Fig.8 Relationship between identified parameters and $E_p$

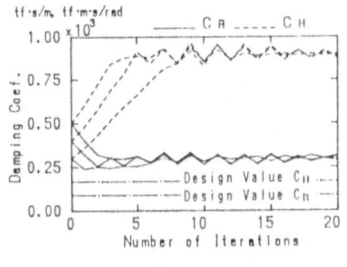

(a) Damping coefficient

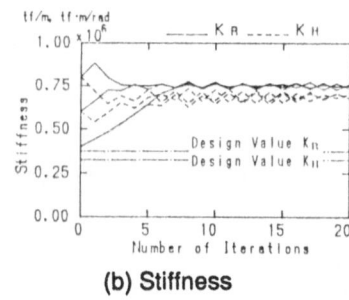

(b) Stiffness

Fig.9 Converging process

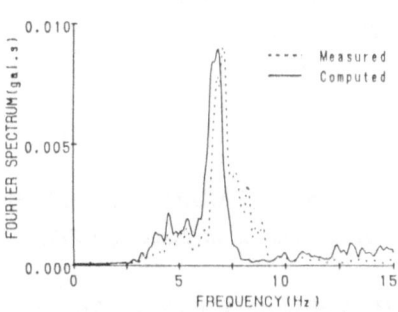

(a) Rocking motion

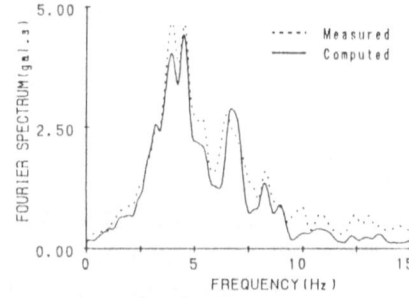

( b) Lateral motion

Fig.10 Comparison of Fourier spectrum

of identification are plotted in Fig.11. Fig.12 shows successfully identified results. Lateral stiffness $K_H$ tends to vary in a wide range compared with rotational stiffness $K_R$

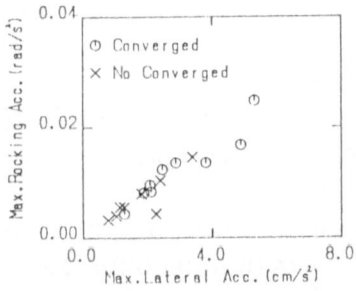

Fig.11 Success and failure of convergency

## Conclusions

Prior to this research we have compared identification procedures and examined the effects of model errors and noises in input and response data by numerical simulations. However it is essential to identify parameters from observed records. After we practiced it, several points have become clear which were not confirmed in the numerical simulation.

1. Both damping coefficient and damping constant vary in a wide range.

2. When a structure is harmonically excited, its frequency should be less than its natural frequency.

3. When a excitation is made by an earthquake, a duration of relatively large input energy should be used instead of entire records.

4. Variation in stiffness parameters of the ground is a lot greater than that in the laboratory model.

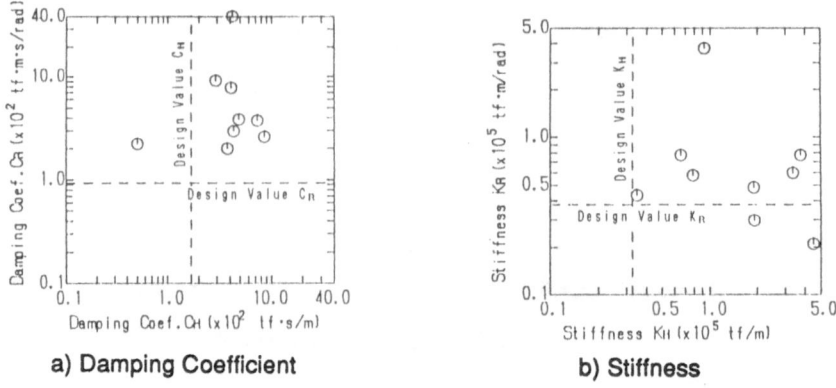

a) Damping Coefficient   b) Stiffness

Fig.12 Identified parameters

## References

1) Hanada, K., Andoh, K., Iwatate, T. and Sawada, Y. : SYSTEM PARAMETER IDENTIFICA-TION FOR A SYSTEM UNDER DYNAMIC LOADS, Journal of Structural Engineering, Vol.32A, pp.725 - 738, 1986.3.(in Japanese)

2) Sawada, T., Tujihara, O., Asega, H. and Kamiya, H.: A FEW REMARKS ON THE IDENTIFI-CATION OF LINEAR CHAIN MODEL WITH MULTI-DEGREE OF FREEDOM IN THE FREQUENCY DOMAIN, Journal of Structural Engineering, Vol.32A, pp. 739 - 748, 1986.3.(in Japanese)

3)Distefano, N. and Panapardo, B. : SYSTEM IDENTIFICATION OF FRAMES UNDER SEIS-MIC LOADS, Journal of Engineering Mechanics Division, ASCE, Vol.102, No.EM2, pp. 313 - 330, April 1976.

4)Hoshiya, M. and Saito, E. : STRUCTURAL IDENTIFICATION BY EXTENDED KALMAN FILTER, Journal of Engineering Mechanics, ASCE, Vol.110, No.12, pp. 1757 - 1770, December, 1984.

5)Matsui, K. and Kurita, T. : STRUCTURAL IDENTIFICATION FROM INPUT AND OUTPUT ACCELERATION RECORDS, Journal of Structural Engineering, Vol.35A, pp. 689 - 698, 1989.3.(in Japanese)

6)Matsui, K. and Kurita, T. : SENSITIVITIES OF PARAMETERS DUE TO MODEL ERROR AND MEASUREMENT NOISES IN STRUCTURAL IDENTIFICATION PROBLEMS, Struct Eng./Earthquake Eng., Vol.7, No.2, 263s - 272s, October 1990.

7)Japan Electric Society : Standard of DENKI-GIJUTSU-KIJUN(JEAG 5003-1980), 1980.5.15.(in Japanese)

# Characterization of the Damping Effect of a Vibrating System

S. Audebert and H. Andriambololona

Electricité de France
Direction des Etudes et Recherches
1 Avenue du Général de Gaulle
92141 Clamart, France

The purpose of the paper is the characterization of the damping effect of a system called a link line, which is attached between electrical lines for safety purpose. A laboratory simplified representation of the system (Fig. 1) is constituted by two free-clamped beams, representing electrical lines, related by a real link line. The aeolian force, which generates the line vibrations in the real situation, is simply represented in the laboratory by a point force, delivered by an excitation device.

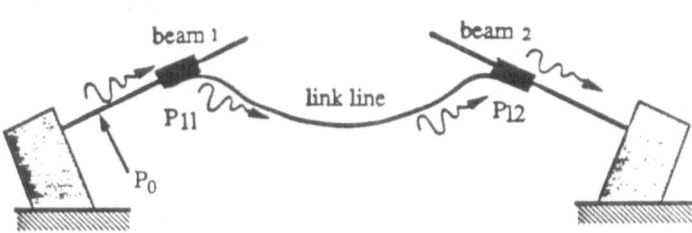

Fig. 1 The considered link line is placed between two free-clamped beams. $P_{11}$ denotes the power transmitted from beam 1 to the link and $P_{12}$ is the power transmitted from the link towards beam 2.

In order to determine the force distribution applied by the link line to each beam, for a given frequency band, two inverse methods are used and compared. These distributions are established from measured displacements at some points on the beams.

The first method uses the general analytical expression of the bending displacements of the beam which can be easily derived. By applying the boundary conditions on the general solution, one can determine the vibration responses of the beam to various excitation distributions.

The second method, more general, uses a finite element discretization technique to determine the response of the beam to various excitation distributions.

In both cases, specific regularisation techniques are used, and especially the stochastic gaussian inversion. Indicators on the quality of the solving are used too.

Finally, an estimation of the power dissipated by the link line is made.

## 1 - PRINCIPLES OF THE IDENTIFICATION OF FORCE DISTRIBUTIONS

We assume that the system is linear.
The beam is supposed to be cylindrical, homogeneous, with a constant section, and with a small diameter compared to its length.

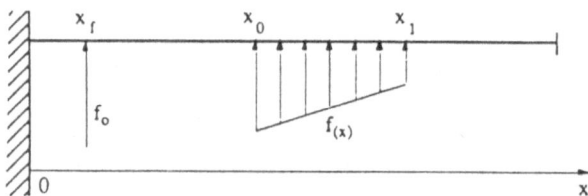

Fig. 2 Distribution of forces acting on the beam

The general configuration of the beams is described on Fig. 2 : the beam 1 is loaded with the point force $f_0$ and the unknown distribution $f_1(x)$ ; the beam 2 is loaded by another force distribution $f_2(x)$.

Analytical model

This method assumes that the general form of displacement u(x) of every point of the beam can be explicitly written. In our case, the flexural vibrations of the beam are governed by the following differential equation :

$$(E I \nabla^4 - \omega^2 \rho S) = f,$$

where E denotes the Young modulus (complex), I the moment of inertia, $\omega$ the pulsation, $\rho$ the density, and S the beam section.
The boundary conditions (clamped at x=0, free at x=L) are :

$$u(0) = u'(0) = u''(L) = u'''(L) = 0$$

Let be $\alpha$, the complex flexural wave number, defined by :

$$\alpha^4 = \omega^2 \frac{\rho}{E} \frac{S}{I}$$

Thus, the complex displacement can be expressed as a function of the abscissa of the beam as the superposition of the general solution of flexural vibrations in a bending beam and of a particular solution due to the force distribution :

$$u(x) = A_1 e^{\alpha x} + A_2 e^{-\alpha x} + A_3 e^{j\alpha x} + A_4 e^{-j\alpha x}$$
$$+ \int_0^x \frac{f(\xi)}{4EI\alpha^3} (e^{\alpha(x-\xi)} - e^{-\alpha(x-\xi)} + je^{j\alpha(x-\xi)} - je^{-j\alpha(x-\xi)}) d\xi$$

The exact response of the beam to excitation f(x) is perfectly determined as soon as the four coefficients $A_i$ and the force distribution f are known. The basic idea used for the identification of f is to project it on a N dimension basis $\phi = [\phi_1, \phi_2, ..., \phi_N]$ such that

$$f(x) = \sum_{i=1}^{N} F_i \phi_i(x) .$$

Therefore, one should obtain N+4 complex magnitudes from at least N+4 complex measurements. A complex linear system is then set :

$$[A]^m_{N+4} \, [F]^{N+4} = [U_{mes}]^m \qquad (1)$$

where $\quad F = \begin{bmatrix} A_1 \\ . \\ A_4 \\ F_1 \\ . \\ . \\ F_N \end{bmatrix} \quad$ and $\quad U_{mes} = \begin{bmatrix} u_{mes}(x_1) \\ . \\ u_{mes}(x_i) \\ . \\ u_{mes}(x_m) \end{bmatrix} \quad ;$

m is the number of measurement points and A is a rectangular matrix whose each line component is constituted of the vibration response of the beam to each unit basis excitation, at a measurement point.

Discretization model

This method uses a discretized representation of the displacement field :

$$u(x) \; = \; \sum_{k=1}^{n} U_k \, e_k(x)$$

in the discretization basis $[e_1, e_2, ..., e_n]$, where the displacement components $U_k$ can be expressed in terms of the excitation function $f(x)$ as follows :

$$U_k = q_k \int_0^x f(\xi) \, e(\xi) \, d\xi$$

where $q_k$ is a given coefficient depending upon the nature of the discretization. Using the same excitation basis as in the first case, one can write the corresponding matrix equation :

$$[B]^m_N \, [F]^N = [U_{mes}]^m \qquad (2)$$

## Resolution

The resolution of systems (1) and (2) can be made classically by the least mean square (LMS) method. Further, it can be improved by using a more robust method, that does not give enormeous errors on results when data are perturbated.

The gaussian stochastic method (Tarantola's method [1], [2], [3]), for instance, is considering data and unknown parameters as random variables following a gaussian law. Thus, we can take into account the confidence we have on each experimental data and quantify eventual a priori information on parameters. The solution of (2) which is, in this case, expressed in terms of mean value and a posteriori covariance matrix, is obtained, for each pulsation $\omega$, by minimising the following quadratic form in terms of F :

$$\left[A\ F\ -\ U_{mes}\right]^{*}[C_D]^{-1}\left[A\ F\ -\ U_{mes}\right]\ +\ \left[F\ -\ F_o\right]^{*}[C_M]^{-1}\left[F\ -\ F_o\right]$$

where $C_D$ and $C_M$ are covariance matrices and $F_o$ an a priori solution.

$F_o$ can be a particular solution, for instance an all-components-equal solution obtained by a LMS method. $C_D$ is defined in terms of the confidence that the measurement quality criteria supply. On the other hand, $C_M$ is estimated in term of a priori knowledge of the shape of the excitation force. For instance, if the force is generated by an external vibrating device, the wavelength in this external device can be used to establish the components of matrix $C_M$. In the case of the linking system that connects the link line to the beams, the large stiffness of this device allows one to assume that the corresponding wavelength is very large and the force distribution slowly varying in space.

## 2 - APPLICATION TO THE LINK LINE

The geometrical and mechanical characteristics of each beam are :

    length : 1 m
    diameter : 0.03 m
    density : 7500 kg.m$^{-3}$
    Young's modulus : $1.4 .10^{11}$ Pa
    damping coefficient : 0.01

The characterization of the damping effect of the link line is studied on the frequency band 5 Hz - 125 Hz that contains the main excitation components of the wind (aeolian) power spectral density. The two first modal frequencies of the beam alone (18 Hz and 114 Hz) are also contained in this bandwidth. The values of density and Young's modulus had been adjusted so that the two first modal frequencies obtained by calculation coincide with those obtained experimentally.

The nature and the number of basis force distributions and their corresponding responses, are first determined from simulated data (obtained either analytically or with a finite element method). A compact sinusoidal basis set of functions, defined on the link line support, is used :

$$\phi_i(x) = \sin [i \ \pi \ \frac{x - x_0}{x_1 - x_0}] \ \text{if} \ x \in [x_0, x_1] \quad \text{and} \quad \phi_i(x) = 0 \ \text{if} \ x \notin [x_0, x_1]$$

Preliminary phase : reconstruction of the point force

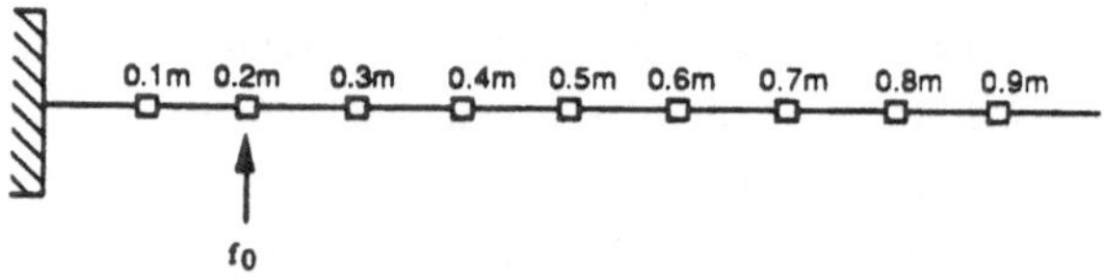

Fig. 3  Location of sensors on the beam

The methods are first verified and compared for the reconstruction of the point force representing the external excitation (known = 1 N), from 9 experimental data. The sensors locations are shown on Fig. 3. The point force to be recognized is decomposed on a small segment $[x_0, x_1]$, the dimension of which is 0.007 m.

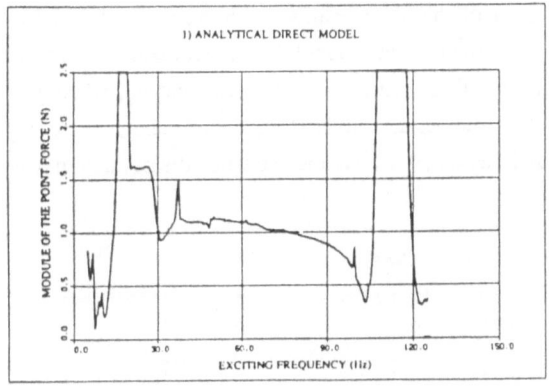

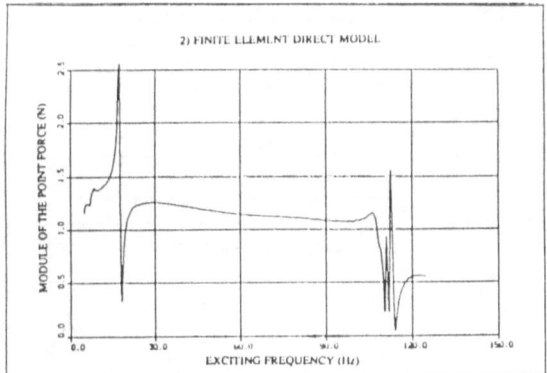

Fig. 4 Identified force modulus function of the exciting frequency (LMS method)

Identification of the force distribution due to the link line

The force distributions on beam 1 and beam 2 are then identified thanks to the gaussian inverse method, from an analytical direct model. Three basis functions are used for the decomposition, in order to be able to quantify the moment induced.

The covariance matrices $C_D$ and $C_M$ are considered as diagonal matrices :

$$C_D = \sigma_D^2 I \qquad \text{and} \qquad C_M = \sigma_M^2 I ,$$

where I denotes the identity matrix, and where $\sigma_D$ and $\sigma_M$ are coefficients respectively related to the precision of the measurement values and to the pulsation ($\sigma_M$ increases with the pulsation).

Fig. 4 shows the results of the identification of the force module 1)by an analytical direct model, 2) by a finite element direct model. The agreement is good, except near the eigen frequencies of the beam. This result is well known ; indeed, when an almost conservative structure is excited on one of its modal frequencies, the response is very close to the corresponding modal deflexion, whatever the excitation distribution may be.

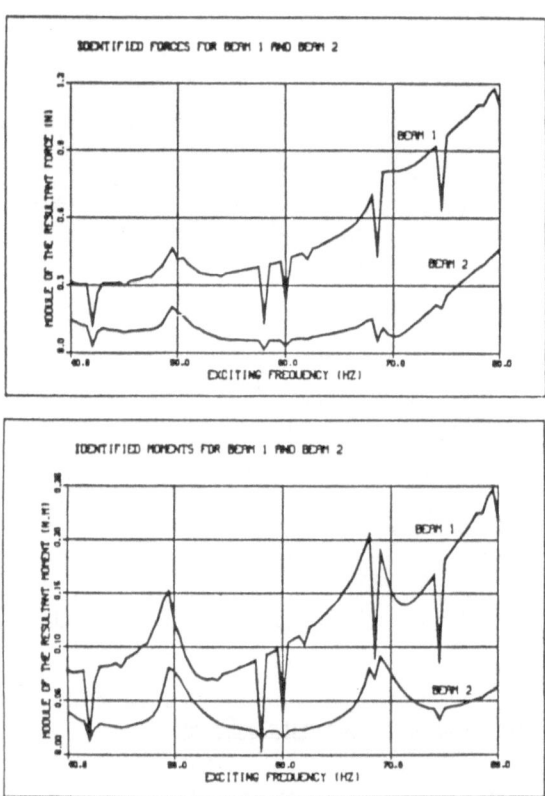

Fig. 5  Resultant identified forces and moments for the two beams, on the frequency band 40 Hz - 80 Hz, that contains two modal frequencies of the system beam + link line.

Fig. 5 shows the modules of the resultant identified forces and moments, relatively to the middle of the link line support, for each beam, relatively to a 1 N excitation point force. Outside the modal frequencies of the system beam + link line, modulus of forces and moments increase with the exciting frequency. Five peaks appear, that may correspond to the modal frequencies of the link line alone.

Determination of power dissipated by link line

The power $P_{11}$ transmitted from beam 1 to the link line is deduced thanks to the classical product $F.v^*$, where v is the velocity field. Similarly, $P_{12}$ is the power transmitted from the link line to beam 2.

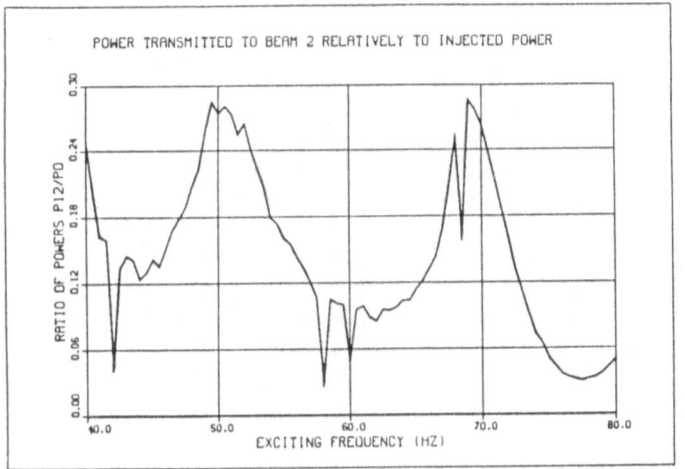

Fig. 6 Power transmitted to beam 2 relatively to injected power on the frequency band 40 Hz - 80 Hz

We can see on Fig. 6 that the power transmitted to beam 2 does not exceed 30 % of the injected power, that is to say that the link line is absorbing at least 70 % of the injected power. The nearest the exciting frequency is compared with the modal frequencies of the system beam + link line, the most important is the damping effect of the link line.

Quality of the resolution

In our case where no real verification is possible, we must try to measure the quality of the inversion process : in the gaussian process, some indicators are able to quantify the solving of the inverse problem. For instance, it has been shown [3] that the quality of the resolution is good when the a posteriori covariance matrix C is small (in norm) compared with the a priori covariance matrix $C_M$. It can be characterized by the number r defined by :

$$r = \frac{\text{Tr} [C\, C_M^{-1} - I]}{\text{Tr} [I]}$$

We can see on Fig. 7 that, excepted near the modal frequencies of the link line, the number r is very close to 1, that indicates a very good accuracy.

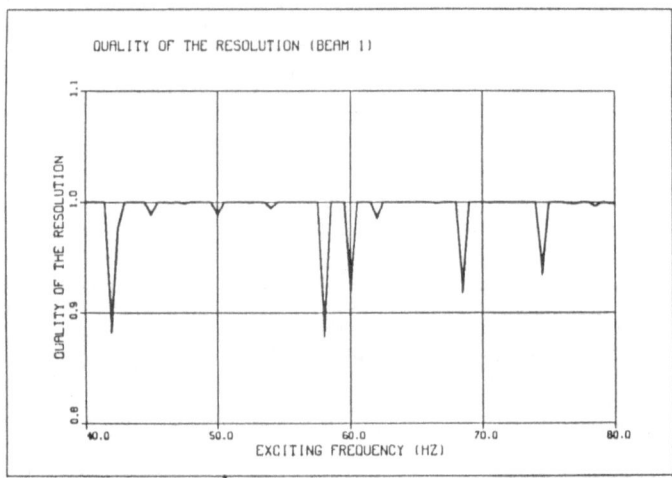

Fig. 7 Quality of the resolution r for the beam 1

3 - CONCLUSION

Inverse techniques have been successfully tested and compared in order to characterize the damping effect of a mechanical system. Two methods, depending on different direct models, are presented. For both methods, the resolution itself can be made thanks to a classical least mean square technique or improved by using the stochastic gaussian inverse technique.

In our case, the results are identical whatever the direct model used. Concerning the solving, the gaussian inverse technique is more sophisticated in so far as it can, for instance, take into account informations on measurement values and, eventually, on unknown parameters ; it also permits to qualify the quality of the solving by defining some adequate indicators.

References
1. Tarantola, A. : Inverse problem theory - Methods for data fitting and Model Parameter Estimation, *Elsevier* (1987).
2. Bonnet, M. : Un aperçu des approches existantes pour la description mathématique et la solution des problèmes inverses, *Rapport EDF (P55L08/1E5240)* (1989).
3. Bonnet, M. : Traitement numérique de problèmes inverses de source en acoustique linéaire, *Rapport EDF (P55L08/1E5240)* (1990).

# Identification of Parameters in Geomechanics Using a Maximum Likelihood Approach

A. Ledesma, A. Gens and E.E. Alonso

School of Civil Engng., Geotechnical Engng. Department.
c/ Gran Capitán s/n, Edif. D2, Technical University of Catalonia.
08034 Barcelona, SPAIN

Summary

The paper presents an statistical formulation based on the maximum likelihood criterion which provides a general framework to pose the backanalysis problem in Geomechanics. The method allows the introduction of prior information on the parameters (i.e. from laboratory tests) in the identification procedure in a systemmatic way. The Finite Element approximation is used to solve the direct problem. On the other hand, the minimization associated problem is solved by means of the Levenberg - Marquardt's algorithm using an 'exact' procedure to compute the sensitivity matrix needed in the formulation. Two examples of application are presented in the paper: a synthetic case and a real one. Both involve excavation problems on rock considered as a linear elastic isotropic material. The error structure of the problem is also identified considered as a new parameter included in the identification procedure.

## Introduction

The estimation of soil or rock parameters from field data has always been an important way of understanding the behaviour of geotechnical structures. Usually, parameters obtained from field instrumentation are more reliable than those from laboratory tests, as they do not include the effect of the macrostructure and non-homogeneity of soils and rocks. Even when representative samples are obtained, the disturbance produced in the sampling and the stress relief will affect the quality of the parameters identified.

Moreover, the recent development of the optimization procedures coupled to numerical techniques provides useful tools to solve the identification problem in a systemmatic way. Therefore the use of such procedures is becoming more frequent in Geomechanics [1]. There are various alternative formulations that can be used to perform backanalysis in Soil and Rock Mechanics, but there are important advantages in formulating the estimation problem within an statistical framework. Following that idea, the maximum likelihood estimator [2] has been used in this work.

The model which relates "state variables" , $x$, and "parameters", $p$, is assumed deterministic, and the best estimation of the parameters is found by maximizing the likelihood,

$L$, of an hypothesis, $p$. As defined by Edwards [2], the Likelihood of an hypothesis is proportional to the joint probability density function $P$ :

$$L(p) = kP(x\ ,\ p) \tag{1}$$

where $k$ is an arbitrary constant. There are some conceptual advantages in this approach [3,4] :

- The model parameters are considered fixed but uncertain due to the lack of information and the error in the measurement process. This allows to introduce the prior information on the parameters in a simple way, using the probability density function of the prior information itself.

- It does not require the model to reproduce the true system exactly.

- The procedure gives information on the reliability of the parameters identified.

The mathematical problem that involves this approach is the minimization of an "objective function" which is the same obtained from a Bayesian approach where the probability density function is maximized. Only the conceptual basis is changed.

The paper presents first the basic formulation of the backanalysis problem using the maximum likelihood approach and the way in which it is coupled with the finite element method used to solve the direct problem. Afterwards, two examples are described, involving the excavation of a tunnel in rock considered as an isotropic elastic material. The first one is a synthetic example and the second one is a real case including prior information.

Maximum Likelihood Formulation

It is assumed that an explicit model $M$ relates parameters $p$ and state variables $x$ as $x = M(p)$. Only "m" state variables are measured and they will be denoted by $x^*$. Although the parameters are unknown there may be some prior information on them that it is convenient to introduce in the formulation. We assume, then, that the probability distribution functions of the error of the measurements and on the error of the prior information are multivariate Gaussian, so:

$$P(x) = \mid C_x \mid^{-1/2} (2\pi)^{-m/2}\ \exp\ \left\{\frac{1}{2}(x^* - x)^t C_x^{-1}(x^* - x)\right\} \tag{2}$$

$$P(p) = \mid C_p^o \mid^{-1/2}(2\pi)^{-n/2}\ \exp\ \left\{\frac{1}{2}(p^o - p)^t C_p^{o-1}(p^o - p)\right\} \tag{3}$$

where $C_x$ is the measurements covariance matrix, $C_p^o$ is the 'a priori' parameter covariance matrix based on the available prior information, "n" is the number of parameters.

$\boldsymbol{x}^*$ is the vector of measurements and $\boldsymbol{p}^\circ$ is the vector which contains the 'a priori' information on the parameters.

If measurements and the prior information available on the parameters are independent, the likelihood (1) becomes:

$$L(\boldsymbol{p}) = kP(\boldsymbol{x})P(\boldsymbol{p}) \tag{4}$$

The problem of parameter estimation is now equivalent to finding the set of parameters, $\boldsymbol{p}^*$ that maximizes (4). This is the same as minimizing the support function $S = -2lnL(\boldsymbol{p})$ :

$$
\begin{aligned}
S = (\boldsymbol{x}^* - \boldsymbol{x})^t \boldsymbol{C_x}^{-1}(\boldsymbol{x}^* - \boldsymbol{x}) \ + \ (\boldsymbol{p}^\circ - \boldsymbol{p})^t \boldsymbol{C_p}^{\circ-1}(\boldsymbol{p}^\circ - \boldsymbol{p}) + \\
+ \ln | \boldsymbol{C_x} | + \ln | \boldsymbol{C_p^\circ} | + n \ \ln(2\pi) + m \ \ln(2\pi) - 2 \ \ln k
\end{aligned}
\tag{5}
$$

If the error structure of measurements and parameters are considered fixed, only the first two terms must be used in the minimization process. Otherwise, only the three last terms can be disregarded.

To solve the identification problem, a suitable algorithm must be used in order to minimize (5) without constraints. In this paper, Levenberg-Marquardt's algorithm has been used [5]. The iterative process starting from initial parameter values follows the expression:

$$\Delta\boldsymbol{p} = \Delta\boldsymbol{p}^\circ + \left[\boldsymbol{A}^t\boldsymbol{C_x}^{-1}\boldsymbol{A} + \boldsymbol{C_p}^{\circ-1} + \lambda\boldsymbol{I}\right]^{-1}\boldsymbol{A}^t\boldsymbol{C_x}^{-1}(\Delta\boldsymbol{x} - \boldsymbol{A}\Delta\boldsymbol{p}^\circ) \tag{6}$$

where $\Delta\boldsymbol{p}^\circ = \boldsymbol{p}^\circ - \boldsymbol{p}$ , $\boldsymbol{I}$ is the identity matrix, $\Delta\boldsymbol{x} = \boldsymbol{x}^* - \boldsymbol{x}$, and $\boldsymbol{A}$ is the sensitivity matrix of the problem:

$$\boldsymbol{A} = \frac{\partial\boldsymbol{x}}{\partial\boldsymbol{p}} \tag{7}$$

$\lambda$ is an scalar that controls the iteration procedure. If $\lambda \to 0$ the algorithm becomes the standard Gauss-Newton algorithm.

Numerical implementation

The identification algorithm represented by equation (6) requires the computation of the sensitivity matrix $\boldsymbol{A}$. This matrix can be estimated using numerical differentiation, but if the number of parameters is significant, it requires a considerable computing effort per iteration. Moreover, the procedure involves numerical and truncation errors associated with numerical differentiation. So an alternative procedure is presented using the Finite Element approximation which is usually applied to solve the direct problem.

Formulating the Finite Element Method in terms of displacements, the basic equation is:

$$Ku = f \tag{8}$$

where $K$ is the global stiffness matrix, $u$ the nodal displacement vector and $f$ the nodal force vector. The classical Finite Element theory provides:

$$K = \int_V B^t D B dV \quad ; \quad f = \int_S N^t \sigma dS \tag{9}$$

where $B$ is the matrix that relates strains and nodal displacements, $D$ is the constitutive matrix, $N$ is the shape function matrix and $\sigma$, for an excavation case, is the vector of stresses acting on the excavation boundary. It is assumed that measurements points may not coincide with nodal points, or in a more general way, that measurement values can be obtained from a linear combination of displacements (i.e. relative displacement between two points). In such case, a matrix relation $R$ between $u$ and $x$ can be found, so

$$x = Ru \quad ; \quad A = \frac{\partial x}{\partial p} = R\frac{\partial u}{\partial p} \tag{10}$$

In order to compute the sensitivity matrix $A$ equation (8) is differenciated with respect to the parameters. After rearranging, we obtain

$$A = R\frac{\partial u}{\partial p} \quad ; \quad \frac{\partial u}{\partial p} = K^{-1}\left[\frac{\partial f}{\partial p} - \frac{\partial K}{\partial p}u\right] \tag{11}$$

assuming that the model is linear, the stiffness matrix does not depend on the displacements, and the geometry is constant.

For instance, if the Young's modulus is going to be identified,

$$\frac{\partial u}{\partial E} = -K^{-1}\frac{\partial K}{\partial E}u \quad ; \quad \frac{\partial K}{\partial E} = \int_V B^t \frac{\partial D}{\partial E} B dV \tag{12}$$

and $\partial D / \partial E$ is straightforward for the elastic case. A similar expression can be deduced for the Poisson modulus. If the identification of the initial stress state is required, the $K_o$ parameter may be used. It represents the ratio between horizontal and vertical initial stresses $K_o = \sigma_h/\sigma_v$ and usually $\sigma_v$ is due to the weight of the rock at the considered point (working in total stresses). So for this particular parameter:

$$\frac{\partial u}{\partial K_o} = K^{-1}\frac{\partial f}{\partial K_o} \quad ; \quad \frac{\partial f}{\partial K_o} = \int_S N^t \frac{\partial \sigma}{\partial K_o} dS \tag{13}$$

where it must be taken into account that $\sigma = (K_o\sigma_v, \sigma_v, 0)$ in two dimensional problems, so $\partial\sigma/\partial K_o = (\sigma_v, 0, 0)$ and the second expression in (13) can be evaluated using basically the subroutines of the program which compute nodal forces.

The procedure outlined allows the 'exact' computation of the sensitivity matrix, using the Finite Element approximation. It can be generalized for the case in which the measurements correspond to various excavation phases [6]. That makes the computation of (13) more complicated. Also the procedure can be extended to non-linear constitutive models whith a one-to-one stress/strain relationship [7].

### Synthetic example

To illustrate the capabilities of the formulation presented, a synthetic example is shown. It simulates the excavation of a circular tunnel in a linear isotropic elastic material under plane strain conditions. The finite element mesh is depicted in figure 1, and because of symmetry, only half of the actual geometry is studied. It will be assumed that horizontal displacements are measured at points 1 to 7 and vertical displacements are measured at points 8 to 12. Usually different measurement equipments are used for horizontal and vertical displacements, i.e. inclinometer and extensometer devices. The horizontal measurements are assumed to have a variance of $\sigma_I^2$ and the vertical ones $\sigma_E^2$, and all of them are independent.

The parameters involved in the problem are the Young's modulus, $E$ , Poisson's ratio, $\nu$ , and the $K_o$ parameter which defines the initial stress state:

$$\sigma_v = \gamma z \quad ; \quad \sigma_h = K_o \gamma z \qquad (14)$$

where $\gamma = 20 KN/m^3$ is the soil density and $z$ is the depth of the considered point. The excavation is supossed to be undrained, so a value of $\nu = 0.49$ has been taken. Appart from the material parameters, there are two variables that are not known 'a priori': the variances of the measurements $\sigma_I^2$ and $\sigma_E^2$. It can be shown that the objective function (5) only depends on the ratio $\mu = \sigma_I^2/\sigma_E^2$, so this variable can be included as a new parameter to be estimated. In this way the error structure of the problem is introduced in the identification procedure. That approach have been already used in hidrology problems [8]. Finally, three parameters have been identified: $E$, $K_o$ and $\mu$.

In order to define the measurement values, the direct problem has been solved using $E = 10 MPa$, $K_o = 1$. The displacements obtained in the points 1 - 12 (figure 1) have been perturbed with a normal noise of zero mean and the following variances:

Case 1. $\sigma_I^2 = 0.06m$ , $\sigma_E^2 = 0.015m$ , $\mu = 16$

Case 2. $\sigma_I^2 = 0.03m$ , $\sigma_E^2 = 0.03m$ , $\mu = 1$

Case 3. $\sigma_I^2 = 0.015m$ , $\sigma_E^2 = 0.06m$ , $\mu = 1/16$

342

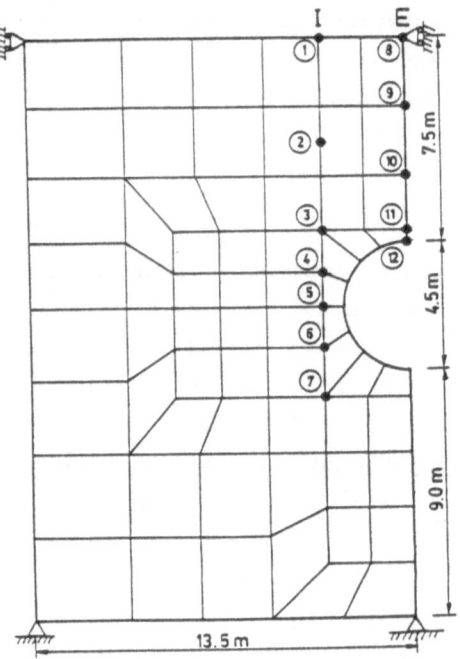

Fig. 1 Synthetic example: finite element mesh and measurement points.

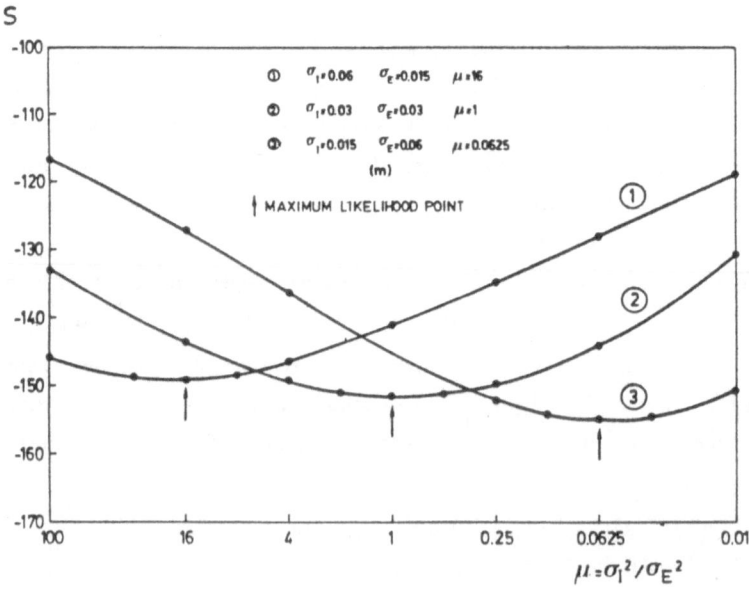

Fig. 2 Synthetic example: variation of objective function S with variance ratio, $\mu$.

Using those displacements as input data, the identification procedure should estimate the correct values of $E$, $K_o$ and $\mu$.

In all cases the procedure gave correct enough values of the parameters, in particular the value of $\mu$ which represents the error structure of measurements. Figure 2 shows the variation of objective function, $S$, with respect to $\mu$. It can be observed that the minimum of $S$ corresponds in every case to the value of $\mu$ of the perturbation applied to the correct measurements, so the variance ratio is correctly identified. On the other hand, maximum likelihood identification analysis leads to good parameter estimates ($E$ and $K_o$), even with an unknown error structure.

Real example

The general identification procedure, including prior information, has been used to back-analyze the excavation of a powerhouse cavern in rock, in the Spanish Pirenées. By analyzing a central section, plane strain conditions can be assumed. The details of the design and construction of the underground opening can be found in [9]. For the backanalysis, two material parameters are considered: the Young's modulus of the rock, $E$ (assumed isotropic linear elastic), and the ratio between initial horizontal and vertical stresses, $K_o$. The Poisson's ratio of the rock, because of its small range of variation and small influence on the results, was fixed to 0.28, derived from laboratory tests.

The measurements correspond to rock displacements obtained from extensometers and convergence devices in the excavation phases. For the purposes of the analysis three excavation phases have been considered, so some of the measurements are only available for a particular stage. That must be taken into account when computing the sensitivity matrix, as was mentioned previously. The finite element mesh and the measurement points are indicated in figure 3. In total, 36 measurements have been used, 8 corespond to the first phase, 7 to the second phase and 21 to the third phase.

Due to the site investigations before the construction of the cavern, there was some prior information on the parameters that could be used in the identification. In particular, the 'in situ' stresses were measured using flat jacks and triaxial cells. In both cases the vertical initial stress was very close to the overburden pressure. However, the $K_o$ value had an important range of variation: 1 – 3. The mean Young's modulus obtained from 'in situ' tests was 15000 MPa.

To obtain a good estimate of the parameters, it is important to include the prior information available in the estimation process. The following values were adopted as 'a priori' estimations of the parameters: $E = 15000 MPa$ ; $K_o = 2$ which are mean

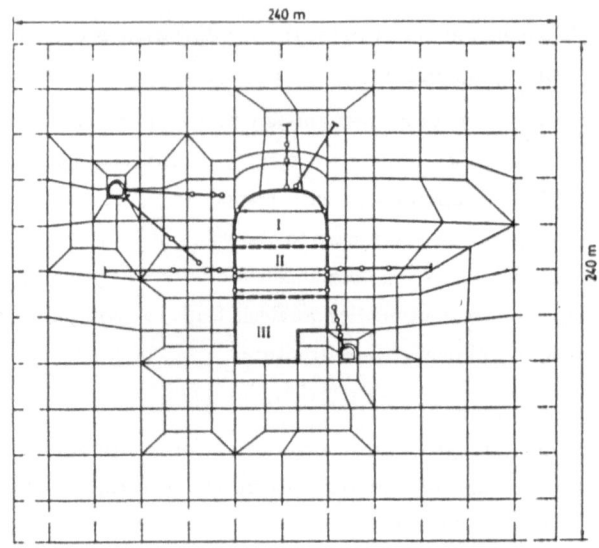

Fig. 3 Real example: finite element mesh, excavation phases and measurement points.

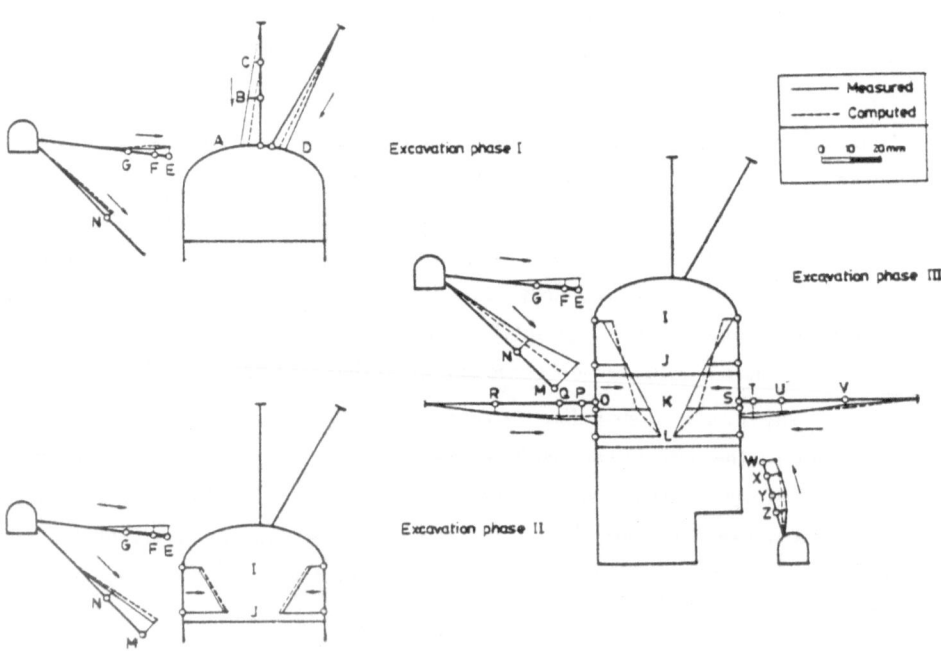

Fig. 4 Real example: comparison between observed and computed displacements.

values obtained in the 'in situ' tests. Still it remains the important decision of giving a standard deviation to those values, to complete the expression of the probability density function of the prior information, that is, to specify $C_p^o$.

These standard deviation are directly related to the relative weight given to the prior information with respect to the field measurements. In order to do that in an objective way, the backanalysis has been performed taking this relative weight as a new parameter to be identified. Therefore the set of unknown parameters are:

$$E \;\; , \;\; K_o \;\; , \;\; \mu_E = \sigma_x^2/\sigma_E^2 \;\; , \;\; \mu_K = \sigma_x^2/\sigma_K^2 \tag{15}$$

because the minimum of the objective function depends on the ratio between standard deviations rather than on the absolute values.

The optimum point obtained as minimum of the objective function corresponds to the values of $\mu_E = 5 \; 10^{-5}$ and $\mu_K = 10^{-5}$. The material parameters associated to these values are: $E = 8200 MPa$ and $K_o = 2.53$. The displacements computed using this set of parameters are compared with the measured ones in figure 4, where a good agreement can be observed. Furthermore, the parameters identified seem to be more realistic. In particular, $E$ has a lower value than 15000 MPa, because it represents the stiffness of the rock mass rather than the rock matrix.

Concluding remarks

In this paper a general methodology to perform backanalysis in geomechanical problems is presented. It is based on the maximum likelihood approach which provides a probabilistic framework for solving the problem. In this way, the error of the measurements is introduced in the procedure, as well as the prior information available on the parameters and its error structure.

The procedure includes a computational method to evaluate 'exactly' the sensitivity matrix, using the finite element approximation. This matrix is needed in the Marquardt's algorithm which has been used to solve the minimization associated problem.

The method is used in two examples concerning the excavation of underground openings in rock, assumed as an elastic isotropic material. Both show the capabilities of the procedure, identifying the elastic modulus, $E$, and the initial stress state defined by the $K_o$ parameter. One of the interesting features of the procedure is that the error structure of the measurements has been included in the first example as a new parameter to be identified. Also, in the second example, where prior information on the parameters was available, the relative weight given to the prior information with respect to the field measurements is included in the identification procedure. In this way the existing prior information is incorporated with a reliability determined by the backanalysis procedure itself.

It is expected that the methodology presented will prove very useful to backanalyze geotechnical structures in a more systematic and rational manner.

Aknowledgements

The support given by the Spanish Research Committee "CICYT" (Projects PB87-0861, PB90-0598) and by FECSA is gratefully acknowledged.

References

1. Gioda, G.; Sakurai, S.: Back analysis procedures for the interpretation of field measurements in Geomechanics, *Int. J. Num. Analyt. Meth. Geomech.*, **11** (1987) 555-583.

2. Edwards, A.W.F.: *Likelihood*, Cambridge University Press, (1972), Cambridge.

3. Beck, J.V.; Arnold, K.J.: *Parameter estimation in engineering science*, Wiley, (1977), New York.

4. Baram, Y; Sandell, N.R.: An information theoretic approach to dynamical systems modelling and identification, *IEEE Transactions on Automatic Control*, **AC-23** (1978) 61-66.

5. Marquardt, D.W.: An algorithm for least-squares estimation of nonlinear parameters, *J. Soc. Industr. Appl. Math.*, **11** (1963) 431-441.

6. Gens, A.; Ledesma, A.; Alonso, E.E.: Back analysis using prior information – Application to the staged excavation of a cavern in rock, *Numerical Methods in Geomechanics, Innsbruck*, (1988), 2009-2016, G. Swoboda ed., A.A. Balkema, Rotterdam.

7. Ledesma, A.; Gens, A.; Alonso, E.E.: Identification of parameters of nonlinear geotechnical models, *Computed methods and advances in Geomechanics, Cairns*, (1991), 1005-1010; G. Beer, J.R. Booker and J.P. Carter eds., A.A. Balkema, Rotterdam.

8. Carrera, J.; Neuman, S.P.: Estimation of aquifer parameters under transient and steady state conditions: 1 Maximum likelihood method incorporating prior information, *Water Resour. Res.*, **22** (1986) 119-210.

9. Wittke, W.; Soria, J.L.: Exploration, design and excavation of the powerhouse cavern Estanygento – Sallente in Spain, *Proc. 5th Congress ISRM, Melbourne* (1983) D167-178.

# Identification Analysis of Distributed-Parameter Systems by Using Kalman Filter-Boundary Element Method

A.Utani * and N.Tosaka **

* Graduate School of Nihon University
  Narashino, Chiba, 275, Japan
** Department of Mathematical Engineering,
  College of Industrial Technology, Nihon University,
  Narashino, Chiba, 275, Japan

## Summary

The purpose of this study is to propose a new approach based on the boundary element method in solving the identification problems of the partial differential equations which are field equations of the problems. In this approach, the boundary element method in conjunction with the Kalman filter theory which is effectively used in the fields of optimal control of dynamic systems is developed. The accuracy and efficiency of the proposed procedure, which is called the Kalman filter-boundary element method, is examined through the numerical examples. They are: 1) identification of the concentrated heat source in steady-state thermal conduction field and 2) identification of material properties and estimation of unknown boundary values in the two-dimensional isotropic elastostatics field.

## Introduction

An increasing interest in the techniques for solving problems of identification analysis has been shown in the fields which is governed by the partial differential equations[1-3]. Kalman filter is one of the powerful methods which allows the error information in the observed data during the solution procedure. There exist identification analyses of the discretized system by using Kalman filter theory in conjunction with the finite element method[4-8]. We developed extended Kalman filtering approach incorporated with boundary element discretization as a new approach[9]. A key ingredient in the identification and estimation analysis by using our proposed procedure is the appropriate selection of boundary element discretizations of the field equations which correspond to the system equations which constitute Kalman filter. In this paper, we show the algorithm of the Kalman filtering as the mathematical basis and apply it to the identification problems of distributed-parameter systems. Then we show the Kalman filter-boundary element formulation for identification analyses, and finally we implement a numerical analysis and discuss the potentiality of our procedure in the fields.

## The Kalman filter approach

The Kalman filtering technique has an on-line data acquisition algorithm which offers successively an optimal presumption to system variables by using input data as follows: 1) dynamic characteristics of the system which generates the signals, 2) statistical characteristics of noises, 3) prior information on initial values of system variables, 4) observation data. In case where the nonlinear system dominates a field equation, we utilize the Taylor series expansion with an estimate of $\hat{z}_{t/t-1}$ for the observation equation in order to avoid nonlinearity. The following system equations constitute the extended Kalman filter.

State equation:

$$z_{t+1} = \boldsymbol{\Phi}_t z_t + \boldsymbol{\Gamma}_t \boldsymbol{\omega}_t \tag{1}$$

Stochastic characteristics on the state noise:

$$E(\boldsymbol{\omega}_t) = \mathbf{0} \ , \ E\left[\boldsymbol{\omega}_t \boldsymbol{\omega}_s^T\right] = \boldsymbol{Q}_t \delta_{ts} \ , \ E\left[\boldsymbol{\omega} z_s^T\right] = \mathbf{0} \quad for \ t \geq s \tag{2}$$

Nonlinear observation equation:

$$\boldsymbol{y}_t = \boldsymbol{m}_t(z_t) + \boldsymbol{\nu}_t \tag{3}$$

Linearized observation equation:

$$\boldsymbol{\eta}_t = \boldsymbol{M}_t z_t + \boldsymbol{\nu}_t \tag{4}$$

$$\boldsymbol{M}_t = \left(\frac{\partial \boldsymbol{m}_t}{\partial z_t}\right)_{z_t = \hat{z}_{t/t-1}} \tag{5}$$

$$\boldsymbol{\eta}_t = \boldsymbol{y}_t - \boldsymbol{m}_t\left(\hat{z}_{t/t-1}\right) + \boldsymbol{M}_t \hat{z}_{t/t-1} \tag{6}$$

Stochastic characteristics on the observation noise:

$$E(\boldsymbol{\nu}_t) = \mathbf{0} \ , \ E\left[\boldsymbol{\nu}_t \boldsymbol{\nu}_s^T\right] = \boldsymbol{R}_t \delta_{ts} \ , \ E\left[\boldsymbol{\nu} z_s^T\right] = \mathbf{0} \quad for \ t \geq s \tag{7}$$

where, $E[\cdot]$ is an average operator, $\boldsymbol{y}_t$ is an observation vector $(p \times 1)$ , $z_t$ is a state vector $(n \times 1)$ , $\boldsymbol{M}_t$ is a sensitivity matrix $(p \times n)$ , $\boldsymbol{\Phi}_t$ is a state transition matrix $(n \times n)$ , $\boldsymbol{\Gamma}_t$ is a driving matrix $(n \times m)$ , $\boldsymbol{\nu}_t$ is an observation noise $(p \times 1)$ and $\boldsymbol{\omega}_t$ is a system noise $(m \times 1)$ . Both noises are 'Gaussian' and 'white'. Now, in case where we discuss the method for identification of distributed-parameter systems, according to the stationary condition of parameters to be identified, the state equation for Kalman filtering can be given by

$$z_{t+1} = \boldsymbol{I} z_t \tag{8}$$

The above state equation expresses a stationary condition that the parameters to be identified should be kept constant in time. Thus, the state transition matrix $\boldsymbol{\Phi}_t$ reduces to a unit matrix $\boldsymbol{I}$ and the system noise $\boldsymbol{\omega}_t$ is not included. The suffix $t$ indicates not the time axis but the iterative axis[4-6].

## Solution procedure

The solution strategy consists of the following filtering scheme. Because of linearity in the fields, an iterative procedure should be adopted under the constant observed data until the convergence of the estimated parameters can be achieved. We review the observation update algorithm of Kalman filtering to complete a numerical procedure for identification problems of distributed-parameter systems as follows:

Filter equation:

$$\hat{z}_{t/t} = \hat{z}_{t/t-1} + K_t \left[ y_t - m_t \left( \hat{z}_{t/t-1} \right) \right] \tag{9}$$

Kalman gain:

$$K_t = P_{t/t-1} M_t^T \left[ M_t P_{t/t-1} M_t^T + R_t \right]^{-1} \tag{10}$$

Estimate error covariance matrix:

$$P_{t/t} = P_{t/t-1} - K_t + M_t P_{t/t-1} \tag{11}$$

Initial condition:

$$\hat{z}_{0/-1} = \bar{z}_0, P_{0/-1} = \Sigma_0 \tag{12}$$

Further details of the filtering algorithm are available in the references[10-11].

## Identification problem in steady-state thermal conduction field

First of all, we consider identification problem of the concentrated heat source in steady-state thermal conduction field.

## Formulation

In this case, the state vector is given by the location of the consentrated heat source and the magnitude as follows:

$$z = (\bar{X}, Q)^T = (X, Y, Q)^T \tag{13}$$

Now, we consider the following boundary-value problem in the steady-state thermal conduction field under the internal heat source:

$$\nabla^2 u(x) = f(x) \quad in \ \Omega \tag{14}$$

$$u = \bar{u} \ on \ \Gamma_u \quad , \quad q = \frac{\partial u}{\partial n} = \bar{q} \ on \ \Gamma_q \tag{15}$$

By taking into consideration of the linearity of Poisson equation (14), we can easily transform the field equation (14) into the following boundary integral equation:

$$\frac{\alpha(y)}{2\pi} u(y) + \int_\Gamma q^*(x, y) u(x) d\Omega(x) = \int_\Gamma u^*(x, y) q(x) d\Gamma(x) + \int_\Omega q^*(x, y) f(x) d\Omega(x) \tag{16}$$

in which u* and q* are the well-known fundamental solution and its derivative, respectively, which are given by

$$u^*(x,y) = \frac{1}{2\pi}ln\left(\frac{1}{r}\right) \quad , \quad q^*(x,y) = \frac{-1}{2\pi r}\frac{\partial r}{\partial n} \quad , \quad ( \ r = \| \ x - y \ \| \ ) \tag{17}$$

In case where the concentrated heat source exists in the internal domain, the inhomogeneous term can be expressed in the form.

$$\int_\Omega f(x)u^*(x,y)d\Omega(x) = \int_\Omega Q\delta(x - \bar{X})u^*(x,y)d\Omega(x) = Qu^*(\bar{X},y) \tag{18}$$

Applying the boundary element discretization to the above boundary integral equation (16), we can get the following well-known discretized system:

$$Hu = Gq + F(z) \tag{19}$$

where the matrices $H$ and $G$ are the so-called influence matrices. By taking into consideration of boundary condition (15), we can easily transform the above discretized system (19) into the following system:

$$AX = BY + F(z) \tag{20}$$

In the end, the odservation equation is derived as

$$X = A^{-1}[BY + F(z)] \tag{21}$$

In this case, the sensitivity matrix is defined as follows:

$$M_t = \left[\frac{\partial X}{\partial X}, \frac{\partial X}{\partial Y}, \frac{\partial X}{\partial Q}\right] \tag{22}$$

These equations correspond to the observation equation (3) if we add the noise term.

Numerical example

As the numerical examples, let us show the identification of the concentrated heat source in thermal conduction model given by Fig.1. The process of identification by using our procedure is shown in Figs.2 (a) and (b). This figure shows the identification process starting from initial values on the magnitude and the location of heat source.

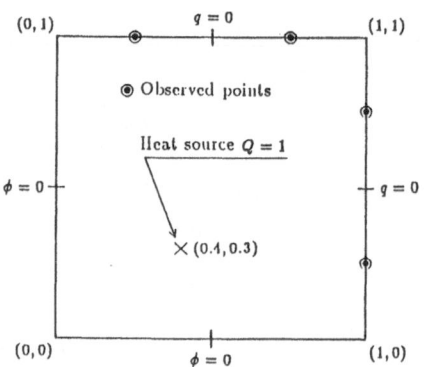

Fig.1 Thermal conduction model

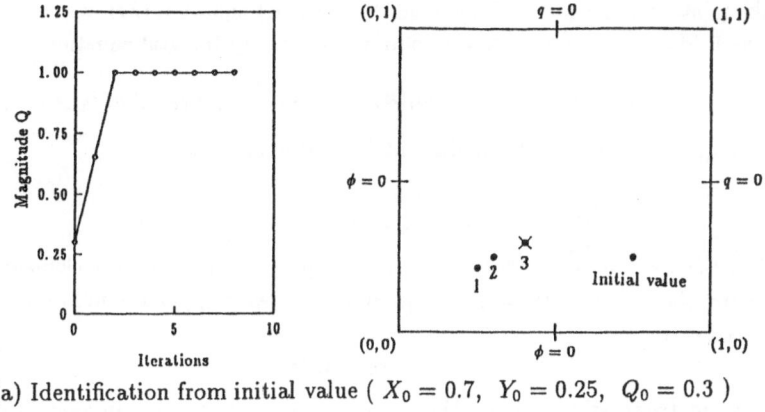

(a) Identification from initial value ( $X_0 = 0.7$, $Y_0 = 0.25$, $Q_0 = 0.3$ )

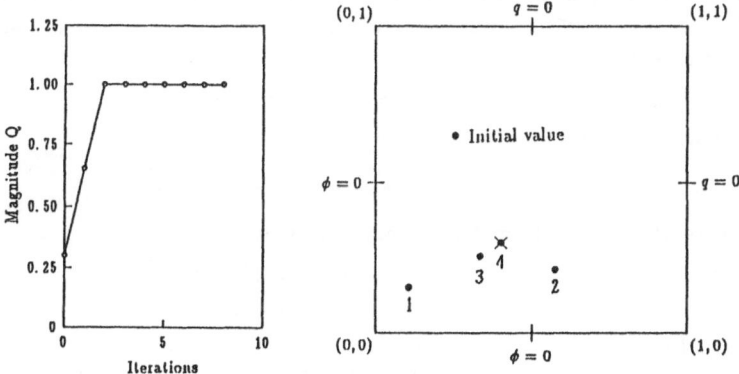

(b) Identification from initial value ( $X_0 = 0.3$, $Y_0 = 0.7$, $Q_0 = 0.3$ )

Fig.2 Identification examples of the concentrated heat source

### Identification problem in linear elastic field

Secondly, we consider identification problem of material properties in the two-dimensional linear elastic field under the plane strain condition.

### Formulation

In this case, the state vector consists of the Young's modulus, E, and Poisson's ratio, $\nu$, in the following form.

$$z = (E, \nu)^T \tag{23}$$

Now, we consider the following boundary-value problem in the linear elastic field:

$$\mu u_{i,jj} + (\lambda + \mu)u_{j,ji} + b_i = 0 \quad in \; \Omega \tag{24}$$

$$u_i = \hat{u}_i \quad on \; \Gamma_u \; , \quad t_i = t_{ij}n_j = \hat{t}_i \quad on \; \Gamma_t \tag{25}$$

By taking into consideration of the linearity of Navier equation (24), we can easily transform the field equation (24) into the following boundary integral equation:

$$\frac{1}{2}\delta_{ij}u_i(\boldsymbol{y}) + \int_{\Gamma} t^*_{ij}(\boldsymbol{x},\boldsymbol{y})u_i(\boldsymbol{x})d\Gamma(\boldsymbol{x}) = \int_{\Gamma} u^*_{ij}(\boldsymbol{x},\boldsymbol{y})t_i(\boldsymbol{x})d\Gamma(\boldsymbol{x}) \tag{26}$$

in which $u^*_{ij}$ is the well-known fundamental solution given by

$$u^*_{ij}(\boldsymbol{x},\boldsymbol{y}) = \frac{-(1+\nu)}{4\pi E(1+\nu)}\{(3-4\nu)\delta_{ij}\cdot lnr - r_{,i}r_{,j}\} \tag{27}$$

and $t^*_{ij}$ is the traction tensor derived with $u^*_{ij}$. Applying the boundary element discretization to the above boundary integral equation (26), we can get the following well-known discretized system:

$$\boldsymbol{H}(z)\boldsymbol{u} = \boldsymbol{G}(z)\boldsymbol{t} \tag{28}$$

where the matrices $\boldsymbol{H}$ and $\boldsymbol{G}$ are the so-called influence matrices in which the unknown material constants (i.e, Lamé's constants) are included. By taking into consideration of boundary condition (25), we can easily transform the above discretized system (28) into the following system:

$$\boldsymbol{A}(z)\boldsymbol{X} = \boldsymbol{F}(z) \tag{29}$$

In the end, observation equation is given in the form.

$$\boldsymbol{X} = \boldsymbol{A}(z)^{-1}\boldsymbol{F}(z) \tag{30}$$

In this case, the sensitivity matrix is defined as follows:

$$\boldsymbol{M}_t = \left[\frac{\partial \boldsymbol{X}}{\partial E}, \frac{\partial \boldsymbol{X}}{\partial \nu}\right] \tag{31}$$

### Numerical example

As the numerical examples, let us show the identification of material properties in the two-dimensional elastic model given by Fig.3. The process of identification is shown in Figs.4 (a) and (b). This figure shows the identification process starting from initial values on the Young's modulus and Poisson's ratio.

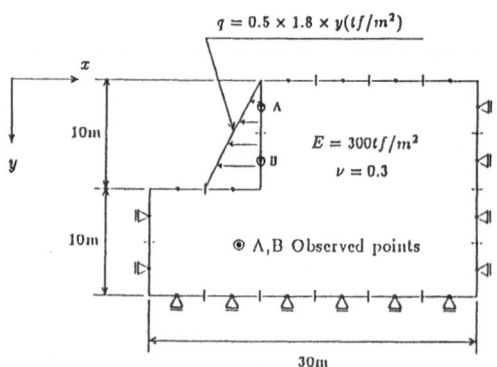

Fig.3 Analytical Model (after Arai et al.[3] and Murakami et al.[5-6])

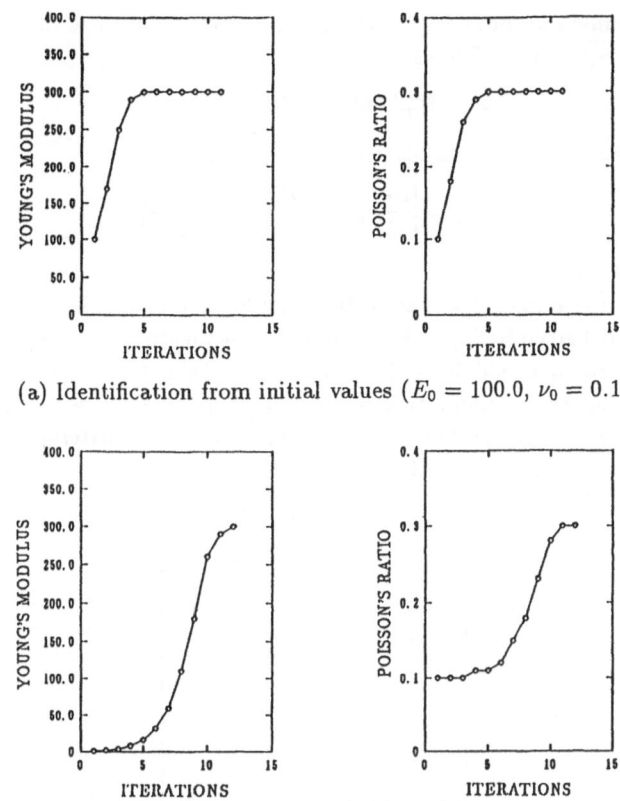

(a) Identification from initial values ($E_0 = 100.0$, $\nu_0 = 0.1$)

(b) Identification from initial values ($E_0 = 1.0$, $\nu_0 = 0.1$)

Fig.4  Identification examples of the Young's modulus and Poisson's ratio

## Another formulation

Here, we consider identification problem of both unknown boundary tractions and material properties. In this case, the state vector is expressed in the following form:

$$z = (E, \nu, t_1^*, \cdots, t_n^*) \tag{32}$$

Now, let us consider again the following boundary element discretized system of the elastostatics:

$$Hu = Gt \tag{33}$$

Here we introduce the following block-partition of the above discretized system by separating the observed displacement $u^*$ from the boundary displacement $u$:

$$
\begin{bmatrix} H_{11} & H_{12} \\ H_{21} & H_{22} \end{bmatrix}
\begin{bmatrix} u^* \\ \hat{u} \end{bmatrix}
=
\begin{bmatrix} G_{11} & G_{12} \\ G_{21} & G_{22} \end{bmatrix}
\begin{bmatrix} \hat{t} \\ t^* \end{bmatrix}
\tag{34}
$$

where $t^*$ is the unknown boundary tractions. Moreover, we arrange the above system as the observation equation as follows:

$$u^* = H_{11}^{-1} \left( G_{11}\hat{t} - H_{12}\hat{u} + G_{12}t^* \right) \tag{35}$$

In this case, the sensitivity matrix is defined as follows:

$$M_t = \left[ \frac{\partial u^*}{\partial \lambda}, \frac{\partial u^*}{\partial \mu}, \frac{\partial u^*}{\partial t_1^*}, \cdots, \frac{\partial u^*}{\partial t_n^*} \right] \tag{36}$$

### Numerical example

As the numerical examples, let us show the identification results in the two-dimensional elastostatics model given by Fig.5. Figure 6 shows the identification process starting from initial values on Young's modulus and Poisson's ratio. Table 1 shows results on estimation of unknown boundary tractions. In this case, we adopt the ones which are obtained by boundary element analysis in the case of the initial values of material properties, $E_0 = 1.0$ and $\nu_0 = 0.1$ as initial values of the tractions.

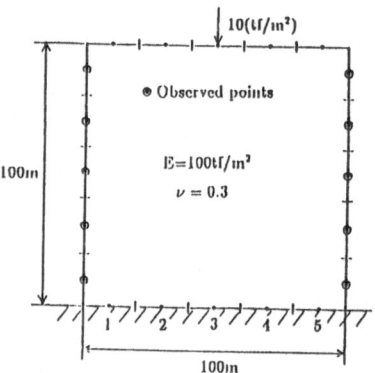

Fig.5  2-D elastostatics model

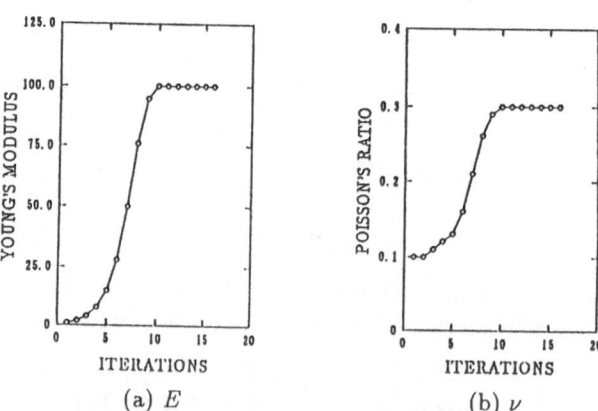

(a) $E$                    (b) $\nu$

Fig.6  Identification example of the Young's modulus and Poisson's ratio

Table 1  Results on estimation of unknown boundary tractions

| Nodal point | | Initial | True | Calculated |
|---|---|---|---|---|
| 1 | Hor. | 0.2910 | 0.7282 | 0.7282 |
|   | Ver. | 2.0803 | 2.1852 | 2.1852 |
| 2 | Hor. | 0.1591 | 0.2842 | 0.2842 |
|   | Ver. | 1.9180 | 1.8506 | 1.8506 |
| 3 | Hor. | 0.0 | 0.0 | 0.0 |
|   | Ver. | 1.9180 | 1.9226 | 1.9226 |
| 4 | Hor. | -0.1591 | -0.2842 | -0.2842 |
|   | Ver. | 1.9180 | 1.8506 | 1.8506 |
| 5 | Hor. | -0.2910 | -0.7282 | -0.7282 |
|   | Ver. | 2.0803 | 2.1852 | 2.1852 |
| Iterations | | | | 11 |

## Conclusion

We have developed the theoretical basis and applicability of the Kalman filter-boundary element method on some identification problem in the fields which is governed by the partial differential equations and discussed several numerical examples for the identification problems. As a result, the accuracy and efficiency of the proposed procedure has been proved through the above numerical examples. The advantages of the method shown herein are that the proposed procedure has the solution algorithm which can consider the observational error as error covariance while renewing the estimate error covariance.

## References

1. S.Kubo; K.Ohnaka; K.Ohji: Identification of Heat-Source and Force Using Boundary Integrals(in japanese), *Proc.JSME*, 54-501(A) (1988) 1329-1334.

2. K.Ohnaka; K.Uosaki: Identification of the external input of distributed-parameter systems by the boundary-element approach, *Int. J. Control*, 43-4 (1986) 1125-1133.

3. K.Arai; H.Ohta; T.Yasui: Simple optimization techniques for evaluating deformation moduli from field observation, *Soils and Foundations*, 23(1) (1983) 107-113.

4. A.Murakami; T.Hasegawa: observational prediction of settlement using Kalman filter theory, *Proc. 5th Int. Corf. on Numerical Methods in Geomechanics*, Nagoya, (1985) 1637-1643.

5. A.Murakami; T.Hasegawa: Back analysis by Kalman filter-finite elements and a determination of optimal observed points location(in Japanese), *Proc.JSCE*, 388 (1987) 227-235.

6. A.Murakami; T.Hasegawa: Back analysis by Kalman filter-finite elements and optimal location of observed points, *Proc. 6th Int. Corf. on Numerical Methods in Geomechanics*, Innsbruck, (1988) 2051-2058.

7. M.Suzuki; K.Ishii: Estimation of spatial variation of soil properties using extended Kalman filter algorithm(in japanese), *Proc.JSCE*, 406 (1989) 71-78.

8. S.Kadota; E.Saito; A.Waku; T.Goto: Identification method for anisotropic mechanical constants of rock mass by local iterated extended Kalman filter and application to excavation control of underground openings(in Japanese), *Proc.JSCE*, 406 (1989) 107-116.

9. A.Utani; N.Tosaka: Identification analysis of elastic constants by extended Kalman filter-boundary element method(in Japanese), *Proc. 8th Symp. on BEM's. JASCOME*, (1991) 23-28.

10. T.Katayama: Applied Kalman Filter(in Japanese), *Asakura-shoten*, (1983).

11. D.E.Catlin: Estimation, Control, and the Discrete Kalman Filter, *Springer-Verlag*, (1989).

12. R.E.Kalman: A new approach to linear filtering and prediction problems, *Trans.ASME, J.Basic Eng.*, 82D(1) (1960) 34-45.

13. R.E.Kalman; R.S.Bucy: New results in linear filtering and prediction theory, *Trans.ASME, J.Basic Eng.*, 83D(1) (1961) 95-108.

14. N.Tosaka; T.Nakayama: Fundation of Boundary Element Method(in Japanese), *Nikkagiren*, (1987).

15. M.Tanaka; T.Matsumoto; M.Nakamura: Boundary Element Method(in Japanese), *Baifu-kann*, (1991).

# Chapter 6
# Elastodynamic Inverse Problems

# Inverse Scattering for Flaw Type Classification

S. Hirose*

* Department of Civil Engineering, Okayama University,
  3-1-1 Tsushima-Naka, Okayama 700, Japan

## Summary

Among various flaws such as voids, inclusions, cracks or interfaces, cracks are the most dangerous defects found in structural materials, since they are obvious sources of unstable failure. It is, therefore, of great importance to distinguish cracks from other volumetric flaws. Our research on the inverse scattering for flaw type classification is motivated by Rose [1]. He has empirically found that the reconstruction of volumetric flaws is found in the real part of the characteristic function obtained by the inverse Born approximation (IBA), while the reconstruction of crack-like flaws is found in the imaginary part of the characteristic function. However, any reasonable explanation for the fact was not given in his paper.

In the present paper, two methods of the inverse Born approximation (IBA) and the inverse Kirchhoff approximation (IKA) are used for the inverse scattering problem of flaw type classification. Volumetric flaws are reconstructed by means of both the IBA and the IKA, while the reconstruction of crack-like flaws is made by the IKA only. Using these properties, we propose an inversion procedure to distinguish between crack-like flaws and other volumetric flaws. Furthermore, the role of the imaginary part of the characteristic function in the IBA, which has been found by Rose, but unexplained, is interpreted on the mathematical basis. In order to confirm the applicability of our inversion method for flaw type classification, numerical examples are presented for a cylindrical cavity, a Griffith crack and their combined flaws.

## Statement of the inverse problem

Consider the scattering of an elastic wave by a flaw in a two-dimensional elastic solid of infinite extent as shown in Fig. 1. It is assumed that the flaw $D$ is a void, for which tractions are zero on the boundary $\partial D$. The domain $D^c$ is assumed to be a homogeneous, isotropic and elastic solid. The inverse scattering problem is to determine the shape of a flaw when the incident wave field and the material constants in the exterior domain $D^c$ are supposed to be known, and the scattered waves are also known within a frequency range on a measurement surface $S$, which surrounds the flaw completely at far-field. There are many possibilities in selecting the arrangement of transmitter and receiver and the wave types of incident and scattered fields. In this paper, the inversion algorithm is developed for the pulse-echo scattering in the frequency domain. So transmitters and

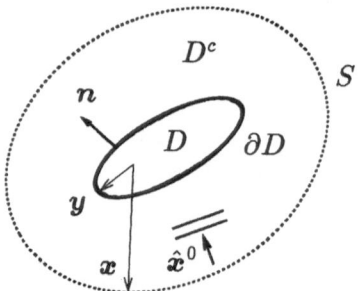

Figure 1: Geometry of elastic wave scattering

receivers coincide and are then moved simultaneously on $S$. Also, the incident wave $\boldsymbol{u}^0$ is assumed to be a plane wave given by

$$\boldsymbol{u}^0(\boldsymbol{y}) = u^0\hat{\boldsymbol{d}}^0 \exp(ik^0\hat{\boldsymbol{x}}^0 \cdot \boldsymbol{y}) \tag{1}$$

where $u^0$, $k^0$, $\hat{\boldsymbol{d}}^0$ and $\hat{\boldsymbol{x}}^0$ are the incident amplitude, the wavenumber, unit vectors defining the directions of particle motion and wave propagation, respectively.

Integral representation for scattered waves

On the assumption that the flaw $D$ is a void with traction-free boundary conditions, the scattered displacement field $\boldsymbol{u}^{sc}$ may be written in the integral form:

$$u_i^{sc}(\boldsymbol{x}) = \int_{\partial D} n_l(\boldsymbol{y})C_{jlmk}\partial U_{im}(\boldsymbol{x},\boldsymbol{y})/\partial y_k u_j(\boldsymbol{y})ds_y, \quad \boldsymbol{x} \in D^c \tag{2}$$

where $\boldsymbol{n}$ is an outward unit vector normal to the boundary $\partial D$, $\boldsymbol{u}$ is the total displacement, and $\boldsymbol{C}$ are elastic constants, which are determined by the Lamé parameters $\lambda$ and $\mu$ for an isotropic material. Also, $\boldsymbol{U}$ is the fundamental solution in 2-D elastodynamics, given by

$$U_{im}(\boldsymbol{x},\boldsymbol{y}) = \frac{i}{4\mu}\left[H_0(k_Tr)\delta_{im} + \frac{1}{k_T^2}\frac{\partial^2}{\partial y_i\partial y_m}\{H_0(k_Tr) - H_0(k_Lr)\}\right] \tag{3}$$

where $r = |\boldsymbol{x}-\boldsymbol{y}|$, $H_0(\cdot)$ is the Hankel function of order zero, $\delta_{im}$ is the Kronecker delta, and $k_T$ and $k_L$ are wavenumbers of transverse and longitudinal waves, respectively. Supposing that the receiver is far away from the flaw relatively to the flaw size, the far-field expression may be obtained by making the approximation of $r \approx |\boldsymbol{x}| - \hat{\boldsymbol{x}} \cdot \boldsymbol{y}$ ($\boldsymbol{x} \in S$, $\boldsymbol{y} \in \partial D$), where $\hat{\boldsymbol{x}} = \boldsymbol{x}/|\boldsymbol{x}|$. Substituting the far-field approximation into eq.(2) yields the integral representation for the scattered far-field as follows:

$$u_i^{sc}(\boldsymbol{x}) = A_i(k_L,\hat{\boldsymbol{x}}^0,\hat{\boldsymbol{x}})\sqrt{\frac{2}{\pi k_L|\boldsymbol{x}|}}e^{i(k_L|\boldsymbol{x}|-\pi/4)} + B_i(k_T,\hat{\boldsymbol{x}}^0,\hat{\boldsymbol{x}})\sqrt{\frac{2}{\pi k_T|\boldsymbol{x}|}}e^{i(k_T|\boldsymbol{x}|-\pi/4)} \tag{4}$$

where the coefficients $A_i$ and $B_i$ are given as

$$A_i(k_L, \hat{\boldsymbol{x}}^0, \hat{\boldsymbol{x}}) = \frac{k_L}{4(\lambda + 2\mu)} \hat{x}_i \hat{x}_m \hat{x}_k C_{jlmk} \int_{\partial D} n_l(\boldsymbol{y}) e^{-ik_L \hat{\boldsymbol{x}} \cdot \boldsymbol{y}} u_j(\boldsymbol{y}) ds_y \tag{5}$$

$$B_i(k_T, \hat{\boldsymbol{x}}^0, \hat{\boldsymbol{x}}) = \frac{k_T}{4\mu} (\delta_{im} - \hat{x}_i \hat{x}_m) \hat{x}_k C_{jlmk} \int_{\partial D} n_l(\boldsymbol{y}) e^{-ik_T \hat{\boldsymbol{x}} \cdot \boldsymbol{y}} u_j(\boldsymbol{y}) ds_y. \tag{6}$$

The coefficients $A_i$ and $B_i$ are the scattering amplitudes of the longitudinal and transverse waves, respectively. The rest of the present paper is concerned with a longitudinal to longitudinal pulse-echo scattering, although the same results can be obtained for transverse to transverse scattering. In this case, the incident wave is supposed to be a longitudinal plane wave with parameters of $k^0 = k_L$, $\hat{\boldsymbol{d}}^0 = \hat{\boldsymbol{x}}^0 = -\hat{\boldsymbol{x}}$.

## Inverse Born approximation (IBA)

The Born approximation consists of replacing the total wave field $\boldsymbol{u}$ in eq.(5) with the incident wave field $\boldsymbol{u}^0$. Although the Born approximation is originally derived for the scattering by a weak scatterer in a low frequency range, in which case the interaction of the incident wave with the scatterer is relatively small, it can still be applied to strong scattering data [1,2]. The longitudinal amplitude $A_i$ using the Born approximation may be expressed by

$$\begin{aligned}
A_i(k_L, -\hat{\boldsymbol{x}}, \hat{\boldsymbol{x}}) &= -\frac{u^0 k_L}{4(\lambda + 2\mu)} \hat{x}_i \hat{x}_j \hat{x}_m \hat{x}_k C_{jlmk} \int_{\partial D} n_l(\boldsymbol{y}) e^{-2ik_L \hat{\boldsymbol{x}} \cdot \boldsymbol{y}} ds_y \\
&= \frac{iu^0 k_L^2}{2(\lambda + 2\mu)} \hat{x}_i \hat{x}_j \hat{x}_l \hat{x}_m \hat{x}_k C_{jlmk} \int_D e^{-2ik_L \hat{\boldsymbol{x}} \cdot \boldsymbol{y}} dV_y \\
&= iu^0 \hat{x}_i k_L^2 / 2 \int_D e^{-2ik_L \hat{\boldsymbol{x}} \cdot \boldsymbol{y}} dV_y \\
&= iu^0 \hat{x}_i k_L^2 / 2 \int_{R^2} \Gamma(\boldsymbol{y}) e^{-2ik_L \hat{\boldsymbol{x}} \cdot \boldsymbol{y}} dV_y
\end{aligned} \tag{7}$$

where $\Gamma$ is the characteristic function defined as

$$\Gamma(\boldsymbol{y}) = \begin{cases} 1 & \text{for } \boldsymbol{y} \in D, \\ 0 & \text{for } \boldsymbol{y} \notin D. \end{cases} \tag{8}$$

The last domain integral on the R.H.S. in eq.(7) corresponds to the $\boldsymbol{K}$-space Fourier transform $\tilde{\Gamma}(\boldsymbol{K})|_{\boldsymbol{K}=2k_L \hat{\boldsymbol{x}}}$ of the characteristic function. If the pulse-echo scattering amplitude is known for all frequencies and for all directions of incidence, then the Fourier component $\tilde{\Gamma}(\boldsymbol{K})$ is completely defined from eq.(7). The characteristic function $\Gamma(\boldsymbol{y})$ can be obtained by an inverse Fourier transform of eq.(7). We have

$$\Gamma(\boldsymbol{y}) = -\frac{2i}{\pi^2} \int_0^{2\pi} \int_0^{\infty} \frac{\hat{x}_i}{u^0 k_L^2} A_i(k_L, -\hat{\boldsymbol{x}}, \hat{\boldsymbol{x}}) e^{2ik_L \hat{\boldsymbol{x}} \cdot \boldsymbol{y}} k_L dk_L d\hat{\boldsymbol{x}}. \tag{9}$$

## Inverse Kirchhoff approximation (IKA)

In a high frequency range, the reflection from a traction-free surface can be analyzed by using the Kirchhoff approximation with conjunction of the ray theory [3]. The total wave field $\boldsymbol{u}$ on the boundary $\partial D$ may be expressed by

$$u_j(\boldsymbol{y}) = \{-u^0 \hat{x}_j e^{-ik_L \hat{\boldsymbol{x}} \cdot \boldsymbol{y}} + \sum_{\alpha=L,T} R^\alpha(\hat{\boldsymbol{x}} \cdot \boldsymbol{n}) u^0 \hat{d}_j^\alpha e^{ik_\alpha \hat{\boldsymbol{x}}^\alpha \cdot \boldsymbol{y}}\} H(\hat{\boldsymbol{x}} \cdot \boldsymbol{n}) \tag{10}$$

where the first term indicates the incident wave and the second term defines the reflected longitudinal ( $\alpha = L$ ) and transverse ( $\alpha = T$ ) waves with the reflection coefficients $R^\alpha$. Also, $H(\cdot)$ is the Heaviside step function, which defines the illuminated side of the boundary $\partial D$. On substitution of eq.(10) into eq.(5), the main contributions to the integral can come from the stationary phase point, where $\nabla_S'[(\hat{\boldsymbol{x}}^\alpha - \hat{\boldsymbol{x}}) \cdot \boldsymbol{y}] = 0$, $\nabla_S'$ denoting the surface gradient with respect to $\boldsymbol{y}$ on $\partial D$. Within this stationary phase approximation, we find that $\boldsymbol{n} = \hat{\boldsymbol{x}}$ is satisfied at the stationary phase point. Thus, the longitudinal amplitude $A_i$ using the Kirchhoff approximation may be written as follows:

$$\begin{aligned} A_i(k_L, -\hat{\boldsymbol{x}}, \hat{\boldsymbol{x}}) &= -\frac{u^0 k_L}{2(\lambda + 2\mu)} \hat{x}_i \hat{x}_j \hat{x}_l \hat{x}_m \hat{x}_k C_{jlmk} \int_{\partial D} H(\hat{\boldsymbol{x}} \cdot \boldsymbol{n}) e^{-2ik_L \hat{\boldsymbol{x}} \cdot \boldsymbol{y}} ds_y \\ &= -u^0 \hat{x}_i k_L/2 \int_{\partial D} H(\hat{\boldsymbol{x}} \cdot \boldsymbol{n}) e^{-2ik_L \hat{\boldsymbol{x}} \cdot \boldsymbol{y}} ds_y \\ &= -u^0 \hat{x}_i k_L/2 \int_{R^2} \gamma_H(\boldsymbol{y}) e^{-2ik_L \hat{\boldsymbol{x}} \cdot \boldsymbol{y}} dV_y \end{aligned} \tag{11}$$

where $\gamma_H$ is the illuminated side singular function defined as

$$\int_{R^2} \gamma_H(\boldsymbol{y}) f(\boldsymbol{y}) dV_y = \int_{\partial D} H(\hat{\boldsymbol{x}} \cdot \boldsymbol{n}) f(\boldsymbol{y}) ds_y. \tag{12}$$

Eq.(11) has the same structure as the integral representation (7) for the scattering amplitude within the Born approximation. The singular function $\gamma_H$ can be reconstructed by an inverse Fourier transform of eq.(11) as follows

$$\gamma_H(\boldsymbol{y}) = -\frac{2}{\pi^2} \int_0^{2\pi} \int_0^\infty \frac{\hat{x}_i}{u^0 k_L} A_i(k_L, -\hat{\boldsymbol{x}}, \hat{\boldsymbol{x}}) e^{2ik_L \hat{\boldsymbol{x}} \cdot \boldsymbol{y}} k_L dk_L d\hat{\boldsymbol{x}}. \tag{13}$$

## Numerical simulation for flaw type classification

As shown in the previous section, the reconstruction of flaws in the IBA is carried out by using the characteristic function $\Gamma$, which is one inside the flaw and zero outside. It may be expected that the IBA works well for volumetric flaws such as cavities, but it can not be applied to crack-like flaws. This is because the characteristic function $\Gamma$ becomes zero everywhere in the domain if the flaw is a crack with no interior volume. On the other

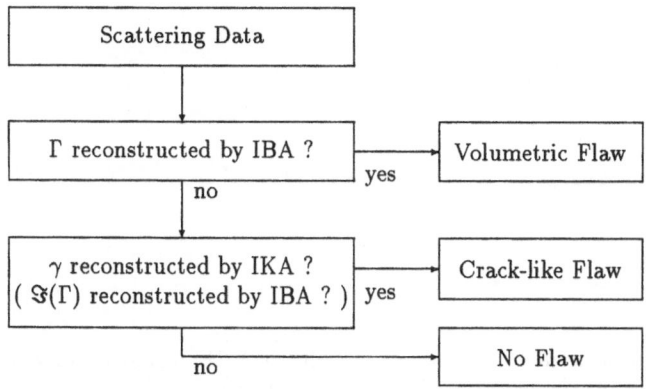

Figure 2: An inversion procedure for flaw type classification

hand, the IKA uses the singular function $\gamma_H$ to reconstruct the boundaries of flaws. As seen in eq.(12), the singular function $\gamma_H$ has singularity at the boundary $\partial D$ of any kind of flaw. Hence, the IKA can be applied to scattering data by both volumetric flaws and crack-like flaws. From the above-mentioned discussion, we propose an inversion procedure for flaw type classification as shown in Fig. 2.

Some numerical examples are presented to show the applicability of our method for discriminating between volumetric flaws and crack-like flaws. Numerical data were calculated, using the boundary integral equation method, for $0 < k_L a < 10$ for six degree increments of the direction angle, where $a$ is a characteristic length of a flaw. Fig. 3 shows the reconstruction of a cylindrical cavity with the radius $a$. Figures (a) and (b) show the characteristic function obtained by the IBA and the singular function using the IKA, respectively. Although the resolution of the singular function is not so good, both inverse approaches are successful in reconstructing the shape of a cylindrical cavity. Fig. 4 shows the similar results for a 2-D Griffith crack with the length of $2a$. The IBA fails to reproduce the crack shape, while the IKA works well in reconstructing the crack. Fig. 5 shows the same results for the combined flaws of a cylindrical cavity and a Griffith crack. As expected, the shape of a cavity is reproduced by both inversion methods, but the crack is clearly reconstructed by the Kirchhoff method only. From these numerical simulations, it is concluded that our inversion procedure shown in Fig. 2 is very useful for the flaw type classification between cracks and other volumetric flaws.

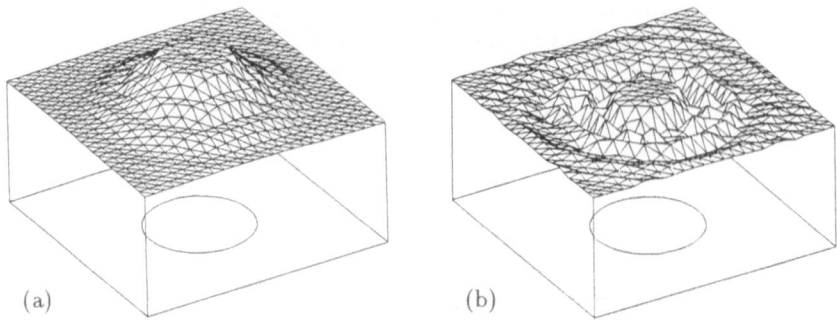

Figure 3: Reconstruction of a cylindrical cavity. (a) Characteristic function $\Gamma$ obtained by the IBA and (b) singular function $\gamma$ obtained by the IKA

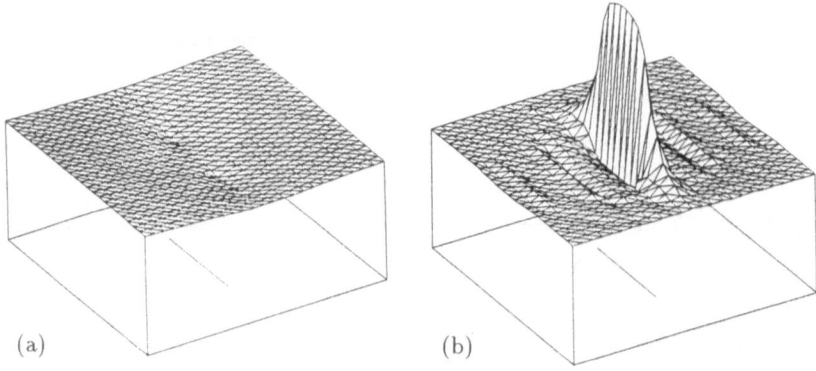

Figure 4: Reconstruction of a Griffith crack. (a) Characteristic function $\Gamma$ obtained by the IBA and (b) singular function $\gamma$ obtained by the IKA

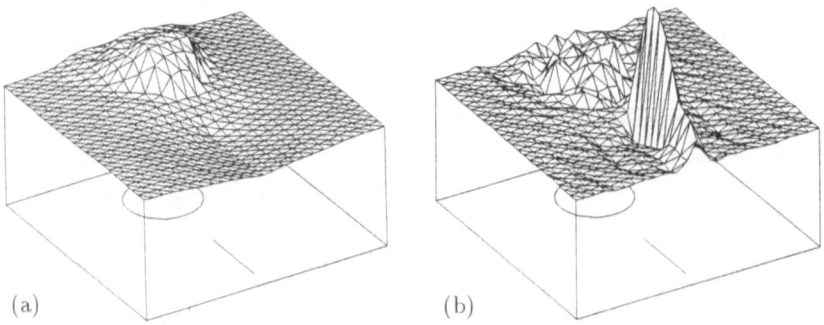

Figure 5: Reconstruction of combined flaws of a Griffith crack and a cylindrical cavity. (a) Characteristic function $\Gamma$ obtained by the IBA and (b) singular function $\gamma$ obtained by the IKA

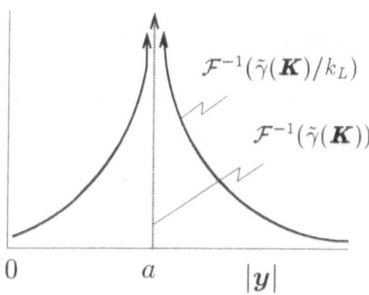

Figure 6: $\mathcal{F}^{-1}(\tilde{\gamma}(\boldsymbol{K}))$ $(= \delta(|\boldsymbol{y}| - a)/|\boldsymbol{y}|)$ and $\mathcal{F}^{-1}(\tilde{\gamma}(\boldsymbol{K})/k_L)$ as a function of $|\boldsymbol{y}|$

Role of the imaginary part of the characteristic function

As seen in eqs.(9) and (13), the difference between the inverse Fourier transforms in the Born and Kirchhoff procedures is only the factor of $i/k_L$. Thus a relationship between the imaginary part of the characteristic function $\Im(\Gamma)$ and the Fourier transformed singular function $\tilde{\gamma}$ is obtained as follows:

$$\Im(\Gamma(\boldsymbol{y})) = -\mathcal{F}^{-1}(\tilde{\gamma}(\boldsymbol{K})/k_L) \tag{14}$$

where $\mathcal{F}^{-1}$ denotes the inverse Fourier transform. For example, we consider the singular function of $\gamma(\boldsymbol{y}) = \delta(|\boldsymbol{y}| - a)/|\boldsymbol{y}|$, which represents the boundary of a cylindrical cavity with the radius $a$. The $\boldsymbol{K}$-space Fourier transform of $\gamma$ can analytically be obtained as $\tilde{\gamma}(\boldsymbol{K}) = 2\pi J_0(k_L a)$, where $J_0$ is the Bessel function of order zero. $\mathcal{F}^{-1}(\tilde{\gamma}(\boldsymbol{K})/k_L)$ can also be calculated in an explicit form of $\mathcal{F}^{-1}(\tilde{\gamma}(\boldsymbol{K})/k_L) = 2/\pi K(a/|\boldsymbol{y}|)$, where $K$ is a complete elliptic integral of the first kind. Fig. 6 shows $\mathcal{F}^{-1}(\tilde{\gamma}(\boldsymbol{K}))$ $(= \gamma(\boldsymbol{y}))$ and $\mathcal{F}^{-1}(\tilde{\gamma}(\boldsymbol{K})/k_L)$ $(= -\Im(\Gamma(\boldsymbol{y})))$ as a function of $|\boldsymbol{y}|$ in a schematic way. The function $\Im(\Gamma)$ becomes singular at the boundary of $|\boldsymbol{y}| = a$ as well as the singular function $\gamma$. This suggests that the imaginary part of the characteristic function has ability for reconstructing the shapes and sizes of both volumetric flaws and crack-like flaws.

Fig. 7 shows the distribution of $\Im(\Gamma)$ obtained by the IBA for (a) a cylindrical cavity, (b) a Griffith crack and (c) their combined flaws corresponding to Figs. 3 to 5. From these figures, it can be seen that the reconstruction of volumetric and crack-like flaws is found in the imaginary part of $\Gamma$. The images reconstructed by $\Im(\Gamma)$ are, however, somewhat smooth and obscure, compared with the results obtained by the IKA as shown in Figs. 3 to 5. This is because the singularity of $\Im(\Gamma)$ is weaker than the singularity of $\gamma$ as seen in Fig. 6.

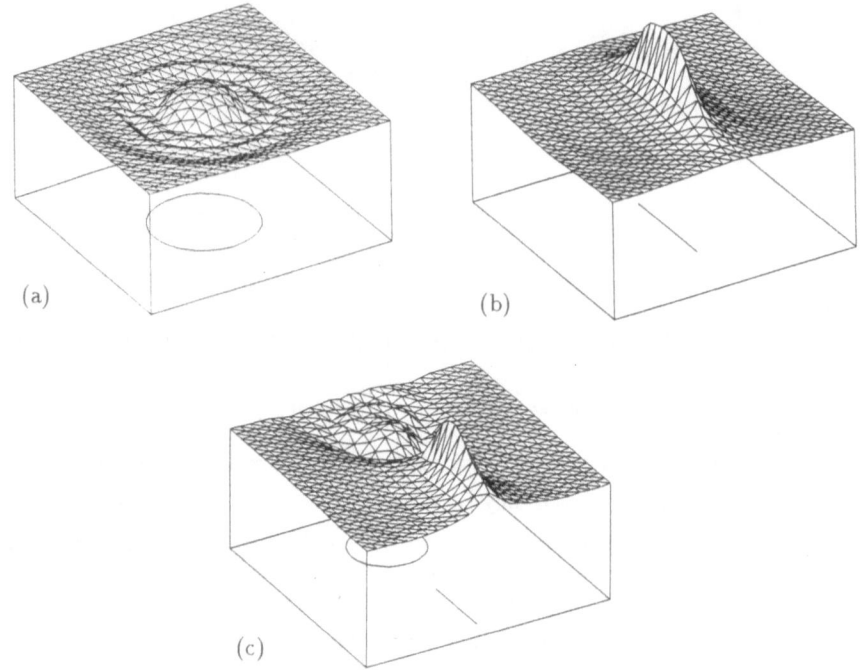

Figure 7: Imaginary part of reconstructed characteristic function $\Gamma$ obtained by the IBA. (a) a cylindrical cavity, (b) a Griffith crack and (c) combined flaws of a cavity and a crack.

References

1. J. H. Rose, Elastic wave inverse scattering in nondestructive evaluation, *J. Pure Appl. Geophys.*, **131** (1989) 715-739.

2. Y. Niwa and S. Hirose, Inverse scattering of elastic waves for a cavity, *Struct. Eng./Earthq. Eng.*, **3** (1986) 267s-275s.

3. J. D. Achenbach and A. K. Gautesen and H. McMaken, *Ray Methods for Waves in Elastic Solids*, Pitman, Boston (1982).

# Shape Inversion from Phase Shifts

M. Kitahara* and K. Nakagawa**

\*    Faculty of Marine Science and Technology
     Tokai University
     3-20-1 Orido, Shimizu, Shizuoka 424, Japan

\*\*   Total System Institute
     1-18-6 Asagaya-minami, Suginami-ku, Tokyo 166, Japan

Summary

The integral representation for the scattered far-field is first derived with the help of the partial wave decomposition of the fundamental solution. The integral representation for the phase shifts is obtained from the integral representation of the scattered far-field. The phase shifts are determined numerically by using the integral representation of the phase shifts. An application of phase shifts to shape inversion is shown. For shape inversion, the coupled system of the integral representation of phase shifts and the integral equation to determine the unknown boundary quantities is used. The numerical procedure to determine the shape of the scatterer is based on the Newton's method.

## Phase shifts

The phase shift analysis is well known in the field of electromagnetics[1] and the atomic collision theory [2 ~ 4]. The point of this paper is to apply the phase shift analysis to the characterization of scattered wave field from flaw and relate the phase shifts to the shape of the scatterer in the sense of the integral representation.

The scattering problem for scalar quantity $u$ is governed by the Helmholtz equation:

$$\Delta u + k^2 u = 0 \qquad \text{in} D \tag{1}$$

where $k = \omega/c$ is the wavenumber, $\omega$ and $c$ are the angular frequency and the wave velocity. The total field $u$ is decomposed into

$$u = u^I + u^S \tag{2}$$

where $u^I$ is the plane incident wave and $u^S$ is the scattered wave. The incident wave is assumed to be a plane wave traveling along the $x_3$ axis :

$$u^I(\mathbf{x}) = e^{ikx_3} \tag{3}$$

and we consider the case that the scatterer has the revolutional symmetry with respect to the $x_3$ axis. Then, in the far ($x \to \infty$) field, the total wave field can be expressed as

$$u(\mathbf{x}) = u^I + u^S \sim e^{ikx_3} + \frac{f(\theta)}{x} e^{ikx}, \qquad (x \to \infty) \tag{4}$$

where, $\mathbf{x} = (x, \theta, \varphi)$ and $x = \mid \mathbf{x} \mid = \sqrt{x_i x_i}$, and the origin of the polar coordinate $(x, \theta, \varphi)$ is located in the center of the scatterer in this expression. In Eq.(4), $f(\theta)$ is the scattering amplitude and $\theta$ is the scattering angle measured from the positive $x_3$ axis (see, Fig.1). The scattering amplitude $f(\theta)$ is relating to the scattered far field $u^S$, and thus, to the shape of the scatterer.

The incident wave in Eq.(4) can be expressed as

$$u^I(\mathbf{x}) = e^{ikx_3} = e^{ikx\cos\theta} = \sum_{l=0}^{\infty}(2l+1)i^l\, j_l\,(kx)\, P_l\,(\cos\theta)$$

$$\sim \sum_{l=0}^{\infty} i^l\,\frac{(2l+1)}{kx}\sin\,(kx-\frac{l\pi}{2})\, P_l\,(\cos\theta) \qquad (5)$$

$$= \sum_{l=0}^{\infty}\frac{(2l+1)}{2ikx}\Big\{e^{ikx}-(-1)^l e^{-ikx}\Big\}\, P_l\,(\cos\theta)$$

in the far-field, where Rayleigh's formula and the asymptotic expression of spherical Bessel function $j_l(\cdot)$ has been used. $P_l(\cdot)$ is the Legendre polynomials. The scattering amplitude $f(\theta)$ is now expanded by the complete system $P_l(\cos\theta)$ as

$$f(\theta) = \frac{1}{k}\sum_{l=0}^{\infty}\frac{2l+1}{2i}a_l P_l(\cos\theta) \qquad (6)$$

where $a_l$ is the unknown coefficients to be determined. Substitution of Eqs.(5) and (6) into Eq.(4) yields following expression for the total field:

$$u(\mathbf{x}) \sim \sum_{l=0}^{\infty}\frac{2l+1}{2ikx}\Big\{(1+a_l)e^{ikx}-(-1)^l e^{-ikx}\Big\} P_l(\cos\theta). \qquad (7)$$

In Eq.(7), the total field is expressed as the sum of the outgoing waves and the incoming waves at the far-field. The energy conservation requires that

$$|1+a_l| = 1 \qquad (8)$$

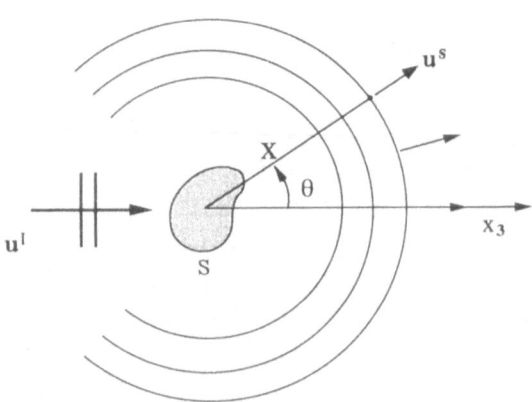

Fig.1 Scattered far-field.

for the coefficients of the outgoing waves and the incoming waves. It suggests to set

$$1 + a_l = e^{2i\delta_l}. \tag{9}$$

Introducing Eq.(9) into Eq.(6), we have the following expressions

$$f(\theta) = \frac{1}{2ik} \sum_{l=0}^{\infty} (2l+1)[e^{2i\delta_l} - 1]P_l(\cos\theta) \tag{10}$$

for the scattering amplitude. In Eq.(10), $f(\theta)$ is now expressed by the unknown quantity $\delta_l$ ($l = 0, 1, 2\cdots$). The $\delta_l$ is called phase shifts in the sense of the following consideration. Equations (7) and (9) lead to the following far field expression for the total field:

$$u(\mathbf{x}) \sim \sum_{l=0}^{\infty} \frac{(2l+1)}{kx} e^{i\delta_l} e^{il\pi/2} \sin\left(kx - \frac{l\pi}{2} + \delta_l\right) P_l(\cos\theta). \tag{11}$$

Comparing this Eq.(11) with the incident wave form in Eq.(5), we know that the quantity $\delta_l$ represents the phase shift in the outgoing component of the total field measured from the incident field.

Scattered far-field

The problem is to determine the phase shifts for the scatterer with an arbitrary shape. The tool what we use is the integral representation. The integral representation for the total field can be written as

$$u(\mathbf{x}) = u^I(\mathbf{x}) + \int_S U(\mathbf{x}, \mathbf{y}) \frac{\partial u(\mathbf{y})}{\partial n_y} dS_y - \int_S \frac{\partial U(\mathbf{x}, \mathbf{y})}{\partial n_y} u(\mathbf{y}) dS_y , \quad \mathbf{x} \in D \tag{12}$$

where $U$ is the fundamental solution for the three dimensional Helmholtz eqation (1) and has the form

$$U(\mathbf{x}, \mathbf{y}) = \frac{1}{4\pi} \frac{e^{ikr}}{r} \tag{13}$$

and $r = |\mathbf{x} - \mathbf{y}|$ is the distance of two points $\mathbf{x}$ and $\mathbf{y}$. This form of the fundamental solution represents the outgoing wave for the time factor $\exp(-i\omega t)$.

The fundamental solution in Eq.(13) can be expanded in the following partial wave form[5]:

$$U(\mathbf{x}, \mathbf{y}) = \frac{e^{ikr}}{4\pi r} = \frac{ik}{4\pi} \sum_{l=0}^{\infty} (2l+1) j_l(ky) h_l^{(1)}(kx) P_l(\cos\chi), \quad (x > y) \tag{14}$$

$$= \frac{ik}{4\pi} \sum_{l=0}^{\infty} (2l+1) j_l(kx) h_l^{(1)}(ky) P_l(\cos\chi), \quad (y > x) \tag{15}$$

where $j_l(\cdot)$ is the spherical Bessel function and $h_l^{(1)}$ is the spherical Hankel function defined as

$$j_l(z) = \sqrt{\frac{\pi}{2z}} J_{l+\frac{1}{2}}(z), \quad n_l(z) = \sqrt{\frac{\pi}{2z}} N_{l+\frac{1}{2}}(z)$$

and

$$h_l^{(1)}(z) = j_l(z) + i n_l(z). \tag{16}$$

The addition theorem of the spherical function is

$$P_l(\cos\chi) = P_l(\cos\theta)P_l(\cos\bar\theta)$$
$$+ 2\sum_{m=1}^{l}\frac{(l-m)!}{(l+m)!}P_l^m(\cos\theta)P_l^m(\cos\bar\theta)\cos m(\varphi-\bar\varphi) \tag{17}$$

where $P_l^m(\cdot)$ is the associated Legendre function, furthermore $\theta, \bar\theta, \varphi$, and $\bar\varphi$ are defined as $\mathbf{x} = (x, \theta, \varphi)$ and $\mathbf{y} = (y, \bar\theta, \bar\varphi)$ as shown in Fig.2. For $x > y$, equation (14) is now written as

$$U(\mathbf{x,y}) = \frac{1}{4\pi}\frac{e^{ikr}}{r} = \frac{ik}{4\pi}\sum_{l=0}^{\infty}(2l+1)j_l(ky)h_l^{(1)}(kx)P_l(\cos\theta)P_l(\cos\bar\theta)$$
$$+ \sum_l\sum_m[\text{terms containing }\cos m(\varphi-\bar\varphi)] \tag{18}$$

from Eq.(17). The normal derivative of the fundamental solution reduces to

$$\frac{\partial U(\mathbf{x,y})}{\partial n_y} = \frac{ik}{4\pi}\sum_{l=0}^{\infty}(2l+1)h_l^{(1)}(kx)[kn_i(\mathbf{y})\frac{\partial y}{\partial y_i}j_l'(ky)P_l(\cos\bar\theta)$$
$$- \sin\bar\theta n_i(\mathbf{y})\frac{\partial\bar\theta}{\partial y_i}j_l(ky)P_l'(\cos\bar\theta)]P_l(\cos\theta)$$
$$+ \sum_l\sum_m[\text{terms containing }\cos m(\varphi-\bar\varphi)]. \tag{19}$$

In the far-field, the spherical Hankel function $h_l^{(1)}(\cdot)$ behaves

$$h_l^{(1)}(kx) \sim (-i)^{l+1}e^{ikx}/kx, \qquad (x\to\infty). \tag{20}$$

Introduction of Eq.(20) into Eqs.(18) and (19) leads to the following far-field expressions of the fundamental solutions:

$$U(\mathbf{x,y}) = \frac{1}{4\pi}\frac{e^{ikr}}{r} \sim \frac{e^{ikx}}{x}(\frac{-1}{4\pi})\sum_{l=0}^{\infty}(2l+1)(-i)^{l+2}j_l(ky)P_l(\cos\bar\theta)P_l(\cos\theta)$$
$$+ \sum_l\sum_m[\text{terms containing }\cos m(\varphi-\bar\varphi)], \qquad (x\to\infty) \tag{21}$$

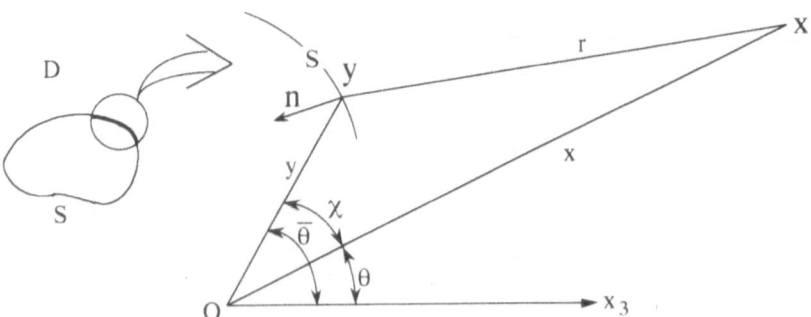

Fig.2 Field point $\mathbf{x}$ and source point $\mathbf{y}$.

$$\frac{\partial U(\mathbf{x}, \mathbf{y})}{\partial n_y} \sim \frac{e^{ikx}}{x} [-\frac{1}{4\pi} \sum_{l=0}^{\infty} (2l+1)(-i)^{l+2}[kn_i(\mathbf{y})\frac{\partial y}{\partial y_i} j'_l(ky) P_l(\cos \bar{\theta})$$

$$- \sin \bar{\theta} n_i(\mathbf{y})\frac{\partial \bar{\theta}}{\partial y_i} j_l(ky) P'_l(\cos \bar{\theta})] P_l(\cos \theta) \tag{22}$$

$$+ \sum_l \sum_m [\text{terms containing } \cos m(\varphi - \bar{\varphi})].$$

From Eqs.(12),(21) and (22), we get the far-field integral representation for the total wave field:

$$u(\mathbf{x}) \sim e^{ikx_3} + \frac{e^{ikx}}{x} \Big[ \sum_{l=0}^{\infty} (2l+1)(-i)^{l+2} \left(\frac{-1}{4\pi}\right) \int_S \Big\{ S(\mathbf{y}) \frac{\partial u(\mathbf{y})}{\partial n_y}$$

$$- D(\mathbf{y}) u(\mathbf{y}) \Big\} dS_y P_l(\cos \theta) \Big), \quad (x \to \infty) \tag{23}$$

where

$$S(\mathbf{y}) = j_l(ky) P_l(\cos \bar{\theta})$$

$$D(\mathbf{y}) = kn_i(\mathbf{y})\frac{\partial y}{\partial y_i} j'_l(ky) P_l(\cos \bar{\theta}) - \sin \bar{\theta} n_i(\mathbf{y})\frac{\partial \bar{\theta}}{\partial y_i} j_l(ky) P'_l(\cos \bar{\theta}). \tag{24a, b}$$

It is to be remarked that the kernels $S(\mathbf{y})$ and $D(\mathbf{y})$ are functions of the source point $\mathbf{y}$ on the surface of the scatterer $S$ and the wave number $k$ of the propagating wave. They are independent of field point $\mathbf{x}$.

Determination of phase shifts

The scattering amplitude has been defined in Eq.(4) and the integral representation of the scattered far-field is given in Eq.(23). Comparison of Eqs.(23) and (4) leads to the integral representation for the scattering amplitude:

$$f(\theta) = \frac{1}{2ik} \sum_{l=0}^{\infty} (2l+1)(-i)^{l+2} \Big[ \frac{-2ik}{4\pi} \int_S \Big\{ S(\mathbf{y}) \frac{\partial u(\mathbf{y})}{\partial n_y} - D(\mathbf{y}) u(\mathbf{y}) \Big\} dS_y \Big] P_l(\cos \theta). \tag{25}$$

Furthermore, the scattering amplitude has been expressed by the phase shifts in Eq.(10). Comparison of Eqs.(25) and (10) leads to the integral representation for the phase shifts:

$$e^{2i\delta_l} - 1 = -(-i)^{l+1} \frac{k}{2\pi} \int_S \Big\{ S(\mathbf{y}) \frac{\partial u(\mathbf{y})}{\partial n_y} - D(\mathbf{y}) u(\mathbf{y}) \Big\} dS_y, \quad (l = 0, 1, 2 \cdots) \tag{26}$$

where $S(\mathbf{y})$ and $D(\mathbf{y})$ have been defined in Eq.(24).

Boundary quantities $u(\mathbf{y})$ and $\partial u(\mathbf{y})/\partial n$ on the right-hand side in Eq.(26) are determined by solving the boundary integral equations for the given boundary condition on the surface of the scatterer. When we know both boundary quantities $u(\mathbf{y})$ and $\partial u(\mathbf{y})/\partial n$, we can determine the phase shifts $\delta_l$ $(l = 0, 1, 2 \cdots)$ from Eq.(26).

From the asymptotic expansion of Bessel functions around the origin, $j_l(ky)$ in Eqs.(26) and (24) behaves

$$j_l(ky) \sim \frac{(ky)^l}{(2l+1)!!}\left(1 - \frac{(ky)^2}{2(2l+3)} + \cdots\right), \quad \text{when} \quad ky \to 0 \qquad (27a)$$

where

$$(2l+1)!! = (2l+1)(2l-1)(2l-3)\cdots 5 \cdot 3 \cdot 1 . \qquad (27b)$$

When the wavelength $\lambda(= 2\pi/k)$ of the incident wave is large compared with the typical dimension of the scatterer, the convergence of the algebraic system in Eq.(26) to determine $\delta_l$ $(l = 0, 1, 2 \cdots)$ is expected to be fairly good.

## Inverse problem

After the determination of phase shifts $\delta_l$ $(l = 0, 1, 2 \cdots)$ in the far field, we want to determine the shape of the scatterer $S$. For example, we consider the case of the Neumann boundary condition $\partial u/\partial n = 0$ on $S$. Then, the integral representation (26) to determine the phase shifts $\delta_l$ reduces to

$$e^{2i\delta_l} - 1 = (-i)^{l+1}\frac{k}{2\pi}\int_S D(\mathbf{y})u(\mathbf{y})dS_y , \qquad (l = 0, 1, 2, \cdots) \qquad (28a)$$

where

$$D(\mathbf{y}) = kn_i(\mathbf{y})\frac{\partial y}{\partial y_i}j_l'(ky)P_l(\cos\bar{\theta}) - \sin\bar{\theta}n_i(\mathbf{y})\frac{\partial\bar{\theta}}{\partial y_i}j_l(ky)P_l'(\cos\bar{\theta}) \qquad (28b)$$

as in defined in Eq.(24b). On the surface of the scatterer $S$ , the following integral equation holds

$$u(\mathbf{x}) + \int_S \frac{\partial U(\mathbf{x},\mathbf{y})}{\partial n_y}u(\mathbf{y})dS_y = u^I(\mathbf{x}), \qquad \mathbf{x} \in S \qquad (29)$$

from Eq.(12). For the given incident field $u^I$ with wavenumber $k$, we suppse that the phase shifts $\delta_l$ $(l = 0, 1, 2, \cdots)$ have been measured at the far field. In this case, Eqs.(28) and (29) can be considered as the coupled integral equations to determine the boundary shape $S$ of the scatterer. Equation (29) is the integral equation to determine the boundary quantity $u$ and Eq.(28) is the equation to determimine the boundary shape $S$ .

In general, the surface $S$ has infinite number of degree of freedom. Equation (28) also has, in principle, infinite number of equations for each $\delta_l$ $(l = 0, 1, 2 \cdots)$ to determine the surface parameters. In practice, the boundary surface $S$ may be represented as a function of $n$ parameters of $\beta_i$ $(i = 1, 2, \cdots, n)$. In this case, Eq.(28) with $n$ phase shifts $\delta_l$ $(l = 0, 1, 2, \cdots, n-1)$ can be considered as $n$ equations to determine the $n$ boundary shape parameters $\beta_i$ $(i = 1, 2, \cdots, n)$. The other choice of $n$ equations is to use Eq.(28) for the zeroth order phase shifts $\delta_0(k_i)$ for $n$ different wavenumers $k_i$ $(i = 1, 2, \cdots n)$ . In this case, it is necessary to calculate $u$ for $n$ different wavenumbers $k_i$ from Eq.(29).

## Newton's method

Now we restrict our attention to the ellipsoidal scatterer which has revolutional symmetry with respect to the $x_3$ axis in the following form (see, Fig.3):

$$\left(\frac{x_1}{a}\right)^2 + \left(\frac{x_2}{a}\right)^2 + \left(\frac{x_3}{b}\right)^2 = 1. \qquad (30)$$

In the polar coordinate,

$$x_1 = r \sin\theta \cos\varphi \, , \;\; x_2 = r \sin\theta \sin\varphi \, , \;\; x_3 = r \cos\theta. \tag{31}$$

Substitution of Eq.(31) into (30) leads to the expression

$$r^2 = 1 / \left\{ \left( \frac{\sin\theta \cos\varphi}{a} \right)^2 + \left( \frac{\sin\theta \sin\varphi}{a} \right)^2 + \left( \frac{\cos\theta}{b} \right)^2 \right\} . \tag{32}$$

Equation (31) with $r$ in Eq.(32) yields the expression of the ellipsoidal surface and it containes two parameters $a$ and $b$ .

To determine two parameters $a$ and $b$ , Newton's method is used. First we consider the following two equations

$$\begin{aligned} F_1(a^*, b^*) &= 0 \\ F_2(a^*, b^*) &= 0 \end{aligned} \tag{33}$$

where $a^*$ and $b^*$ are exact parameters. We now assume that

$$\begin{aligned} F_1(a + \Delta a, b + \Delta b) &= 0 \\ F_2(a + \Delta a, b + \Delta b) &= 0 \;\; . \end{aligned} \tag{34}$$

Then it follows that

$$\begin{aligned} \frac{\partial F_1}{\partial a}\Delta a + \frac{\partial F_1}{\partial b}\Delta b &= -F_1(a, b) \\ \frac{\partial F_2}{\partial a}\Delta a + \frac{\partial F_2}{\partial b}\Delta b &= -F_2(a, b) \end{aligned} \tag{35}$$

for the small increments of $\Delta a$ and $\Delta b$ . These $\Delta a$ and $\Delta b$ can be obtained as

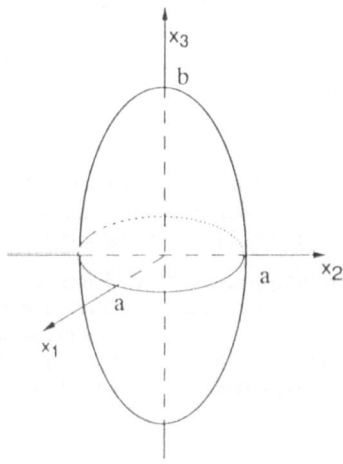

Fig.3 Ellipsoidal boundary.

$$\left\{ \begin{matrix} \Delta a \\ \Delta b \end{matrix} \right\} = - \left[ \begin{matrix} \dfrac{\partial F_1}{\partial a} & \dfrac{\partial F_1}{\partial b} \\ \dfrac{\partial F_2}{\partial a} & \dfrac{\partial F_2}{\partial b} \end{matrix} \right]^{-1} \left\{ \begin{matrix} F_1(a,b) \\ F_2(a,b) \end{matrix} \right\} \tag{36}$$

from Eq.(35) .

For the determination of two parameters $a$ and $b$ , we adopt two zeroth order phase shifts $\delta_0(k_1)$ and $\delta_0(k_2)$ for two different wave numbers $k_1$ and $k_2$ . Then the integral equations (28) and (29) to determine the boundary shape parameters and the boundary quantity $u$ can be written as

$$F_1(a,b) = \frac{(-i)k_1}{2\pi} \int_{S(a,b)} D(\mathbf{y})u(\mathbf{y})dS_y - [e^{2i\delta_0(k_1)} - 1]$$

$$F_2(a,b) = \frac{(-i)k_2}{2\pi} \int_{S(a,b)} D(\mathbf{y})u(\mathbf{y})dS_y - [e^{2i\delta_0(k_2)} - 1] \tag{37}$$

$$u(\mathbf{x}) + \int_{S(a,b)} \frac{\partial U(\mathbf{x},\mathbf{y})}{\partial n_y} u(\mathbf{y})dS_y = u^I(\mathbf{x}) \qquad \text{for } k = k_1 \text{ and } k_2$$

It is necessary to evaluate $\partial F_1/\partial a$ , $\partial F_1/\partial b$ , $\partial F_2/\partial a$ , $\partial F_2/\partial b$ to determine the increments $\Delta a$ and $\Delta b$ in Eq.(36) . It can be done numerically. For example , $\partial F_1/\partial a$ can be evaluated at $(a_0, b_0)$ as

$$\frac{\partial F_1(a_0, b_0)}{\partial a_0} = \frac{(-i)k_1}{2\pi} \left[ \int_{S(a_0+\Delta\alpha, b_0)} D(\mathbf{y})u(\mathbf{y})dS_y - \int_{S(a_0, b_0)} D(\mathbf{y})u(\mathbf{y})dS_y \right] / \Delta\alpha \tag{38}$$

where $\Delta\alpha$ is the small parameter for the numerical differentiation. In this expression, $u$ on the surfaces $S(a_0 + \Delta\alpha, b_0)$ and $S(a_0, b_0)$ can be calculated from integral equations

$$u(\mathbf{x}) + \int_{S(a_0+\Delta a, b_0)} \frac{\partial U(\mathbf{x},\mathbf{y})}{\partial n_y} u(\mathbf{y})dS_y = u^I(\mathbf{x})$$

$$u(\mathbf{x}) + \int_{S(a_0, b_0)} \frac{\partial U(\mathbf{x},\mathbf{y})}{\partial n_y} u(\mathbf{y})dS_y = u^I(\mathbf{x}) \tag{39}$$

for the given wave number $k_1$ .

## Shape inversion

Tables 1 shows the phase shifts $\delta_l$ $(l = 0, 1, 2 \cdots)$ for the ellipsoid calculated from Eq.(26). They are listed for two different nondimensional wavenumbers $ak = 0.5$ and $1.0$. The boundary condition is $\partial u/\partial n = 0$ on the surface. The aspect ratio of the ellipsoid is $b/a = 0.5$.

A result of the shape inversion from phase shifts is shown in Fig.4. In the system of Eq.(37), two zeroth order phase shifts for $ak = 0.5$ and $1.0$ in Table 1 were chosen as $\delta_0(k_1)$ and $\delta_0(k_2)$. This Fig.4 summarizes the convergence process for the ellipsoid of $a = 1$ and $b = 0.5$. The initial values for $a$ and $b$ are chosen to be $a^0 = b^0 = 2.5$. At the 6th iteration, Newton's method based on Eqs.(36) and (37) converges to the final values of $a = 1$ and $b = 0.5$ within an acceptable error bound.

Table 1 Phase shifts $\delta_l$ ($l = 0, 1, 2, \cdots$) for ellipsoid.

| $\ell$ | ak=0.5 $\delta_\ell$ | ak=1.0 $\delta_\ell$ |
|---|---|---|
| 0 | -9.83E-1 | -6.70E+0 |
| 1 | 7.64E-1 | 5.90E+0 |
| 2 | 2.18E-3 | 7.46E-2 |
| 3 | 2.28E-3 | 7.22E-2 |
| 4 | E-6 | E-4 |

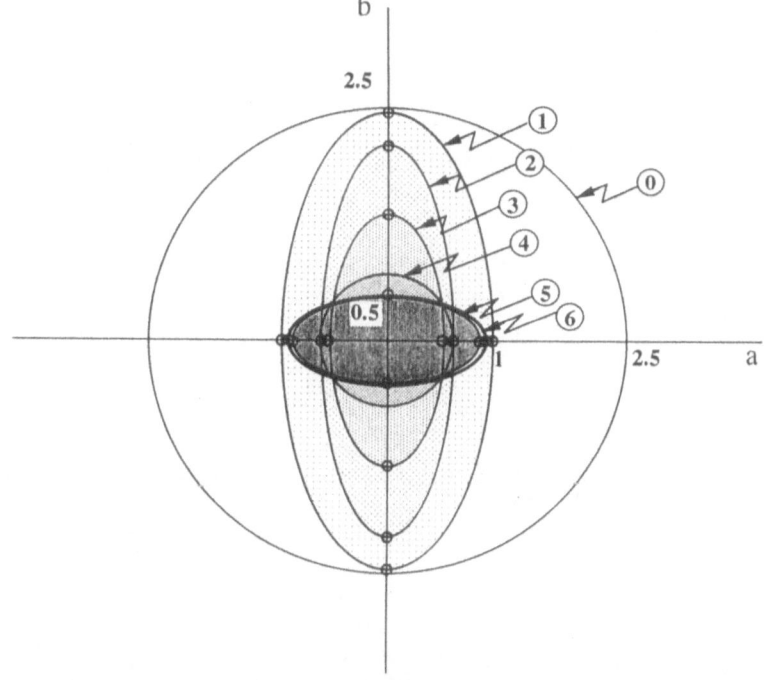

Fig.4 Convergence process for the ellipsoid of $a = 1$ and $b = 0.5$
with Neumann boundary condition $\partial u/\partial n = 0$
(Initial values of $a$ and $b$ are $a^0 = b^0 = 2.5$).

References

1. Sunakawa, S.: Theoretical Electromagnetics, 2nd Ed., Chap.8, Kinokuniya,Tokyo (1973).
2. Bohm, D.: Quantum Theory, Part 5, Prentice-Hall (1951).
3. Schiff, L.I.: Quantum Mechanics, Chap.5, McGraw-Hill (1955).
4. Goldberger, M.L. and Watson, K.M.: Collision Theory, Chap.6, John Wiley & Sons (1964).
5. Gradshteyn, I.S. and Ryzhik, I.M.: Table of Integrals, Series, and Products, Corrected and Enlarged Edition, Academic Press (1980).

# Determination of Shapes, Sizes and Material Compositions of Large Inclusions in Elastic Media from Scattered Waves

Y. M. Chen

Department of Applied Mathematics and Statistics
State University of New York at Stony Brook
Stony Brook, New York 11794-3600, U. S. A.

Summary

The problem of determining the shapes, sizes and material compositions of large inclusions embedded in elastic media from scattered elastic waves leads to the inverse scattering problems of the anisotropic linear elastic wave equations. Generalized Pulse-Spectrum Technique (GPST), a class of versatile and efficient Newton-like iterative algorithms with Tikhonov regularization, is used to solve this extremely difficult multi-parameter inverse problem. The summary of past and present applications of GPST to elastic inverse scattering problems are presented; in particular, the determination of shapes and sizes of large rigid near-spherical inclusions in an elastic medium from the scattered time-harmonic waves is described in detail. The future development of GPST and its tremendous potential for applications are discussed.

## Introduction

The general anisotropic linear elastic wave equations are

$$\underline{L}(\underline{e}, \rho) \bullet \underline{u} \equiv \partial[e_{ijmn}(\underline{x}) \, \partial u_m(\underline{x},t)/\partial x_n]/\partial x_j - \rho(\underline{x}) \, \partial^2 u_i(\underline{x},t)/\partial t^2 = f_i(\underline{x},t),$$

$$\underline{x} = (x_1, x_2, x_3) \, \epsilon \, \Omega, \quad t > 0, \quad i, j, m, n = 1, 2, 3, \tag{1}$$

with given initial conditions,

$$u_i(\underline{x},0) = g_i(\underline{x}), \quad \partial u_i(\underline{x},0)/\partial t = h_i(\underline{x}), \quad \underline{x} \, \epsilon \, \Omega, \quad i = 1, 2, 3, \tag{2}$$

and a non-reflective boundary condition or a proper radiation condition as $|\underline{x}| \rightarrow \infty$,

$$\underline{B} \bullet \underline{u} = 0, \quad \underline{x} \, \epsilon \, \partial\Omega, \quad t > 0, \tag{3}$$

where $\Omega$ is a finite domain in the three-dimensional space, $\partial\Omega$ is the boundary of $\Omega$, $\underline{u} = (u_1, u_2, u_3)$ is the displacement vector, $\underline{f} = (f_1, f_2, f_3)$ is the external force with its compact support in both t and $\underline{x} \, \epsilon \, \Omega$, $\underline{e} = e_{ijmn}$, i, j, m, n = 1, 2, 3, and $\rho$ are the elastic moduli tensor and mass density respectively whose spatial values characterize the shape, size and material composition of the inclusion or scatterer.

Let the measurements of the scattered elastic waves at P spatial locations $\underline{x}_p$, p = 1, 2, 3, ....., P, and Q time steps $t_q$, q = 1, 2, 3, ....., Q, be $\underline{u}^*(\underline{x}_p, t_q)$.

In general, the standard direct problem is to solve for $\underline{u}$ from (1) - (3) with given $e_{ijmn}$, i, j, m, n = 1, 2, 3, (36 of them), $\rho$, $\underline{f}$, $\underline{g}$, and $\underline{h}$. The multi-parameter inverse scattering problem is to solve for $e_{ijmn}$, i, j, m, n = 1, 2, 3, and $\rho$ from (1) - (3) with the given $\underline{f}$, $\underline{g}$, $\underline{h}$ and measured data $\underline{u}^*(\underline{x}_p, t_q)$, p = 1, 2, 3, ....., P, q = 1, 2, 3, ....., Q.

## Generalized Pulse-Spectrum Technique (GPST)

The GPST inversion algorithm begins by setting

$$\underline{u}_{r+1} = \underline{u}_r + \delta\underline{u}_r, \quad \underline{e}_{r+1} = \underline{e}_r + \delta\underline{e}_r, \quad \rho_{r+1} = \rho_r + \delta\rho_r, \quad r = 0, 1, 2, 3, ..........., \tag{4}$$

where $\underline{e}_0$ and $\rho_0$ are the initial guesses of $\underline{e}$ and $\rho$ respectively, and $\delta$ is a small number such that the $\delta$-terms are smaller than their corresopnding non-$\delta$-terms in some norms for numerical convergence.

Upon substituting (4) into (1) - (3) and neglecting terms of order $\delta^2$ and higher, one obtains the system for $\underline{u}_r$ same as (1) - (3),

$$\underline{L}(\underline{e}_r, \rho_r) \bullet \underline{u}_r = \underline{f}, \qquad \underline{x} \in \Omega, \qquad t > 0, \qquad r = 0, 1, 2, 3, .........., \tag{5}$$

with $\qquad \underline{u}_r(\underline{x}, 0) = \underline{g}(\underline{x}), \qquad \partial\underline{u}_r(\underline{x}, 0)/\partial t = \underline{h}(\underline{x}), \qquad \underline{x} \in \Omega, \tag{6}$

and $\qquad \underline{B} \bullet \underline{u}_r = 0, \qquad \underline{x} \in \partial\Omega, \qquad t > 0, \tag{7}$

and a similar system for $\delta\underline{u}_r$,

$$\underline{L}(\underline{e}_r, \rho_r) \bullet \delta\underline{u}_r = -\underline{L}(\delta\underline{e}_r, \delta\rho_r) \bullet \underline{u}_r, \qquad \underline{x} \in \Omega, \quad t > 0, \qquad r = 0, 1, 2, 3, .........., \tag{8}$$

with $\qquad \delta\underline{u}_r(\underline{x}, 0) = \partial\delta\underline{u}_r(\underline{x}, 0)/\partial t = 0, \qquad \underline{x} \in \Omega, \tag{9}$

and $\qquad \underline{B} \bullet \delta\underline{u}_r = 0, \qquad \underline{x} \in \partial\Omega, \qquad t > 0. \tag{10}$

By using the method of Green's function, setting $(\underline{x}, t)$ to $(\underline{x}_p, t_q)$, p = 1, 2, 3, ....., P, q = 1, 2, 3, ....., Q, and replacing $\underline{u}_{r+1}$ by the measurements $\underline{u}^*$, one obtains a system of Fredholm integral equations of the first kind for the unknowns $\delta\underline{e}_r(\underline{x})$ and $\delta\rho_r(\underline{x})$,

$$\int_0^T \int_\Omega \underline{G}_r(\underline{x}_p, t_q; \underline{x}', t') \bullet [\underline{L}(\delta\underline{e}_r, \delta\rho_r) \bullet \underline{u}_r] \, d\underline{x}' \, dt' = \underline{u}_r(\underline{x}_p, t_q) - \underline{u}^*(\underline{x}_p, t_q), \qquad (11)$$

$$p = 1, 2, 3, \ldots, P, \qquad q = 1, 2, 3, \ldots, Q, \qquad r = 0, 1, 2, 3, \ldots\ldots,$$

where $\underline{G}_r$ is the Green's matrix of (8) - (10).

Now, Eqs. (4) - (7) and (11) form the basic theoretical structure for each iteration of the GPST inversion algorithm which is clearly a fast convergent Newton-like method. One has the option to solve Eqs. (5) - (7) and (11) as they are or their Fourier or Laplace transforms (with respect to "t"). One also has the option to solve them numerically or analytically; we prefer the purely numerical approach.

For solving the multi-parameter inverse problems of small isotropic (only two $e_{ijmn}$ are needed in this case) elastic inclusions or scatterers (computationally, coarse grids are sufficient), the GPST inversion algorithm has been quite successful [1-8]. However, for solving the multi-parameter inverse problems of large anisotropic elastic inclusions (computationally, fine grids are needed), the existing GPST inversion algorithm is just not efficient enough to handle this difficult task. For example, the most CPU-time consuming task in the existing GPST at each iteration is to invert a 37N × 37N (N is the number of grid points) full matrix come from the discretization of (11) and Tikhonov regularization method. For large inclusions, it is common to have N = $O(10^6)$. Since the number of floating point arithmetic operations (FLO) of a standard numerical method for inverting 37N × 37N full matrix is of $O\{(37N)^3\} = O(10^{22})$, the CPU-time on the super computer CRAY X-MP/4 (it can execute $10^8$ FLOPS) to perform this task will be of $O(10^{14})$ sec. = $O(10^6)$ yr.

To overcome the above-mentioned difficulty, a simple hierarchical parallelism in parameter-space [6,8-10] reduces the effort of solving a multi-parameter inverse problem to that of solving a one-parameter inverse problem. Domain decomposition in the spatial domain divides a large domain into many small sub-domains and it is possible to calculate the unknown parameters in each sub-domain on an individual processor in parallel; thus it has the potential to reduce the effort of solving a large inverse problem to that of solving the corresponding small inverse problem. Algorithm re-structuralization and hierarchical multigrid [9, 11] reduce the full matrix to a banded sparse matrix which can be inverted with FLO count of $O(N)$. For the above example, the original domain can be divided into $10^3$ sub-domains each with N = $O(10^3)$. If there are 37 × $10^3$ processors available, each with $10^6$ FLOPS capability, with the implementation of all of the present and future improvements (massive parallelism), the CPU-time needed to perform the same inversion task will be of $O(10^{-3})$ sec. per iteration, a possible improvement of $O(10^{17})$!

## Determination of Shapes and Sizes of Large Rigid Near-spherical Inclusions in Isotropic Elastic Media from Scattered Time-harmonic Waves

On the other hand, to determine the shapes and sizes of large rigid near-spherical inclusions embedded in isotropic elastic media from scattered time-harmonic single-frequency elastic wave data at many spatial locations, a simple special case, can be achieved by using a hybrid of analytical and numerical methods. In this case, the Fourier transform of (1) and (2) can be reduced to

$$\underline{u} = \underline{u}_i + \underline{u}_s, \qquad \underline{x} = (r, \theta, \phi) \, \epsilon \, \Omega, \tag{12}$$

$$\alpha^2 \, \nabla \, (\nabla \cdot \underline{u}_s) - \beta^2 \, \nabla \times \nabla \times \underline{u}_s + \omega^2 \, \underline{u}_s = 0, \qquad \underline{x} \, \epsilon \, \Omega, \tag{13}$$

$$\underline{u}_s = - \, \underline{u}_i, \qquad \underline{x} \, \epsilon \, \partial\Omega, \tag{14}$$

and $\underline{u}_s$ satisfies a proper radiation condition at $r \to \infty$,

where $\underline{u}_i$ and $\underline{u}_s$ are the displacement vectors of the incident (a plane compression wave propagating along $-z$ - axis) and scattered waves respectively, $\alpha = [(\lambda+2\mu)/\rho]^{1/2}$ is the compressional wave velocity, $\beta = (\mu/\rho)^{1/2}$ is the shear wave velocity, $\omega$ is the angular frequency of $\underline{u}_i$, $k_1 = \omega/\alpha$, $k_2 = \omega/\beta$, $\Omega$ is the exterior of the rigid near-spherical inclusion, $\partial\Omega$ is the boundary of $\Omega$ and is represented by the shape function $\underline{R}(\theta, \phi) = a \cdot [1 + \epsilon f(\theta,\phi)]$, $|\epsilon f(\theta,\phi)| < 1$, "a" is the radius of the mean sphere (Fig. 1) and $\epsilon$ is a small non-dimensional parameter.

The inverse problem here is to determine $\underline{R}(\theta,\phi)$ from $\underline{u}_s$ measured at few discrete locations far away from the rigid scatterer $\partial\Omega$.

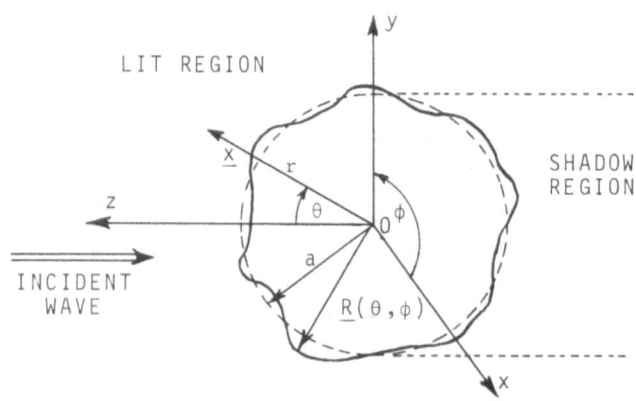

Fig. 1   The geometry of a rigid inclusion in an elastic medium.

Upon using the boundary perturbation method [12], the above boundary value problem is solved analytically in the form,

$$\underline{u}_s = \sum_{j=0}^{\infty} \epsilon^j \underline{u}_{sj} = \underline{u}_{s0} + \epsilon \, \underline{u}_{s1} + O(\epsilon^2). \tag{15}$$

Next, $\underline{u}_s$ is expanded asymptotically [13] for large $k_i a$ and $k_i r$, $i = 1, 2$, to become

$$\underline{u}_s(\underline{x}) \sim \underline{u}_{s0}^{g}(\underline{x}, a) + \underline{u}_{s0}^{d} + \epsilon \, \underline{u}_{s1}^{g}[\underline{x}, a, f(\theta, \phi)] + \epsilon \, \underline{u}_{s1}^{d}[\underline{x}, a, f(\theta, \phi)] + O(\epsilon^2), \tag{16}$$

where the superscripts "g" and "d" denote the contributions from the geometric optics and diffracted waves respectively.

Since the contributions from the geometric optics waves are asymptotically large than those from the diffracted waves, it is reasonable to omit the contributions from the diffracted waves. Finally,

$$\underline{u}_s(r, \theta, \phi) \sim \underline{u}_s^{g,c}[r, \theta, \phi, \epsilon, a, f(\theta, \phi)] + \underline{u}_s^{g,s}[r, \theta, \phi, \epsilon, a, f(\theta, \phi)] + O(\epsilon^2), \tag{17}$$

where the superscripts "c" and "s" denote the compression and shear wave components respectively,

$$\begin{aligned} \underline{u}_s^{g,c} &= \underline{e}_r \, (a/2r) \, R_c \exp \{-ik_1[r - 2a \cos(\theta/2)]\}, && \underline{x} \text{ in lit region}, \\ &= 0, && \underline{x} \text{ in shadow region}, \end{aligned} \tag{18}$$

$$\begin{aligned} \underline{u}_s^{g,s} &= \underline{e}_\theta \, (a/r) \, [N \cos(\theta-\theta^*) \sin \theta^*]^{1/2} \, / \, \{[1 + (\cos \theta^*/ N \cos(\theta-\theta^*))] \sin \theta\}^{1/2} \\ &\quad \bullet \, R_s \exp \{-ik_2[r - a \cos(\theta-\theta^*) - (a/N) \cos \theta^*]\}, && \underline{x} \text{ in lit region}, \\ &= 0, && \underline{x} \text{ in shadow region}, \end{aligned} \tag{19}$$

the effective reflection coefficient of P or C wave,

$$\begin{aligned} R_c = \; & \{[N^2 - \sin^2(\theta/2)]^{1/2} \cos(\theta/2) - \sin^2(\theta/2)\} \, / \\ & \{[N^2 - \sin^2(\theta/2)]^{1/2} \cos(\theta/2) + \sin^2(\theta/2)\} \\[6pt] & + \, i2k_1 a \, \epsilon \, f(\theta/2, \phi) \, \{[N^2\cos^2(\theta/2) - \sin^2(\theta/2)] \cos(\theta/2)\} \, / \\ & \qquad \{[N^2 - \sin^2(\theta/2)]^{1/2} \cos(\theta/2) + \sin(\theta/2)\}^2, \end{aligned} \tag{20}$$

the effective reflection coefficient of S wave,

$$R_s = - \, (2 \cos \theta^* \sin \theta^*) \, / \, [N \cos(\theta-\theta^*) \cos \theta^* + \sin^2 \theta^*]$$

$$- i2k_1 a \, \epsilon \, f(\theta^*, \, \phi) \, \{[\cos \theta^* + N \cos(\theta - \theta^*)] \sin \theta^* / \sin \theta\}^{1/2} \, [\cos \theta^* / \cos^3(2\theta^* - \theta)]$$
$$\bullet \, [(1 - N^2)\sin \theta^* \cos 2\theta^* \cos(\theta - \theta^*) - N\sin \theta \cos \theta^* \cos(\theta - \theta^*) - N^2 \sin \theta \sin^2 \theta^*], \quad (21)$$

$$\theta^* = \tan^{-1}[N \sin \theta \, / \, (1 + N \cos \theta)], \tag{22}$$

$N = k_2/k_1$, and $\underline{e}_r$ and $\underline{e}_\theta$ are the unit vectors in the directions of r and $\theta$ respectively. Clearly, if the mean radius "a" is known, the deviation function $\epsilon f(\theta, \, \phi)$ in the lit region can be obtained in a straightforward manner from either $\underline{u}_s^{g,c}$ [(18) and (20)] or $\underline{u}_s^{g,s}$ [(19), (21) and (22)]. Similarly, an additional incident wave propagating along +z - axis is needed to determine $\epsilon f(\theta, \, \phi)$ in the original shadow region. However, simultaneous measurements of both $\underline{u}_s^{g,c}$ and $\underline{u}_s^{g,s}$ are needed to determine "a". For example if $\underline{u}_s^{g,s}(r_0, \, \theta_0, \, \phi_0)$ is measured in addition to $\underline{u}_s^{g,c}\{r_0, \, 2 \tan^{-1}[N \sin \theta_0 \, / \, (1 + N\cos \theta_0)], \, \phi_0\}$, then from (18) - (22) one obtains a system of two nonlinear equations for two unknowns, "a" and $\epsilon f\{\tan^{-1}[N \sin \theta_0 \, / \, (1 + N \cos \theta_0)], \, \phi_0\}$, which can be solved by using the simplest version of GPST, a Newton-like iterative algorithm.

## References

1. Chen, Y.M.; Xie, G.Q.: A numerical method for simultaneous determination of bulk modulus, shear modulus and density variations for nondestructive evaluation, *Nondestructive Testing Communications*, 21 (1984) 125-135.
2. Xie, G.Q.; Chen, Y.M.: A modified Pulse-Spectrum Technique for solving inverse problems of two-dimensional elastic wave equation, *Appl. Numer. Math.*, 1 (1985) 217-237.
3. Chen, Y.M.: Generalized Pulse-Spectrum Technique, *Geophysics*, 50 (1985) 1664-1675.
4. Qiu, C.H.; Chen, Y.M.: Inverse problems for elastic plates with variable flexural rigidity, *Int. J. Solids & Structures*, 22 (1986) 901-908.
5. Chen, Y.M.; Xie, G.Q.: An iterative method for simultaneous determination of bulk and shear moduli and density variations, *J. Comput. Phys.*, 62 (1986) 143-163.
6. Chen, Y.M.: Parallelism in hierarchy of GPST inversion algorithm for elastic wave equation, *Appl. Numer. Math.*, 4 (1988) 83-95.
7. Wang, S.L.; Chen, Y.M.: An efficient numerical method for exterior and interior inverse problems of Helmholtz equation, *Wave Motion*, 13 (1991) 387-399.
8. Chen, Y.M.: Multi-parameter inversion by GPST with hierarchical parallelism and multigrid, <u>Geophysical</u> <u>Inversion</u>, ed. J. Bee Bednar, SIAM (1991).
9. Chen, Y.M.; Liu, M.S.: Efficiency improvement of GPST inversion algorithm, *J. Comput. Phys.*, 72 (1987) 372-382.
10. Chen, Y.M.: High level parallelism in hierarchy of GPST algorithm for parameter identification in engineering mechanics, *Computers & Struct.*, 30 (1988) 821-829.
11. Chen, Y.M.; Zhang, F.G.: Hierarchical multigrid strategy for efficiency improvement of the GPST inversion algorithm, *Appl. Numer. Math.*, 6 (1989/90) 431-446.
12. Fan, W.; Chen, Y.M.: Scattering of a plane longitudinal elastic wave by a large convex rigid object with a statistically corrugated surface. I. Perturbation solutions, *J. Math. Phys.*, 15 (1974) 831-841.
13. Fan, W.; Chen, Y.M.: Scattering of a plane longitudinal elastic wave by a large convex rigid object with a statistically corrugated surface. II. Far field solution, *J. Math. Phys.*, 15 (1974) 950-953.

# Estimation of Unknown Boundary Values by Inverse Analysis with Elastodynamic Boundary Element Method

M. Tanaka*, M. Nakamura* and R. Ochiai**

*  Faculty of Engineering, Shinshu University
   500 Wakasato, Nagano 380, Japan
** Graduate School of Shinshu University
   500 Wakasato, Nagano 380, Japan

## Summary

This paper is concerned with inverse analysis for estimating unknown boundary values in a steady elastodynamic state by using the boundary element method. It is assumed that the measured values of dynamic responses at some points on the boundary can be used for the inverse analysis. The inverse problem is cast into an optimization problem in which the parameters describing the unknown boundary values should be determined such that an objective function is minimized by the standard optimization technique. Numerical experiment is carried out to confirm the effectiveness of the proposed procedure. Influences of measurement errors are also studied in some detail.

## Introduction

There are many inverse problems in which the unknown values which are difficult to measure directly should be estimated. Recently, a large number of investigations have been done to analyze these inverse problems by means of computational software which has been so far developed for the corresponding direct problems. Such inverse analyses aim at estimation of unknown boundary values, material constants, or part of the boundary shape including, for example, crack or defect shape identification[1-10].

It is a difficult task in practice to measure directly the displacement or traction distributions on the contact boundary of elastic bodies. In this study, we propose an inverse analysis method for estimating unknown boundary values on some boundary part of the elastic body in a steady elastodynamic state. It is assumed that some of measured data are available for inverse analysis. The inverse problem under consideration is cast into a parameter identification problem for which the boundary element method can be used together with the standard optimization procedure. Since in general the inverse problem is ill-posed, estimation results are much influenced by errors included in the measured data. Therefore, the influence of the measurement errors on the estimation results should be investigated to develop a reliable, efficient procedure for inverse analysis. In the present paper, numerical simulation is also carried out for the estimation of unknown boundary values when the dynamic responses include measurement errors.

## Inverse Analysis Procedure

We shall consider the inverse problem in which unknown boundary values in a steady elastodynamic state should be estimated by inverse analysis. The schematic view of this inverse problem is given in Fig.1. The elastic body is subjected to a time-harmonic exciting force. The traction and/or displacements on some part of the boundary are not known, while the other boundary conditions are prescribed. It is assumed that displacement and/or strain responses at some points on the boundary are measured and available for inverse analysis. The whole geometry and the material constants of the elastic body are known in advance.

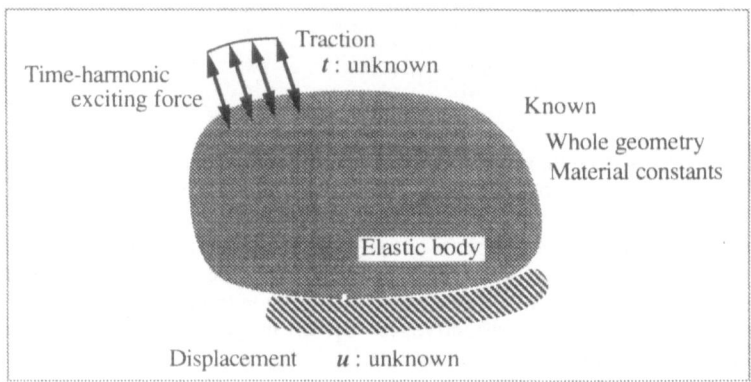

Fig. 1 Schematic view of the inverse problem

In this study, we propose a method of inverse analysis in which the unknown boundary values are estimated directly or approximated by a function with parameters. The problem under consideration can be reduced to an optimization problem in which the parameters describing unknown boundary values should be determined by minimizing an objective function using the standard optimization procedure.

The objective function can be chosen as a square sum of residuals between the measured dynamic responses at selected points on the boundary and the corresponding values computed by the boundary element method assuming the unknown parameters. The objective function can be written as follows:

$$W = \sum_{n=1}^{N} \sum_{i=1}^{D} \left( u_i^n - \overline{u}_i^n \right)^2 \tag{1}$$

where $\overline{u}_i^n$ denotes the measured displacement at point $n$ in the $i$-direction and $u_i^n$ the computed one at the same point which can be obtained from boundary element analysis of the steady-state elastodynamic problem. $N$ is the number of measuring points on the boundary, $D$ is the number of spatial dimensions. The conjugate gradient method[11] is employed in this study to determine an optimal set of parameters minimizing the objective function.

In the direct method, the unknown values of displacement $u(x)$ at some nodes on the boundary are directly taken into the parameters of the optimization problem. In the method of approximate functions, the parameters to be determined are as follows. If the polynomial expression is employed, we can approximate the distribution of displacement $u(x)$ as

$$u_i(x_j) = \sum_{k=0}^{m} a_{ki}\, x_j^k \tag{2}$$

where $u_i(x_j)$ denotes the displacement value in the $i$-direction at point $x_j$ on the boundary. $a_{ki}$ are the unknown coefficients which are the parameters to be determined by the optimization procedure. On the other hand, if we approximate the distribution of displacement $u(x)$ by the $M$th order B-spline functions[12,13], we have

$$u_i(x_j) = \sum_{k=1}^{m} C_{ki}\, B_{ki}(x_j) \tag{3}$$

where $B_{ki}(x_j)$ denote the $M$th order B-spline functions. In this case, the optimization parameters are coefficients $C_{ki}$. In the same manner, the traction $t(x)$ can be treated.

### Estimation Procedure

The algorithm of parameter estimation is illustrated as a flow chart in Fig. 2. The estimation procedure consists of the following steps:

*STEP* 1. Input the data concerning the geometry shape, material constants of the elastic body and the measured data of displacement and/or strain responses.

*STEP* 2. Assume the initial parameters of the unknown boundary values.

*STEP* 3. Approximate the unknown boundary values by using the direct method or the method of approximate functions.

*STEP* 4. Carry out boundary element analysis of the steady-state elastodynamic problem and calculate the objective function. Modify the parameters by using the optimization technique.

*STEP* 5. Check convergence. If the solution converges, stop iterations. Otherwise, go to *STEP* 3.

In this study, as the convergence criterion of iterative computations, we use either of the following two inequalities:

$$Z^k < \eta_1, \quad |Z^k - Z^{k-1}| < \eta_2 \tag{4}$$

where $\eta_1$ and $\eta_2$ are given tolerance limits, and $Z^k$ is the non-dimensional expression of the objective function at the $k$th iteration, which can be expressed as

$$Z^k = \frac{1}{2}\log (w^k / w^0) \tag{5}$$

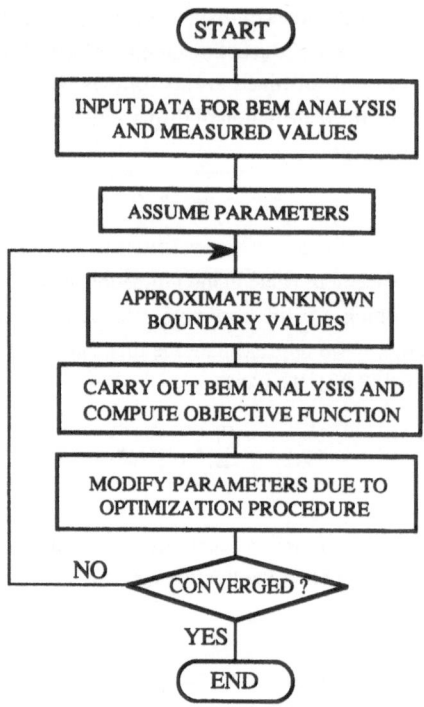

Fig. 2  Main flow of inverse analysis

## Numerical Results

We now carry out numerical experiment for estimation of unknown boundary values to check if the proposed inverse analysis method works. In this numerical experiment we consider the rectangular plate model ABCD of 200mm × 300mm as shown in Fig. 3. The material constants of the plate model are shown in Table 1. It is assumed that the displacement on the side AB and the traction on a part of the side BC are not known. Six measuring points are assigned almost uniformly on the boundary except for the side AB. The values of displacement responses at the measuring points are computed by boundary element analysis of the steady-state elastodynamic problem under the exact displacement and traction distributions. Then, the measurement data for numerical experiment is produced by rounding the computed displacements in consideration of the three cases of measurement accuracy in 1 $\mu$ m, 10 $\mu$ m and 100 $\mu$ m. Figure 4 shows the boundary element discretization for this computation, where 20 isoparametric quadratic boundary elements are used. The same discretization is employed for the iterative computations in the optimization process.

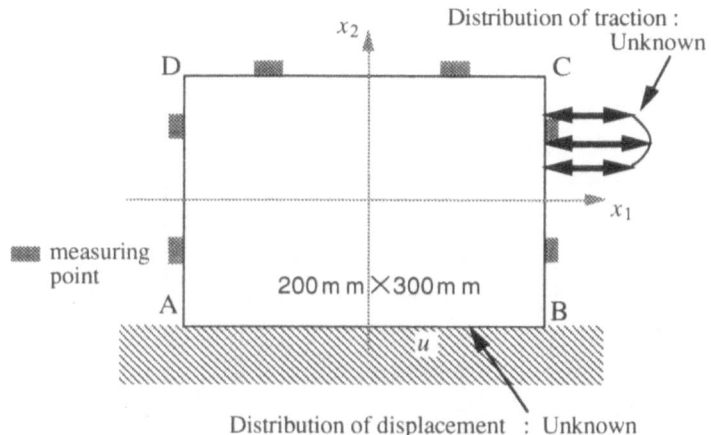

Fig. 3 Numerical experiment model

Table 1 Material constants

| Young's modulus $E$ | 210 GPa |
|---|---|
| Poisson's ratio $\nu$ | 0.3 |
| Mass density $\rho$ | $7.85 \times 10^3 \text{kg/m}^3$ |

Fig. 4 Boundary element discretization

Now, we show some of the estimation results obtained by the present inverse analysis. The first example is the estimation of traction distribution. In this example, it is assumed that the value of the traction distribution on the side BC is not known, while the displacement distribution on the side AB is known. The traction applied on the part of the side BC is assumed to be expressed as a quadratic function of $x_2$ coordinate, the distribution of traction component is approximated by nodal values if the direct method is used. Figure 5 shows the estimation results obtained by the direct method over the whole side BC. Small circles denote the estimation values obtained by the inverse analysis and the solid line the exact distribution of traction. Figure 6 shows the estimation results obtained when the traction distribution is assumed to be not zero only in the limited part in consideration of the estimation results in Fig.5. Figures 7 and 8 show the estimation results obtained using the method of approximate functions in which the distribution of traction is approximated by the 3rd order B-spline functions. As shown in Figs. 5 to 8, when the measurement accuracy is assumed as $1\,\mu$ m or $10\,\mu$ m, the accurate results can be obtained. However, if the accuracy is assumed to be $100\,\mu$ m, no successful estimation can be made.

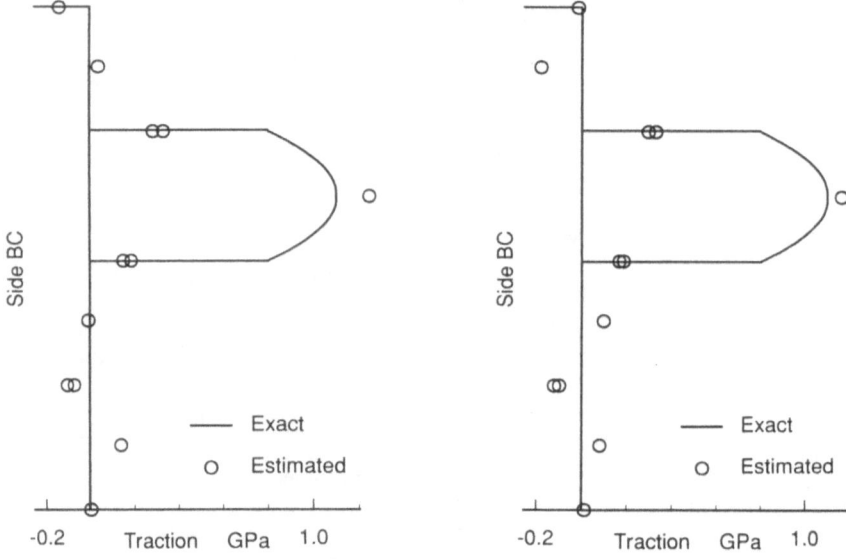

(a) 10 $\mu$ m accuracy of measurement  (b) 1 $\mu$ m accuracy of measurement

Fig. 5  Estimation results of traction distribution on side BC by direct method

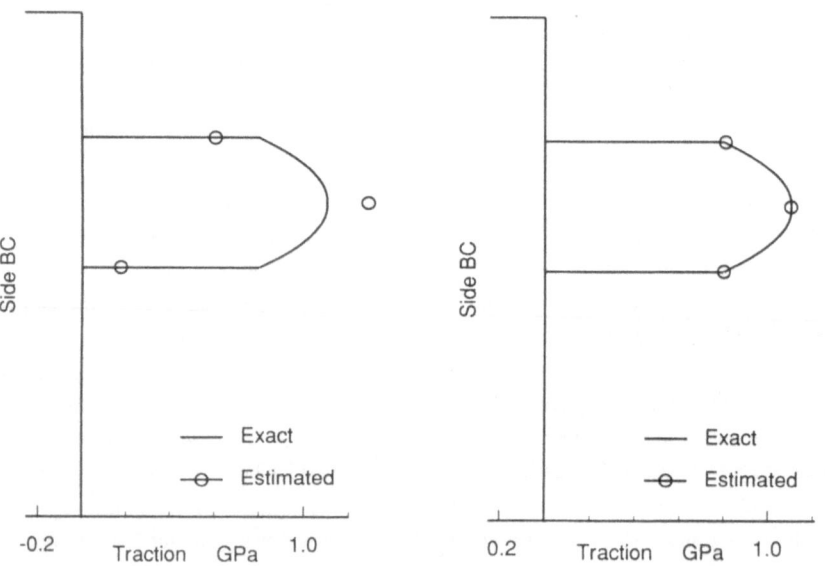

(a) 10 $\mu$ m accuracy of measurement  (b) 1 $\mu$ m accuracy of measurement

Fig. 6  Estimation results of traction distribution by direct method when traction is applied
only on a limited part of side BC

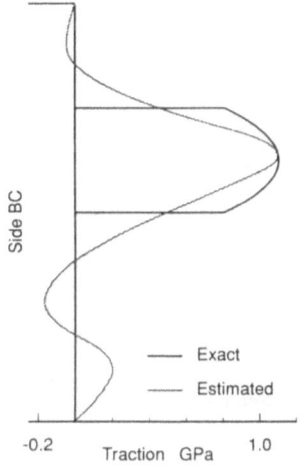
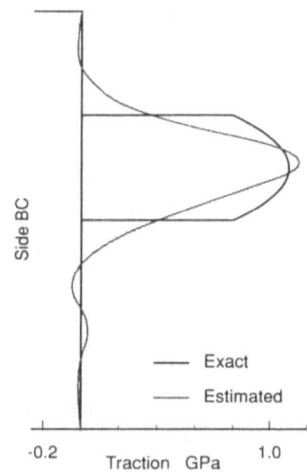

(a) 10 $\mu$ m accuracy of measurement      (b) 1 $\mu$ m accuracy of measurement

Fig. 7 Estimation results of traction distribution on side BC by method of approximate functions using B-spline functions

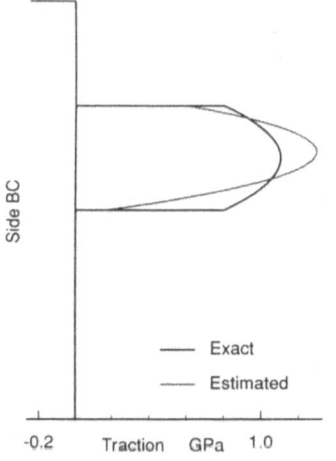
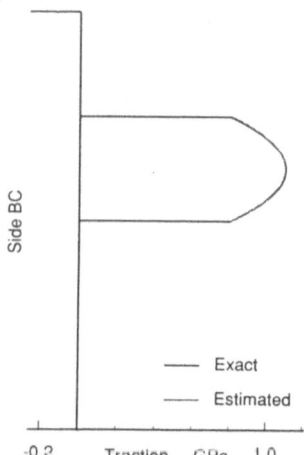

(a) 10 $\mu$ m accuracy of measurement      (b) 1 $\mu$ m accuracy of measurement

Fig. 8 Estimation results of traction distribution by method of approximate functions using B-spline functions if traction is applied in a limited part of side BC

The second example is the estimation of displacement distribution on side AB. In this case, the displacement components are assumed to be approximated by the 3rd order B-spline functions. The exact values of displacement distribution at the side AB are given in terms of quadratic function of $x_1$ coordinate. Figure 9 shows the estimation results of displacement distribution. The solid line denotes the exact distribution and the broken line the estimation result. The inverse analysis can provide satisfactory estimations in this example.

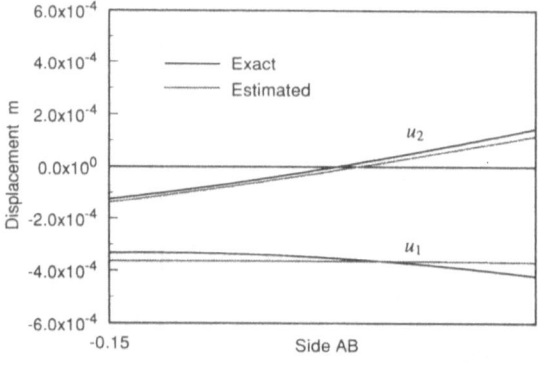

(a) 100 $\mu$ m accuracy of measurement

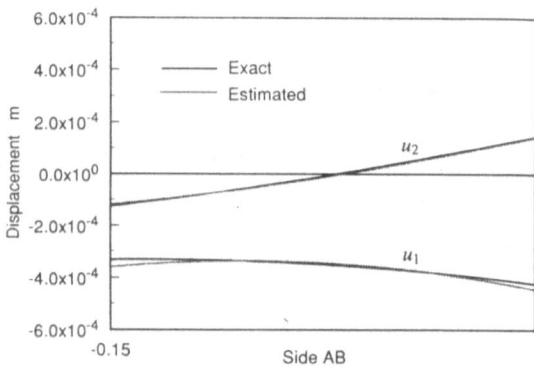

(b) 10 $\mu$ m accuracy of measurement

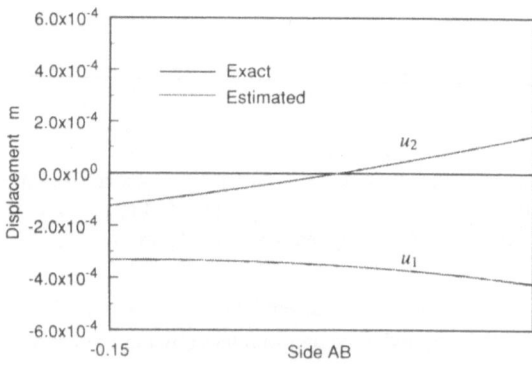

(c) 1 $\mu$ m accuracy of measurement

Fig. 9 Estimation results of displacement distribution approximated by B-spline functions

Finally, we shall show numerical experiment for simultaneous estimation of unknown displacement and traction distributions. In this case, the method of approximate functions using the 3rd order B-spline functions is used for displacement distribution and the direct method for traction distribution. In the optimization process, we modify alternatively 2 groups of optimization parameters, the group of spline coefficients and the group of boundary nodes. Successful estimation can be made for the cases of $1 \mu$ m and $10 \mu$ m accuracy of measurement.

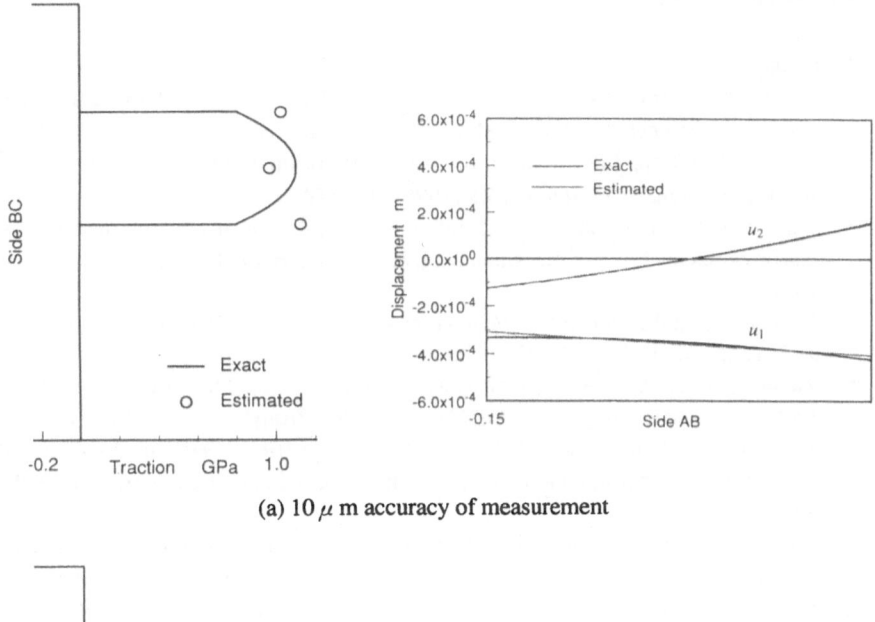

(a) $10 \mu$ m accuracy of measurement

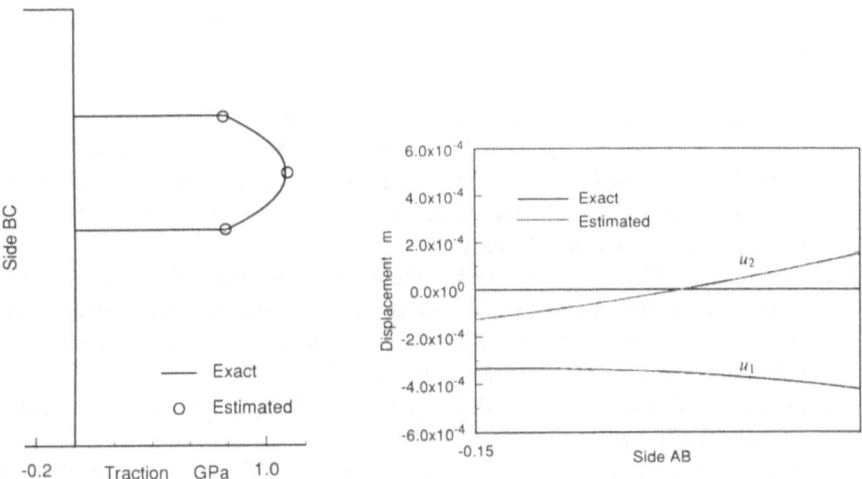

(b) $1 \mu$ m accuracy of measurement

Fig. 10 Simultaneous estimation of traction and displacement distributions

## Conclusion

In this paper we proposed the method of inverse analysis to estimate the unknown boundary values in a steady-elastodynamic state based on the measured data by using jointly the boundary element analysis and the optimization procedure. Numerical experiment was carried out for the two-dimensional cases where the values of traction and/or displacement on the boundary are not known. The proposed method was also applied to the case where the measured data included measurement errors. It was demonstrated that even to these complicate cases the method could be successfully applied.

## References

1. Tanaka, M. : Some recent advances in boundary element research for inverse problems, *Boundary Elements X*, ed. by Brebbia, C.A. Springer-Verlag, **1** (1988) 567-582.

2. D'Cruz, J., Crisp, J.D.C. and Ryall, T.G. : Determining a force acting on a plate-An inverse problem, *AIAA Journal*, **29** (1990) 464-470.

3. Schnur, D.S. and Zabaras, N. : Finite element solution of two-dimensional inverse elastic problems using spatial smoothing, *Int. J. Numerical Methods in Eng.*, **30** (1990) 57-75.

4. Maniatty, A. and Zabaras, N. : Method for solving inverse elastoviscoplastic problems, *J. Eng. Mech.*, **115** (1990) 2216-2231.

5. Zabaras, N. and Liu, J.C. : An analysis of two-dimensional linear inverse heat transfer problems using an integral method, *Numerical Heat Transfer*, **13** (1988) 527-533.

6. Beck, J.V., Litkouchi, B. and St. Clair Jr., C.R. : Efficient sequential solution of the nonlinear inverse heat conduction problem, *Numerical Heat Transfer*, **5** (1982) 275-286.

7. Reinhardt, H.-J. : A numerical method for the solution of two-dimensional inverse heat conduction problems, *Int. J. Numerical Methods in Eng.*, **32** (1991) 363-383.

8. Flach, G.P. and Özisik, M.N. : Inverse heat conduction problem of simultaneously estimating spatially varying thermal conductivity and heat capacity per unit volume, *Numerical Heat Transfer, Part A*, **16** (1989) 249-266.

9. Tanaka, M., Nakamura, M. and Ishikawa, H. : Identification of defects by the elastodynamic boundary element method using noisy additional information, *Boundary Elements XIII*, ed. by Brebbia, C.A. and Gipson, G.S., Computational Mechanics Publications and Springer-Verlag, (1991) 799-810.

10. Tanaka, M., Nakamura, M. and Nakano, T. : Detection of cracks in structural components by the elastodynamic boundary element method, *Boundary Elements XII*, Vol. 2, ed. by Tanaka, M., Brebbia, C.A. and Honma, T., Computational Mechanics Publications and Springer-Verlag, (1991) 413-424.

11. Fox, R.L. : *Optimization Methods for Engineering Design*, Addison-Wesley Publishing Co., (1971) 38-116.

12. Rivlin, J.T. : *An Introduction to the Approximation of Functions*, Dover Publications, (1969).

13. Rice, J.R. : *The Approximation of Functions*, Addison-Wesley Publishing Co., (1964).

# Source Inversion of High-Frequency Strong-Motion Records in the Near-Source Region

M. Iida[*] and S. Hartzell[**]

* Earthquake Research Institute, University of Tokyo
  1-1-1 Yayoi, Bunkyo-ku, Tokyo 113, Japan
** United States Geological Survey, MS 966, Box 25046
  Denver Federal Center, Denver, CO 80225, USA

Summary
One of the most controversial topics on high-frequency strong seismic motion prediction is earthquake source effects which are related to an earthquake's damage and intensity. While reviewing recent strong-motion source inversion methods, effectiveness of an inversion method for high-frequency components is investigated. Implications of high-frequency source inversion are summarized.

## 1. Introduction

Probably, the best way to understand the nature of high-frequency strong seismic ground motion is interpretation of the whole strong-motion waveform. It has been undoubtedly verified by numerous studies that high-frequency strong motion is heavily affected by local effects near the station, including soil nonlinear behavior. For example, this situation was summarized by Aki [1]. Recently, we come to recognize that source effect, path effect, and site effect jointly affect strong-motion seismograms. This might be exemplified by the 1985 Michoacan, Mexico earthquake (e.g., [2]). Considering these situations, one of the most controversial topics is source effects which are related to damage and intensity patterns even in the high-frequency range. Importantly, source effects of an earthquake influence all the observation stations. Near-source records which are not much influenced by local site effects should be analyzed to minimize propagation path effects.

## 2. General source inversion

The most effective way to investigate source effects for a large earthquake is to perform a source inversion for the retrieval of the temporal behavior of slip on an extended fault. An elast-dynamic representation theorem is used in order to relate the dislocation across the fault surface to the radiated elastic field. The discretization of the representation theorem leads to a system of linear equations relating the unknown slip distribution vectors with the recorded ground motion data vectors, through a matrix of the Green's functions for the medium.

Trifunac's [3] formulation established a basic methodology of modern source inverse problems. In his study, the prescribed fault plane was divided into a small number of subfaults. Ground motion from each subfault was calculated using the Haskell model [4], which gives a synthetic displacement in an infinite homogeneous medium. A least-squares fit to displacement time histories was used to determine slip on each subfault on the condition that the hypocenter location and the rupture mode were prescribed.

Physically reasonable constraints were added to the Trifunac's formulation in order to obtain the practical, spatio-temporal characteristics of the inferred slip on the fault plane [5,6]. The constraints include (a) linear smoothing constraints to stabilize the solutions and (b) positive constraints on the solutions to obtain positive dislocation. The third constraint is (c) minimum constraints on the solutions to uniquely define which solutions are to be chosen.

Different types of formulations are an iterative deconvolution method [7], a frequency-domain inversion method [8], and a differential array analysis [9]. In the iterative deconvolution method, constituent events of a multiple shock sequence are determined one by one in the decreasing order of the event size using an iterative procedure with a least-squares criterion. This method seems to work well in the cases of large earthquakes with separate concentrations of slip in the low-frequency range. The frequency-domain inversion method has an advantage that performing the temporal deconvolution in the frequency domain allows the spatial dependence of slip to be computed independently at each frequency. This greatly reduces the computational effort and allows the grid spacing to be chosen sufficiently fine. The differencial array analysis is suitable for high frequencies, which are very difficult to process in usual source inversion frameworks. The technique makes use of a difference in arrival times of distinguishable phases in body-wave seismograms obtained from a source region. Recently, non-linear inversion methods of determining an optimal location of the rupture front are being developed (e.g., [10,11]).

## 3. High-frequency source inversion

The suitable way to estimate source effects in the high-frequency range seems to treat a system of linear equations with physically reasonable constraints, as Olson and Apsel [5] or Hartzell and Heaton [6] performed. The first reason is that reliable solutions are expected over the whole fault plane because of linear smoothing and positive constraints. Secondly, since the whole fault plane is subdivided into a large number of subfaults, the rupture process can be analyzed in detail. Thirdly, this method has been applied to many earthquakes with actual velocity models, confirming the effectiveness of the method. However, source effects that are closely related to damage and intensity patterns have not been demonstrated although some studies have been conducted in the high-frequency range. The most likely reason is luck of well-instrumented earthquakes. Also, detailed damage and intensity patterns have not been surveyed in most earthquakes.

We apply a modified version of Hartzell and Heaton's method [6] to the 1987 Whittier Narrows, California, earthquake, one of the best instrumented earthquakes to date [11]. We note the unusual damage and intensity patterns of this earthquake [12] (Fig. 1); In other words, Whittier, which is not located in the source area, experienced the greatest damage. Also, the northwestern side of the fault had more damage than the surrounding areas. Such patterns could not be interpreted by only local site effects [13].

The pattern of 17 near-source stations that form good azimuthal coverage of the source is expected to give good resolution (Fig. 1) [14,15]. To minimize propagation path effects, only stations within 15 km of the epicenter are used. With this cutoff, the station ranges are comparable to or less than the source depths, which emphasizes the direct body waves. Only horizontal components are used, which are dominated by S wave energy. Band-pass filtered velocity records from 0.2 to 3.0 Hz are used. This frequency range is responsible for much of an earthquake's damage and intensity.

## 4. Inversion method

Our inversion method was discussed in detail by Hartzell and Iida [11], and is roughly reviewed here. The model fault is a square planar region 10 km on an edge. We fix the strike of our model fault at $280°$. Two different values of dip, $30°$ and $40°$ were tried, and a dip of $30°$ gave a marginally better fit to the data. The hypocenter is located at the center of the fault plane at a depth of 14.6 km. The fault plane is divided up into small

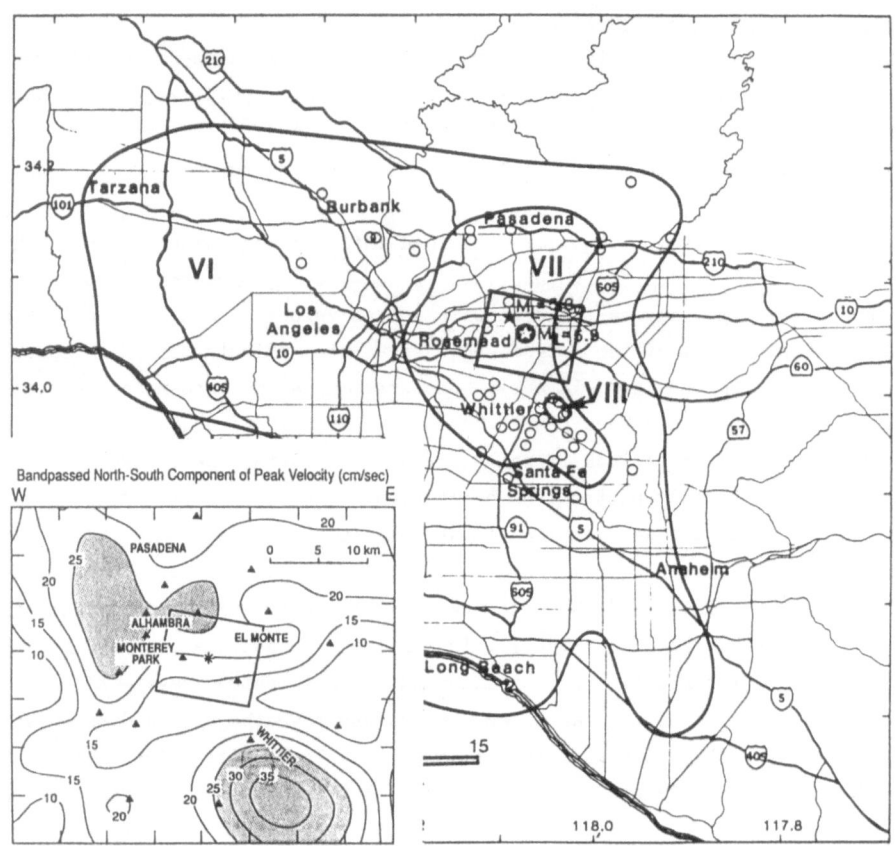

Fig. 1 Regional Modified Mercalli intensity isoseismals of the Los Angeles area in the earthquake of October 1, 1987. The surface projection of the model fault plane is indicated by the rectangular box. Open circles represent the center of the census tracts surveyed. The circled star is the main shock epicenter (After Leyendecker, et al.). Predicted peak velocities (cm/sec) for the model (b) of Fig. 2 in the bandpass 0.2 to 3.0 Hz are shown on the same scale at the lower left corner. Values contoured are peak whole record velocities in the north-south component of motion. The stations used are indicated by solid triangles.

square regions of equal area. Each subfault is $1km^2$. The Green's function that includes all theoretical arrivals is calculated for each subfault and station pair, using the discrete wave number/finite element method of Olson et al. [16], which is applicable to a 1-dimensional velocity gradient model. The velocity model is characterized by slow surfave velocities of steep gradients in the upper 5 km, typical of the Los Angeles basin. Actually, a cross section of the basin does not show uniform underground structure horizontally [17]. Although no local effects are taken into account, abnormal site effects are not recognized as far as we visually inspect the seismograms.

If we wish to solve for the slip amplitudes for a prescribed rupture velocity, the problem is linear. The observed records and the subfault synthetic records then form an overdetermined system of linear equations,

$$A \; x \cong b$$

where A is the matrix of synthetics, b is the data vector, and x is the solution vector of the subfault dislocation weights. Because instability of the solution arises, the problem is stabilized by appending linear constraints giving

$$\begin{pmatrix} Cd^{-1} A \\ \lambda_1 S \\ \lambda_2 M \end{pmatrix} x \cong \begin{pmatrix} Cd^{-1} b \\ 0 \\ 0 \end{pmatrix}$$

S is a matrix of smoothing constraints. M is a matrix of minimization constraints. $\lambda_1$ and $\lambda_2$ are linear weights. Cd is an a priori data covariance matrix. The solution vector is solved for using a Householder reduction method that invokes a positivity constraint on the solution [18]. If we wish to solve simultaneously for the magnitude of the slip and the rupture initiation time for each subfault, the problem is nonlinear and is solved in an iterative manner. We then have an overdetermined problem which is solved using a least squares criterion for a model parameter perturbation vector.

Three different types of waveform inversion are performed. The first and simplest approach assumes a constant rupture velocity with each subfault rupturing once. The second formulation also uses a fixed rupture velocity, but each subfault is allowed to rupture twice with a small time interval, to allow for a more complex source-time function. The third type of inversion allows each subfault to rupture once, but the rupture velocity, in other words, the rupture time of each subfault may vary.

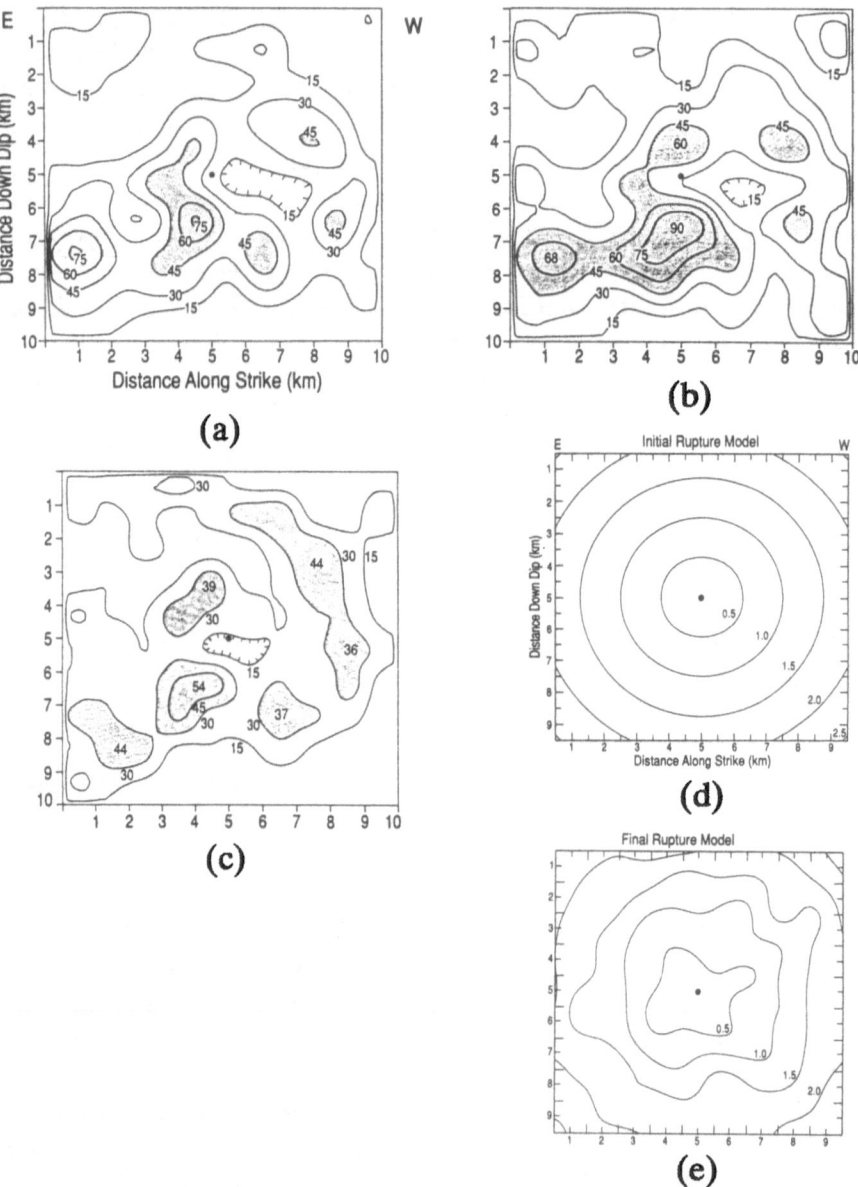

Fig. 2 Contours of slip in centimeters. (a) Each subfault ruptures once at a constant rupture velocity of 2.5 km/s. (b) Each subfault is allowed to rupture twice with a certain small time interval in order to allow for a more complex source-time function. The two ruptures progress at a fixed speed of 2.5 km/s. The rupture mode of (a) and (b) is shown in (d). (c) Each subfault ruptures once, but the rupture time for each subfault is free to vary. Starting with the initial rupture mode of (d), the final rupture mode of (e) is obtained.

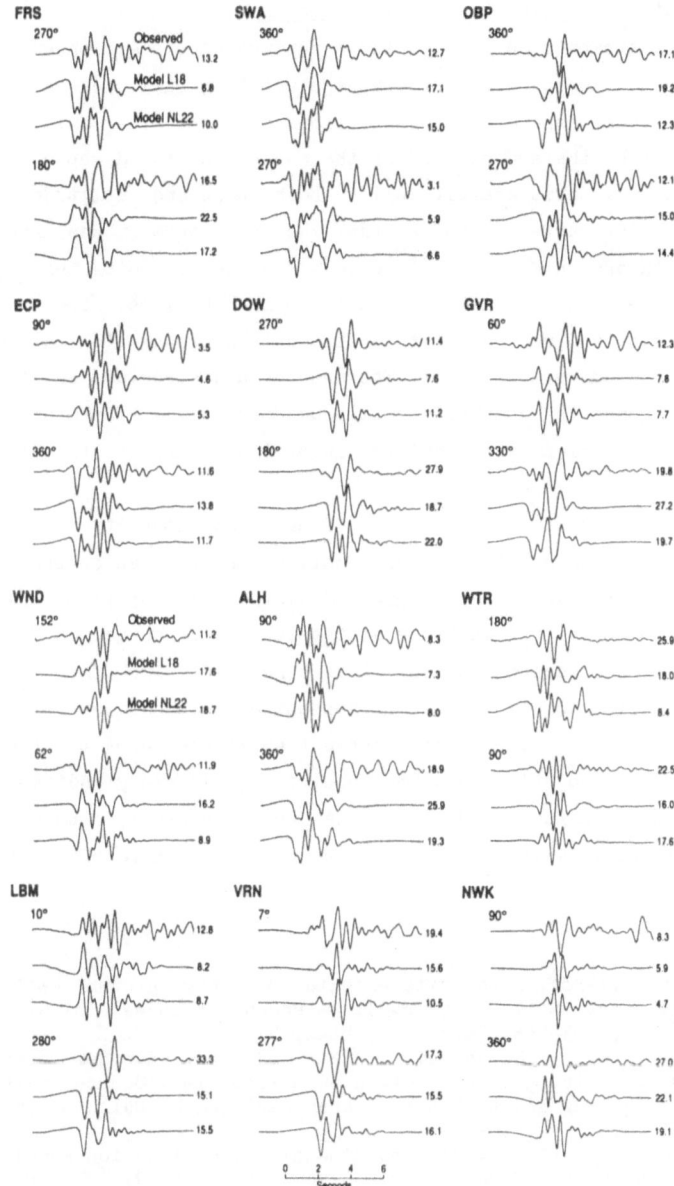

Fig. 3  Comparison of the observed strong-motion velocity records (first trace) with the synthetic ground motion records for the model (b) (second trace) and the model (c) (third trace) of Fig. 2.

## 5. Implications of high-frequency inversion

The slip distributions for the three rupture models are shown in Fig. 2. They indicate a complex rupture process within a small source volume, with at least four separate concentrations of slip. The data records are compared with the synthetics for the second and third rupture models in Fig. 3. The waveforms are fit well in both shape and amplitude, especially at the earlier parts of the records. The same tendency was recognized in previous inversion studies [5,6,10]. We should note that the large-amplitude sections are not influenced so much by local site effects. The later parts of the records most likely contain propagation path effects not included in our simple model. The underground structure in the Whittier Narrows region is not horizontally uniform [17]. Therefore, Iida and Spudich are developing a scatterer inversion to explain the later phases seen at observed seismograms [19].

In order to interpret the unusual damage and intensity patterns for the Whittier earthquake, the ground motion in the epicentral region is predicted on the basis of the inferred distribution of slip. The result for the model (b) of Fig. 2 is shown at the lower-left corner of Fig. 1, where peak velocities are contoured. It is consistent with the intensity and damage distributions; The area of highest expected velocities is near the town of Whittier. The second largest amplitudes are to the west and northwest of the epicenter. The results of this study indicate that the ground motion in the source region can be explained by considering source effects coupled with the same averaged propagation path effects to each strong motion station.

## References

1. Aki, K. (1988). Local site effects on strong ground motion, Proc. Earthq. Eng. Soil Dynamics II -- Recent Advances in Ground Motion Evaluation, 103-155, Park City, Utah, USA.
2. Campillo, M., J.C. Gariel, K. Aki and F.J. Sanchez-Sesma (1989). Destructive strong ground motion in Mexico City: Source, path, and site effects during great 1985 Michoacan earthquake, Bull. Seism. Soc. Am. 79, 1718-1735.
3. Trifunac, M.D. (1974). A three-dimensional dislocation model for the San Fernando, California, earthquake of February 9, 1971, Bull. Seism. Soc. Am. 64, 149-172.
4. Haskell, N.A. (1969). Elastic displacements in the near-field of a propagation fault, Bull. Seism. Soc. Am. 59, 865-908.
5. Olson, A.H. and R.J. Apsel (1982). Finite faults and inverse theory with applications to the 1979 Imperial Valley earthquake, Bull. Seism. Soc. Am. 72, 1969-2001.

6. Hartzell, S.H. and T.H. Heaton (1983). Inversion of strong motion and teleseismic waveform for the fault rupture history of the 1979 Imperial Valley, California, earthquake, Bull. Seism. Soc. Am. 73, 1553-1583.
7. Kikuchi, M. and H. Kanamori (1982). Inversion of complex body waves, Bull. Seism. Soc. Am. 72, 491-506.
8. Olson, A.H. and J.G. Anderson (1988). Implications of frequency-domain inversion of earthquake ground motions for resolving the space-time dependence of slip on an extended fault, Geophysical Journal 94, 443-455.
9. Spudich, P. and E.Cranswick (1984). Direct observation of rupture propagation during the 1979 Imperial Valley earthquake using a short baseline accelerometer array, Bull. Seism. Soc. Am. 74, 2083-2114.
10. Beroza, G.C. and P. Spudich (1988). Linearized inversion for fault rupture behavior: Application to the 1984 Morgan Hill, California, earthquake, J. Geophys. Res. 93B, 6275-6296.
11. Hartzell, S. and M. Iida (1990). Source complexity of the 1987 Whittier Narrows, California, earthquake from the inversion of strong motion records, J. Geophys. Res. 95B, 12475-12485.
12. Leyendecker, E.V., L.M. Highland, M. Hopper, E.P. Arnold, P. Thenhaus and P. Powers (1988). The Whittier Narrows, California earthquake of October 1, 1987 -- Early results of isoseismal studies and damage surveys, Earthquake Spectra 4, 1-10.
13. Kawase, H. and K. Aki (1990). Topography effect at the critical SV wave incidence: Possible explanation of damage pattern by the Whittier Narrows, California earthquake of October 1, 1987, Bull. Seism. Soc. Am. 80, 1-22.
14. Iida, M., T. Miyatake and K. Shimazaki (1988). Optimum strong-motion array geometry for source inversion, Earthq. Eng. Struct. Dyn. 16, 1213-1225.
15. Iida, M. (1990). Optimum strong-motion array geometry for source inversion -- II, Earthq. Eng. Struct. Dyn. 19, 35-44.
16. Olson, A.H., J. Orcutt and G. Frazier (1984). The discrete wavenumber/finite element method for synthetic seismograms, Geophys. J.R. Astron. Soc. 77, 421-460.
17. Davis, T.L., J. Namson and R.F. Yerkes (1989). A cross section of the Los Angeles area: Seismically active fold and thrust belt, the 1987 Whittier Narrows earthquake, and earthquake hazard, J. Geophys. Res. 94B, 9644-9664.
18. Lawson, C. and R. Hanson (1974). Solving least squares problems, Prentice-Hall, Englewood Cliffs, N.J.
19. Iida, M. and P. Spudich (1992). Scatterer inversion based on microearthquake seismograms, Proc. Int. Symp. Effects Surface Geology Seismic Motion I, 239-244, Odawara, Japan.

# Optimum Strong-Motion Station-Array Geometry for Earthquake Source Studies

M. Iida

Earthquake Research Institute, University of Tokyo
1-1-1, Yayoi, Bunkyo-ku, Tokyo 113, Japan

Summary
Near-source seismic strong-motion records are controlled by source effects.
Using a source inversion method, however, the true solution is not
recovered, and the goodness of recovery much depends upon station array.
The resolving power of station array is measured, and the optimum array
geometry is determined.

## 1. Introduction

In almost all earthquakes, the largest damage occurs in the earthquake
source area. Near-source strong-motion records are mainly affected by
source effects and local site effects because propagation path effects are
minor. Recently, we come to recognize that consideration of source and path
effects is required in addition to local site effects [1]. While local site
effects have been verified by numerous studies, source effects remain
unknown. Exactly speaking, whereas low-frequency seismic records follow
seismic source theory, source effects are not confirmed in the high-
frequency range of primary concern in the earthquake engineering community.
The most likely reason is luck of well-instrumented earthquakes for source
studies. Applying a source inversion method to one of the best instrumented
earthquakes to date, Hartzell and Iida [2] demonstrated source effects on
high-frequency strong-motion records . In other words, they showed strong-
motion seismograms were basically explained by source theory.

Certainly, a source inversion method is the most powerful tool to
estimate source effects on strong-motion records. However, we recognize
that relatively large inconsistencies have been seen among inversion
results for an identical earthquake, as exemplied by the results for the
1979 Imperial Valley earthquake [3-5]. The basic reason seems to be that,
because uniqueness of the inversion solution is not guaranteed, the true
solution is not recovered. Olson and Anderson (1988) [6] showed that the
assumed solution was not recovered by their inversion method with a minimum

norm for the uniqueness of the solution, and that the goodness of recovery was heavily dependent upon the station array (Fig. 1). On the other hand, Iida et al. (1988) [7] measured effects of array configuration by using a single parameter, the accuracy of a source inversion. Although these two studies used too simple Green's functions in a homogeneous half space, effectivess of strong-motion array is examined.

## 2. Resolving power of strong-motion arrays

At present, we need to measure the resolving power of strong-motion arrays. The twofold purposes are to understand the limitations of current source inversion studies imposed by available strong-motion arrays, and to provide the guidelines of strong-motion array installations for source studies.

Since our method was explained in a previous study [8], we will give here a brief summary. We define the resolving power by the accuracy of a waveform source inversion, and estimate it efficiently without calculating the values of faulting parameters on the basis of the Wolberg's prediction analysis [9]. We divide the entire fault into many subfaults and use the displacement waveform representation for each subfault. A complete Green's function in a semi-infinite elastic space is used. A common, simple source time function is assumed for each subfault. The seismic moment and the rupture onset time for each subfault are chosen as unknown parameters. The unknown parameters are determined using a least-squares criterion. Here, uncertainties are assumed for several known independent variables. We estimate the accuracy of the source inversion, $\sigma$ by the maximum standard deviation of seismic moments of all subfaults, normalized by the seismic moment.

The theoretical waveform is a function of known and unknown parameters. The known parameters whose uncertainties are taken into account (the dip angle, the strike direction, the slip angle, the wave amplitude and the arrival time) are denoted by $x_p$ (p=1, ...., Np). We denote by $a_i$ (i=1, ...., Nu) the unknown parameters, the seismic moment and rupture onset time for each subfault. The wave amplitude for the j-th time point at the k-th station is expressed as $u^k(t_j) = f_j^k(x_{1kj}, \ldots, x_{N_pkj}; a_1, \ldots, a_{Nu})$. Two residuals, $Ru_{kj}$ and $Rx_{pkj}$ are defined as the differences between the observed and calculated values,

$$Ru_{kj} = U^k(t_j) - u^k(t_j) \qquad (1)$$

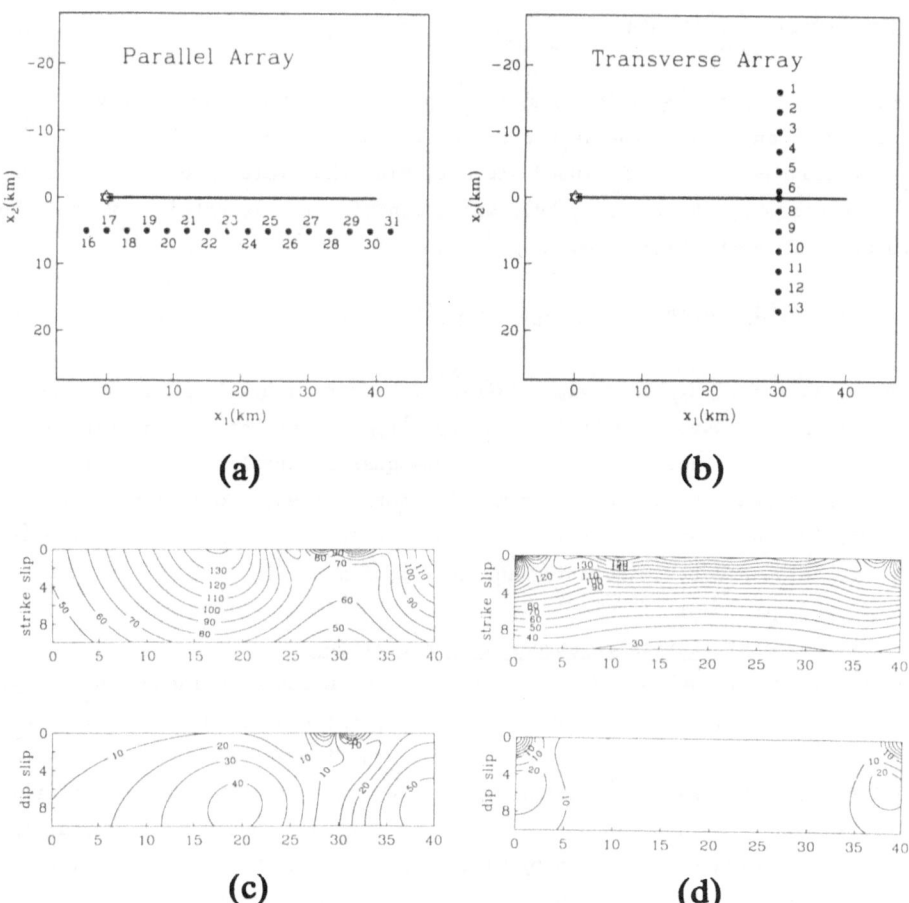

Fig. 1 (a) A transverse array and (b) a parallel array used as test cases. The vertical fault is shown as a solid line with a star at the epicenter. The input model was a uniform 100 cm in strike-slip offset and 0 cm in dip-slip offset. (c) Static slip solution by inversion of synthetic data from the transverse array and (d) that from the parallel array (After Olson and Anderson).

$$Rx_{pkj} = X_{pkj} - x_{pkj} \qquad (2)$$

where we denote by $U^k(t_j)$ and $X_{pkj}$ (p=1, ...., Np) the observed wave amplitude and the true values for known parameters, respectively. The least-squares method is used to determine the values of the unknown parameters, $a_i$ (i=1, ...., Nu), which minimize the weighted sum of the squares of the residuals, S:

$$S = \sum_k \sum_j (Wu_{kj} Ru_{kj}{}^2 + \sum_p^{Np} Wx_{pkj} Rx_{pkj}{}^2) \qquad (3)$$

where $Wu_{kj} = 1/\sigma u_{kj}^2$ and $Wx_{pkj} = 1/\sigma x_{pkj}^2$ . We denote by $\sigma u_{kj}$ the standard deviation of wave amplitude and by $\sigma x_{pkj}$ that of known parameter. Generally, the solution for the least-squares problem is obtained by solving normal equations in a matrix form. Then, following Wolberg's prediction analysis [8], the uncertainty of the i-th unknown parameter $\sigma a_i$ can be estimated by calculating an inverse of a matrix.

## 3. Optimum distribution of strong-motion stations

At the International Workshop on Strong-Motion Earthquake Instrument Arrays held in Honolulu, Hawaii in 1978 [10], a desirable array configuration was proposed on the basis of empirical judgement for each of three typical types of earthquake faults (Fig. 2). However, to date, little quantitative analysis has been done as to how the array stations should be deployed. The exceptional studies were done by Olson and Anderson (1988) [6] and Iida et al. (1988) [7]. But they did not use realistic Green's functions. Through physical wave simulations [8], we have confirmed that our current Green's function is adequate.

The optimum strong-motion array geometry is determined for each of three types of earthquake faults: strike-slip, dip-slip and offshore subduction thrust [11], which were considered at the International Workshop. By 'optimum', we mean that the solution of the source inversion becomes the most accurate for the same process of fault rupturing and with the same number of array stations. By assuming an unilateral rupture mode and fixing the number of array stations at 16, we perform it by trial and error. A wide variety of array configurations is tested for the respective faults. In some cases of the offshore fault simulation, hypothetical strong-motion ocean bottom seismographs are tested.

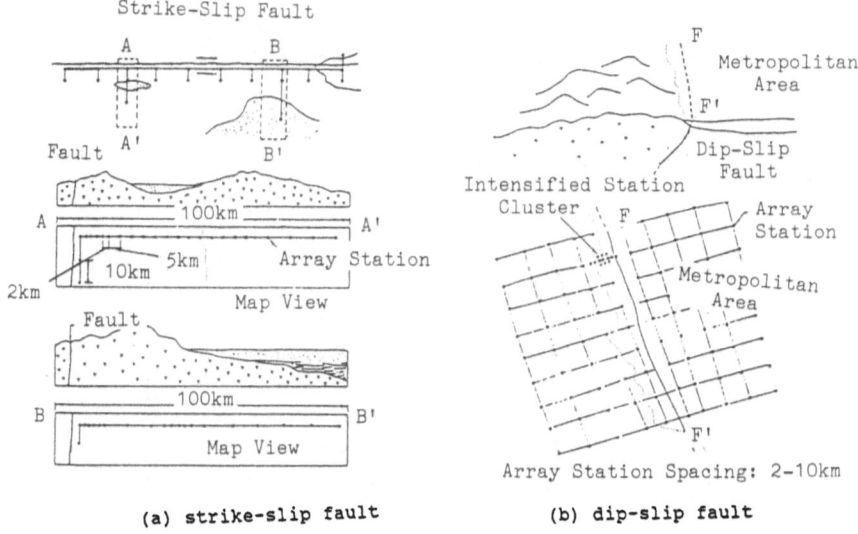

(a) strike-slip fault     (b) dip-slip fault

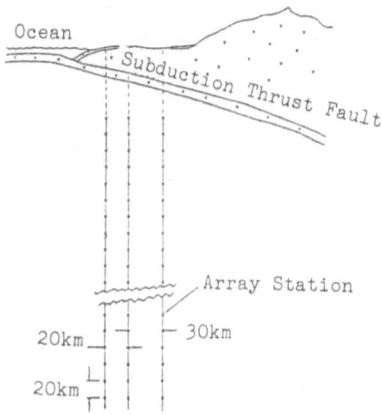

(c) subduction thrust fault

Fig. 2  Array configurations for source mechanism and wave propagaion studies recommended at the 1978 International Workshop on Strong-Motion Earthquake Instrument Arrays: (a) strike-slip; (b) dip-slip; (c) subduction thrust fault.

408

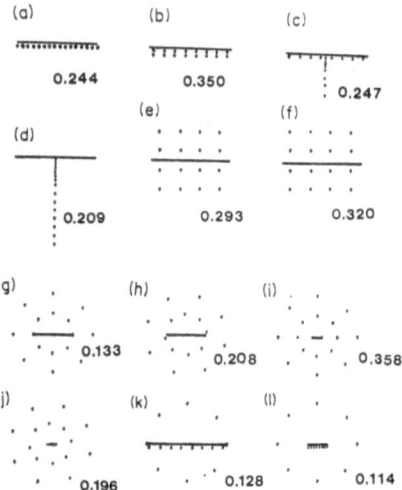

Fig. 3  Array configurations (plan views) tested for a vertical strike-slip
fault.  The horizontal bar shows the fault and the dots indicate  stations.
The  numerical value indicates the accuracy of the inversion solution, σ.
For example, 0.244 means 24.4 per cent error.

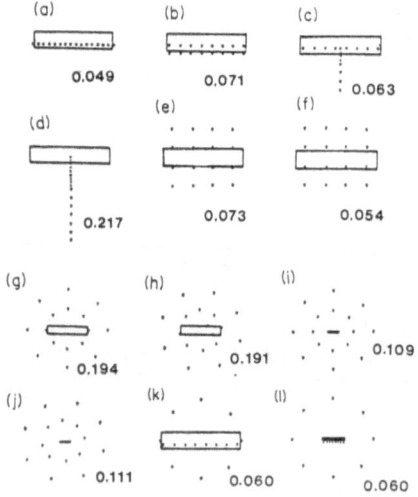

Fig. 4  Array configurations (plan views) tested for an inclined  dip-slip
fault.  The rectangle shows the fault and the dots indicate  stations.  The
numerical value indicates the accuracy of the inversion solution, σ.

## 4. Discussion

The array configurations examined for a strike-slip fault and a dip-slip fault are shown in Figs. 3 and 4. The numerical values in these figures show the accuracy of the source inversion. The results suggest that the optimum strong-motion array for a strike-slip fault is characterized by stations well distributed in azimuth, while the optimum array for a dip-slip event has stations arranged in a grid-shaped form. For a vertical strike-slip fault of producing close phase arrivals at most stations, stations immediately above the fault plane, which are robust at the vertical resolution, are needed. An inclined dip-slip fault favors a grid pattern of stations that appears to help many phases to be separated. In the case of the offshore fault, the slight dependence of the inversion uncertainty on the surface array configuration is due to the restricted array geometry. We find that ocean bottom stations are effective for

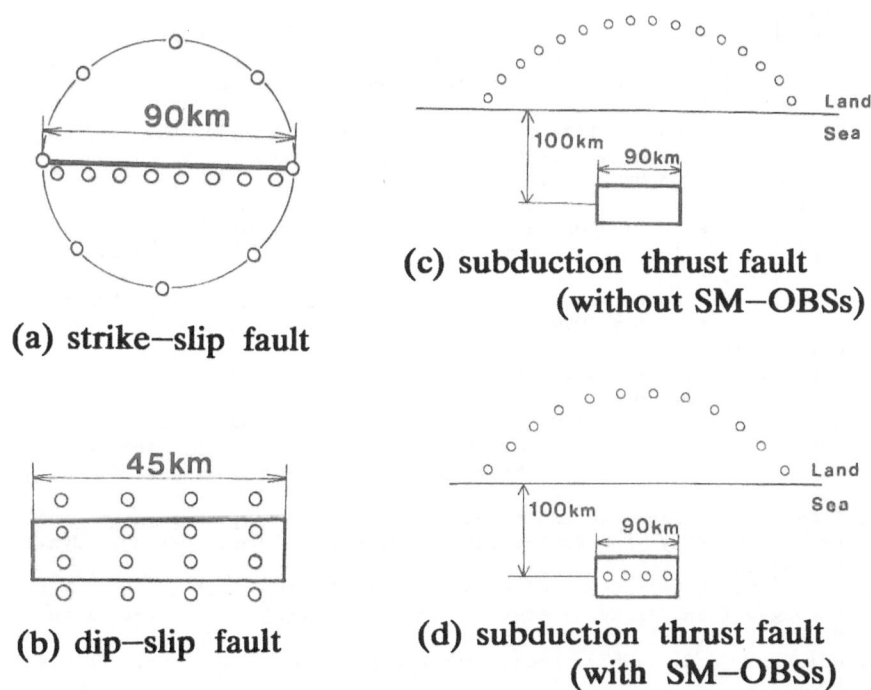

**(a) strike−slip fault**

**(b) dip−slip fault**

**(c) subduction thrust fault (without SM−OBSs)**

**(d) subduction thrust fault (with SM−OBSs)**

Fig. 5 The optimum array configurations for (a) a vertical strike-slip fault, (b) an inclined dip-slip fault, and a subduction thrust fault ((c) without and (d) with hypothetical ocean bottom seismographs).

further improvment of the inversion uncertainty. The optimum array configuration for each fault geometry is summarized in Fig. 5. The optimum array geometries are consistent with the ones proposed at the 1978 International Workshop (Fig. 2).

In a previous study [7], contribution of different types of stations during the whole rupture process was examined in the same strike-slip fault using only the far-field S waves. The study indicated that stations surrounding the fault area with good azimuthal coverage resolve the earlier stage of the rupturing process, while near-source stations resolve the later stage. By comparing the previous and present studies, we note that use of a more realistic Green's function reduces the dependency of the inversion solution on the array configuration. In other words, the simultaneous use of different kinds of waves in source studies is desirable.

References

1. Campillo, M., J.C. Gariel, K. Aki and F.J. Sanchez-Sesma (1989). Destructive strong motion in Mexico City: Source, path, and site effects during great 1985 Michoacan earthquake, Bull. Seism. Soc. Am. 79, 1718-1735.
2. Hartzell, S. and M. Iida (1990). Source complexity of the 1987 Whittier Narrows, California, earthquake from the inversion of strong motion records, J. Geophys. Res. 95B, 12475-12485.
3. Hartzell, S. and T. Heaton (1983). Inversion of strong ground motion and teleseismic waveform data for the fault rupture history of the 1979 Imperial Valley, California, earthquake, Bull. Seism. Soc. Am. 73, 1553-1583.
4. Olson, A.H. and R. Apsel (1982). Finite faults and inverse theory with applications to the 1979 Imperial Valley earthquake, Bull. Seism. Soc. Am. 72, 1969-2001.
5. Archuleta, R.J. (1984). A faulting model for the 1979 Imperial Valley earthquake, J. Geophys. Res. 89, 4559-4585.
6. Olson, A.H. and J.G. Anderson (1988). Implications of frequency-domain inversion of earthquake ground motions for resolving the space-time dependence of slip on an extended fault, Geophysical Journal 94, 443-455.
7. Iida, M., T. Miyatake and K. Shimazaki (1988). Optimum strong-motion array geometry for source inversion, Earthq. Eng. Struct. Dyn. 16, 1213-1225.
8. Iida, M., T. Miyatake and K. Shimazaki (1990). Relationship between strong-motion array parameters and the accuracy of source inversion, Bull. Seism. Soc. Am. 80, 1533-1552.
9. Wolberg, J.R. (1967). Prediction analysis, Van Nostrand, Princeton, N.J.
10. Iwan, W.D. (Editor) (1978). Strong-motion earthquake instrument array, Proc. Int. Workshop Strong-Motion Earthquake Instrument Arrays, Honolulu, Hawaii.
11. Iida, M. (1990). Optimum strong-motion array geometry for source inversion --- II, Earthq. Eng. Struct. Dyn. 19, 35-44.

# Chapter 7
# Ultrasonic Nondestructive Evaluation

# Neural Network Approach to the Inverse Problem of Crack-Depth Determination from Ultrasonic Backscattering Data

M. Takadoya*, M. Notake*, M. Kitahara**,
J.D. Achenbach***, Q.C. Guo*** and M.L. Peterson***

*  Mathematical Engineering Dept.
   Mitsubishi Research Institute
   3-6 Otemachi 2-Chome, Chiyoda-ku, Tokyo 100, Japan

**  Faculty of Marine Science and Technology
    Tokai University
    3-20-1 Orido, Shimizu, Shizuoka 424, Japan

***  Center for Quality Engineering and Failure Prevention
     Northwestern University
     Evanston, IL 60208, U.S.A.

## Introduction

A neural network approach has been developed to determine the depth of a surface breaking crack in a steel plate from ultrasonic backscattering data. The network is trained by the use of a feedforward three-layered network together with a back-propagation algorithm for error corrections[1,2]. Synthetic data are employed for network training. The signal used for crack insonification is a mode converted 45° transverse wave. The plate with a surface breaking crack is immersed in water and the crack is insonified from the opposite uncracked side of the plate. A numerical analysis of the backscattered field is carried out based on elastic wave theory, by the use of the boundary element method. The numerical data are calibrated by comparison with experimental data. The numerical analysis provides synthetic data for the training of the network. The training data have been calculated for cracks with specified increments of the crack depth. The performance of the network has been tested on other synthetic data and experimental data which are different from the training data.

Other recent studies on the use of neural networks for the classification of ultrasonic data have been reported in Refs. [3] and [4].

## Neural Network Approach to the Inverse Problem

When ultrasonic wave is incident on a structure which has defects inside, the interactive with the defects generate backscattered waves. It is called a "forward problem" to calculate the backscattered wave forms when the parameters of the defects, for example, sizes, locations, and so on, are given. On the other hand, it is called an "inverse problem" to estimate the parameters of defects from the observed backscattered waves. Determining parameters of defects is a typical inverse problems and various numbers of analysis have been proposed. The forward problem as well as inverse one can be regarded as a quantitative mapping problem which relates the signal information (the backscattering waves) and the characteristics of the defects (sizes or locations).

A neural network is sometimes called a mapping network if it is able to compute some functional relationship between its input and its output. For example, if the input to a neural network is the value of an angle, and the output is the cosine of that angle, then the network performs the mapping $\theta \rightarrow \cos(\theta)$. For such a simple function, we do not need a neural network; however, we might want to perform a complicated mapping where we do not know how to describe the functional relationship in advance, but we do know examples of the correct mapping. Therefore, if we make the neural network learn by using some examples for the relationship between the signal information and the characteristics of defects,then we are able to represent this nonlinear and continuous mapping relationship as a pair of the network weight and the offset value. By its ability of generalization, the neural network can make an estimation, even for that were not pare of the training set. It may take a time to train the neural network. Once the training has been completed, a solution can, however, be obtained quickly, because only sum and product operations have to be performed.

## Experimental Data and Theoretical Calculations
A surface-breaking crack with depth a in a steel plate of thickness h is considered. The plate is immersed in a water bath as shown in Fig.1. Ultrasound is generated by an immersed piezoelectric transducer. The angle of incidence on the insonified top face of the plate is taken to be 18.9° with the normal to the plate face. This angle of incidence exceeds the critical angle, and the ultrasonic beam is primarily converted into a beam of transversely polarized ultrasound in the plate, which is incident under an angle of 45° on the insonified face of the crack. The back-scattered ultrasonic signal is utilized for crack-depth determination.

The general forms of the experimental and theoretical back-scattered signals are briefly discussed in this section. For details we refer to Ref.[5]. In the frequency domain, the experimentally obtained back-scattered transducer signal may be expressed as

$$Y_{exp}(\omega) = T_0 H_w H_b H_{ws} H_{crack}^{exp} H_{sw} H_w T_r \ . \tag{1}$$

The response functions in this expression represent the effects of

| | |
|---|---|
| $T_0(\omega)$: transducer output, | $H_w(\omega)$: water path, |
| $H_b(\omega)$: beam spreading, | $H_{ws}(\omega)$: water $\rightarrow$ solid interface, |
| $H_{sw}(\omega)$: solid $\rightarrow$ water interface, | $T_r(\omega)$: transducer reception, |

and

$H_{crack}^{exp}(\omega)$: interaction with crack in solid.

For the corresponding theoretical results, the expression is exactly the same except for the response of the crack:

$$Y_{theory}(\omega) = T_0 H_w H_b H_{ws} H_{crack}^{BEM} H_{sw} H_w T_r \ . \tag{2}$$

In equation (2), $H_{crack}^{BEM}(\omega)$ represents the interaction with the crack of the incident wave as calculated by the boundary element method(BEM). The BEM calculation is based on two-dimensional elastodynamic theory for an elastic body with a surface-breaking crack.

To uncouple the theoretical signal in Eq.(2) from the response functions, the signal for a corner

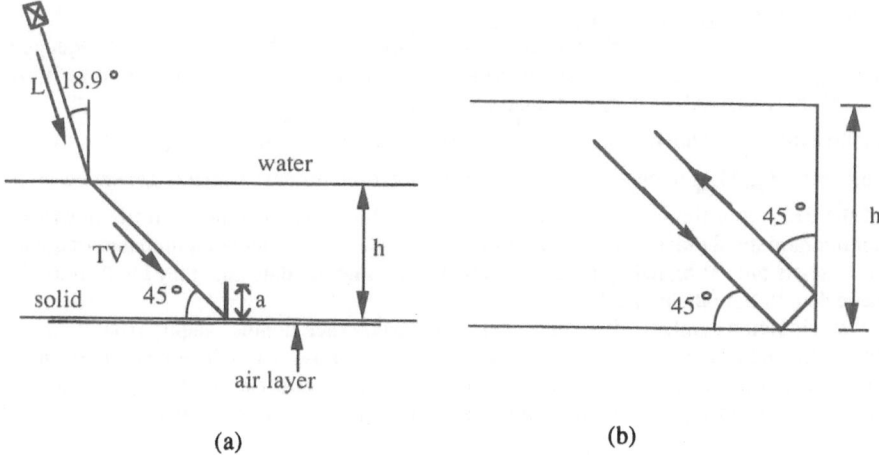

Fig.1   Surface breaking crack of depth a in a steel plate (a),
and corner reflection of the reference signal (b).

reflection is introduced as the reference signal, see Fig.1.  For the same transducer angle, the
same water paths, and the same specimen but with a rectangular corner, this reference corner
signal can be written as

$$X_{ref}(\omega) = T_0 H_w H_b H_{ws} H_{cor}(\omega) H_{sw} H_w T_r \ ,$$  (3)

where $H_{cor}(\omega)$ represents the corner reflections in the solid.  For a solid-air interface, the term
$H_{cor}(\omega)$ can be expressed in simple form as shown in Ref.[5].  The formal deconvolution of the
theoretical signal of Eq.(2) by the reference signal of Eq.(3) yields

$$\frac{Y_{theory}(\omega)}{X_{ref}(\omega)} = \frac{H_{crack}^{BEM}(\omega)}{H_{cor}(\omega)} \ .$$  (4)

Thus, the theoretical signal can be expressed as

$$Y_{theory}(\omega) = \frac{X_{ref}(\omega)}{H_{cor}(\omega)} H_{crack}^{BEM}(\omega) \ ,$$  (5)

where the term $X_{ref}/H_{cor}$ accounts for the beam paths in the water and across the solid-water
interface.  The theoretical signal in Eq.(5) is a convolved signal of the water path, $X_{ref}/H_{cor}$,
and the elastodynamic interaction, $H_{crack}^{BEM}$, with a crack in the solid.  We call the signal,
$Y_{theory}(\omega)$, of Eq.(5) the theoretical signal, which may be directly compared with the
experimental signal, $Y_{exp}(\omega)$, of Eq.(1).  The numerical calculations of the elastodynamic
interaction term, $H_{crack}^{BEM}(\omega)$, have been discussed in detail by Zhang and Achenbach[6].

## The Neural Network System Scheme

The neural network system is schematically depicted in Fig.2. In the neural network strategy of this paper, the theoretical signals of Eq.(5) are used for the training of the network. In the theoretical analysis, the boundary element calculation is carried out in the frequency domain to evaluate the interaction term $H_{crack}^{BEM}$. The numerical result is subsequently convolved with the term $X_{ref}/H_{cor}$ to obtain $Y_{theory}(\omega)$. Next the time domain signal is generated by the use of the FFT algorithm. The back scattered signals have been calculated for both cases of perfect mathematical cracks and notches of 0.5mm width. The calculated time and frequency domain signals are calibrated by comparison with experimental data and they are then used as synthetic data to train the network.

First, we calculated synthetic data for a total of nineteen crack depths ranging from 0.6mm to 2.4mm, with equal increments of 0.1mm. We used ten of those data with equal increments of 0.2mm as the training data for a neural network (the time and frequency domain data for notches are shown in Fig.3) and remaining data for the evaluation of the network performance.

## Neural Network and its Training

The three-layered feedforward network shown in Fig.4 is employed. An input signal vector $\{O_1^1, O_2^1, \cdots O_N^1\}^T$ is applied to the sensory units of the first layer, where N is the number of sensory units. The sensory units distribute the signals to the association units in the next layer according to the following relation

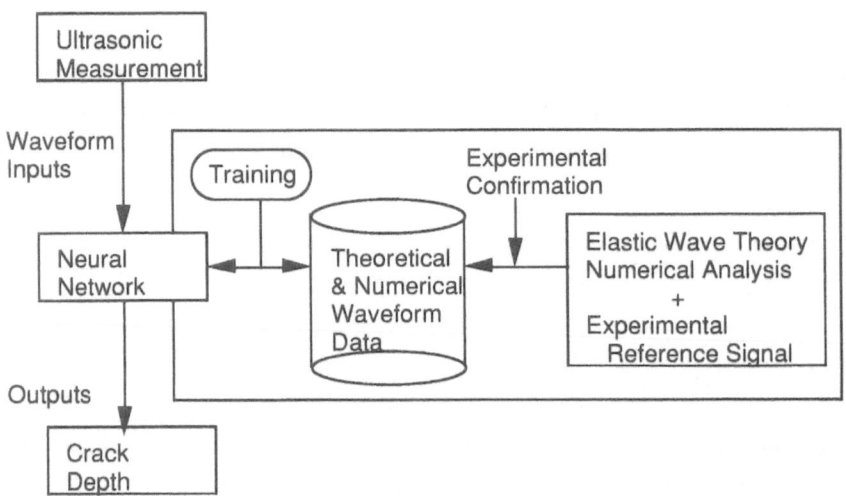

Fig.2 Neural network system

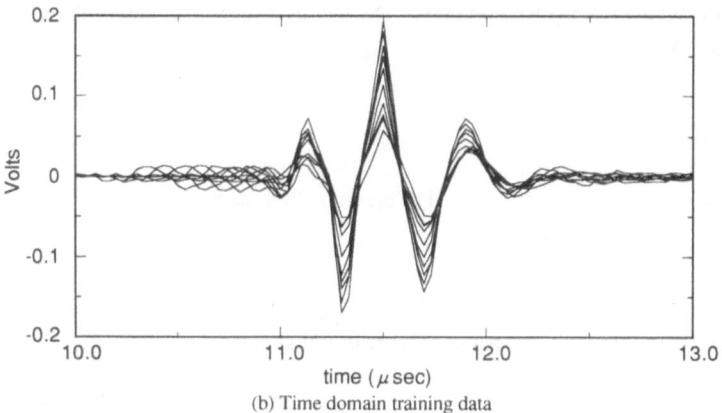

(b) Time domain training data

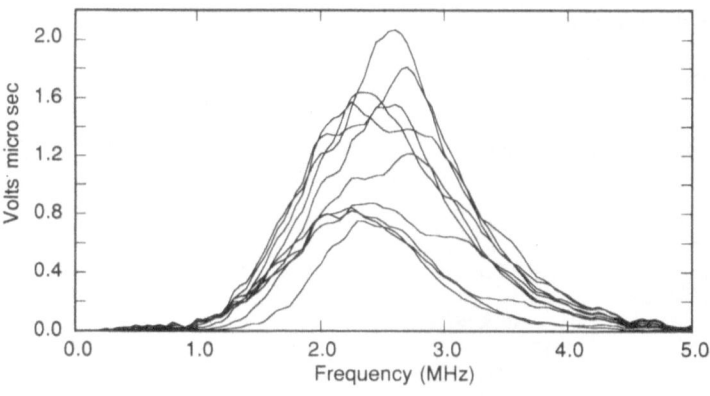

(a) Frequency domain training data

Fig.3 Theoretical training signals for ten notches, with depth
ranging from 0.6mm to 2.4mm in increments of 0.2mm.

$$I_j^2 = \sum_{i=1}^{N} W_{ij}^{12} O_i^1 + \theta_j^2 , \qquad (j=1,2,\cdots,L),$$

(6)

where $W_{ij}^{12}$ is the connection weight from the $i$th unit of the 1st. (sensory) layer to the $j$th unit of
the 2nd (association) layer, and $\theta_j^2$ is the threshold firing value of the $j$th unit in the association
layer. In equation (6), the superscript refers to the layer, the subscript refers to the unit in each

layer, and L is the number of association units. The output from the $j$th unit of the association layer is assumed to be

$$O_j^2 = f(I_j^2) ,$$
(7)

where the transformation function f is defined as the sigmoidal function

$$f(x) = \frac{1}{1 + e^{-x}} .$$
(8)

The input for the $k$th unit in the last (response) layer is defined by

$$I_k^3 = \sum_{j=1}^{L} W_{jk}^{23} O_j^2 + \theta_k^3 , \qquad (k=1,2,\cdots,M),$$
(9)

where the output is assumed as

$$O_k^3 = f(I_k^3) .$$
(10)

Here M is the number of response units.

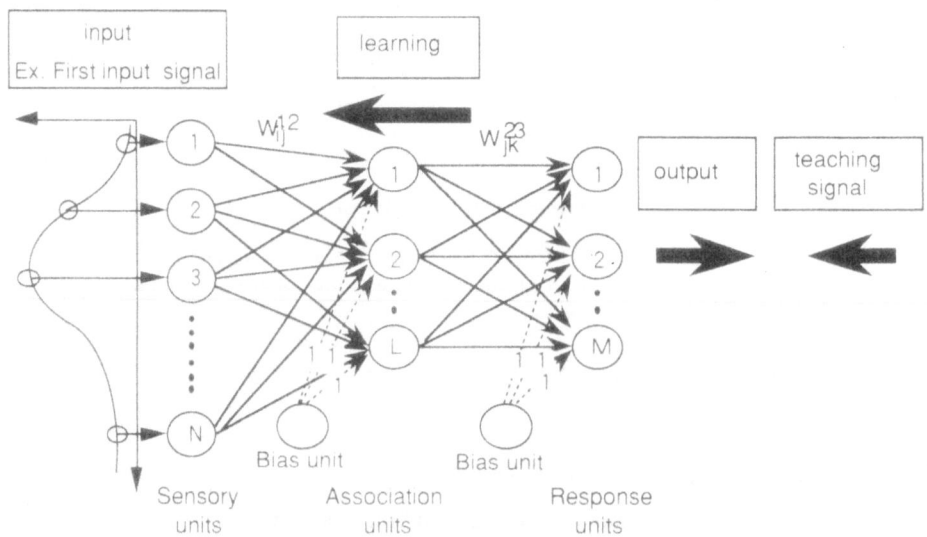

Fig.4 Three-layered feedforward network.

The initial weight values, $w_i^{k-1}{}_j^k$, and threshold values, $\theta_j^k$, are chosen as random numbers. For a given input signal vector $O_i^1$ (i=1,2,...,N), the network proceeds to forward according to Eqs.(6)-(10). At the response units, a teaching signal which has unit value in the kth unit is assigned to recognize the kth input signal.

To train the network, an error-backpropagation algorithm is adopted to adjust the connection weights. The weight-correction $\Delta w$ is expressed as

$$\Delta w_i^{k-1}{}_j^k(t+1) = -\varepsilon d_j^k O_i^{k-1} + \alpha \Delta w_i^{k-1}{}_j^k(t) , \tag{11}$$

where t is the number of iterations, $\varepsilon$ is the learning gain factor with a value in the range of (0,1), and $\alpha$ is the momentum coefficient which is in [0,1). For the response unit, the modified teaching signal $d_k^3$ in Eq.(11) is expressed as

$$d_k^3 = (O_k^3 - y_k) f'(I_k^3) , \qquad (k=1,2,\cdots,M) , \tag{12}$$

where $y_k$ is the earlier assigned teaching signal. For the association unit, $d_j^2$ takes the form

$$d_j^2 = (\sum_{k=1}^{M} w_{jk}^{23} d_k^3) f'(I_j^2) . \tag{13}$$

The function f in Eqs.(12) and (13) is given by Eq.(8) and the expression for $f'(\cdot)$ reduces to

$$f'(I_j^k) = O_j^k(1 - O_j^k) . \tag{14}$$

To adjust the threshold value $\theta_j^k$ in the process of correcting the connection weight $w_i^{k-1}{}_j^k$, a bias unit is attached to each layer. The bias unit provides a fictitious unit input value to each unit, as shown in Fig.4. Inspection of Eqs.(6) and (9) shows that the threshold value can be treated exactly like a connection weight, provided that the input signal from the bias unit is kept at unit value. The threshold values for each unit are now determined like the other connection weights.

## Network Architecture for Crack-Depth Determination

In order to construct a neural network which can estimate crack depths, we prepared two types of network structure:

(a) Binary output type

This type of network has, for example, ten discrete response units. For each of these, a unit corresponds to a certain depth of the crack.

For any given input wave form, this network should activate one and only one of the ten response units to represent ten wave forms. Therefore, in the Eq.(12), the teaching signal is given as $y_k = \{0, 0, \cdots, 1(k\text{th unit}), 0, \cdots 0\}^T$ to recognize the k-th input signal.

(b) Analog output type

This type of network has only one response unit. The output value ranges from 0.0 to 1.0 so that we map the actual crack depth on (0.0,1.0). As a result, the response unit has analog values. In our study, the upper limit of the crack depth is 2.4mm, but it is normalized at 2.5mm because the value of the response unit never reaches 1. Therefore, the actual crack depth is 1.0mm, the teaching signal $y_1 = 1.0 / 2.5 = 0.4$ is given.

Table 1 summarizes the network structures used in this study.

Table 1  Structure of the neural network  - numbers of unit

| | layer | sensory | association | response |
|---|---|---|---|---|
| (a) Binary output | time | 91 | 15 | 10 |
| | freq. | 101 | 15 | 10 |
| (b) Analog output | time | 91 | 5 | 1 |
| | freq. | 101 | 10 | 1 |

## Network Performance

(a) Binary output type

Table 2 summarizes the network performance when experimental backscattered data for cracks of depths 1.05mm, 1.49mm and 2.19mm are entered into the network as input data for crack-depth characterization. The outputs from the response units are listed in the table.The first row of the table lists the crack depths for the synthetic data used to train the network. Three sets of two rows each list the response numbers for the three sets of experimental input data. The label's time and frequency indicate that the output numbers in the labeled rows were obtained from time domain and frequency domain data, respectively. Thus the response numbers 0.97 and 0.98  for the time domain and frequency domain inputs of the 2.19mm crack show that for both sets of data the network indicates a crack depth of 2.2mm.

Table 2   Network performance for experimental data inputs.  - Binary output type (1)

| exp. data | output unit | 0.60 | 0.80 | 1.00 | 1.20 | 1.40 | 1.60 | 1.80 | 2.00 | 2.20 | 2.40 |
|---|---|---|---|---|---|---|---|---|---|---|---|
| 1.05mm | time | - | 0.22 | 0.97 | - | - | - | - | - | - | - |
| | freq. | - | - | - | 0.99 | - | - | - | - | - | - |
| 1.49mm | time | - | - | - | - | 0.52 | 0.45 | - | - | - | - |
| | freq. | - | - | - | - | 0.29 | 1.00 | - | - | - | - |
| 2.19mm | time | - | - | - | - | - | - | - | - | 0.97 | - |
| | freq. | - | - | - | - | - | - | - | - | 0.98 | - |

Table 3  Network performance for experimental data inputs. - Binary output type (2)

| exp. data | output unit | 1.30 | 1.40 | 1.50 | 1.60 | 1.70 | 1.80 |
|-----------|-------------|------|------|------|------|------|------|
| 1.49mm | time | - | - | 0.95 | - | - | - |
| | freq. | - | - | 0.86 | 0.09 | - | - |

The response numbers for the frequency domain data are also quite conclusive for the 1.49mm crack. For this case the time domain results may be interpreted as indicating a crack of depth in-between 1.4 and 1.6mm. On the other hand the frequency domain data suggest a crack depth quite close to 1.6mm. For the 1.05mm case the time domain data suggest a crack depth close to 1mm, where the frequency domain data indicate a crack depth close to 1.2mm.

Table 3 shows the outputs from the response units when the network was trained by the synthetic data in the range of 1.3mm and 1.8mm, with the increment of 0.1mm. The input signal for the network is again experimental backscattered signals for the crack with the depth 1.49mm. The network is indicating the crack of 1.5mm depth.

(b) Analog output type

For this type of network, the output value of the response layer directly indicates the crack depth if the output is multiplied by the factor 2.5. Table 4 summarizes the the network performance when the remaining synthetic data are entered into the network.In this case, the first row of the table indicates the crack depth for the synthetic data used to evaluate the network performance. The second row and the third one are output values from the time domain and frequency domain data, respectively. The results show that the network estimates quite well the crack depths for these synthetic data.

Table 5 summarizes the network performance when the experimental data for crack depths 0.67mm, 1.05mm, 1.49mm and 2.19mm are entered into the network as input data for crack-depth characterization. The first row of the table lists the crack depth for the experimental data used to evaluate the network performance. The results correlate well with the actual crack depths.

Table 4  Network performance for synthetic data inputs.  - Analog output type

| synthetic data (mm) | 0.70 | 0.90 | 1.10 | 1.30 | 1.50 | 1.70 | 1.90 | 2.10 | 2.30 |
|---------------------|------|------|------|------|------|------|------|------|------|
| time domain (mm) | 0.63 | 0.90 | 1.08 | 1.29 | 1.49 | 1.68 | 1.94 | 2.17 | 2.33 |
| freq. domain (mm) | 0.63 | 0.85 | 1.14 | 1.36 | 1.47 | 1.75 | 1.90 | 2.09 | 2.34 |

Table 5  Network performance for experimental data inputs.   - Analog output type

| experimental data | 0.67mm | 1.05mm | 1.49mm | 2.19mm |
|---|---|---|---|---|
| time domain | 0.82 | 1.00 | 1.23 | 2.06 |
| frequency domain | 0.79 | 0.96 | 1.57 | 2.27 |

## Summary

A method for quantitative evaluation of defects  (crack-depth determination) using a neural network has been discussed.

Synthetic data based on the elastic wave theory have been employed for the training of the neural network .

As a representation of outputs of the neural network, binary type output and analog type output have been tested.

- The network size for the analog type is smaller than that of binary type, because of the difference of the response unit numbers.
- However the actual learning time is not different, because the analog type needs more learning times to converge.
- It is difficult to interpret the output from the binary type when data for intermediate crack depths are entered into the network.
- Both networks show good performances on the intermediate synthetic or experimental data.

## Reffereces

1. D.E. Rumelhart, G.E. Hinton, and R.J. Williams, in Parallel Distributed Processing, edited by D.E. Rumelhart and J.L. McClelland (MIT Press, Cambridge, 1986), Vol.1, p.318.
2. J.A. Freeman and D.M. Skapura, Neural Networks: Algorithms, Applications, and Programming Techniques (Addison-Wesley Publishing Co., 1991).
3. J.J. Thomsen and K. Lund, Materials Evaluation, 49(5), 594(1991).
4. D.O. Thompson and D.E. Chimenti (eds.), Review of Progress in Quantitative NDE (Plenum Press, New York, 1991), Vol.10A, Chap.3, Sec.A.
5. M. Kitahara, J.D. Achenbach, Q.C. Guo, M.L. Peterson, T.Ogi and M. Notake, in Review of Progress in Quantitative NDE, edited by D.O. Thompson and D.E. Chimenti (Plenum Press, New York, 1991), Vol.10A, p.689.
6. Ch.Zhang and J.D. Achenbach, Ultrasonics, 26, 132(1988).

# Simulation of Directivity Synthesis for Ultrasonic Transducers

Takao TSUCHIYA*, Hiroshi KHONO** and Yukio KAGAWA*

* Faculty of Engineering, Okayama University
 3-1-1 Tsushima-naka, Okayama 700, Japan

** Daihen Corp., 5-1 Minami-senrioka, Settsu city
 Osaka 569, Japan

**Summary**

A numerical simulation is carried out on the directivity synthesis for ultrasonic transducer applications. Two types of sound source configurations are discussed: one is a line array consisting of point sound sources (point source array) and another is a distributed surface sound source (surface source array). To satisfy the specification, such as the radiation beam width and the side-lobe suppression level, the least means square method (LMS method) is practiced by means of an iterative scheme. For acceptable convergence to be achieved, an error control is introduced. Axisymmetric field is considered for the point source array and two-dimensional field is considered for the surface source array for which boundary element approach is used. Some numerical simulations are demonstrated.

## 1. Introduction

Ultrasonic transducers are widely used for the ultrasonic imaging systems for medical diagnosis and nondestructive testing. The resolution of the ultrasonic imaging systems depends on the directional capability of the transducers. The synthesis of the desired directivity is a key factor for the optimal design.

The directivity of the ultrasonic transducer is determined by the amplitude and phase distributions of the velocity over the transducer's vibratory surface. The desired directivity can be achieved by properly adjusting these distributions. This is an inverse problem in the sense that the amplitude and phase distributions must reversely be determined from the directivity given. Direct inversion is not possible and an iterative scheme must be used on the basis of the least mean square method (LMS method) [1, 2]. Directional pattern is first calculated for certain amplitude and phase distributions assumed over the sound source, which is compared with the desired or specified directivity to give the error function. The amplitude and phase distributions are then corrected so as to minimize the error function. Minimum of the mean square error or norm is searched by iteration until a convergence is reached. To satisfy the specification, such as the radiation beam width and the side-lobe suppression level, an error control is introduced.

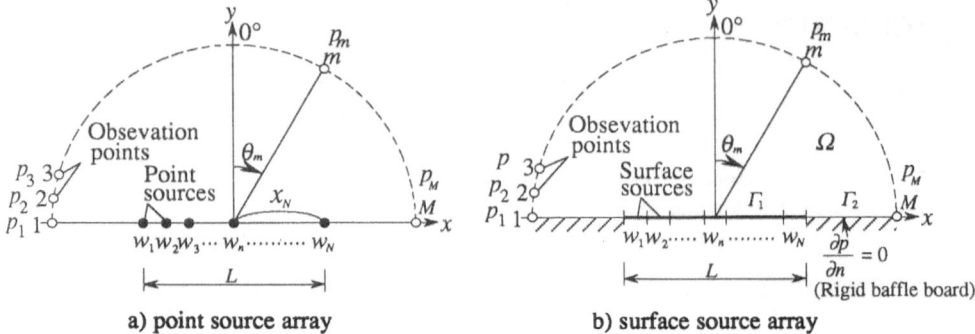

a) point source array         b) surface source array

Fig.1 Ultrasonic array transducers considered.

Two types of sound source configurations are considered: one is a line array consisting of point sound sources (point source array) as shown in Fig.1 a) and another is a distributed surface sound source (surface source array) as shown in Fig.1 b). Axisymmetric field is considered for the point source array and two-dimensional field is considered for the surface source array for which boundary element approach is used. To examine the validity of the approach, some numerically simulated results are demonstrated.

## 2. Point source array

First, we consider a point source array of length $L$ consisting of $N$ point sources as shown in Fig.1 a). Sound wave is sinusoidal with angular frequency $\omega$ which is to radiate from the point sources to the homogeneous medium. Sound pressure is measured at $M$ discrete observation points remote from the array. The sound pressure $p_m$ at the $m$-th observation point is given as

$$p_m = \sum_{n=1}^{N} s_{mn} w_n \qquad (m=1,2,...,M),$$ (1)

where

$$s_{mn} = e^{jkx_n \sin\theta_m},$$ (2)

$$w_n = A_n e^{-j\phi_n}.$$ (3)

$A_n$ and $\phi_n$ are the amplitude and phase of the $n$-th point source, $x_n$ is the distance from the center of line array to position of the $n$-th point source, $k$ is wave number (=$\omega/c$, $c$ sound speed), $M$ and $N$ are number of total observation points and point sources, and $\theta_m$ is the angle from the central line to the direction of the $m$-th observation point. Discrete directivity function $\{D\}$ is defined in terms of sound pressure as

$$\{D\} = \{p_1 \ p_2 \ ... \ p_M\}^{\mathrm{T}} = [S]\{W\},$$ (4)

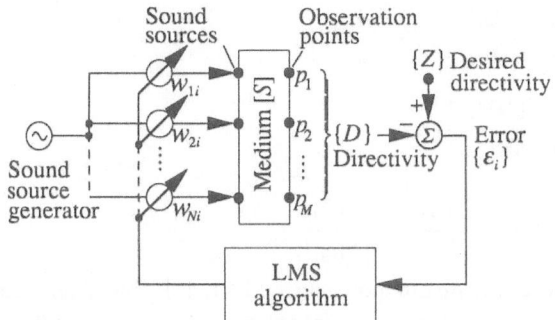

Fig.2 Schematic diagram for directivity sysnthesis by LMS method.

where

$$[S] = \begin{bmatrix} s_{11} & \cdots & s_{1N} \\ \vdots & & \vdots \\ s_{M1} & \cdots & s_{MN} \end{bmatrix} \quad \text{and} \quad \{W\} = \{w_1 \ w_2 \ \dots \ w_N\}^T, \tag{5}$$

$[S]$ is a transfer matrix which relates the source distribution to the directivity and $\{W\}$ is a source distribution function which represents the amplitude and phase distributions over the array. When the transmission space is given, the directivity function $\{D\}$ is under the control of the source distribution function $\{W\}$. Thus directivity can be synthesized by properly adjusting the source distribution function. The problem is to reversely determine the amplitude and phase distributions of the sound source from the sound field distribution given.

## 3. Directivity synthesis

To meet the specification, such as the radiation beam width and the side-lobe suppression level given, the least means square method (LMS method) is used based on an iterative scheme. A schematic diagram for synthesizing the desired directivity by means of LMS method is illustrated in Fig.2. Directivity $\{D_i\}$ is first calculated for certain amplitude and phase distributions given or the distribution function $\{W_i\}$, and the result is then compared with the desired or specified directivity $\{Z\}$ to give the error function $\{\varepsilon_i\}$ as follows

$$\{\varepsilon_i\} = \{Z\} - \{D_i\} \ , \tag{6}$$

where the subscript $i$ denotes the $i$-th iteration. The mean square error $E_i^2$ is defined by

$$E_i^2 = \{\varepsilon_i\}^T \{\varepsilon_i\}/M = (\{Z\}^T\{Z\} - 2\{D_i\}^T\{Z\} + \{D_i\}^T\{D_i\})/M, \tag{7}$$

where $M$ is again the number of the total observation points. The vector $\{W_i\}$ must be corrected so as to minimize the mean square error $E_i^2$. It follows that

$$\{W_{i+1}\} = \{W_i\} - \mu\{\nabla_i\}, \tag{8}$$

where $\{\nabla_i\}$ is a gradient vector

$$\{\nabla_i\} = \{\frac{\partial E_i^2}{\partial w_{1i}} \quad \frac{\partial E_i^2}{\partial w_{2i}} \quad \cdots \quad \frac{\partial E_i^2}{\partial w_{Ni}}\}^T,$$ (9)

where

$$\frac{\partial E_i^2}{\partial w_{ni}} = -\frac{2}{M}\frac{\partial\{|D|_i\}^T}{\partial w_{ni}}\{\varepsilon_i\},$$ (10)

and

$$\frac{\partial\{|D|_i\}^T}{\partial w_{ni}} = \{\frac{d_1\bar{s}_{1n}}{|d_1|_i} \quad \frac{d_2\bar{s}_{2n}}{|d_2|_i} \quad \cdots \quad \frac{d_M\bar{s}_{Mn}}{|d_M|_i}\},$$ (11)

$w_{ni}$ is a component of the distribution matrix $\{W_i\}$, $|\ |$ and $\bar{\phantom{x}}$ respectively represent the absolute value and complex conjugate, and $\mu$ is a scaling factor to adjust the step size. For stable convergence, it is necessary and sufficient condition if the followings is satisfied

$$0 < \mu < 1/\text{tr}[R], \qquad \text{where } [R] = [S]^T[S],$$ (12)

and tr[ ] represents the trace of matrix.

## 4. Surface source array

We consider a surface sound source of two-dimension as shown in Fig.1 b). Here we derive the transfer matrix of the transmission space for the surface source array by means of boundary element method. The field is governed by the following Helmholtz equation

$$\nabla^2 p + k^2 p = 0, \qquad (\text{in } \Omega)$$ (13)

where $p$ is sound pressure. On the boundary $\Gamma\ (=\Gamma_1+\Gamma_2)$, the following conditions are prescribed

$$q = \hat{q} \qquad (\text{on } \Gamma_1),$$
$$q = 0 \qquad (\text{on } \Gamma_2),$$ (14)

where $q$ is flux and $\wedge$ represents the prescribed value. The process of the weighted residuals leads to the following boundary integral expression[3]

$$c_j p_j + \int_\Gamma pq^* d\Gamma = \int_\Gamma qp^* d\Gamma,$$ (15)

where $p_j$ is the pressure at the point of interest $j$, $c_j$ is the angle around $j$ at which the region $\Omega$ is looked. Providing the fundamental solution and its derivative with respect to the normal to the boundary for $p^*$ and $q^*$ respectively and dividing the boundary $\Gamma$ into constant elements, we now have a simultaneous equation of the form

$$[H]\{p\} = [G]\{q\}.$$ (16)

The driving boundary $\Gamma_1$ consists of $N$ surface sources each of which is divided into several boundary elements. Assuming that the amplitude and phase are uniform on a single surface source, a source driving vector $\{q\}$ is expressed as

$$\{q\} = \hat{q}[Q]\{W\},$$ (17)

where $[Q]$ is a transfer matrix which relates the amplitude and phase on the surface source to those on the boundary elements. It is assumed that surface source array is built in an infinite rigid baffle board. Without loss of accuracy, the baffle board is truncated at a certain finite distance from the source since the sound pressure and the particle velocity become small rapidly as the distance increases.

The sound pressure $p_m$ at the $m$-th observation point is given by following equation

$$p_m = \sum_{j=1}^{n_e} q_j \, G'_{mj} - \sum_{j=1}^{n_e} p_j \, H'_{mj} \, , \tag{18}$$

where $p_j$ and $q_j$ are the solution given on the boundary and $n_e$ is the number of elements. The directivity function $\{D\}$ is now given by

$$\{D\} = [G']\{q\} - [H']\{p\} \, , \tag{19}$$

Eliminating the pressure term $\{p\}$ from equation (19) by the help of equations (16) and (17), the directivity function $\{D\}$ can be rewritten as

$$\{D\} = \hat{q} \, ([G'] - [H'][H]^{-1}[G'])[Q]\{W\} \, , \tag{20}$$

Comparing equations (20) with (4), the transfer matrix $[S]$ for the surface source array can be obtained as

$$[S] = ([G'] - [H'][H]^{-1}[G'])[Q] \tag{21}$$

The procedure to follow is the same as in the previous case.

## 5. Numerical examples

### 5.1 Simple point source array

A simple numerical example is first demonstrated. A point source array consisting of 26 point sources is considered. Length of the array is $5\lambda$ where $\lambda$ is the wavelength. The point sources are placed at same interval. The desired directivity is chosen to be that of a line source with length $5\lambda$, excited at uniform amplitude and phase distribution. For the initial

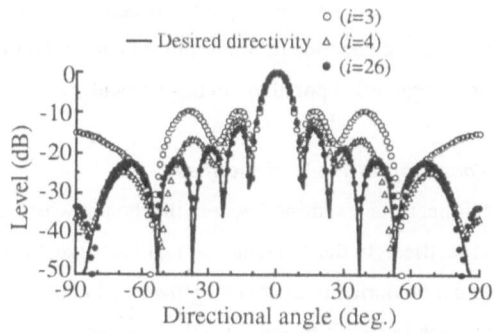

Fig.3 Convergence process of iteration. (Point source array)

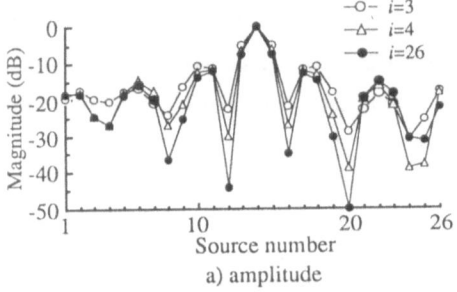

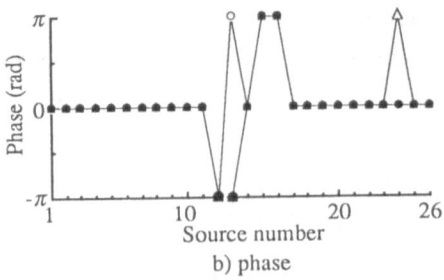

a) amplitude

b) phase

Fig. 4 Convergence process of iteration for amplitude and phase. (Point source array)

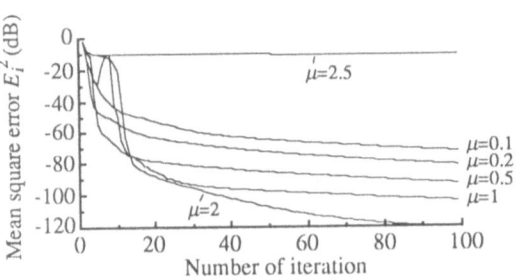

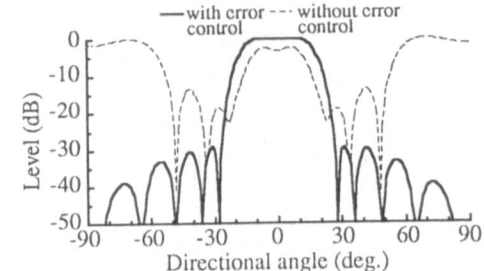

Fig.5 Convergence process for various scaling factor. (Point source array)

Fig.6 Synthesized results for the beam width and the side-lobe level specified. (Point source array)

condition, the amplitude of one of the point sources is set to be of a unit intensity at the center of the line. The number of the observation points is chosen to be 181 and $\mu$ is set to be 1. The directivity synthesized is shown in Fig.3 including the ones at some stages to convergence, in which plots indicate the cases at the 3rd, 4th and 26th iteration and solid line indicates the directivity of the line source set as the desired one. The directivity of the point source array approaches to that of the line source as iteration advances and the reasonable agreement is achieved at the 26th iteration. The amplitude and phase distributions over the array are shown in Fig.4. The convergence process is shown in Fig.5. It can be seen that the number of iterations required for convergence depends on the scaling factor $\mu$ and there is an optimal value for $\mu$ at which rapid convergence is possible. In this particular case, it is around 2.

### 5.2 Beam width and side-lobe suppression level specified

The directivity synthesis is examined when the beam width and the side-lobe suppression level are both specified. In the conventional LMS scheme, the desired or specified directional characteristic must numerically be given. However in most cases of engineering design, the specification is given in such a manner that the side-lobe suppression level is more

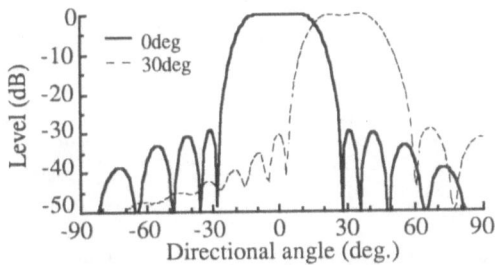

Fig.7 Synthesized results for beam steering. (Point source array)

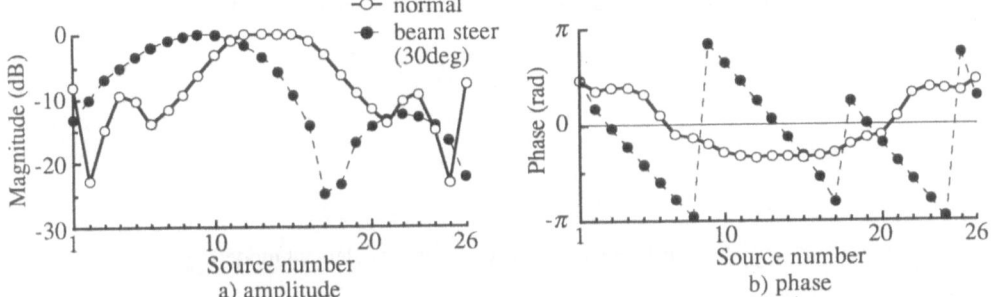

a) amplitude    b) phase

Fig.8 Amplitude and phase distributions for beam steering given in Fig.7.

than 30dB together with the beam width required for the main beam. It will be difficult that the side-lobe is suppressed as much as just 30dB though it could be more than 30dB, at which mean square error might be larger. To cope with this situation, the error control is introduced into the scheme in which if such a case occurs the value of the error is weighted by, say, a half to evaluate the gradient to follow. Numerical results when the beam width and the side-lobe suppression level are both specified to be 40° and more than 30dB are shown in Fig.6. The array considered is the same as in the previous example. 300 iterations are made for $\mu=1$ chosen. No convergence is reached when the conventional LMS scheme is simply used as illustrated by dashed line.

The beam steering is next examined. The synthesized result when the center of the main-lobe turns as much as 30° keeping other specification fixed. The amplitude and phase distributions over the array are respectively shown in figures 7 and 8 to achieve this. The error control is again incorporated.

430

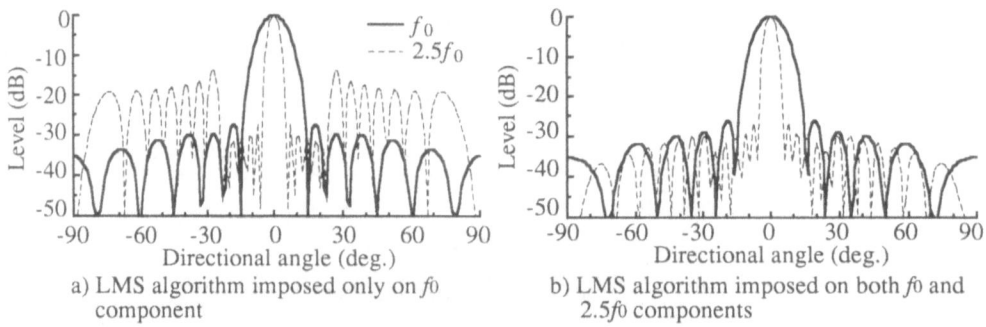

a) LMS algorithm imposed only on $f_0$
component

b) LMS algorithm imposed on both $f_0$ and
2.5$f_0$ components

Fig.9 Synthesized results for multiple frequency operation.

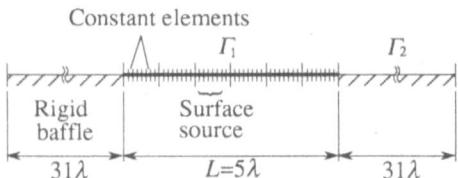

Fig.10 Surface source array. (Boundary element model)

*5.3 Multiple frequency operation*

Ultrasonic transducers are often operated at multiple frequency to provide various resolution capability and observation range. Here the directivity synthesis for the transducer operated at two different frequencies, $f_0$ and 2.5$f_0$, is considered. The beam width and the side-lobe suppression level are both specified to be less than 30° and more than 30dB. The result when the specification is imposed only on frequency $f_0$ is shown in Fig.9 a). The side-lobe level increases when the array is operated at 2.5$f_0$ as illustrated by dashed line while the specification is satisfied for frequency $f_0$. To consider multiple frequency operation, the transfer matrix [$S$] must be obtained for every frequency component to which the directivity is calculated. The LMS scheme is then applied so as to minimize the total mean square error for every frequency component. The result when the specification is satisfied for both frequencies is shown in Fig.9 b) in which the beam width at 2.5$f_0$ is much narrower.

*5.4 Surface source array*

Lastly, the directivity synthesis for the two-dimensional surface source array is demonstrated. The sound field is evaluated by means of the boundary element method. An array transducer consisting of 6 surface sources is shown together with the element division in Fig.10. The amplitude is assumed to be uniform within each surface source. Total length of the array is again 5$\lambda$ which is divided into 38 constant elements. The array is built in the rigid

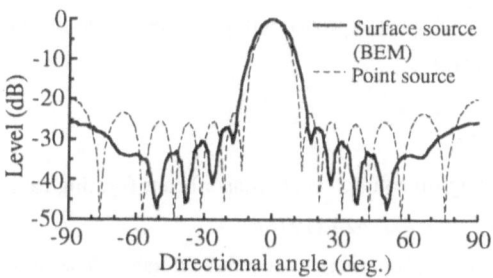

Fig.11 Surface source array and point source array.

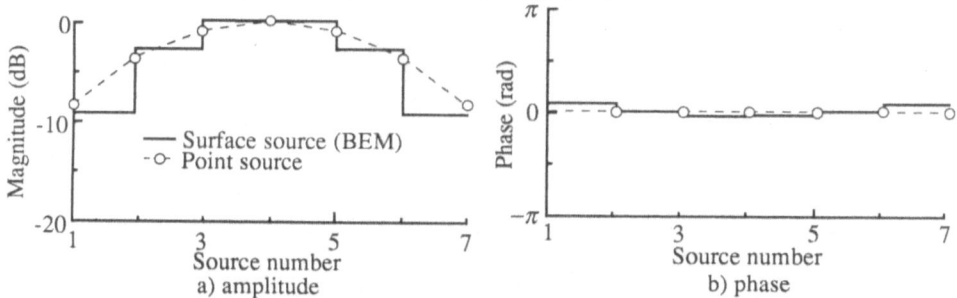

a) amplitude               b) phase

Fig.12 Amplitude and phase distributions for the directivity given in Fig.11.

baffle board whose length is $31\lambda$ in each side which is divided into 95 elements. The observation points are placed at the distance of $30\lambda$ from the center of the array. The beam width and the side-lobe suppression level are both set to be 30° and more than 30dB. The results after the convergence is reached are shown in figures 11 and 12. In the figures, the results for the point source array consisting of 7 point sources are also shown for the comparison. As this is an axisymmetric around the array, comparison is only possible for its cross section. With this particular point source array, the side-lobe suppression can not properly be possible as the point sources are too small in number.

## 6. Concluding remarks

The directivity synthesis for ultrasonic transducers was treated by means of LMS method. Some numerical simulation was carried out. The numerical results provided reasonable and typical solutions. Following conclusions might be drawn:

1) The error control is effective and necessary for acceptable convergence to be achieved when the beam width and the side-lobe level are both specified.

2) The present method can also be applied to the synthesis for the transducers at multiple frequency operation.

We hope that the present approach will pave the way to the computer-aided design of ultrasonic array transducers.

## References

1) Homer P.Bucker; "High-resolution cross-sensor beam-forming for a uniform line array," J. Acoust. Soc. Am. **63**(2), pp.420-424 (1978).

2) H.Date; "Digital signal processing -advanced course-," The 30th-lecture of Acoustical Society of Japan (1987) (*in Japanese*).

3) C.A.Brebbia; "The boundary element method for engineers," Pentech Press. (1978).

# Inversion Using Multi-Tip Echoes of Elastic Waves

+          ++          +++          +++
K. Harumi,  M.Uchida,  T.Miyajima  and  Y.Ogura

+    Tokyo University of Information Sciences
     1200 Yatou-chou, Chiba, Chiba, Japan
++   Nihon University, Narashino, Chiba, Japan
+++  Hitachi Construction Machinary Co. Ltd.,
     Kandatsu, Tsuchiura, Ibaraki, Japan

Summary

In applications  tip echoes generated  from the tips and corners
of defects are considered as the most effective way of inversion
( defect sizing ).    However,  the theoretical approach has been
done  mainly  by the use of continuous wave  theory,  but  these
proposals  are  sophisticated  and too  expensive  in  practice.
Numerical  experiments reveal that the tip waves are dominant in
the  reflection of elastic waves. [1]  Recently, low noise  full
digital  flaw detector have been developed,  and it  has  become
very easy to catch several tip waves simultaneously.   By the use
of  this detector and with the help of numerical experiment,  it
is  possible to distinguish each  experimental  echo.  Multi-tip
echo  method gives good results for sizing of an inclined crack,
1.5 mm height,  on a free surface.  Multi-tip waves method gives
more  information  than the conventional method,  and  it  is  a
simple and economical way of ultrasonic inversion in a solid.

## Introduction

The  tip echo method is recognised as  the most effective way of
ultrasonics  inversion(defect sizing),  but tip waves have  very
weak  intensity,  about  -20  or  30 dB below  the  main  echo.
Therefore,  the  application  of  the  method  needs  skill  and
experience. In the conventional method, we must make preliminary
inspection  and  then  use  a  combination  of  several  methods
explained in the book " Ultrasonic Defect Sizing ".  [2]  Several
kinds  of  low noise digital flaw detectors have  been  developed
recently  and these are able to catch tip echoes easily.   In the
first  stage,  we  have  done experiments using a  full  digital
detector DT-2000 to find  the reflection of the transverse  wave
by  an inclined crack on a free surface,  and caught several tip
echoes  simultaneously.   Using  these  experimental  values  of
distances of several tip echoes,  we have obtained crack height,
inclination  angle,  and crack width.   Good results  have  been
obtained from only three echo distances.  If we use N distances,
we  can  obtain N unknown values,  by using  several  receiving
transducers, and so invert complicated shaped cracks. The multi-
tip  wave method will be the most effective inverse  method,  as
the tip waves are dominant in the reflection of elastic waves in
a solid.

## REFLECTION FOR Y2 DIRECTION

### Comparison of Experiments with Numerical Experiments

Experiments have been done from both side of crack as shown in Fig. 1, Y2 and Y1 directions; the results are compared with numerical experiments, and theoretical consideration. Experiments on the reflection of transverse wave by an inclined crack were made using a full digital flaw detector HITACHI-DT-2000 with a 5C10x10A45 wide-band probe. The inclination angle of crack is given for every 10° between 10° and 40° , and the incident angle is 45 degrees. A scanning graph of echo amplitude is recorded as shown in Fig. 2. Noise level of this detector proved to be about -60 dB lower than that from a 3 mm dia transverse hole, so that we were able to catch tip echoes easily without noise as shown in Fig. 2. A,B, and C in the upper left graph in Fig. 2 are the main echoes: A is the first tip echo generated when the incident wave IT passes the crack tip, B is the corner echo, and C is the second tip echo caused by RT1 in Fig. 3-a when it passes the crack tip. Each echo corresponds to the reflected waves A,B, and C in Fig. 2. Preceding these main echoes we observe several echoes in Fig. 3-b, which are the mode converted longitudinal waves RL in Fig. 3-b. When the sensitivity is increased, about 25 dB higher than that of the main echoes in Fig. 2, main echoes A,B, and C are out of the upper right figure of Fig. 2, and several tip echoes D,E,and F having almost equal distances follow the main echo. These echoes correspond to D,E, and F in Fig. 2. Details of these tip echoes will be explained in the next section. Thus, these tip echoes are found clearly using full digital flaw-detector DT-2000. As shown in the vector diagram Fig. 2, crack height is 1.5 mm, the inclination angle of the crack is 30 degrees, and the crack width is 0.2 mm. The thickness of the test piece is 20 mm, and the incident angle of 45 degrees is denoted by an arrow in Fig. 2. Thus, distinction of each echo in a scanning graph, which will appear in experiments, is possible using numerical experiments. Vector diagram and scanning graph similar to those of Fig. 2, for the reflection by a 40 degree inclined crack, are shown in Fig. 3. A, B, C, D, and E correspond to those in Fig. 2, and F is the corner wave which is generated as the tip wave E passes the right corner of the inclined crack. Because of the page restriction, similar results for the incident 10°and 20° angles are excluded, and only the results will be described in another section.

### Detailed Explanation of Reflection

The first reflection is made on the crack tip, and generates tip wave A, having propagating center at the crack tip, when the incident transverse wave IT passes the tip. Incident wave IT is reflected on the free surface, and generates RT1 as shown in Fig. 3-a. After passing the corner, RT1 travels along the inclined crack right surface, and generates tip wave C as it passes the crack tip as shown in Fig. 3-b. Corner echo B is generated as IT and RT1 passes the corner, and comes back to the incident direction shown by an arrow. Tip wave C goes around the left side of the crack, after reaching the left corner, then comes back to the tip and generates the tip wave E in Figs. 3-c. Tip wave C is also reflected on the bottom surface, and generates tip wave D shown in Fig.3-c. Another corner wave $\bar{D}$, generated when the tip wave C passes the right corner, is also observed in Fig. 3-c. This wave $\bar{D}$ is similar wave to D, since it is a reflection of tip wave C on a free surface. The mode converted longitudinal wave RL and the head wave H is observed in Fig. 3-c, and the head wave H becomes the main echo B. Several echoes F and etc. are observed in Fig. 2, found as the tip or corner echoes, as the reflected waves go back and forth from the corner to the crack tip.

## Theoretical Estimation of Crack Size

In the 30 degree inclination crack shown in Figs.1 and 2, we can estimate crack inclination angle θ, crack height L, and crack width d shown in Fig. 4, A rough estimation of crack length is $\overline{ab}$ = ( L / cos θ ). Incident wave IT of 45 degree incidence propagates to the corner after passing $\bar{c}$ in Fig. 4, is reflected back to $\bar{c}$, and becomes a corner reflected wave B in Fig. 2. The length $\overline{ac}$ is given approximately by $\overline{ab}$ sin(45° - θ). The distance of AB in Fig. 2 is given by $\overline{ab}$ sin(45°- θ )=(1.5/cos θ ) sin(45°- θ )= 1.732 sin15°= 0.448 mm. Reflected wave RT1 passes $\bar{a}$ at the same time as IT passes $\bar{a}$, propagates along the right side of crack, goes around the round shape crack tip, and generates tip wave C, The distance of BC in Fig. 2 is given by BC = $\overline{ab}$(1- sin15°) = 1.2838 mm, and a half of this value becomes 0.6419 mm for the distance BC in the scanning graph in Fig. 2, After generating the tip wave C, it go around the crack left side to the left corner, and come back to the tip, generating the tip wave E in Fig. 2. Therefore, the distance CE in Fig. 2 is given approximately by twice the length $\overline{kh}$ + $\overline{gh}$ in Fig. 4. Here $\overline{kh}$ = ($\overline{ag}$ - 0.2tan θ ) = 3.233 mm, and $\overline{gh}$ is the circular tip correction length, namely ( $\pi$ (135°+ θ )/360°)d. Then, the total length is 3.233+1.1187d = 4.3517 mm, it means 2.176 mm in the scanning graph in Fig. 2, The length CD is estimated by graphical method, and is given by 1.25mm for the middle point of D$\bar{D}$.

## Comparison of Theoretical and Numerical Values with Experiments

The distances between these echo peaks are tabulated in the Table 1 to 4. For the 30° inclination crack, numerical values ΛB,BC,CD,CE, and EF are measured from the length of each point in Fig. 2, and they are 0.36 mm, 0.66 mm, 1.25 mm, 2.05 mm, and 0.91 mm. Experimental values of these values are AB = 0.367 mm (S.D.= 0.117 mm), BC = 0.658 mm(S.D.= 0.118 mm), CD = 1.24 mm (S.D.= 0.124 mm), CE = 1.945 mm(S.D.= 0.115 mm), and EF = 0.98 mm(S.D.= 0.124 mm), where S.D. is the standard deviation of experimental values. As observed in Fig. 5, experimental values have distribution from the mean values for BC. This result is summarized as BC = 0.658 mm(S.D.= 0.118 mm). Other experimental values have similar deviations, and are summarized as shown. Table 1 shows the comparison of theoretical, numerical, and experimental values for a 30° inclination crack.

Table 1 : Comparison of theoretical, numerical, and experimental
          values for 30° inclination crack(mm).

|            | AB | BC | CD | CE | EF |
|------------|------|------|------|------|------|
| Theory     | 0.448 | 0.642 | 1.25 | 1.778(1.933) | |
| Numerical  | 0.358 | 0.665 | 1.25 | 2.050 | 0.912 |
| Experiment | 0.367(0.117) | 0.658(0.118) | 1.24(0.124) | 1.945(0.12) | 0.98 |

Table 2,3 and 4 is for the 10° ,20° ,and 40° inclined cracks.

Table 2 : Comparison of theoretical, numerical, and experimental
          values for 10° inclination crack(mm).

|            | AB | BC | CD | CE | EF |
|------------|------|------|------|------|------|
| Theory     | 0.874 | 0.325 | 1.27 | 1.688(1.835) | |
| Numerical  | 0.756 | 0.353 | 1.00 | 1.566 | 0.974 |
| Experiment | 0.771(0.007) | 0.314(0.064) | 0.925 | 1.543(0.073) | 0.97 |

Table 3 : Comparison of theoretical, numerical, and experimental
          values for 20° inclination crack(mm).

|            | AB | BC | CD | CE | EF |
|------------|------|------|------|------|------|
| Theory     | 0.675 | 0.461 | 1.25 | 1.707(1.856) | |
| Numerical  | 0.504 | 0.462 | 1.25 | 1.796 | 0.974 |
| Experiment | 0.585(0.095) | 0.418(0.072) | 1.24 | 1.548(0.06) | 0.78 |

Table 4 : Comparison of theoretical, numerical, and experimental
          values for 40° inclination crack(mm).

|            | AB | BC | CD | CE | EF |
|------------|------|------|------|------|------|
| Theory     | 0.171 | 0.894 | 1.25 | 1.922(2.089) | |
| Numerical  | 0.170 | 0.873 | 1.25 | 2.01 | 1.138 |
| Experiment | 0.300(0.049) | 0.822(0.17) | 1.59(0.057) | 2.058(0.12) | 0.90 |

## Estimation of Crack Size Using Experimental Values

Since  the length $2\bar{a}\bar{c}$ is denoted as $\bar{a}\bar{c}$ in Fig.2 (in the scale of
flaw  detector), the length in  Fig. 2 should be  multiplied by 2
for real length.  Therefore,  the  theoretical  estimation gives
$AB = 2(L/\cos\theta)\sin(45°-\theta) = 2\bar{a}\bar{c}$,  but $AB = \bar{a}\bar{c}$ in the length  of
Fig. 1-a, and $2BC = 2(L/\cos\theta)[1-\sin(45°-\theta)] = 2(L/\cos\theta) - AB$.
Therefore,  $2BC + AB = 2(L/\cos\theta) = 2ab$, but $\bar{ab}$  in  the  scale
of DT-2000 in Fig. 2 and Table 1, that means  $2BC + AB = ab$, and
$\bar{a}\bar{c} = \bar{ab}\sin(45°-\theta)$.   Approximately,  two values  $\bar{ab}$ and $\bar{a}\bar{c}$ are
obtained from these equations.  On  the  other  hand  the  crack
width  d  is  estimated  from  $CE = 2[\bar{ab}-d\tan\theta] + \bar{gh}$, $CE-\bar{ab} =$
$(\pi(135°+\theta)/360°- \tan\theta)d$, where   CE and $\bar{ab} = 2BC + AB$  have
already been obtained, and we can obtain d from these equations.
Thus,  the experimentally estimated value $L, \theta$, and d are given
as  follows  for  10° ,20° ,  30° ,  and 40° in  Table  5.  These
estimated values will be helpful for theoretical evaluation.

Table 5 : Estimated crack sizes

| Inclination | 10° | 20° | 30° | 40° | |
|-------------|-------|-------|-------|-------|--------|
| L(1.5 mm)   | 1.389 | 1.365 | 1.499 | 1.573 | mm |
| θ           |       | 7.8   | 18.0  | 31.0  | 35.1 degree |
| d(0.2 mm)   | 0.129 | 0.126 | 0.261 | 0.159 | mm |

REFLECTION FOR Y1 DIRECTION
Figure 6 shows results similar to those in Fig. 2: the
correspondence of several echoes in the scanning graph of DT-
2000 to the reflected waves in the vectors diagram of 30°
inclination. A is the tip echo from the upper crack tip, C is
the corner reflection of the main echo, B is the bottom
reflection of the tip wave A. These waves are observed as three
pulses in the left part in the scanning graph shown at upper
left figure in Fig. 6. D is the bottom reflection of C, and E is
the bottom reflection of C having the propagation center at the
right corner of the crack.

## Comparison of Experiments with Numerical Experiments

As explained for the Y2 direction we used a DT-2000 with a
5C10x10A45 wide-band probe for the Y1 direction. Inclination
angles are given for every 10 degree between 10° and 40° and the
incident angle is 45 degrees. The results of 10° inclination
are explained for Y1 in Table 6, the comparison of theoretical,
numerical, and experimental values for 20° , 30° , and 40°
inclination are given in Tables 7, 8, and 9.

## Comparison of Theoretical and Numerical Values with Experiments

Distances of these echoes are illustrated in Tables 6 to 10.

Table 6 : Comparison of theoretical, numerical, and experimental
values for 10° inclined crack.

|  | AB | BC | CD | CE |
|---|---|---|---|---|
| Theory | 1.06 | 0.16 | | |
| Numerical | 0.88 | 0.366 | 1.354 | 2.225 |
| Experiment | 0.976 | 0.298 | 1.600 | 2.73 |
|  | (0.14) | (0.057) | (0.163) | (0.287) |

Table 7 : Comparison of theoretical, numerical, and experimental
values for 20° inclined crack.

|  | AB | BC | CD | CE |
|---|---|---|---|---|
| Theory | 1.06 | 0.332 | | |
| Numerical | 1.01 | 0.52 | 1.354 | 2.19 |
| Experiment | 0.99 | 0.47 | 1.66 | |
|  | (0.15) | (0.046) | (0.019) | |

Table 8 : Comparison of theoretical, numerical, and experimental
values for 30° inclined crack.

|  | AB | BC | CD | CE |
|---|---|---|---|---|
| Theory | 1.06 | 0.47 | 1.06 | |
| Numerical | 1.04 | 0.69 | 0.99 | 1.98 |
| Experiment | 0.90 | 0.83 | 1.05 | |
|  | (0.045) | (0.102) | (0.185) | |

438

Table 9 : Comparison of theoretical, numerical, and experimental
          values for 40° inclined crack.

|            | AB     | BC      | CD      | CE     |
|------------|--------|---------|---------|--------|
| Theory     | 1.06   | 0.64    |         |        |
| Numerical  | 1.08   | 0.65    | 0.41    | 1.15   |
| Experiment | 0.88   | 0.78    | 0.60    | 1.06   |
|            | (0.04) | (0.063) | (0.105) | (0.05) |

## Estimation of Crack Size Using Experimental Results

We have no time to examine this case in detail as was done for
Y-2 direction, and we have made a rather rough estimation. A
multiple reflection between the bottom plane and the left side
of crack is observed and BC becomes long and complicated.
However, from simple geometric considerations, the crack
height L is given approximately by $L = (2)^{1/2} \cdot AB$; therefore the
theoretical value AB is 1.06 as shown in each table. The
experimental value are as follows, after small corrections for
inclination and tip circular effects, and after averaging with L
in Table 5 gives A.V.L.

Table 10 :

| Inclination     | 10°  | 20°  | 30°  | 40°  |
|-----------------|------|------|------|------|
| Experimental  L | 1.41 | 1.44 | 1.32 | 1.24 |
| A. V. L.        | 1.40 | 1.40 | 1.41 | 1.41 |

Combining Tables 5 and 10 gives the estimated crack size as

Table 11 : Estimated crack sizes

| Inclination | 10°   | 20°   | 30°   | 40°   |
|-------------|-------|-------|-------|-------|
| L(1.5mm)    | 1.399 | 1.40  | 1.41  | 1.405 |
| θ           | 7.8   | 18.0  | 31.0  | 35.1  |
| d(0.2mm)    | 0.376 | 0.188 | 0.170 | 0.183 |

These values are acceptable, and will be able to improve
further using other values.

## Experiments  for Fatigue Crack on a Free Surface

Several experiments have been executed for the reflections by a
fatigue crack on a free surface using DT-2000.  The tip echoes
and the corner echoes were found easily than by the conventional
method as shown in Fig. 7.  Therefore, we expect the multi-tip
wave (MTW) method will be applied to the fatigue crack on a free
surface.  Especially, it will be effective for a small crack on
a free surface, since the shape of the small fatinue crack will
be uncomplicated. Moreover, it will be rather easy to use higher
frequencies; we are using 15 MH in the detection by DT-2000.
This means it will be easy to measure the small crack hight of
about 0.5 mm or less size.

CONCLUSION
We have shown that numerical experiments using the one wave
model shows very good agreement with experiment, if we use wide
band pulse, and the estimated crack size is proved good
agreement with true values. Therefore, we shall be able to use
numerical experiments as a powerful tool in distinguishing many

tip waves, and therefore as a tool of " Ultrasonics Defect Sizing ". Until now the detection of tip waves is considered rather difficult. However, the difficulties have been greatly reduced using the low noise full-digital flaw detector and numerical experiment. Therefore any signal in a scanning graph of full digital flaw detector is meaningful, and we shall be able to obuain more information than the conventional method. Recently, X. M. Wu and D. K. Mak [3] extended the multi-tip-method, and obtained good result for the inversion of inclined crack on a free surface up to the 40 degree inclination angle using two angle probes. We hope this method will be a new practical and ecconomical way for the inversion of elastic waves in a solid.

ACKNOWLEDGEMENT
The computation of this work was done by the use of HITAC S820-80 in Tokyo University.

References

1) K. Harumi, M. Uchida, S. Ogura, T. Miyajima. : Motion Picture of Tip Waves, Proc. 7th Asian Pacif. Conf. NDT(1990) pp.18.1-18.7
2) 210 and 202 sub - committee. : Ultrasonic Defect Sizing, Translated by M. Moles of reference 4, ASNT, 1991.
3) X. M. Wu and D. K. Mak. : A New Method of Sizing Surface-Breaking Inclined Cracks Using Multi-Tip-Waves, Metals Technology Laboratories in Canada Report MTL91-58(J),1991.
4) 210 and 202 sub-committee WG, : Tip Echo Handbook, Japanese Soc. NDI edited by Harumi, 1987, ( in Japanese )

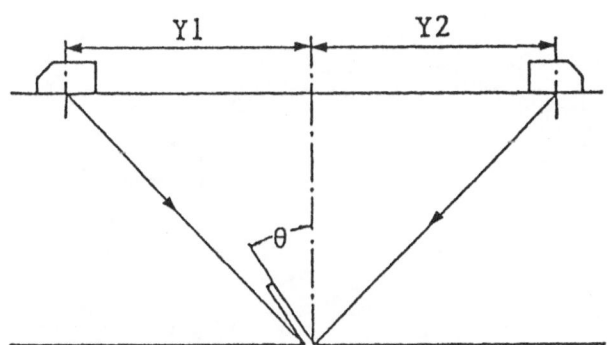

Fig. 1   Configuration of an inclined crack and transducer.

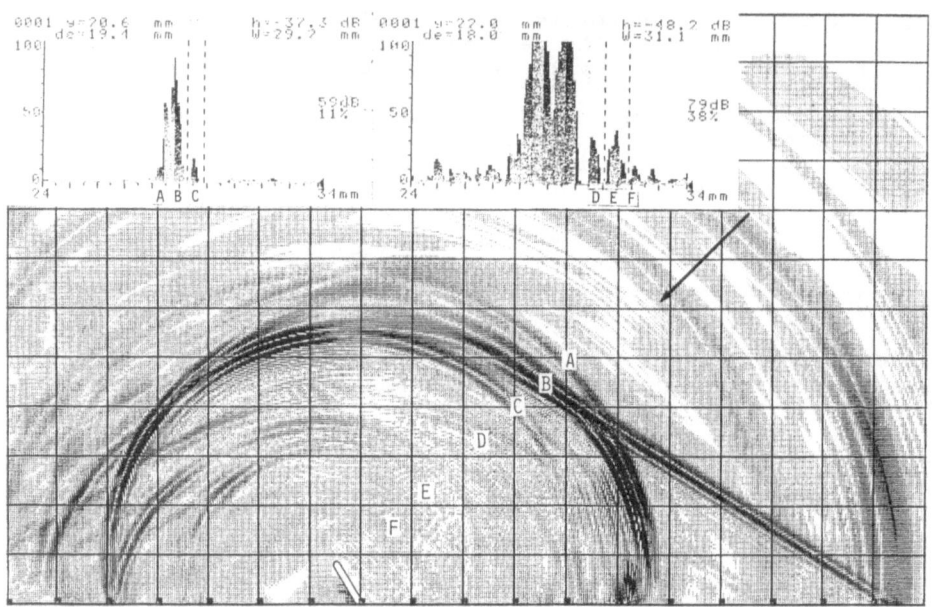

Fig. 2    Vectors diagram of reflection by a 30° inclined
          crack, and its scanning graph of DT-2000.( Y2 )

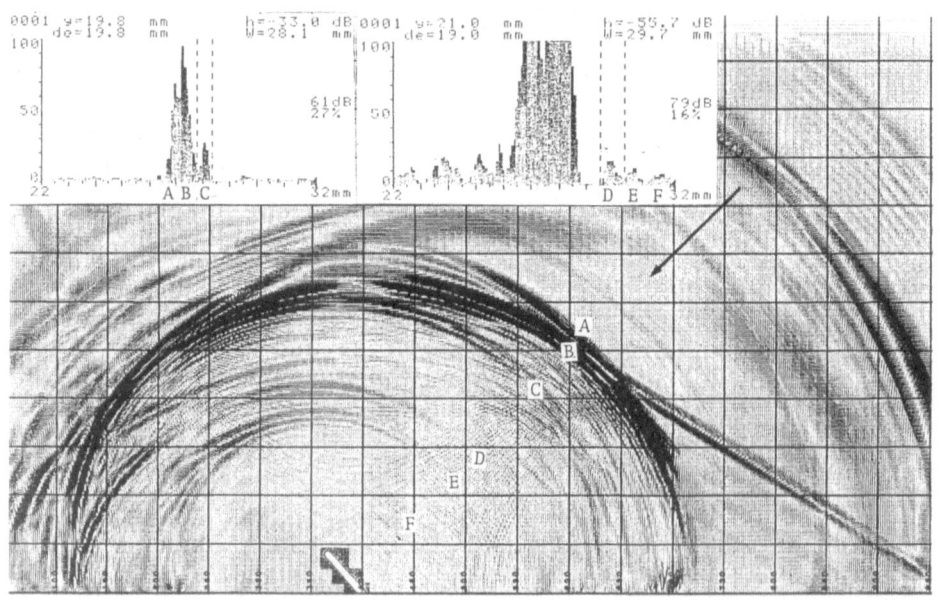

Fig. 3    Vectors diagram of reflection by a 40° inclined
          crack, and its scanning graph of DT-2000.( Y2 )

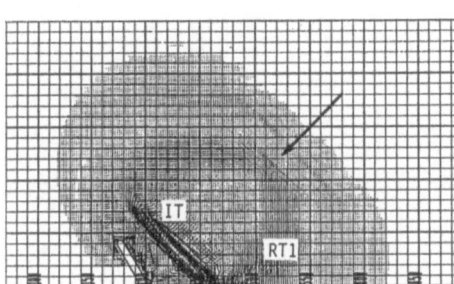

Fig. 3-a  Vectors diagram of
incident wave IT,
and reflected wave RT1.

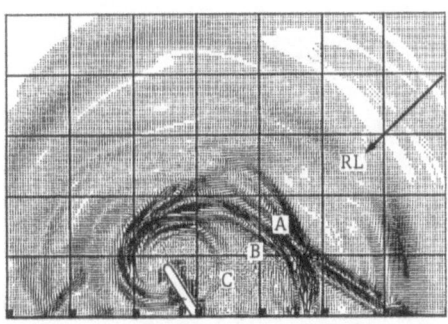

Fig. 3-b Main echo B, and tip
echoes A, C, and D. Mode
converted logitudinal wave RL.

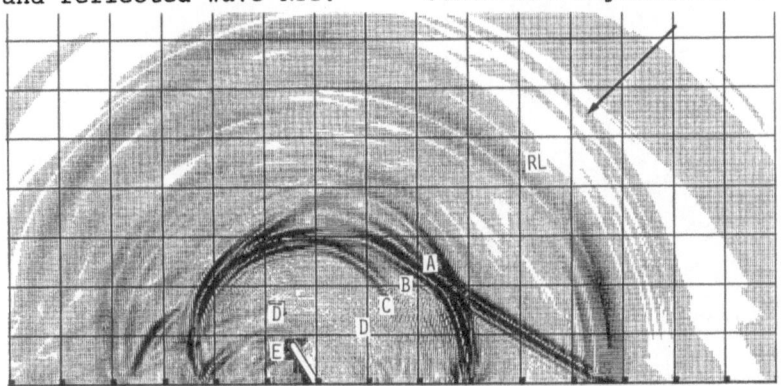

Fig. 3-c  Tip echoes A, C, E and main echo B. Reflected wave
of C on a free surface; right side wave D̄, left side wave D.

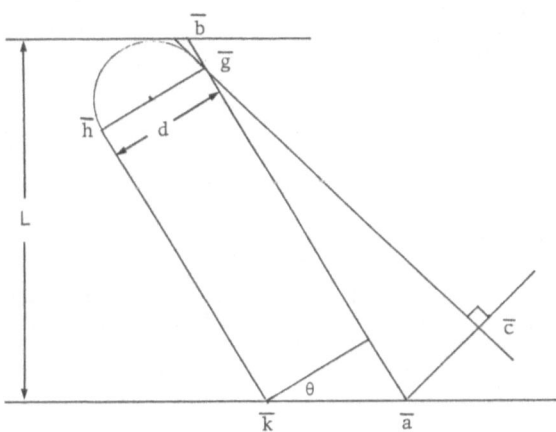

Fig. 4   Schematic figure of crack. L, crack height;
θ , crack inclination angle; d, crack width.

442

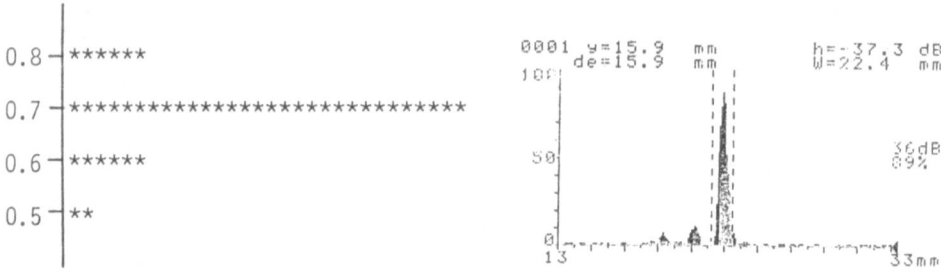

Fig. 5  Experimental distribution
of length BC.

Fig. 7  Scanning graph of a
fatigue crack by DT-2000.

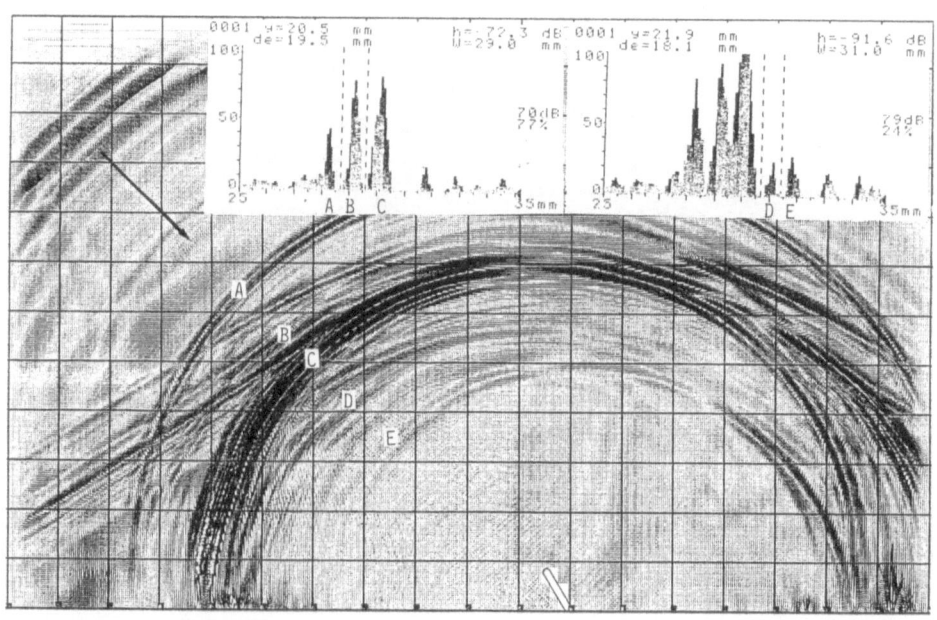

Fig. 6  Vectors diagram of reflection by a 30° inclined crack,
and its scanning graph of DT-2000.( Y1 )

# Surface Reconstruction of a Three-Dimensional Ultrasonic Flaw

Lat S. Koo
Idaho National Engineering Laboratory
EG&G Idaho, Inc.
P.O. Box 1625
Idaho Falls, ID 83415-2209
U.S.A.

Summary
In three-dimensional inverse scattering problems, the reconstruction of a solid scatterer is often difficult, if not impossible, and computationally expensive due to the dimensionality. To obtain only the geometrical information, a surface reconstruction algorithm is naturally more desirable since no additional knowledge can be gained from doing the solid reconstruction and the computation is reduced to two dimensions. With the application of the first Born approximation, this paper proposes a simple surface reconstruction technique for a three-dimensional target. In general, this method is ill-posed. However, the numerical instability part of the ill-posedness is removable when the surface has a two-fold symmetry with respect to a plane. To demonstrate this approach, three analytical examples are shown.

Introduction

Reconstruction of the geometry of an embedded flaw has been one of the major areas in nondestructive evaluation (NDE) research. Some of the algorithms published in the last decade were based on the elastic wave inverse (first) Born approximation, which is essentially a low frequency approximation to an isolated weak scatterer in an otherwise isotropic and homogeneous medium whose material properties are similar to the host's [1,2]. A key factor in the development of these inverse algorithms is the spatial-frequency function "shape factor" that appears in the formulation of the inverse Born approximation. This shape factor embodies all information about the size, shape, and orientation of the scatterer and, therefore, one can develop inverse procedures to extract the geometry in terms of a characteristic function (which has the value 1 inside the scatterer and 0 outside). The direct approach to recovering this characteristic function requires three-dimensional information in the spatial-frequency domain [3], which is computationally expensive. However, the basic equation of the general three-dimensional inverse Born approximation can be further cast to form a set of equations for the surface of a three-dimensional flaw. As a consequence, one can develop a "surface reconstruction" algorithm based upon these equations. Since the surface of a three-dimensional flaw is two-dimensional, the algorithm requires only knowledge of a two-dimensional hyper-plane of the shape factor instead of a complete three-dimensional domain, which reduces the dimensionality by one. The present algorithm can be reduced to a special case that was suggested by R. M. Lewis [4]. For flaws having very dissimilar material properties (strong scatterers), the aforementioned methods may still be applicable since studies have shown that the inverse Born approximation is useful not only in sizing weak scatterers but also strong ones [5]. This was recently shown when Chen [6] applied the elastodynamic ray theory to an arbitrary void in an isotropic and homogeneous medium and found that the inverse Born approximation accurately predicted both the front surface echo and the constant term in the asymptotic expansion of the impulse response as long as the host material has a Poisson ratio of 1/3. Therefore, for a host material having a Poisson ratio about 1/3 (this is true for most structural materials), the early-arriving portion of the response signal is expected to agree with the Born prediction.

The next section describes the theory behind the surface reconstruction method (a more detailed discussion of this technique will be reported in a future publication [7]). Then analytical examples of reconstruction of three scatterers are given. After the examples, the reconstruction algorithms for a finite spatial-frequency bandwidth are given.

Theory
Based on the inverse (first) Born approximation, for a pulse-echo test with a longitudinal impulsive incident wave, the longitudinal scattering amplitude A can be written as [1,5]

$$A = \frac{k^2}{4\pi} B \ S(\mathbf{q}) \tag{1}$$

where k is the longitudinal wave number and B, which is dependent upon material properties of both the flaw and the host, is

$$B = \frac{\delta\rho}{\rho} + \frac{\delta\lambda + 2\delta\mu}{\lambda + 2\mu} \tag{2}$$

where $\lambda$, $\mu$, and $\rho$ are respectively the Lamé elastic constants and the mass density of the host and $\delta\lambda = \lambda_{flaw} - \lambda$, etc. In this paper, a bold-typed letter denotes a vector (e.g. $\mathbf{q}$). In (1), q = $|\mathbf{q}|$ = 2k where $|\mathbf{q}|$ denotes the magnitude of $\mathbf{q}$. $S(\mathbf{q})$ is the shape factor, which embodies all the information about the size, shape, and orientation of the scatterer and forms a Fourier transform pair with the three-dimensional characteristic function $\gamma(\mathbf{x})$ (which has the value 1 inside the scatterer and 0 outside) [5]

$$S(\mathbf{q}) = \int_{-\infty}^{\infty} \int_{-\infty}^{\infty} \int_{-\infty}^{\infty} dV(\mathbf{x}) \ \gamma(\mathbf{x}) \ e^{i\mathbf{q}\cdot\mathbf{x}}$$

$$= \int_{-\infty}^{\infty} \int_{-\infty}^{\infty} dxdy \ e^{i(q_x x + q_y y)} \int_{-\infty}^{\infty} dz \ \gamma(\mathbf{x}) \ e^{iq_z z}. \tag{3}$$

In (3), $dV(\mathbf{x})$ denotes the differential volumetric element with the position vector $\mathbf{x} = (x,y,z)$ in the infinite spatial domain. From (1), $S(\mathbf{q})$ can be expressed as

$$S(\mathbf{q}) = C\frac{A}{k^2}. \tag{4}$$

Here the proportionality $C = 4\pi/B$ is a function of material properties only. Once $S(\mathbf{q})$ is known, $\gamma(\mathbf{x})$ can be obtained by applying the inverse-Fourier transformation to (3). In (3), when one isolates $\gamma(\mathbf{x})$ along the z-direction, then

$$S(\mathbf{q}) = \int_{-\infty}^{\infty} \int_{-\infty}^{\infty} dxdy\, e^{i(q_x x + q_y y)}\, g(u,v;q_z) \tag{5}$$

where $u(x,y)$ and $v(x,y)$ are the upper and lower surfaces, respectively, of the flaw along the z-axis in the spatial domain (Fig. 1). Obviously, u and v depend on which plane is chosen as $z =$ constant. For convenience, sometimes the coordinate is designated such that $z = 0$ cuts through the centroid of the flaw and separates u and v. These two surface functions are assumed to be *single-valued*. In (5), the unknown $g(u,v;q_z)$ is a nonlinear function of u and v with a constant $q_z$

$$g(u,v;q_z) = \int_{v(x,y)}^{u(x,y)} dz\, e^{iq_z z}. \tag{6}$$

Notice that $q_z =$ constant is the only hyper-plane in the shape factor domain (**q**) we will extract information from. In [4], $q_z = 0$ was used, here $q_z \neq 0$ is assumed. If $q_z = 0$, $g(u,v;0)$ becomes the thickness function of the scatterer along the z-direction and one must seek a second equation in order to resolve u and v. To acquire the second equation, Lewis [4] differentiated (8) with respect to $q_z$ and then set $q_z = 0$ while the present proposed technique uses a plane $q_z \neq 0$ and resolves both u and v from (8) alone without any differentiation. In the following discussion, the numerical instability is shown to be related to the nonsymmetry of the scatterer surface and thus can be eliminated when symmetry is present. A two-dimensional version of the above nonlinear Fredholm integral equation of the first kind (5) was considered by both Kogan and Lopes [8] and Vogel [9]. While Kogan and Lopes used a guess-and-iterate scheme to calculate u and v for a double-rectangular box with simulated geometric scattering data, Vogel applied a nonlinear optimization scheme similar to the Levenberg-Marquardt algorithm [9] to recover u and v for a peanut-nut-shaped flaw with simulated scattering data. In the present paper, we will demonstrate a much simpler algorithm derived from (5) and (6) for a three-dimensional target. When u and v are symmetrical to each other, the proposed algorithm becomes numerically stable. Since $g(u,v;q_z)$ forms a two-dimensional Fourier transform pair with $S(\mathbf{q})$, $g(u,v;q_z)$ can be also written as

446

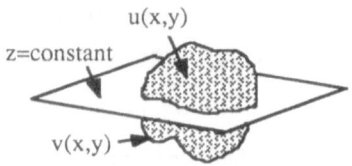

Figure 1. Defining the upper, u(x,y), and lower, v(x,y), surfaces.

$$g(u,v;q_z) = \frac{1}{(2\pi)^2} \int_{-\infty}^{\infty} \int_{-\infty}^{\infty} dq_x dq_y \, S(q) \, e^{-i(q_x x + q_y y)}.$$
(7)

Notice that the right hand side of (7) is a measurable quantity. For convenience, this quantity will be rendered as $d(x,y;q_z)$ hereafter. That is,

$$g(u,v;q_z) = d(x,y;q_z)$$
(8)

where u and v are functions of x and y. By further evaluating the integral in (6), we find

$$u = \frac{1}{q_z}[\tan^{-1}\frac{Im(d)}{Re(d)} + \sin^{-1}\frac{q_z|d|}{2} + N_u\pi]$$
(9a)

$$v = \frac{1}{q_z}[\tan^{-1}\frac{Im(d)}{Re(d)} - \sin^{-1}\frac{q_z|d|}{2} + N_v\pi]$$
(9b)

where $Re(d)$ and $Im(d)$ denote the real and imaginary parts of $d(x,y;q_z)$, respectively, $q_z \neq 0$, and $N_u$, $N_v = 0, \pm1, \pm2, \pm3, ....$ Contrary to the original surfaces, the above u and v, seemingly, are functions not only of x and y, but also of $q_z$ and the integers $N_u$ and $N_v$. However, from the three examples which will be shown in the following section, the $q_z$ term is cancelled by the first two terms in both (9a) and (9b) such that it can be arbitrary and the integers, $N_u$ and $N_v$, are always zero. The ill-posedness in (6) is clearly demonstrated in the first terms in (9). However, when u and v are symmetrical to each other with respect to z = constant, these terms become zero. That is, if u = -v, then $Im(d) = 0$ and (9) becomes a numerically stable algorithm

$$u = \frac{1}{q_z}[\sin^{-1}\frac{q_z|d|}{2} + N\pi]$$
(10a)

$$v = -u$$
(10b)

where $N = 0, \pm1, \pm2, \pm3, ....$

Analytical examples

For any finite right cylinder with the major axis aligned along the z-direction, its three-dimensional characteristic function $\gamma(x)$ becomes separable as

$$\gamma(x) = \gamma(z) \, \gamma(x,y).$$
(11)

Substituting (11) into (3) and then applying the result into (7), the measurable quantity $d(x,y;q_z)$ becomes

$$d(x,y;q_z) = \int_{-\infty}^{\infty} \gamma(z) \, e^{-iq_z z} \, dz \, \gamma(x,y)$$

$$= \frac{2\sin(q_z a_z)}{q_z} \, e^{iq_z(1-p_z)a_z} \, \gamma(x,y) \qquad (12)$$

where $-p_z a_z$ is the distance between the bottom surface $v(x,y)$ and the origin with $2a_z$ being the axial length. As a consequence of (12) and (9) together with $N_u = N_v = 0$, the top and bottom surfaces can be recovered exactly as

$$u(x,y) = (2 - p_z) \, a_z \, \gamma(x,y) \qquad (13a)$$
$$v(x,y) = -p_z \, a_z \, \gamma(x,y). \qquad (13b)$$

The expression of the two-dimensional characteristic function $\gamma(x,y)$ depends upon not only the coordinates on the x-y plane but also on the location of the origin, which can be anywhere inside or outside the scatterer. Examples of a rectangular parallelepiped (Fig. 2a) and a finite right circular cylinder (Fig. 2b) given below will verify the above results (13). In addition, equations (10) will be tested on a third scatterer: a spheroid (Fig. 2c). Only results are given here and all the details of the calculations will be reported in [7].

1. Rectangular Parallelepiped

The semi-lengths of this target are assumed to be $(a_x, a_y, a_z)$ along the Cartesian (x, y, z) directions respectively (Fig. 2a). The origin of the coordinate is assumed to be $(p_x a_x, p_y a_y, p_z a_z)$ away from the sides on the negative (x, y, z) axes respectively. Equivalently, the origin is $((2-p_x)a_x, (2-p_y)a_y, (2-p_z)a_z)$ away from the sides on the positive (x, y, z) axes. The ratios $p_x$, $p_y$, and $p_z$ are all real numbers and $0 \le p_x, p_y, p_z \le 2$. Correspondingly, the shape factor (3) becomes

$$S(q) = \frac{8}{q_x q_y q_z} \sin(q_x a_x) \sin(q_y a_y) \sin(q_z a_z) \, e^{iq_x(1-p_x)a_x} \, e^{iq_y(1-p_y)a_y} \, e^{iq_z(1-p_z)a_z}. \qquad (14)$$

Substituting (14) into (8), d is

$$d(x,y;q_z) = \frac{2\sin(q_z a_z)}{q_z} \, e^{iq_z(1-p_z)a_z} \cdot [H(x + p_x a_x) - H(x - (2-p_x)a_x)]$$
$$\cdot [H(y + p_y a_y) - H(y - (2-p_y)a_y)] \qquad (15)$$

where H is the Heaviside unit function. Consequently, from (9) with $N_u = N_v = 0$,

$$u(x,y) = (2-p_z)a_z \, [H(x + p_x a_x) - H(x - (2-p_x)a_x)]$$
$$\cdot [H(y + p_y a_y) - H(y - (2-p_y)a_y)] \qquad (16a)$$
$$v(x,y) = -p_z a_z \, [H(x + p_x a_x) - H(x - (2-p_x)a_x)]$$
$$\cdot [H(y + p_y a_y) - H(y - (2-p_y)a_y)] \qquad (16b)$$

which are exactly the upper and lower surfaces of the rectangular parallelepiped.

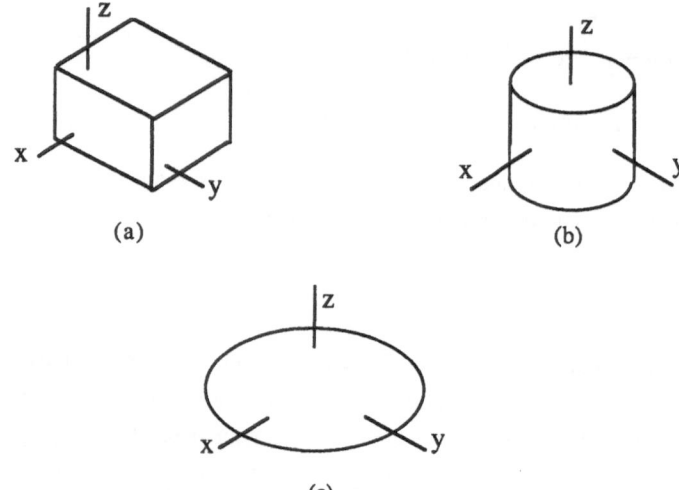

Figure 2. Three scatterers: (a) a rectangular parallelepiped, (b) a finite right circular cylinder, and (c) a spheroid with circular symmetry on the x-y plane.

## 2. Finite Right Circular Cylinder

The major axis of the cylinder is aligned with the z-axis and the origin is placed $-p_z a_z$ away from the bottom surface (Fig. 2b). Its radius and semi-length are assumed to be $(a_x, a_z)$ respectively. Correspondingly, the shape factor (3) becomes

$$S(q) = \frac{4\pi a_x \sin(q_z a_z) J_1(b a_x)}{b q_z} e^{i q_z(1-p_z)a_z} \tag{17}$$

where $b = \sqrt{(q_x)^2+(q_y)^2}$ and $J_1$ is the Bessel function of the first kind and first order. Substituting (17) into (8), d is

$$d(x,y;q_z) = \frac{2\sin(q_z a_z)}{q_z} e^{i q_z(1-p_z)a_z} H(a_x - r) \tag{18}$$

where r is the radial length on the x-y plane. Consequently, from (9) with $N_u = N_v = 0$,

$$u(x,y) = (2-p_z)a_z\, H(a_x - r) \tag{19a}$$
$$v(x,y) = -p_z a_z\, H(a_x - r) \tag{19b}$$

which are exactly the upper and lower surfaces of the right circular cylinder.

## 3. Spheroid

The spheroid is centered at the origin with its semi-axes $(a_x, a_x, a_z)$ along the (x, y, z) directions respectively (Fig. 2c). Correspondingly, the shape factor (3) becomes

$$S(q) = 4\pi a_x a_x a_z \frac{\sin(K) - K\cos(K)}{K^3} \tag{20}$$

where $K = \sqrt{(a_x q_x)^2 + (a_x q_y)^2 + (a_z q_z)^2}$ . Substituting (20) into (8), d is

$$d(x,y;q_z) = \frac{2}{q_z} \sin(q_z a_z \sqrt{1 - (\frac{r}{a_x})^2}) H(a_x - r) \qquad (21)$$

where r is the radial length on the x-y plane. Consequently, from (10a) with $N = 0$,

$$u(x,y) = a_z \sqrt{1 - (\frac{r}{a_x})^2} H(a_x - r) \qquad (22)$$

which is exactly the upper surface of the spheroid and the lower surface can be obtained by using (10b).

## Theory for Finite Spatial-Frequency Bandwidth

The bandlimited frequency domain (the three-dimensional shape factor $q$-space) is considered to be a spherical shell with $q_a$ and $q_b$ as the inside and outside radii respectively. In general, the interception between this spherical shell and the infinite hyper-plane $q_z$ = constant, D(q), can be either an annular disk, or a circular disk depending on whether $q_z < q_a$, or $q_a \leq q_z \leq q_b$. In this paper, D(q) is chosen to be at $q_z = q_a$ [7] where the circular disk has the largest size, as is shown in Fig. 3. With such a choice, D(q) becomes

$$D(q) = \{q \mid [q_z = q_a] \cap [(q_x)^2 + (q_y)^2 \leq (q_D)^2]\} \qquad (23)$$

where $q_D = \sqrt{q_b^2 - q_a^2}$ is the radius of D(q). The window function K(q) for D(q) is

$$K(q) = \delta(q_z - q_a) H(q_D - \sqrt{q_x^2 + q_y^2}) \qquad (24)$$

where $\delta(q_z - q_a)$ is the Dirac delta function. The corresponding Fourier transform, k(x), of K(q) is

$$k(x) = \frac{1}{(2\pi)^3} \int_{-\infty}^{\infty} \int_{-\infty}^{\infty} \int_{-\infty}^{\infty} K(q) e^{-iq \cdot x} dq$$

$$\approx \frac{1}{2\pi} \delta(x)\delta(y) e^{-iq_a z} \qquad (25)$$

where $q_b$ is assumed to be sufficiently large [7]. Applying this pair of window functions, k(x) and K(q), to the Fourier transform pair, $\gamma(x)$ and S(q) from Eq. (3), one has

$$g(u,v;q_a) = d_D(x,y;q_a) \qquad (26)$$

where the measurable set $d_D(x,y;q_a)$ is obtained from the disk D(q)

$$d_D(x,y;q_a) = \frac{1}{(2\pi)^2} \int \int_{D(q)} S(q_x,q_y;q_a) e^{-i(q_x x + q_y y)} dq_x dq_y. \qquad (27)$$

Since Eq. (27) is similar to (8), the surfaces u and v can be determined by the previous

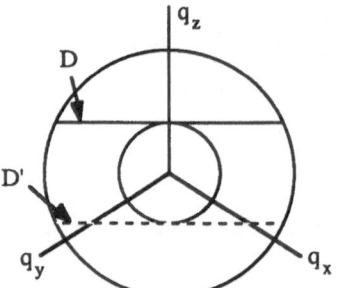

Figure 3. The data disk $D(\mathbf{q})$ and its image $D'(\mathbf{q})$ in the spherical shell whose inside and outside radii are $q_a$ and $q_b$ respectively.

analysis. Thus,

$$u = \frac{1}{q_a}[\tan^{-1}\frac{Im(d_D)}{Re(d_D)} + \sin^{-1}\frac{q_a|d_D|}{2} + N_u\pi] \tag{28a}$$

$$v = \frac{1}{q_a}[\tan^{-1}\frac{Im(d_D)}{Re(d_D)} - \sin^{-1}\frac{q_a|d_D|}{2} + N_v\pi] \tag{28b}$$

where $N_u$, $N_v = 0, \pm 1, \pm 2, \pm 3, ....$ Obviously when $u = -v$, results similar to Eq. (10) can be obtained. To determine the appropriate values of the ambiguous constants $N_u$ and $N_v$, future numerical studies are needed.

Another method of analysis for the finite bandwidth case is to replace the Heaviside function, as is shown in Eq. (24), by a Gaussian function in the window function $K(\mathbf{q})$. In this approach, *no assumption of high frequency approximation is required*. This new window function $K_G(\mathbf{q})$ is given as

$$K_G(\mathbf{q}) = \delta(q_z - q_a)\, e^{-(q_{xy}/(pq_D))^2} \tag{29}$$

where $q_{xy} = \sqrt{q_{\tilde{x}}^2 + q_{\tilde{y}}^2}$ and p is an arbitrary nonzero positive real number. Since in practice a Gaussian function must be truncated at the edge of the disk $D(\mathbf{q})$, the parameter p is thus included to control the amplitude at the truncation. For this new window function, its Fourier transform $k_G(\mathbf{x})$ is [7]

$$k_G(\mathbf{x}) = \frac{1}{(2\pi)^3}\int_{-\infty}^{\infty}\int_{-\infty}^{\infty}\int_{-\infty}^{\infty} K_G(\mathbf{q})\, e^{-i\mathbf{q}\cdot\mathbf{x}}\, d\mathbf{q}$$

$$= \frac{1}{2\pi}\, e^{-iq_a z}\frac{1}{4\pi}\,(pq_D)^2 e^{-(1/4)(r_{xy}pq_D)^2} \tag{30}$$

where $r_{xy} = \sqrt{x^2 + y^2}$. Applying this pair of window functions, $k_G(x)$ and $K_G(q)$, to the Fourier transform pair, $\gamma(x)$ and $S(q)$ from Eq. (3), one has

$$g(u,v;q_a) = d_D^W(x,y;q_a) \tag{31}$$

where the measurable function $d_D^W(x,y;q_a)$ is the two-dimensional inverse Weierstrass transformation [10] of the data set $d_D^G(x,y;q_a)$ [7] and

$$d_D^W(x,y;q_a) = \frac{-1}{4\pi} \int_{\sigma_{y'}-i\infty}^{\sigma_{y'}+i\infty} \int_{\sigma_{x'}-i\infty}^{\sigma_{x'}+i\infty} d_D^G(x',y';q_a)\, e^{[(x'-x)^2 + (y'-y)^2]/4}\, dx'dy' \tag{32}$$

$$d_D^G(x,y;q_a) = \frac{1}{(2\pi)^2} \int_{-\infty}^{\infty} \int_{-\infty}^{\infty} S(q_x,q_y;q_a)\, e^{-(q_{xy}/(pq_D))^2}\, e^{-i(q_x x + q_y y)}\, dq_x dq_y. \tag{33}$$

with $\sigma_{x'}$ and $\sigma_{y'}$ being real variables [10]. Since Eq. (32) is similar to (8), the surfaces u and v can be determined by the previous analysis. Thus,

$$u = \frac{1}{q_a}[\tan^{-1}\frac{\text{Im}(d_D^W)}{\text{Re}(d_D^W)} + \sin^{-1}\frac{q_a|d_D^W|}{2} + N_u\pi] \tag{34a}$$

$$v = \frac{1}{q_a}[\tan^{-1}\frac{\text{Im}(d_D^W)}{\text{Re}(d_D^W)} - \sin^{-1}\frac{q_a|d_D^W|}{2} + N_v\pi] \tag{34b}$$

where $N_u$, $N_v = 0, \pm1, \pm2, \pm3, ....$ Again when $u = -v$, results similar to Eq. (10) can be obtained. To determine the appropriate values of the ambiguous constants $N_u$ and $N_v$, future numerical studies are needed.

Discussion and Conclusion

This paper has proposed a surface reconstruction technique that requires only two-dimensional hyper-plane data in the spatial-frequency domain (q) and thus significantly improves on the traditional three-dimensional approach [3]. For the above studies, whether with infinite or finite frequency spectra, all the algorithms share a similar expression as can be seen in Eqs. (9), (28) and (34). The differences are found in the selection of the hyper-plane $q_z$ = constant and in the data processing of the shape factor set S(q) to acquire the appropriate "data function" $d(x,y;q_z)$, $d_D(x,y;q_a)$, or $d_D^W(x,y;q_a)$. In general, this technique is ill-posed, as is shown in Eqs. (9), (28) and (34), which is inherent with most inverse problems. The numerical instability in this algorithm, however, is identified with the nonsymmetry of the surface of a general target. Hence, whenever a scatterer has a two-fold symmetric surface with respect to the assigned plane, z = constant, the algorithms become stable. The validity of this method has been demonstrated by three analytical examples.

## Acknowledgment

The author is thankful to Dr. K. L. Telschow, Mr. B. A. Barna, Dr. V. G. Kogan and Prof. L. W. Schmerr for their encouragement and discussions of the present work. This work was sponsored by the U.S. Department of Energy, Office of Energy Research, Office of Basic Energy Sciences. The initial work was performed at the Ames Laboratory under DOE Contract No. W-7405-ENG-82; later work was performed at the Idaho National Engineering Laboratory under DOE Idaho Field Office Contract No. DE-AC07-76ID01570.

## References

1. Gubernatis, J.E.; Domany, E.; Krumhansl, J.A.; Hubermann, M.: The Born approximation in the theory of the scattering of elastic waves by flaws, *J. of Appl. Phys.*, **48** (1977) 2812-2819.

2. Hudson, J.A.; Heritage, J.R.: The use of the Born approximation in seismic scattering problems, *Geophys. J. Royal Astron. Soc.*, **66** (1981) 221-240.

3. Koo, L.S.; Wormley, S.J.; Hsu, D.K.; Thompson, D.O.: Fourier method approaches to three-dimensional ultrasonic flaw reconstruction, in *Review of Progress in QNDE*, edited by D.O. Thompson and D.E. Chimenti (Plenum Press, New York), **10A** (1991) 67-74.

4. Lewis, R.M.: Physical optics inverse diffraction, *IEEE Trans. on Ant. and Prop.*, **17** (1969) 308-314.

5. Rose, J.H.: Elastic wave inverse scattering in nondestructive evaluation, *Pure and Appl. Geophys.*, **131** (1989) 715-739.

6. Chen, J.S.: Elastodynamic ray theory and asymptotic methods for direct and inverse scattering problems, Ph.D. dissertation, Iowa State Univ., 1987.

7. Koo, L.S.: Theoretical surface reconstruction of a three-dimensional ultrasonic flaw based on the Born approximation, to appear.

8. Kogan, V.G.; Lopes, E.F.: On the Born approximation for weak uniform scatterers, *Inverse Prob.*, **1** (1985) 331-338.

9. Vogel, C.R.: Numerical solution of a nonlinear ill-posed problem arising in inverse scattering, *Inverse Prob.*, **1** (1985) 393-403.

10. Zemanian, A.H., *Generalized Integral Transformations*. New York: Dover Publications 1987.

# Generalized Ray Theory and its Application to Long-Distance Ultrasonic Testing

D. J. Chinn, H. A. Dieterman

Faculty of Civil Engineering
Delft University of Technology
2628 CN Delft, The Netherlands

Summary
The theory of generalized rays is investigated for use in modeling the detection of vertical defects in plates using long-distance ultrasonic testing. The theory can be extended to analytically verifying SAFT-generated images of defects from experimental data. Used together, generalized ray theory and long-distance ultrasonic testing allow crack detection without knowledge from previous measurements.

Purpose

Ultrasonic testing is commonly used for local defect detection in steel plate members. The localized aspect of ultrasonic testing is necessary for precise characterization of defects. One disadvantage of localized ultrasonic testing is the small inspection area of a single test that renders it time-consuming for large members. Further, to detect defects locally the defect must be accessible to the equipment as well as to the inspector.

To increase the inspection area and facilitate accessibility at hard-to-reach connections, long-distance ultrasonic testing (LDUT) has been under investigation. In this method, ultrasonic testing is performed at distances of 0.3 to 1.0 meters from the detection area. An angled probe generates a wave field that reaches the detection area after several reflections from the plate faces. The probe is chosen so that a reduced wave field, consisting only of shear and surface waves, is generated. Wavelengths are kept sufficiently small in relation to the plate thickness so that no Lamb waves are generated. Scanning at offset intervals of 5 - 10 wavelengths is necessary to locate a defect. Fig. 1 illustrates the concept of LDUT.

Generalized ray theory principles

Generalized ray theory (GRT) can be used to analytically validate experimental data from LDUT. The theory was originally developed and is now widely used in the field of seismics to analyze wave propagation in layered media [1,2]. It has been applied in plates to describe the wave field generated from acoustic emission sources [3-5]. GRT superposes the wave fields from all possible wave paths, or rays, giving an exact solution of the wave field in the time period examined. The theory is used ideally when the effects of many rays must be considered, as is the case with LDUT. The theory is briefly described in this section.

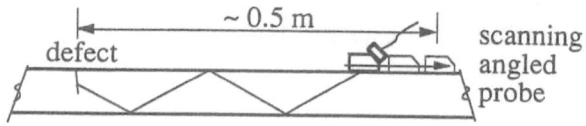

Fig. 1. Long-distance ultrasonic testing

The response in a plate at location **x** due to a point force **f**(t) at **x** = 0 acting at time t = 0 can be expressed as:

$$u(x,t) = \int_0^t G(x,t) \times f(t-\tau) \, d\tau \tag{1}$$

where G(x,t) is the Green's transfer function for the plate medium due to an impulse force on the plate at t = 0.

Written as a convolution, eqn. (1) becomes:

$$u(x,t) = G(x,t) * f(t) \tag{2}$$

GRT is one of several methods available to find G(x,t). Normal modes theory and various integral transform methods may also be used. With GRT, the complete Green's function is found by summing the transfer functions resulting from all possible rays arriving at the source within the time period examined. In Fig. 2, three possible ray paths are shown for a source at x = 0 and receiver at x. The first path corresponds to a surface wave between the source and receiver. The second ray is reflected once from the bottom of the plate, and the third ray is reflected three times before arriving at the receiver. Each ray path has a different Green's transfer function denoted by $G_1$, $G_2$ and $G_3$. The complete Green's function is then:

$$G(x,t) = \sum_{i=1}^{n} G_i(x,t) \tag{3}$$

where n is the number of rays whose first arrival time occurs before the end of the time period examined, t. Using the notation of ref. [3], the Green's function for each ray can be described as:

$$G_i = G(S, \Pi_i, D, x, t) \tag{4}$$

where S is a function based on the source type, $\Pi_i$ is a function of the number and type of reflections in the ray path i, and D is a function of the receiver type. Full descriptions of the derivation and solution of Green's transfer function for each type of ray using GRT are described in refs. [3] and [5].

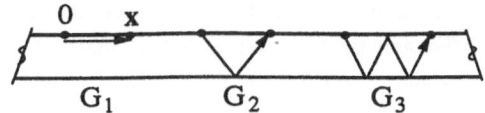

Fig. 2. Three different ray paths between 0 and x in a plate

Generally, normal modes theory is more accurate than GRT in describing the wave field in a plate at large source-receiver offset distances since the slight numerical errors in calculating each ray with GRT are compounded with large numbers of ray superpositions or long time periods [6,7]. While GRT superposes all possible rays to describe the wave field, the theory of normal modes superposes all possible vibrational modes of wave propagation in the plate. However, for high frequency (5 Mhz) testing as used in LDUT, many modes of vibration must be considered to cover the frequency range present. Further, in LDUT the wave field is limited by the highly directional source which generates only shear waves. The rays that must be considered are limited to those generated by the source. The decreased set of rays in the LDUT wave field produces a GRT solution with reduced numerical error. Therefore, GRT is appropriate for modeling the long-distance problem here.

LDUT experimental results

Experimental work has shown that LDUT is successful in detecting defects from long distances [8]. Both pulse-echo and tandem probe configurations are used. The synthetic aperture focusing technique (SAFT) is used to reconstruct images of the detection area from scanned long-distance data. This technique uses the different arrival times of waves recorded at different receiver positions to focus on a source of reflection. Scanning simulates a large-aperture, highly-focused probe. The simulated probe is focused on different locations within the area of observation to obtain a cross-sectional image of the area. SAFT has been used successfully in localized ultrasonic testing to obtain a detailed image of the defect area [9]. At large offsets however, SAFT has proven to be suitable only for defect detection. Defect characterization from long distance scanning is highly dependent on probe configuration and crack orientation.

A 35 mm thick plate containing a vertically-oriented, surface-breaking fatigue crack was scanned using LDUT from a distance of 0.3 - 0.5 m. Growth of the fatigue crack from 5 to 10 mm depth can be seen with the aid of the SAFT image. It was shown in ref. [8] that the small amount of diffraction energy from the crack tip is not visible in the presence of reflection energy from the vertical crack face. In the time period examined, four unique ray paths resulting from reflection from the crack face are possible with the test configuration used. They are the 1-2, 1-4, 2-3, and 3-4 paths shown in Fig. 3. A 1-2 reflection path has 1 reflection on the path to the defect and 2 reflections on the return path. On all the to and from paths a perfect reflection is assumed without scattering from the top and bottom faces of the plate, as shown in Fig. 4a. (A 1-2 reflection path has the same travel time as a 2-1 reflection path. Thus, a 1-2 reflection path here denotes both the 1-2 and the 2-1 paths.)

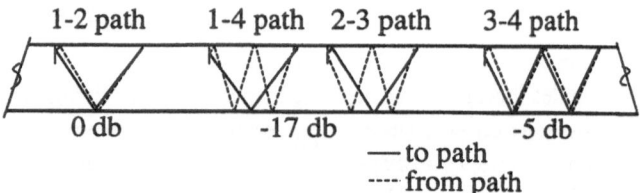

Fig. 3. Relative energy of 1-2, 1-4, 2-3 and 3-4 ray paths

For each ray path reflected by the crack, a different amount of energy is returned to the probe. Since a single ray path is used to generate a SAFT image, the relative energy of each ray can be estimated from the values in their respective images. Fig. 3 shows the different energy levels of the four paths relative to the 1-2 path. Because of the crack configuration, the 1-4 and 2-3 ray paths have exactly the same travel time and cannot be distinguished from one another. The largest amount of energy returned to the probe comes from the 1-2 path. The 3-4 path has 5 db less energy than the 1-2 path. The 1-4 and 2-3 paths, however, have 17 db less energy.

Fig. 4. Reflection surfaces (a) perfect reflector (b) scattering reflector

In all the ray paths, we assume a finite source-receiver where the offset between the source and receiver may range from 0 - 10 wavelengths. Using this assumption, the reflection from the crack face in the 1-2 and 3-4 paths occurs as if the crack were a perfect reflector as in Fig. 4a, such that $\theta_0 \sim \theta_1$. This type of reflection returns the maximum amount of energy to the receiver. For the case of the 1-4 and 2-3 paths, the angle of propagation in the to-path is quite different from the angle of propagation in the from-path. Although we continue to assume a finite source-receiver, the reflection angle at the crack of the 1-4 and 2-3 paths must result from scattering at the crack surface as shown in Fig. 4b. The energy in the 1-4 and 2-3 paths is thus a result of scattering at the crack in the direction required for the return path to the receiver. This explains the large difference in energy between the paths. For cracks whose surface roughness is much smaller than the wavelength of the incident wave, most of the energy comes from the ray paths which reflect from the crack as perfect reflectors.

## Modeling LDUT with GRT

In LDUT, a signal is sent over an extended distance to a probed area. If a reflector is present, a response is returned, again traveling over the distance. The long-distance propagation of the LDUT signal both to and from the probed area can be modeled with GRT. A simplified model of pulse-echo LDUT of a surface-breaking, vertically-oriented crack is shown in Fig. 5 for a 1-2 ray path. The probe-source is modeled with a directional point force $f(t)$, the probe-receiver is modeled with a directional displacement $u(t)$ at offset distance $2x$. The Green's function for the 3-reflection ray path between the source and receiver is found from eqn. (4) and is solved using numerical integration techniques found in ref. [3]. The displacement $u(t)$ is found from eqn. (2).

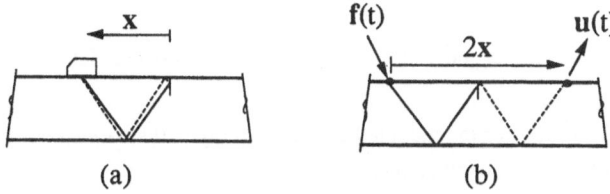

Fig. 5. 1-2 reflection in pulse-echo LDUT (a) experimental path
(b) simplified GRT model

In the simplified model, the vertical crack face is assumed to be a perfect reflector . Only with this assumption can a single generalized ray between two points model the pulse-echo path to and from the crack. Ray paths which require scattering at the crack face cannot be modeled this way. It was shown in the previous section, however, that most of the energy detected from a crack of this configuration returns along a path which requires near perfect reflection at the crack face. If all the ray paths that have near perfect reflection at the crack face are included in the GRT model, a large part of the actual energy returned from the crack will be included in the theoretical model.

Within the constraints of a finite source-receiver, the simplified model also requires that the reflection from the crack face occur very close to the surface of the plate. This is the only way that the offset distance of the to and from paths can be approximately equal as is required in pulse-echo testing.

An experimental pulse-echo signal from a 10 mm deep vertically-oriented, surface-breaking fatigue crack in a 35 mm thick plate is shown in Fig. 6a. The radial distance between the crack and the probe is 350 mm. The response using a GRT model of the same configuration is shown in Fig. 6b. Only the 1-2 and 3-4 ray paths were modeled.

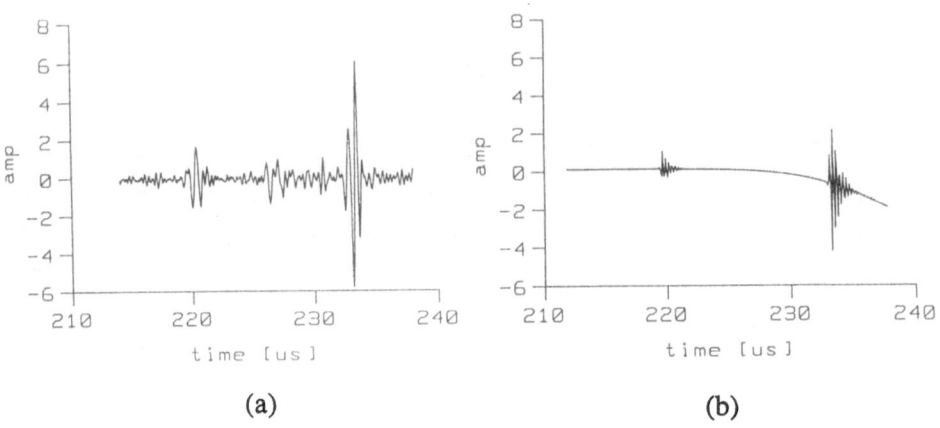

Fig. 6. Pulse-echo response at x = 350 mm  (a) experimental data   (b) GRT model

The arrival times of the modeled 1-2 and 3-4 ray paths correspond very well with the arrival times of the experimental data. However, the pulse shapes of the responses are not in agreement. The source forcing function used in the GRT model, f(t), is evidently not accurate enough to model the actual probe output signal. A discussion of the source function is included later.

SAFT imaging using the GRT model
SAFT imaging of scan data is carried out entirely in the time domain. The technique uses the arrival time of the ray to focus on possible reflector points. In the focusing, the actual pulse shape of the received signal is of less importance than the consistency of the pulse shape over different scan positions. Therefore, despite the difference in pulse shapes between the real pulse-echo response and GRT-simulated responses, the GRT model can still be used in SAFT imaging if the same source function is used at each scan position. Fig. 7 shows SAFT images generated using an experimental LDUT 3-4 reflection pulse-echo scan from 0.3 - 0.5 m and its corresponding GRT-simulated scan. The ambiguity of a pulse-echo scan is seen in both experimental and simulated data images. However, the GRT simulated scan data verifies the lateral offset of the defect.

Tandem testing has been shown in ref. [8] to give better characterization of cracks from long-distances than pulse-echo testing. The GRT model can also be applied to tandem testing. Once again, only perfect reflection at the crack face is modeled. Unlike the pulse-echo model, the tandem model does not require reflection from the crack at a point close to the top surface.

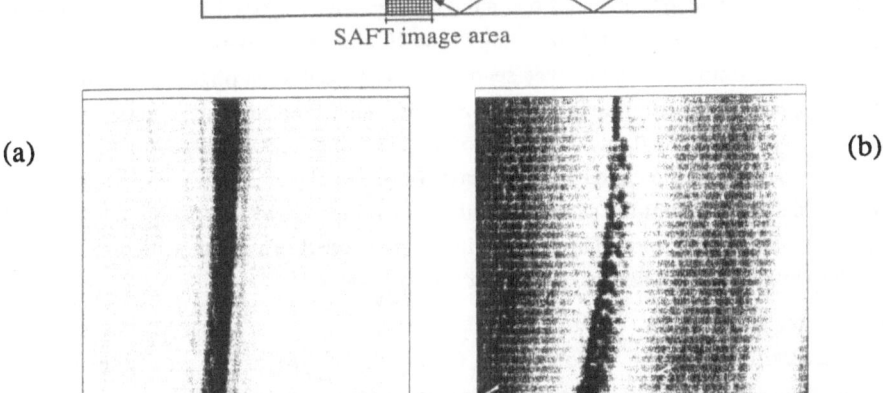

Fig. 7. SAFT images using 3-4 reflection in pulse-echo testing
(a) from experimental LDUT scanning  (b) from GRT model scan

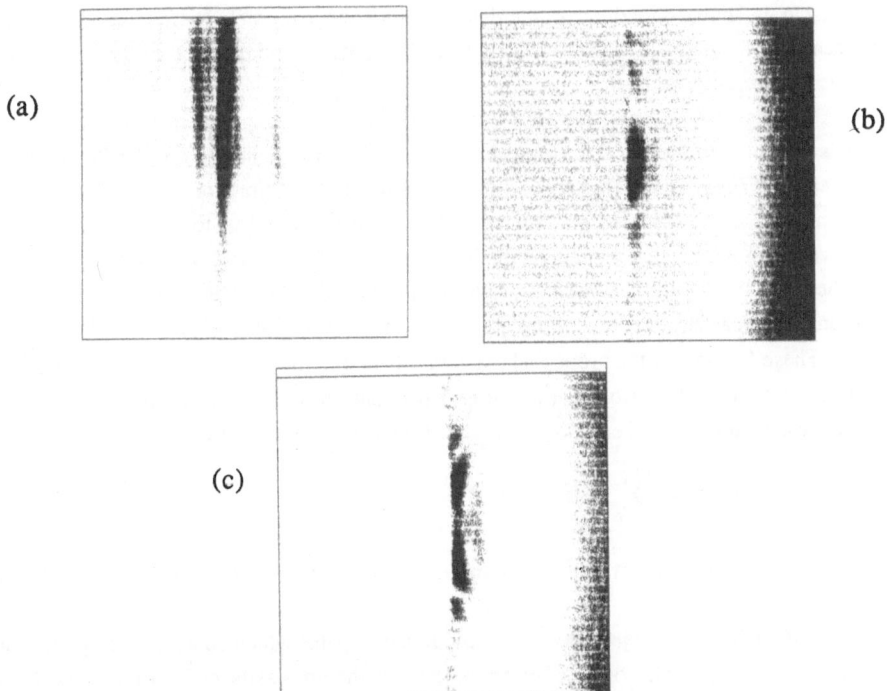

Fig. 8. SAFT image using 1-4 reflection in tandem testing  (a) from
experimental LDUT scanning of 10 mm crack  (b) from GRT model
of 10 mm crack scan  (c) from GRT model of 5 mm crack scan

By limiting reflections to a certain, simulated crack depth, a comparison can be made between relative energy levels of real and GRT-generated SAFT images. Fig. 8a shows an image from experimental tandem testing of the 10 mm vertically-oriented, surface-breaking, fatigue crack using a 1-4 path. The send probe was fixed at an offset of 150 mm while the receiving probe scanned from 300 - 500 mm. Figs. 8b and 8c show the corresponding images from the GRT model of tandem testing of 10 and 5 mm cracks, respectively. The energy increases 5 db as the simulated crack increases from 5 to 10 mm. For the experimental tandem testing, the energy from the same 1-4 reflection path increased 8 db as the crack grew from approximately 5 to 10 mm. The increase in energy in the SAFT image with crack growth is a result of reflection from the increased crack surface area.

## Discussion

The general objective of modeling with GRT is to be able to use a numerical model for comparison of data instead of comparison to earlier measurements. There are two parts to this general objective. First, the response $u(t)$ received from long-distance probing of a crack can be verified. Second, the end result of LDUT, the detection of detects, can be verified. It has been shown above that with SAFT, the lateral location of a defect can be verified despite an inaccurate source function. SAFT processing fills the void in the model resulting from the lack of an accurate source function. In the detection of a vertical crack, there is good agreement between experimental and GRT-simulated data. Thus, the second part of the objective has been achieved for vertical cracks.

The most significant parameter in the verification of the experimental response $u(t)$ with GRT is the source function $f(t)$. The source in LDUT is a circular 1" dia. 5 Mhz transducer mounted on a 70° wedge. This probe was chosen for its high directionality and clarity in the resultant wave field. With a highly directional probe, the energy loss to spreading of the wave field over long-distance propagation is minimized. By choosing a propagation angle larger than 32°, the wave field is limited to shear waves in all the paths. In addition, by propagating at a large angle such as 70°, less reflections are required in traveling to and from the defect.

These features of the probe make it very difficult to model. The source function $f(t)$ of each ray path emanating from the probe is dependent on its angle of propagation [8]. The frequency and amplitude of each ray is unique. Eqn. (2) is more accurately written as:

$$u(x,t) = \sum_{i=1}^{n} G_i(x,t) * f(t) \tag{5}$$

Work is in progress on a more accurate transfer function for the probe and will be reported at a later date.

The simplified GRT model includes only those reflected paths which contain the largest amount of energy. For vertically-oriented defects, these paths are easily chosen since the largest amount of energy results from perfect reflectors back to the source. For non-vertical defects, a model for the crack must be included. The largest amount of energy will still result from a perfect reflection however the return path is not as clear. In this case, the to and from paths must be modeled separately in GRT. The crack orientation determines the direction of the from path.

## Conclusions

A simplified model is presented for modeling the detection of vertically-oriented cracks from a long distance. The model does not absolutely agree with experiment because of the lack of an accurate source function. SAFT allows the model to overcome this deficiency by using spatial and temporal differences in the response. In LDUT, it was found that only defect detection was possible. Defect characterization is limited to very special cases. The same can be said about the GRT model presented here. No less information is derived from GRT-simulated data making it suitable for verification of LDUT detection. Together, the two analyses give an analytical tool to determine the existence of cracks from collected data, without information from previous measurements.

## Acknowledgements

The authors acknowledge the help of Prof. W. Sachse in providing the computer code developed at Cornell University for the Green's function. This work is supported by the Technology Foundation (STW) in the Netherlands.

## References

1. Ewing, W.M.; Jardetsky, W.S.; Press, F.: *Elastic Waves in Layered Media*, McGraw-Hill (1957).
2. Kennett, B.L.N.: *Seismic Wave Propagation in Stratified Media*, Cambridge University Press (1983).
3. Pao, Y.H.; Gajewski, R.R.: The generalized ray theory and transient responses of layered elastic solids, in *Physical Acoustics*, ed. Mason, W.P., Thurston, R.N., Academic Press, 13 (1977) 183-265.
4. Michaels, J.E.; Pao, Y.H.: The inverse source problem for an oblique force on an elastic plate, *J. Acoustical Society of America*, 77 (1985) 2005-2011.
5. Eitzen, D.G. et. al.: *Fundamental Developments for Quantitative Acoustic Emission Measurements*, EPRI NP-2089, Electric Power Research Institute (1981).
6. Weaver, R.L.; Pao, Y.H.: Axisymmetric elastic waves excited by a point source in a plate, *J. Applied Mechanics*, 49 (1982) 821-836.
7. Santosa, F.; Pao, Y.H.: Transient axially asymmetric response of an elastic plate, *Wave Motion*, 11 (1989) 271-295.
8. Chinn, D.J.; Dieterman, H.A.: Defect imaging using long-distance ultrasonic testing, *Review of Progress in Quantitative NDE*, 11B (1991) 1901-1908.
9. Lorenz, M.; van der Wal, L.F.; Berkhout, A.J.: Ultrasonic imaging with multi-SAFT, *Nondestructive Testing and Evaluation*, 6 (1991) 149-177.

# A New Approach to Ultrasonic Image Reconstruction

M.YAMANO* and S.R.GHORAYEB**

*   System Engineering Division
    Sumitomo Metal Industries,Ltd.
    Amagasaki 660 Japan
**  Dept. of Electrical Engineering and Computer Engineering
    Iowa State University
    Ames,IA.,50011,U.S.A.

Summary
For quantitative nondestructive evaluation(QNDE) with using ultrasound, a new
approach to improve spatial resolutions is presented in this paper.  The newly
implemented technique is compared with  the synthetic aperture focusing technique(SA
FT) and the Wiener filtering  in the resolution  and the signal-to-noise ratio.  The
approach provides with a much higher resolution than the SAFT   and comparable resolution to the Wiener filtering, while it requires only one PSF.  The images reconstructed by  the technique,however,have a lower signal-to-noise ratio compared  with
that of  the others.

Introduction
Techniques of  quantitative  nondestructive  evaluation(QNDE), which  characterize a
defect in size,shape and orientation, have been required  in quality/process control
of materials, in-service monitoring  and  estimation of residuary life expectancy of
structures.  The conventional pulse-echo method, which is the most popular method in
ultrasonic NDE,has the lateral and depth resolutions  that depend on  the transducer
size,the duration time of a transmitted pulse and the transducer's position relative
to a defect,so that usually it doesn't provide with enough resolution for the defect
characterization.
    Various  ultrasonic  image  reconstruction  techniques  have  been developed[1]-[7]
[9]-[12] to improve  resolution  in lateral  and/or  depth  direction,  and to characterize a defect.  The synthetic aperture focusing  technique(SAFT)  and the Wiener
filtering  have  been used   for the purpose  described above.   Although the SAFT
improves resolution, the reconstructed SAFT images are still degraded by the factors
such as the pulse duration time, limited synthetic aperture length and discrete data
sampling. The factors  are described by  a point spread function(PSF).  On the other
hand the Wiener filtering is  one of the deconvolution techniques  that restore  the
degraded images with  using appropriate  PSF. In ultrasonic NDE applications,however,
any particular PSF of raw B-scan data  is valid over  relatively narrow depth  range
because  the ultrasonic beam spread varies with beam propagation[1][2]. The requirement of the depth variant PSF makes the Wiener filtering applications difficult.

464

A new approach for improving the spatial resolution in ultrasonic NDE applications is presented in this paper. This paper describes the image reconstruction algorithm that we proposed and subsequent results. The newly implemented technique is compared with the SAFT and the Wiener filtering in the resolution and the signal-to-noise ratio. The effect of the resolution improvement is shown on simulated data numerically.

## Reconstruction Algorithm

Fig.1 shows a geometry and notation of the ultrasonic NDE imaging. A 2-D defect in a specimen is shown being illuminated by a cylindrical wave $p_0(x_t,t)$ transmitted from a transducer at a point$(x_t,0)$. The wave scattered by the defect function $f(x,z)$ is received by the same transducer as a received signal $p(x_t,t)$. As the transducer is scanned along a line above the specimen, successive raw A-scan data are collected.

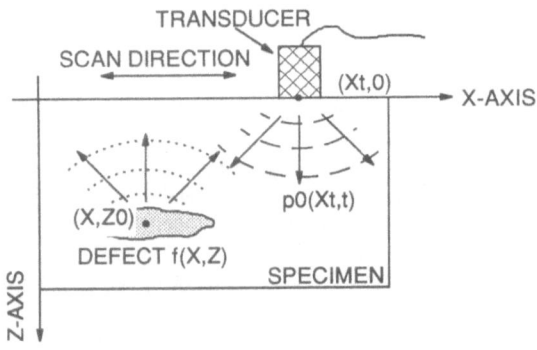

Fig.1 Geometry and notation of the ultrasonic NDE imaging.

To characterize the interaction of the defect with the ultrasonic field, one must solve the inhomogeneous Helmholtz wave equation

$$(\nabla^2+k^2)P(x,z,f) = k^2f(x,z)P(x,z,f) \qquad (1)$$

where $k=2\pi f/C_0$ is the wavenumber with $C_0$ being the longitudinal wave velocity in the specimen and $f(x,z)$ is defect function. Here $P(x,z,f)$ denotes the temporal Fourier transform of the ultrasonic field $p(x,z,t)$; i.e.

$$P(x,z,f) = \int_{-\infty}^{+\infty} p(x,z,t) \, exp[-j2\pi ft] \, dt \qquad (2)$$

Although it is not possible, in general, to solve the exact ultrasonic field from Eq.(1), the first Born approximation and the Rytov approximation provide the approximate solutions. With the first Born approximation, we obtain the following solution as the received signal.

$$P(x_t, f) = A \cdot \frac{jf}{C_0} \cdot P_0(f) \cdot$$

$$\iint\limits_{-\infty}^{+\infty} \frac{f(x,z)}{R} exp[j2(2\pi f/C_0)R] \, dx \, dz \qquad (3)$$

where $P_0(f)$ is the temporal Fourier transform of the transmitted pulse $p_0(x_t, t)$, $R = \sqrt{(x-x_t)^2 + z^2}$ and $A$ is a constant. In deriving Eq.(3), in addition to the first Born approximation, the following approximation of the Hankel function is made use of:

$$G(\vec{r} \mid \vec{r_0}) = \frac{j}{4} \cdot H_0^{(1)}(k \mid \vec{r} - \vec{r_0} \mid)$$

$$\fallingdotseq \left(\frac{j}{8\pi kR}\right)^{\frac{1}{2}} \cdot exp(jk \mid \vec{r} - \vec{r_0} \mid) \qquad (4)$$

where $G(\vec{r} \mid \vec{r_0})$ is the 2-D free-space Green's function, $H_0^{(1)}(k \mid \vec{r} - \vec{r_0} \mid)$ is the zeroth order Hankel function of the first kind.
Letting

$$\phi(x_t, f) = \frac{P(x_t, f)}{P_0(f) \cdot f} \qquad (5)$$

and $\Phi(u, f)$, the spatial Fourier transform of $\phi(x_t, f)$; that is

$$\Phi(u, f) \equiv \int_{-\infty}^{+\infty} \phi(x_t, f) exp[-j2\pi u x_t] \, dx_t \qquad (6)$$

Substituting (3) and (5) in (6), we obtain

$$\Phi(u, f) = \frac{A}{C_0 \sqrt[4]{(2f/C_0)^2 - u^2}} F(u, -\sqrt{(2f/C_0)^2 - u^2}) \qquad (7)$$

where

$$F(u, v) = \iint\limits_{-\infty}^{+\infty} f(x,z) exp[-j2\pi(ux+vz)] \, dx \, dz \qquad (8)$$

The discrete Fourier transform(DFT), in numerical calculation, provides with $\Phi(u, f)$ only at discrete sampling points, so that $F(u, v)$, the Fourier transform of the defect function, is obtained only at the points along circular arcs $\sqrt{(2f/C_0)^2 - u^2}$ as shown in Fig.2. In going from this circular grids to the rectangular grids required by the 2-D inverse DFT algorithm, which reconstruct defect function $f(x, z)$, one will make use of the interpolation algorithm such as nearest-neighbor and bilinear interpolation. According to the investigation by Kak et al.[3], one can obtain reconstructions of quality by using bilinear interpolation which is comparable to that obtained by the filtered-backpropagation algorithm proposed by Devaney[4], whereas the former approach reduces the CPU processing time. After interpolation in frequency domain, we finally obtain defect function $f(x,z)$;

$$f(x,z) = \int\int_{-\infty}^{+\infty} F(u,v) \, exp[j2\pi(ux+vz)] \, du \, dv \qquad (9)$$

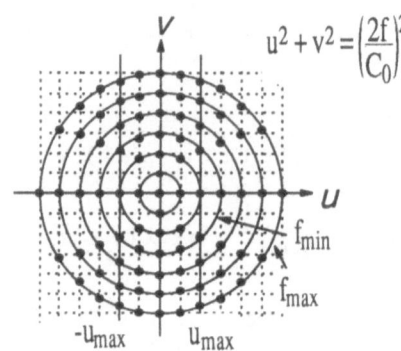

Fig.2 Circular arcs of f=constant in the (u,v) plane including a set of points at which F(u,v) is obtained. ● is calculated from Eq.(7).

The SAFT algorithm described above is the Fourier domain reconstruction(FDR) algorithm, which has been proposed by Nagai[5]. The SAFT improve spatial resolution in lateral and depth directions. The improvement is,however, limited to

$$dr_{lat} = max\left(\frac{C_0}{f} \cdot \frac{\sqrt{z_0^2+(D_x/2)^2}}{2D_x}, \Delta x_t\right) \qquad (10)$$

$$dr_{dep} = \frac{1}{v_{max}-v_{min}} \qquad (11)$$

where $dr_{lat}$ and $dr_{dep}$ denote the lateral and depth resolution, respectively[5]. And Dx is the synthetic aperture length, $z_0$ is the averaged depth of the defect ,$\Delta x_t$ denotes the spatial sampling interval and

$$v_{min} = \frac{2f_{min}}{C_0} \qquad (12)$$

$$v_{max} = \sqrt{\left(\frac{2}{C_0}f_{max}\right)^2 - u^2_{max}} \qquad (13)$$

That limitation results from the limited-length synthetic aperture and/or the band-limited transmitted(received) pulse. Furthermore the limitation of the ultrasonic beam spread and the discrete data sampling in scan(lateral) and time(depth) axis tend to degrade the quality of reconstruction images.

Since the SAFT is considered as a technique that *backfocuses* the received signals (diffracted signals by defects) into the specimen, the defect image reconstructed by the SAFT doesn't depend on the depth where the defect is located. This depth invariance of the defect images allows the use of the Wiener filtering to restore the SAFT images degraded by the factors described above.

The newly implemented algorithm applies the Wiener filtering on the SAFT images to obtain a higher resolution. We find the Wiener filter

$$M(u,f) = \frac{H(u,f)^{*}}{|H(u,f)|^{2}+\eta} \tag{14}$$

where $H(u,f)$ is the 2-D Fourier transform of the PSF $h(x,z)$ and $*$ denotes the complex conjugation. $\eta$ is a constant related to the signal-to-noise ratio. The PSF $h(x,z)$ is a SAFT image of a point scatterer,which may be at an arbitrary depth. After the Wiener filtering is applied on the SAFT images, the 2-D inverse Fourier transform provides with the restored defect function $f_+(x,z)$:

$$f_+(x,z) = \iint_{-\infty}^{+\infty} \{F(u,f)M(u,f)\} exp[j2\pi(ux+fz)]dudf \tag{15}$$

Fig.3 shows a block diagram of this reconstruction algorithm.

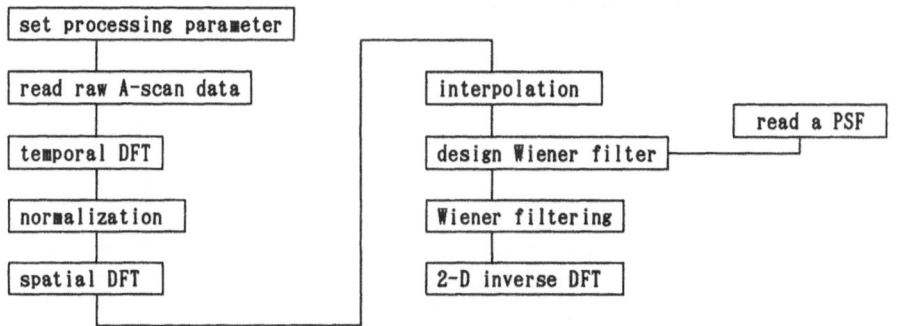

Fig.3 Block diagram of the new reconstruction algorithm

Computer Simulation

Computer simulation with using a simple model shown in Fig.4, 5 and 6, instead of measuring experimental data, is used for generating unrectified RF B-scan data in order to exercise the reconstruction algorithm. A point-focus transducer shown in Fig.4 is used in the model for illuminating a specimen with highly broad beam. The ultrasonic beam converges on the specimen's surface and then diverges at a wide

angle in the specimen. The assumption of the transducer directivity and the transmitted waveshape are shown in Fig.5 and 6, respectively. The waveshape is the
Gaussian modulation of the 5MHz sine function.

A 2-D test specimen contains four double point-defects spaced $\lambda$ at 10mm to 40mm
in depth(Fig.4). A referential single point-defect at a depth of 10mm is included
to obtain a PSF required for the Wiener filtering. The transducer is scanned along
a line above the specimen with a interval $\Delta x_t = \lambda/5$.

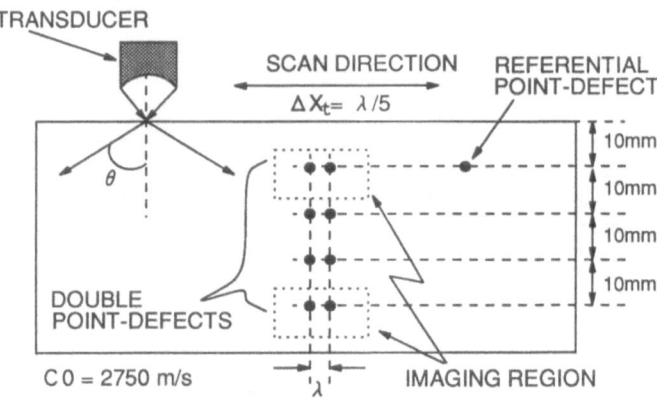

Fig.4 Schematic cross-sectional illustration of a 2-D test specimen. Specimen contains four double point-defects with a referential point-defect.

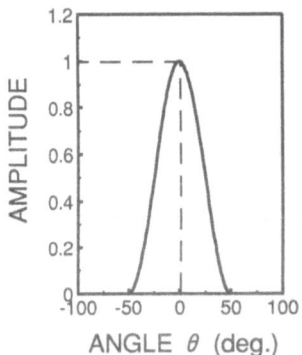

Fig.5 Transducer's directivity

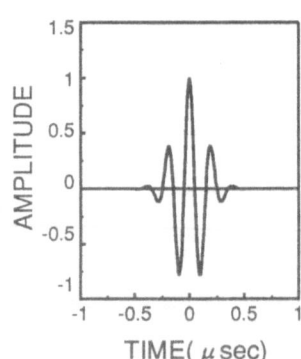

Fig.6 Transmitted waveshape

Results
The newly implemented technique is made a comparison with the SAFT and the Wiener
filtering to verify the effect of the resolution improvement. Fig.7 shows ex-

tracted images of the raw B-scan data after envelop detection, generated from the model. The double point-defect at depths of 10mm and 40mm is centered on the extracted image in Fig.7a and Fig.7b, respectively.  Figures are 13.2mm in lateral (horizontal) direction and 8.8mm in depth(vertical) direction. Any pixel whose reconstructed amplitude is above -6dB level of the peak amplitude are coded with black and contours at -12dB and -18dB level are plotted. In both images, the responses from the individual point-defects overlap in lateral direction and consequently the two responses have merged to give only one identifiable response.

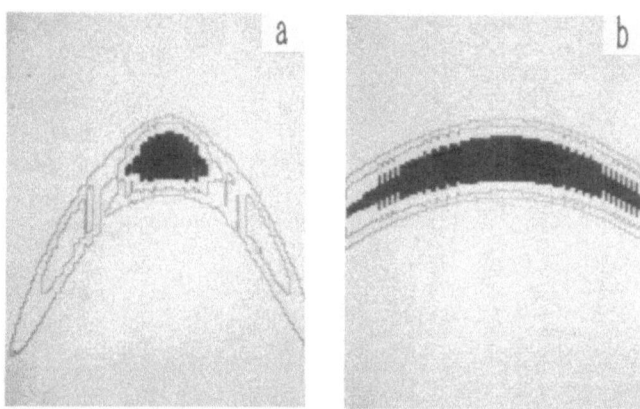

Fig.7   Raw B-scan data generated by computer simulation after envelop detection. The responses from individual point-defects overlap in lateral direction.


The corresponding B-scan images after the SAFT processing are shown in Fig.8a and Fig.8b.  Black represents the pixel above -6dB level, and a contour at -12dB level is plotted. Not only the improvement of the lateral resolution but also depth invariance of the defect images is clearly evident. The responses from the individual point-defects,however,are not clearly discriminated, because the SAFT images are degraded by several factors. The resolution, that is estimated by the full width of the image at -6dB level,is about a wavelength in both lateral and depth directions.
The Wiener filtering is applied on the SAFT results shown in Fig.8 to restore them and obtain a much higher resolution.  The results(see Fig.9a and Fig.9b) apparently show the improved lateral resolution compared with the SAFT images, while depth resolution is almost same.  The lateral resolution is a fifth of the wavelength and the depth is a wavelength.  The responses from the double point-defect were resolved into two separated signals.  The PSF for the Wiener filtering was derived from the SAFT image of the referential point-defect at 10mm depth. In case of being applied on unrectified RF B-scan data,as shown later,the Wiener filtering requires the depth variant PSF, but here only one PSF is required,  because the SAFT has given approximately depth invariant defect images(see Fig.8).   However,the signal-to-noise ratio of Fig.9 is relatively lower than that of the SAFT(Fig.8) and the Wiener filtering (Fig.10a).

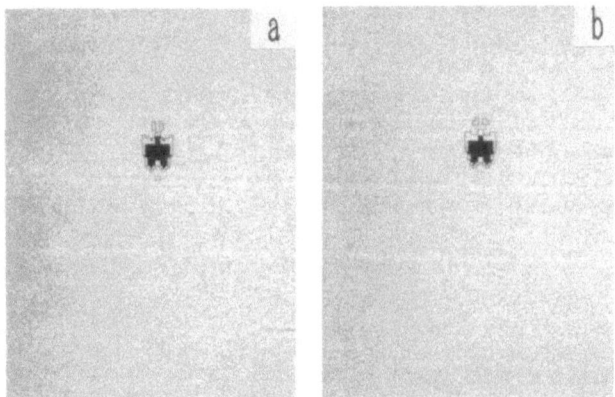

Fig.8 Corresponding B-scan images after the SAFT processing.

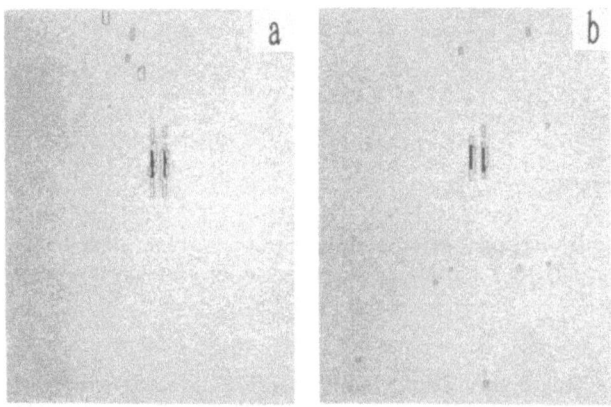

Fig.9 Images after applying the Wiener filtering on the SAFT images.    The responses
     from the double point-defect were resolved into two separated signals


   Fig.10a and Fig.10b show  the results from  applying  the Wiener filtering on unrectified RF B-scan data.  The PSF was derived from  the computer simulated response
of the referential point-defect.  The greatest improvement in both lateral and depth
resolution, a fifth and a third of the wavelength  respectively, is shown in Fig.10a
and the clearest distinction is made  between individual point-defects.  Because the
defect in Fig.10a is  at a same depth  as the referential point-defect. On the other
hand, an erroneous image is shown in Fig.10b  because the double point-defect is at a
depth of 40mm and has a different depth with that of the referential point-defect.
The results show apparently  that the signals  from defects  having different depths
must be  restored  separately  in  a case  of applying  the Wiener filtering  on unrectified RF B-scan data.

Table 1 shows the spatial resolution of three different techniques; SAFT, Wiener filtering and a new approach proposed in this paper.

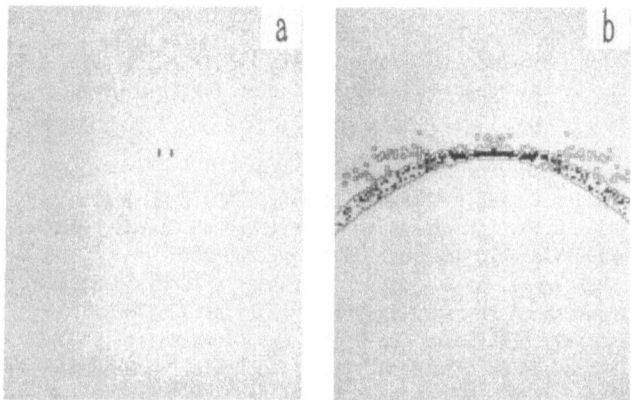

Fig.10 Images after applying the Wiener filtering on the unrectified RF B-scan data

Table 1

| method | resolution | advantage | disadvantage |
|---|---|---|---|
| SAFT | $\lambda^1$ $\lambda^2$ | ·no PSF required ·depth invariant result | ·not high resolution comparatively |
| Wiener filtering | $\lambda/5^1$ $\lambda/3^2$ | ·highest resolution in both directions | ·depth variant PSF required |
| SAFT + Wiener filtering | $\lambda/5^1$ $\lambda^2$ | ·high lateral resolution ·only one PSF required | ·lower signal-to-noise ratio |

1; lateral resolution    2; depth resolution

## Conclusions

A new approach to ultrasonic image reconstruction has been proposed and compared with the SAFT and the Wiener filtering in the spatial resolution and the signal-to-noise ratio. Comparing images, this technique is better in the lateral resolution than the SAFT and comparable to the Wiener filtering. Furthermore it is more useful in NDE applications than the Wiener filtering because of requiring only one PSF, while signals from defects having different depths must be restored with different PSFs in the Wiener filtering. A disadvantage of this approach is that the reconstructed image has a relatively lower signal-to-noise ratio, which may cause incorrect evaluation of defects. Now we have been engaged in the research to improve the signal-to-noise ratio. Because the technique should be preferred for improving spatial resolution in ultrasonic NDE applications if the signal-to-noise ratio can be improved. And also we've tried to exercise the algorithm on some experimental data.

Acknowledgement
The authors wish to express their sincere thanks to Palmer Prof.W.Lord of the Iowa State University for his help and useful advices for this work.

References
[1]  S.F. Burch,Ultrasonics(1987),25,pp.259-266
[2]  M. Fatemi and A.C. Kak,Ultrasonic Imaging(1980),2,pp.1-47
[3]  S.X.Pan and A.C.Kak,IEEE Trans. on Acoust,Speech,Signal Processing,ASSP-31,5
[4]  A.J.Devaney,Ultrasonic Imaging(1982),4,pp.336-350
[5]  K. Nagai,IEEE Trans. on Sonics and Ultrasonics(1985),SU-32,4,pp.531-536
[6]  J.Johnson,Review of Progress in QNDE(1982),1,pp.735-752
[7]  J.Johnson et al.,IEEE Trans. on Sonics and Ultrasonics(1983),SU-30,5,pp.283-294
[8]  J.Seydel,Research Tech. Nondest. Test. Academic Press(1982),6,pp.1-47
[9]  K.J.Langenberg et al.,NDT International(1986),19,2,pp.177-188
[10] E.E.Hundt et al.,IEEE Trans. on Sonics and Ultrasonics(1980),SU-27,5,pp.249-252
[11] N.R.Chapman et al.,Geophys,J.R. astr.Soc.(1983),72,pp.93-100
[12] H.Schomberg et al.,Ultrasonic Imaging(1983),5,pp.38-54

# Chapter 8
# Thermal Inverse Problems

# Identification of Heat Conduction Coefficient: Application to Nondestructive Testing

# Identification of Heat Conduction Coefficient: Application to Nondestructive Testing.

**Marc BONNET, Huy Duong BUI, Hubert MAIGRE**

Laboratoire de Mécanique des Solides

(centre commun CNRS–Polytechnique–Ecole des Mines–Ecole des Ponts et Chaussées),

Ecole Polytechnique, 91128 Palaiseau Cedex, FRANCE.

**Jacques PLANCHARD**

Electricité de France, Direction des Etudes et Recherches,

1 av. du Général de Gaulle, 92141 Clamart cedex, FRANCE.

## 1 Introduction

Thermographic Non-Destructive Testing methods for the evaluation of materials and structures are of great interest. They are attractive because of rapid-scanning capabilities available today, Balageas et al [1]. The principle of photothermal methods are well known. The heat flux is applied on the sample by a laser pulse. Then, by means of a Infra-Red camera, one makes a full time record of the whole surface temperature field. There are very rich informations saved in the surface time history $\theta(\mathbf{x}, t)$, which can be used for the reconstruction of unknown thermal conduction coefficient $\eta(\mathbf{x})$, defects, cracks etc...For example, the temperature response of an homogeneous body subjected to a pulse heating $Q$ is $\theta(t) = Q/e\sqrt{t}$ , where $e = (\eta\rho c)^{1/2}$ is the effusivity. The slope $-1/2$ of the response in the log-log scale means the absence of defects. All perturbations of the straight line are indications of non-homogeneous conduction of heat. Qualitative and quantitative evaluations of defects or inhomogeneities from the knowledge of the surface temperature field belongs to the class of mathematically ill-posed problems.

The numerical solution of thermal inverse problems gives rise to a growing interest in the recent literature, see e.g. [2], [3], [7], [15]. Other different physical contexts share the same mathematical basic structure, notably electric conductivity inverse problems, [6], [8], [9], [10].

Generally, solutions to inverse problems are based on two steps. Firstly, objective functions may be derived for the optimized fitting between experimental and theoretical data. Secondly, one makes use of regularization methods for solving the ill-posed problems.

The paper by H.G. Natke [18] is devoted specifically to the second step, which is also dealt with e.g. in [13], [16], [17], [21], [22]. [23]. The present paper will concentrate on the first problem, which consists of deriving the so-called observation equation. We shall make a review of some mathematical methods for establishing the non-linear relation, which links the thermal coefficient $\eta$ to the surface temperature field, in transient as well as in stationary problems. In particular, we shall discuss problem of determining the optimal correction $\delta\eta$ of the model, in a sensitivity approach of the inverse problem.

## 2    The inverse heat conduction problem

Consider the heat conduction equations with inhomogeneous thermal conduction coefficient $k(\mathbf{x})$ in a three-dimensional solid $\Omega$ of external boundary $S$ ($S \neq \partial\Omega$ in case of a void defect).

$$\frac{\partial\theta}{\partial t} - \mathrm{div}(k(\mathbf{x})\nabla\theta) = 0 \qquad \mathbf{x} \in \Omega \tag{1}$$

$$\theta(\mathbf{x}, t \leq 0) = 0 \qquad \text{(initial conditions)} \tag{2}$$

with boundary conditions to be specified.

The direct problem is stated as follows : given the thermal coefficient $k(\mathbf{x})$ such that $k > 0$ and the boundary condition

$$k(\mathbf{x}^0)\frac{\partial\theta}{\partial n} = f(\mathbf{x}^0, t) \qquad\qquad (\mathbf{x}^0 \in \partial\Omega, t \geq 0) \tag{3}$$

find the temperature field $\theta(\mathbf{x}, t)$ inside the solid. We shall use the following notations:

- $\theta(\mathbf{x}, t)$ : the temperature field satisfying the heat equation (1) with unknown coefficient $k(\mathbf{x})$
- $u^\eta(\mathbf{x}, t)$ : the temperature field satifying the heat equation (1) with a given coefficient $\eta(\mathbf{x})$.

We consider the inverse problem of determining the unknown heat conduction coefficient $k(\mathbf{x})$, from boundary data which consist of the heat flux (3) and the surface temperature $\theta(\mathbf{x}, t)$. We suppose also that the heat conduction coefficient $k(\mathbf{x}) > 0$ is known on the surface $\partial\Omega$ of the solid. Then the simplest formulation of the inverse problem is the minimization of a least-squares best-fit objective function $J(\eta) = J(u^\eta, \eta)$, where:

$$J(u, \eta) = \frac{1}{2}\int_0^T \int_S \{u(\mathbf{x}, t) - \theta(\mathbf{x}, t)\}^2 \, dS_{\mathbf{x}} \, dt \tag{4}$$

which is a functional of the difference between the surface temperature of the model $u^\eta$ and the experimental data $\theta \mid_S$

Another related inverse problem is the identification of a defect of known thermal characteristics but unknown location and shape.

First, two methods leading to integral equations on the coefficient perturbation $\delta\eta = k(\mathbf{x}) - \eta(\mathbf{x})$ are discussed (sections 3 and 4). They allow discussion of uniqueness and stability with respect to the temperature data, as well as linearization methods in the case of small perturbation $\delta\eta(\mathbf{x})$.

Then we consider (section 5) the derivation of the gradient of data-fit functionals using the calculus of variations. This is very useful for many practical applications, since the actual numerical solution of the inverse problem involves a nonlinear optimization problem, which, after discretization, is best solved using conjugate gradient, BFGS or similar algorithms. All these algorithms make use of gradient information, and its computation using finite difference approaches is less efficient and less accurate than using an analytical derivation. Application to the shape identification problem is also discussed in section 6.

# 3   The Fréchet derivative method

Without loss of generality, and for simplicity reason, we consider in this section the particular geometry of a half space which is representative of most experiments in thermal inspections of materials. We shall then use the following additional notations:

- $S : \mathbf{x}^0 \mid x_3^0 = 0$

- $\Omega : \mathbf{x} \mid \mathbf{x}_3 < 0$

- $V^\eta$ : the fundamental solution of the heat equation with coefficient $\eta(\mathbf{x})$.

In order to apply the sensitivity method, one needs the calculation of the functional derivative $D_\eta J$ of the objective function, or the functional derivative $D_\eta u$ of the surface value $u^\eta(\mathbf{x}^0, t)$ with respect to $\eta$. The last derivative may be interpreted as the linear map $D_\eta u$: $\delta\eta \rightarrow \delta u$, or the Fréchet derivative defined by

$$u^{\eta+\delta\eta}(\mathbf{x}^0, t) - u^\eta(\mathbf{x}^0, t) = D_\eta u^\eta(\mathbf{x}^0, t)\,\delta\eta + O(\delta\eta^2).$$ (5)

To determine explicitly the Fréchet derivative $D_\eta u$, let us consider the fundamental impulse heat flux solution $V^\eta$ with coefficient $\eta(\mathbf{x})$, associated to a given point $\mathbf{x}^0 \in S$, which solves:

$$\frac{\partial V}{\partial t} - \mathrm{div}(\eta(\mathbf{x})\nabla V) = 0 \qquad \mathbf{x} \in \Omega$$ (6)

$$V(\mathbf{x}, t \le 0) = 0 \qquad \text{(initial conditions)}$$ (7)

$$k(\mathbf{x}^0)\frac{\partial V}{\partial n}(\mathbf{x}^0, t) = \delta(y^0 - x^0)\delta(t) \qquad \mathbf{x}^0 \in S, t \ge 0$$ (8)

$$V \rightarrow 0 \text{ as } \| \mathbf{x} \| \rightarrow +\infty$$ (9)

This solution is identical to the solution of the instantaneous point heat source located at $\mathbf{x}^0$ in an infinite 3D solid, which can be found in textbooks (e.g. Carslaw and Jaeger [5]).

The model solution $u^\eta(\mathbf{x}, t)$ satisfies the heat equation with the same boundary condition (3), with $k(\mathbf{x}^0) = \eta(\mathbf{x}^0)$:

$$\frac{\partial u}{\partial t} - \mathrm{div}(\eta(\mathbf{x})\nabla u) = 0 \qquad \mathbf{x} \in \Omega$$ (10)

$$u(\mathbf{x}, t \le 0) = 0 \qquad \text{(initial conditions)}$$ (11)

$$k(\mathbf{x}^0)\frac{\partial u}{\partial n}(\mathbf{x}^0, t) = f(\mathbf{x}^0, t) \qquad \mathbf{x}^0 \in S, t \ge 0$$ (12)

The field $u^{\eta+\delta\eta}(\mathbf{x}, t)$ satisfies eqns. (10)-(12) with coefficient $\eta + \delta\eta$, and the same boundary conditions (12). Therefore, $R^\eta(\mathbf{x}, t; \delta\eta) = u^{\eta+\delta\eta}(\mathbf{x}, t) - u^\eta(\mathbf{x}, t)$ satisfies the equations

$$\frac{\partial R}{\partial t} - \mathrm{div}(\eta(\mathbf{x})\nabla R) - \mathrm{div}(\delta\eta(\mathbf{x})\nabla u) = 0 \qquad \mathbf{x} \in \Omega$$ (13)

$$R(\mathbf{x}, t \le 0) = 0 \qquad \text{(initial conditions)}$$ (14)

$$k(\mathbf{x}^0)\frac{\partial R}{\partial n}(\mathbf{x}^0, t) = 0 \qquad \mathbf{x}^0 \in S, t \ge 0$$ (15)

$$R \rightarrow 0 \text{ as } \| \mathbf{x} \| \rightarrow +\infty$$ (16)

Taking the time convolution of (13) with the fundamental solution $V^\eta$, and observing that $\delta\eta \equiv 0$ on $S$, we obtain the formula

$$\delta u(\mathbf{x}^0, t; \eta) \equiv D_\eta u\,\delta\eta = -\frac{1}{k(\mathbf{x}^0)}\int_\Omega \delta\eta(\mathbf{x})\nabla u^\eta(\mathbf{x}, t) * \nabla V^\eta(\mathbf{x}, t; x^0)\, dV_\mathbf{x}$$ (17)

where $(*)$ denotes the time convolution of scalar product. Eq. (17) gives the modification of the surface temperature at point $\mathbf{x}^0$, at time $t$, from perturbation $\delta\eta(\mathbf{x})$. It confirms the "golden rule" according to which the determination of unknown $\delta\eta(\mathbf{x})$ defined in a 3D subset of $\Omega$ requires data $\delta u$ defined also on a 3D subspace, here $S \times [0, T]$.

Eq. (17) provides a linear relation of the form $A\delta\eta = \delta u$, which can be used for determining $\delta\eta$ which makes the best fit between data $\delta u$ and the prediction $A\delta\eta$. However the linear system $A\delta\eta - \delta u = 0$ is a Fredholm integral equation of first kind. Hence it belongs to the class of ill-posed problems [22], and cannot be used directly in the inverse problem. Instead of solving exactly this equation, one can determine the steepest direction $\delta\eta$ for minimizing the function $J$. Since

$$\delta J = -\int_0^T \int_S \int_\Omega \frac{1}{k(y)}(u^\eta(y,t) - \theta(y,t))\nabla u^\eta(\mathbf{x},t) * \nabla V^\eta(\mathbf{x},t;y)\delta\eta(\mathbf{x})\, dt dS_\mathbf{y}\, dV_\mathbf{x} \qquad (18)$$

One can see from (18) that $\delta J = (\beta, \delta\eta)$ is a linear functional of $\delta\eta$. Therefore, for all perturbation of the same norm, the perturbation $\delta\eta$ corresponding to the maximum variation of $\delta J$ is proportional to $\beta$. Consequently

$$\delta\eta(\mathbf{x}) = \lambda\int_0^T \int_S \frac{1}{k(y)}(u^\eta(y,t) - \theta(y,t))\, \nabla u^\eta(\mathbf{x},t) * \nabla V^\eta(\mathbf{x},t;y)\, \delta\eta(\mathbf{x})\, dt dS_\mathbf{y} \qquad (19)$$

This result allows to calculate explicitly the steepest gradient of the objective function.

## 4 The adjoint fields method

For simplicity reason, we study the inverse problem for identifying the heat coefficient $\eta(\mathbf{x}) = 1 + h(\mathbf{x})$, with $h(\mathbf{x}^0) = 0$ on the surface $S$ of the solid. The perturbation $h(\mathbf{x})$ represents physically the damaged zone in the solid. Generally, damage by microcraking may change the heat coefficient, but neither the density $\rho$ nor the specific heat $c$. Normalizing constants and variables, we assume that the unperturbed heat coefficient is equal to unity.

Let us consider a family $(\lambda)$ of experiments on the actual solid, with the unknown coefficient $k(\mathbf{x}) = 1 + h(\mathbf{x})$, and the heat flux $f_\lambda(\mathbf{x}^0, t)$. The response solution to the boundary condition is denoted by $\theta_\lambda(\mathbf{x}, t)$, which satisfies the following equations

$$\frac{\partial}{\partial t}\theta_\lambda(\mathbf{x}, t) - \operatorname{div}(k(\mathbf{x})\nabla\theta_\lambda(\mathbf{x}, t)) = 0 \qquad x \in \Omega \qquad (20)$$

$$\theta_\lambda(\mathbf{x}, t)(\mathbf{x}, t \le 0) = 0 \qquad \text{(initial condition)} \qquad (21)$$

$$\frac{\partial}{\partial n}\theta_\lambda(\mathbf{x}^0, t)(\mathbf{x}, t) = f_\lambda(\mathbf{x}^0, t) \qquad \mathbf{x}^0 \in S,\, t \ge 0 \qquad (22)$$

where $h(\mathbf{x}^0) = 0$ (i.e. $k(\mathbf{x}^0) = 1$) on $S$ has been taken into account. The inverse problem under consideration is to determine $h(\mathbf{x})$ from surface measurements of $f_\lambda(\mathbf{x}^0, t)$ and $\theta_\lambda(\mathbf{x}^0, t)$, during the time interval $[0, T]$. Here, we have much more data than necessary, so that the inverse problem is overdetermined.

Let us introduce two auxiliary fields $\phi_\lambda$ and $\psi_\mu$. The first field, for each $\lambda$, is solution of the unperturbed problem with the unit heat coefficient and the same surface data (22).

$$\frac{\partial}{\partial t}\phi_\lambda(\mathbf{x}, t) - \operatorname{div}(\nabla\phi_\lambda(\mathbf{x}, t)) = 0 \qquad x \in \Omega \qquad (23)$$

$$\phi_\lambda(\mathbf{x}, t)(\mathbf{x}, t \le 0) = 0 \qquad \text{(initial condition)} \qquad (24)$$

$$\frac{\partial}{\partial n}\phi_\lambda(\mathbf{x}^0, t)(\mathbf{x}, t) = f_\lambda(\mathbf{x}^0, t) \qquad \mathbf{x}^0 \in S,\, t \ge 0 \qquad (25)$$

The second field depending on another family of parameter ($\mu$) is solution of the adjoint problem

$$-\frac{\partial}{\partial t}\psi_\mu(\mathbf{x}, t) - \text{div}(\nabla\psi_\mu(\mathbf{x}, t)) = 0 \qquad x \in \Omega \tag{26}$$

$$\psi_\mu(\mathbf{x}, t \geq T) = 0 \qquad \text{(\textit{final} condition)} \tag{27}$$

$$\frac{\partial}{\partial n}\psi_\mu(\mathbf{x}^0, t) = g_\mu(\mathbf{x}^0, t) \qquad \mathbf{x}^0 \in S, t \leq T \tag{28}$$

The unperturbed direct problems (23)-(25) and the unperturbed adjoint problems (26)-(28) are well-posed problems, which can be solved by the finite element method. We assume that the auxiliary fields are known for all $\lambda$ and $\mu$. Multiplying (20) by $\psi_\mu$ and (26) by $\theta_\lambda$, then combining the results, yields

$$\int_0^T \int_\Omega \text{div}\{\theta_\lambda\nabla\psi_\mu - \psi_\mu\nabla\theta_\lambda\}\, dV_\mathbf{x}\, dt - \int_0^T \int_\Omega \psi_\mu\text{div}\{h\nabla\theta_\lambda\}\, dV_\mathbf{x}\, dt = 0 \tag{29}$$

Upon suitable application of the divergence formula and taking into account the boundary conditions (22), (28), eq. (29) can be arranged in the following form

$$\int_\Omega h(\mathbf{x}) K^{NL}(\mathbf{x}; \lambda, \mu, T)\, dV_\mathbf{x} = B(\lambda, \mu, T) \tag{30}$$

where

$$K^{NL}(\mathbf{x}; \lambda, \mu, T) = \int_0^T \nabla\psi_\mu \cdot \nabla\theta_\lambda\, dt \tag{31}$$

$$B(\lambda, \mu, T) = \int_0^T \int_S (\theta_\lambda g_\mu - \psi_\mu f_\lambda)\, dt\, dS_\mathbf{x} \tag{32}$$

The integrand in the right-hand side of (32) is a known surface quantity. But a look at (31) shows that the kernel $K^{NL}$ depends on the unknown field $\theta_\lambda$, hence eq. (31) is non-linear with respect to $h(\mathbf{x})$ (the superscript NL means non-linear kernel).

An alternative expression of (32) may be derived as follows:

$$\int_0^T \int_S \psi_\mu f_\lambda\, dS_\mathbf{x}\, dt = \int_0^T \int_S \psi_\mu \frac{\partial}{\partial n}\phi_\lambda\, dS_\mathbf{x}\, dt$$

$$= \int_0^T \int_\Omega \text{div}\psi_\mu\nabla\phi_\lambda\, dV_\mathbf{x}\, dt$$

$$= \int_0^T \int_\Omega \left\{\phi\frac{\partial}{\partial t}\psi_\mu + \nabla\phi_\lambda \cdot \nabla\psi_\mu\right\}\, dV_\mathbf{x}\, dt$$

$$= \int_0^T \int_S \phi_\lambda\frac{\partial}{\partial n}\psi_\mu\, dS_\mathbf{x}\, dt + \int_0^T \int_\Omega \frac{\partial}{\partial t}\psi_\mu\phi_\lambda\, dV_\mathbf{x}\, dt$$

$$= \int_0^T \int_S \phi_\lambda\frac{\partial}{\partial n}\psi_\mu\, dS_\mathbf{x}\, dt \tag{33}$$

$$B(\lambda, \mu, T) = \int_0^T \int_S (\theta_\lambda - \phi_\lambda)g_\mu\, dt\, dS_\mathbf{x} \tag{34}$$

In particular, application of (34) to the special case $\psi_\mu(\mathbf{x}, t) = V(\mathbf{x}, T - t; x_\mu)$ (with $x_\mu) \in S$, $t \in [0, T]$) gives:

$$B(\lambda, \mu, T) = \theta_\lambda(\mathbf{x}, t) - \phi_\lambda(\mathbf{x}, t) \tag{35}$$

Then, (34) and (37) provide a generalization of the Frechet derivative (17) for perturbations $h(\mathbf{x})$ of arbitrary amplitude:

$$\theta_\lambda(\mathbf{x}, t) - \phi_\lambda(\mathbf{x}, t) = \int_\Omega h(\mathbf{x}) \int_0^T \nabla\psi_\mu \cdot \nabla\theta_\lambda\, dt\, dV_\mathbf{x} \tag{36}$$

The adjoint method is also interesting when the perturbation $h(\mathbf{x})$ is small enough to allow the linearization process $\theta = \phi + O(h)$, or $\theta \sim \phi$. Substituting $\theta_\lambda$ by $\phi_\lambda$ in the kernel (31) yields the linear equation for $h(\mathbf{x})$:

$$\int_\Omega h(\mathbf{x}) K(\mathbf{x}; \lambda, \mu, T) \, dV_\mathbf{x} = B(\lambda, \mu, T) \tag{37}$$

where $B$ is still given by (32) and $K$ is a known kernel:

$$K(\mathbf{x}; \lambda, \mu, T) = \int_0^T \nabla \psi_\mu \cdot \nabla \phi_\lambda \, dt \tag{38}$$

Equations (37), (38) is nothing but the generalization to transient heat problem of the method suggested by Calderon (see [6], [10]) for stationary problem. Furthermore, eq. (36) then becomes identical to (17) (apart from notation differences).

## 5  Variational formulation of the inverse conductivity problem

Let us reformulate the model problem (10)-(12) in the weak form. Set the following definitions and notations

$$\mathcal{V} = \{v \in H^1(\Omega),\ v(\mathbf{x}, 0) = 0\ (\mathbf{x} \in \Omega),\ v(\mathbf{x}^0, t) = 0\ (\mathbf{x}^0 \in S,\ t \geq 0)\} \tag{39}$$

$$u' = \frac{\partial u}{\partial t} \tag{40}$$

$$\langle w, v \rangle_\Omega = \int_\Omega w(\mathbf{x}) v(\mathbf{x}) \, dV_\mathbf{x} \tag{41}$$

$$\langle f, v \rangle_S = \int_S f(\mathbf{x}) v(\mathbf{x}) \, dS_\mathbf{x} \tag{42}$$

$$a_\Omega^\eta(u, v) = \int_\Omega \eta(\mathbf{x}) \nabla u \cdot \nabla v \, dV_\mathbf{x} \tag{43}$$

Then the model problem with coefficient $\eta$ is the solution of the variational problem ('state equation')

$$\int_0^T \langle u', v \rangle_\Omega \, dt + \int_0^T a_\Omega^\eta(u, v) \, dt = \int_0^T \langle f, v \rangle_S \, dt \qquad \forall v \in \mathcal{V} \tag{44}$$

The optimization problem ($\min_\eta J(\eta)$) with respect to $\eta$ belongs to the optimal control theory for partial derivative equations, cf. Lions [14]. It consists of minimizing $J(u; \eta)$ with the constraints on $u$ which satisfies the variational equation (44). The classical approaches to this minimization problem, with constraints, make use of the Lagrange multiplier $\psi(\mathbf{x}, t)$ and the Lagrangian $\mathcal{L}$

$$\mathcal{L}(u; \eta, \psi) = J(u; \eta) + \int_0^T \{\langle u', \psi \rangle_\Omega + a_\Omega^\eta(u, v) - \langle f, \psi \rangle_S\} \, dt \tag{45}$$

It is clear that $\mathcal{L} \equiv J$ when (44) is satisfied. The optimal solution $u$ minimizing $J$ with the constraints (44) is the stationary point of $\mathcal{L}$. One observes that the stationarity condition $\delta \mathcal{L} = 0$ under fixed $u$ and $\eta$, and arbitrary $\delta \psi$ yields the variational equation ((44)) with $v \equiv \delta \psi$.

Consider now the variation of $\mathcal{L}$ due to $\delta u$ and $\delta \eta$:

$$\delta \mathcal{L} = \frac{\partial \mathcal{L}}{\partial u} \delta u + \frac{\partial \mathcal{L}}{\partial \eta} \delta \eta \tag{46}$$

$$\frac{\partial \mathcal{L}}{\partial u} \delta u = \int_0^T \{\langle u - \theta, \delta u \rangle_S + \langle \delta u', \psi \rangle_\Omega + a_\Omega^\eta(\delta u, \psi)\} \, dt \tag{47}$$

$$\frac{\partial \mathcal{L}}{\partial \eta} \delta \eta = \int_0^T a_\Omega^{\delta \eta}(u, \psi) \, dt. \tag{48}$$

We now restrict the choice of $\psi$ in such a way that $\delta\mathcal{L} = 0$ for $\delta\eta \equiv 0$, that is, we put:

$$\frac{\partial\mathcal{L}}{\partial u}\delta u = 0 \qquad \forall\delta u \in \mathcal{V} \tag{49}$$

By integrating the time integral by parts in (47), we obtain

$$\int_0^T \{\langle u - \theta, \delta u\rangle_S + a_\Omega^\eta(\delta u, \psi) - \langle \delta u, \psi'\rangle_\Omega\} \, dt + \langle \delta u, \psi'\rangle_\Omega \, |_0^T = 0 \qquad \forall\delta u \in \mathcal{V} \tag{50}$$

Since $\delta u(\mathbf{x}, 0) = 0$, one arrives at the adjoint *backward* heat equation for the lagrangian multiplier $\psi(\mathbf{x}, t)$

$$-\frac{\partial\psi}{\partial t} - \text{div}(\eta(\mathbf{x})\nabla\psi) = 0 \qquad x \in \Omega \tag{51}$$

$$\psi(\mathbf{x}, t \geq T) = 0 \qquad (\textit{final condition}) \tag{52}$$

$$k(\mathbf{x}^0)\frac{\partial\psi}{\partial n}(\mathbf{x}^0, t) = (u^\eta - \theta)(\mathbf{x}^0, t) \qquad \mathbf{x}^0 \in S, t \leq T \tag{53}$$

where $u^\eta$ is the solution of the state equation (44). This system corresponds to the back diffusion equation, with the final condition (52) and boundary condition (53), which is a well-posed problem. Its variational formulation reads:

$$\int_0^T -\langle\psi', w\rangle_\Omega \, dt + \int_0^T a_\Omega^\eta(\psi, w) \, dt = \int_0^T \langle u^\eta - \theta, w\rangle_S \, dt \qquad \forall w \in \mathcal{V}' \tag{54}$$

where $\mathcal{V}'$ is a test function space similar to $\mathcal{V}$ but with a *final* condition $w(\mathbf{x}, T) = 0$, $\forall w \in \mathcal{V}'$.

Finally, this particular choice $\psi^\eta(\mathbf{x}, t)$ for the Lagrange multiplier $\psi(\mathbf{x}, t)$ yields the following formula

$$\delta J(\eta) = \delta\mathcal{L}(u^\eta; \psi^\eta, \eta) = \int_0^T a_\Omega^{\delta\eta}(u^\eta, \psi^\eta) \, dt \tag{55}$$

Comparison with formula (18) shows that $\psi$ is related to the mismatch $(u^\eta - \theta)$ of surface data by the fundamental solution $V$. Although formulae (55) and (18) are different in their presentation, they represent esentially the same result.

## Some remarks about the result (55)

Equation (55) expresses analytically the gradient of $J(\eta)$ with respect to $\eta(\mathbf{x})$. It is therefore a very valuable tool for usual nonlinear optimization strategies, since:

1. It is an exact expression (provided the state and adjoint equations are solved exactly).

2. The whole gradient of $J$ is computed using only *one* adjoint equation per objective function considered, whereas a direct differentiation (either analytical or numerical) would require to set up as many auxiliary boundary/initial value problems as the number of design variables describing the conductivity field $k(\mathbf{x})$.

3. The adjoint variational equation (54) involves the same bilinear forms $\langle\cdot, \cdot\rangle_\Omega$ and $a_\Omega^\eta(\cdot, \cdot)$ than the state equation. In a finite element discretization approach, it means that the corresponding matrices are assembled only once. The numerical setting up of the adjoint equation needs only the building of a new right-hand side.

# 6  Variational formulation of the defect shape inverse problem

Let us consider the case where the physical nature of the defect is known (e.g. cavity, inclusion of a known material...). For simplicity of presentation, the present discussion is restricted to volumic defects, although other types may be considered (e.g. interface defects). Accordingly we put $\Omega = \Omega_1 \cup \Omega_2$, where $\Omega_1$ and $\Omega_2$ (the defect) have known thermal conductivities $k_1(\mathbf{x})$ and $k_2(\mathbf{x})$ respectively. We consider a variant of the variational approach of section 5 above, in which the unknown is the defect boundary $\Gamma = \partial \Omega_2$ instead of $k(\mathbf{x})$. Accordingly, the least-squares functional (4) is considered as a function of $\Gamma$: $J \equiv J(\Gamma)$.

As in the previous section, the minimization of $J(\Gamma)$ is constrained: $u$ must solve

$$\frac{\partial u}{\partial t} - \operatorname{div}(\eta_i(\mathbf{x})\nabla u) = 0 \qquad \mathbf{x} \in \Omega_i, \; i = 1,2 \tag{56}$$

$$u(\mathbf{x}, t \le 0) = 0 \qquad \text{(initial condition)} \tag{57}$$

$$k(\mathbf{x}^0)\frac{\partial u}{\partial n}(\mathbf{x}^0, t) = f(\mathbf{x}^0, t) \qquad \mathbf{x}^0 \in S, \, t \le T \tag{58}$$

$$[\![u]\!](\mathbf{x}, t) = [\![\eta\frac{\partial u}{\partial n}]\!](\mathbf{x}, t) = 0 \qquad \mathbf{x} \in \Gamma \text{ (continuity accross } \Gamma) \tag{59}$$

where $[\![\cdot]\!](\mathbf{x}) = (\cdot)_2(\mathbf{x}) - (\cdot)_1(\mathbf{x})$ ($\mathbf{x} \in \Gamma$) denotes the jump accross $\Gamma$. The lagrangian $\mathcal{L}$ for the problem under consideration accordingly reads:

$$\mathcal{L}(u; \psi, \Gamma) = J(u; \Gamma) + \int_0^T \{\langle u', \psi \rangle_\Omega + a_\Omega^\eta(u, v) - \langle f, \psi \rangle_S\} \, dt \qquad \psi \in \mathcal{V} \tag{60}$$

where $\mathcal{V}$ is defined by (40) (in particular, each $\psi \in \mathcal{V}$ is continuous accross $\Gamma$).

The derivative of $\mathcal{L}(u; \psi, \Gamma)$ with respect to $\Gamma$, necessary for the minimization of $J(\Gamma)$ using standard algorithms, is provided by the *shape differentiation approach* [19]. Let $\Gamma$ denote the current location of the unknown boundary during the minimization process, and assume a further evolution of the surface described by means of a time-like parameter $\tau$ and a normal 'velocity' field $v_n$:

$$\Gamma(\tau) = \Gamma + v_n \mathbf{n} \tau \tag{61}$$

while the external boundary $S$ remains fixed ($v_n(\mathbf{x}) = 0 \; \forall \mathbf{x} \in S$). Then the derivative $\frac{d\mathcal{L}}{d\tau}$ of a functional $\mathcal{L}$ is a linear form of the field $v_n$. In what follows, all derivatives with respect to $\tau$ are implicitly taken for $\tau = 0$.

Various formulas are given in the literature (see e.g. [19]) for the derivative of integrals with respect to variable volumes $\Omega$ or surfaces $\Gamma$, among which:

$$\frac{d}{d\tau}\int_\Omega a(\mathbf{x}, \tau)\, dV_\mathbf{x} = \int_\Omega \frac{\partial}{\partial\tau} a(\mathbf{x}, \tau)\, dV_\mathbf{x} + \int_{\partial\Omega} a(\mathbf{x}, \tau)v_n(\mathbf{x})\, dS_\mathbf{x} \tag{62}$$

$$\frac{d}{d\tau}\int_\Gamma a(\mathbf{x}, \tau)\, dS_\mathbf{x} = \int_\Gamma \left\{\frac{\partial}{\partial\tau}a(\mathbf{x}, \tau) + \left(\frac{\partial}{\partial n}a(\mathbf{x}, \tau) - 2K(\mathbf{x})a(\mathbf{x}, \tau)\right)v_n(\mathbf{x})\right\} dS_\mathbf{x} \tag{63}$$

where $K(\mathbf{x})$ denotes the mean curvature at $\mathbf{x} \in \Gamma$. Equation (63) holds only for a closed smooth surface, while in equation (62) $v_n$ refers to the unit normal $\mathbf{n}$ directed towards the exterior of $\Omega$. Application of formulas (62)-(63) above gives:

$$\frac{d}{d\tau}\mathcal{L} = \frac{\partial\mathcal{L}}{\partial u}\frac{\partial u}{\partial \tau} + \frac{\partial\mathcal{L}}{\partial\Gamma}v_n \tag{64}$$

$$\frac{\partial\mathcal{L}}{\partial u}\frac{\partial u}{\partial \tau} = \int_0^T \left\{ \langle u - \theta, \frac{\partial u}{\partial \tau}\rangle_S + \langle\frac{\partial u'}{\partial \tau}, \psi\rangle_\Omega + a_{\Omega_1}^{\eta_1}(\frac{\partial u}{\partial \tau}, \psi) + a_{\Omega_2}^{\eta_2}(\frac{\partial u}{\partial \tau}, \psi) \right\} dt \tag{65}$$

$$\frac{\partial\mathcal{L}}{\partial\Gamma}v_n = \int_0^T \langle [u'\psi + \eta\nabla u \cdot \nabla\psi], v_n\rangle_\Gamma \, dt \tag{66}$$

where the condition $v_n \equiv 0$ on $S$ has been taken into account.

Now the choice of $\psi$ is restricted in such a way that $\frac{d}{d\tau}\mathcal{L} = 0$ for $v_n \equiv 0$, that is, we put:

$$\frac{\partial\mathcal{L}}{\partial u}\frac{\partial u}{\partial \tau} = 0 \qquad \forall\frac{\partial u}{\partial \tau} \in \mathcal{V} \tag{67}$$

By integrating the time integral by parts in (65), (67) and (65) give:

$$\left(\forall\frac{\partial u}{\partial \tau} \in \mathcal{V}\right) \int_0^T \left\{ \langle u - \theta, \frac{\partial u}{\partial \tau}\rangle_S + a_{\Omega_1}^{\eta_1}(\frac{\partial u}{\partial \tau}, \psi) + a_{\Omega_2}^{\eta_2}(\frac{\partial u}{\partial \tau}, \psi) - \langle\frac{\partial u}{\partial \tau}, \psi'\rangle_\Omega \right\} dt + \langle\frac{\partial u}{\partial \tau}, \psi'\rangle_\Omega \big|_0^T = 0 \tag{68}$$

The statement of the (backward) adjoint problem readily follows:

$$-\frac{\partial\psi}{\partial t} - \text{div}(\eta_i(\mathbf{x})\nabla\psi) = 0 \qquad \mathbf{x} \in \Omega_i, \, i = 1, 2 \tag{69}$$

$$\psi(\mathbf{x}, t \geq T) = 0 \qquad (\textit{final condition}) \tag{70}$$

$$k(\mathbf{x}^0)\frac{\partial\psi}{\partial n}(\mathbf{x}^0, t) = (u_\Gamma - \theta)(\mathbf{x}^0, t) \qquad \mathbf{x}^0 \in S, t \leq T \tag{71}$$

$$[\psi](\mathbf{x}, t) = [\eta\frac{\partial\psi}{\partial n}](\mathbf{x}, t) = 0 \qquad \mathbf{x} \in \Gamma \, (\text{continuity accross } \Gamma) \tag{72}$$

Finally, the jump conditions (59) on $\Gamma$ imply that

$$[u_\Gamma'\psi + \eta\nabla u_\Gamma \cdot \nabla\psi] = \eta_1\frac{\partial u_\Gamma}{\partial n}[\frac{\partial\psi}{\partial n}] \tag{73}$$

and, as a result, the derivative $\frac{d}{d\tau}J(\Gamma)$ is given by:

$$\frac{d}{d\tau}J(\Gamma) = \frac{d}{d\tau}\mathcal{L}(u_\Gamma; \Gamma, \psi_\Gamma) = \int_0^T \langle\eta_1\frac{\partial u_\Gamma}{\partial n}[\frac{\partial\psi}{\partial n}], v_n\rangle_\Gamma \, dt \tag{74}$$

## Some remarks about the result (74)

1. Expression (74) gives at once the whole gradient of $J(\Gamma)$, and (linearly) depends upon the design variables which describe the current surface $\Gamma$ through the normal 'velocity' $v_n$.

2. Contrarily to the result (55) of the previous section, the derivative of $J(\Gamma)$ is expressed using only boundary integrals, as is always the case in shape differentiation approach [19].

3. It is worth noticing that, in the case of *piecewise constant* material properties (i.e. $\eta_i(\mathbf{x}) \equiv \eta_i$, $i = 1, 2$), the temperature field $u^\eta(\mathbf{x}, t)$ and the adjoint field $\psi^\eta(\mathbf{x}, t)$ may be conveniently solved using boundary integral equations (BIE). This, combined to the 'boundary only' character of (74), allows a 'boundary only' treatment of the shape identification problem. The BIE formulation is well-known (see [4] among many references) and will not be repeated here.

# 7  Use of the constitutive equation error

The so-called 'constitutive equation error' $E$ is a special type of objective function which has been considered e.g. in [11] for electric conductivity inverse problems or in [20] for elastic FEM model updating. For thermal problems, let us assume that the boundary condition (3) holds and that the temperature field of the solid with defect $\delta\eta$ is measured on the entire domain $\Omega$: $u^{\eta+\delta\eta} = U$ on $\Omega$. The following functional $J(u, q; \eta)$ is introduced:

$$
\begin{aligned}
J(u, q; \eta) &= \frac{1}{2}\int_0^T \left\{ \int_\Omega \frac{1}{\eta}(q - \eta\nabla u)\cdot(q - \eta\nabla u)\,dV_{\mathbf{x}} + \frac{\gamma}{2}\int_\Omega \eta\nabla(u - U)\cdot\nabla(u - U)\,dV_{\mathbf{x}} \right\} dt \\
&= \frac{1}{2}\langle\frac{1}{\eta}(q - \eta\nabla u), q - \eta\nabla u\rangle_\Omega + \frac{\gamma}{2}a_\Omega^\eta(u - U, u - U)
\end{aligned}
\tag{75}
$$

where $u, q$ denote 'admissible' temperature and heat flux fields respectively: $u \in \mathcal{V}$ and $q \in \mathcal{B}$, with

$$
\mathcal{B} = \{q \mid \mathrm{div}(\eta q) - u' = 0, \; q\cdot n(\mathbf{x}) = f(\mathbf{x}), \mathbf{x} \in S\}
\tag{76}
$$

and $\gamma$ is an adjustable weighing constant, which expresses the expected degree of accuracy of the measured field $U$. Let $u_\Omega$, $q_\Omega$ and $J(\eta)$ be defined as:

$$
\begin{aligned}
J(\eta) &= \min_{u\in\mathcal{V}, q\in\mathcal{B}} J(u, q; \eta) \\
(u_\Omega, q_\Omega) &= \mathrm{Arg}\,\min J(u, q; \eta)
\end{aligned}
\tag{77}
$$

Then the constitutive equation error $E(\eta)$ is defined as:

$$
E(\eta) = \frac{1}{2}\int_0^T \left\{ \langle\frac{1}{\eta}(q_\Omega - \eta\nabla u_\Omega), q_\Omega - \eta\nabla u_\Omega\rangle_\Omega \right\} dt
\tag{78}
$$

Let us examine how $E(\eta)$, together with the gradient of $J$ with respect to $\eta$ are computed in practice. Due to the constraint $q \in \mathcal{B}$, the following lagrangian $\mathcal{L}$, with multiplier field $w(\mathbf{x}, t)$, is introduced:

$$
\mathcal{L}(u, q, w; \eta) = J(u, q; \eta) + \int_0^T \left\{ \langle u', w\rangle_\Omega + b_\Omega(q, w) - \langle f, w\rangle_S \right\} dt
\tag{79}
$$

$$
b_\Omega(q, w) = \int_\Omega q\cdot\nabla w\,dV_{\mathbf{x}}
\tag{80}
$$

Its variation is given by:

$$
\delta\mathcal{L} = \frac{\partial\mathcal{L}}{\partial q}\delta q + \frac{\partial\mathcal{L}}{\partial u}\delta u + \frac{\partial\mathcal{L}}{\partial\eta}\delta\eta
\tag{81}
$$

$$
\frac{\partial\mathcal{L}}{\partial q}\delta q = \int_0^T \left\{ b_\Omega(\delta q, w) + \langle\frac{1}{\eta}(\delta q, q - \eta\nabla u)\rangle_\Omega \right\} dt
\tag{82}
$$

$$
\frac{\partial\mathcal{L}}{\partial u}\delta u = \int_0^T \left\{ \gamma a_\Omega^\eta(u - U, u - U) + \langle\delta u', w\rangle_\Omega - b_\Omega(q - \eta\nabla u, \delta u) \right\} dt
\tag{83}
$$

$$
\frac{\partial\mathcal{L}}{\partial\eta}\delta\eta = \int_0^T \langle\delta\eta(\nabla u - \frac{q}{\eta}), \nabla u - \frac{q}{\eta}\rangle_\Omega a_\Omega^{\delta\eta}(\nabla u - \frac{q}{\eta}, \nabla u + \frac{q}{\eta}))\,dt
\tag{84}
$$

First, $J(\eta) = \mathcal{L}(u_\Omega, q_\Omega, w_\Omega; \eta)$, where $(u_\Omega, q_\Omega, w_\Omega)$ solve

$$
\forall\delta q \in \mathcal{B}, \; \forall T > 0 \qquad \frac{\partial\mathcal{L}}{\partial q}\delta q = 0
\tag{85}
$$

$$
\forall\delta u \in \mathcal{V}, \; \forall T > 0 \qquad \frac{\partial\mathcal{L}}{\partial u}\delta u = 0
\tag{86}
$$

and the constraint (76). This leads to the following equations:

$$q = \eta \nabla (u - w) \tag{87}$$

and

$$\forall \delta u \in \mathcal{V} \quad \gamma a_\Omega^\eta (u - U, \delta u) - \langle \delta u', w \rangle_\Omega + b_\Omega (\eta \nabla w, \delta u)$$
$$\forall v \in \mathcal{V} \quad \langle u', v \rangle_\Omega + a_\Omega^\eta (u, v) - \langle f, w \rangle_S = a_\Omega^\eta (w, v)$$

The last two equations above can be rewritten as follows, putting $u = u^\eta + \Delta u$, $U = u^\eta + \Delta U$ with $u^\eta$ the solution of the unperturbed direct problem:

$$\forall \delta u \in \mathcal{V} \quad \gamma a_\Omega^\eta (\Delta u - \Delta U, \delta u) + \langle \delta u, w' \rangle_\Omega + b_\Omega (\eta \nabla w, \delta u)$$
$$\forall v \in \mathcal{V} \quad \langle \Delta u', v \rangle_\Omega + a_\Omega^\eta (\Delta u, v) = a_\Omega^\eta (w, v)$$

or, using 'stiffness' and 'mass' operators $\mathbf{K}$, $\mathbf{M}$ (e.g. within the finite element method framework):

$$\mathbf{K}w - \mathbf{M}w' = -\gamma \mathbf{K}(\Delta u - \Delta U) \tag{88}$$
$$\mathbf{K}w = \mathbf{K}\Delta u + \mathbf{M}\Delta u' \tag{89}$$

Then $\Delta u$ is expressed in terms of $w$ in terms of $w$ using (89):

$$\Delta u = \Delta U - \frac{1}{\gamma} \left( w - \mathbf{K}^{-1} \mathbf{M} w' \right) \tag{90}$$

and the result is inserted in (88), giving:

$$\mathbf{K}w + \frac{1}{\gamma} \left( \mathbf{K}w - \mathbf{M}\mathbf{K}^{-1}\mathbf{M}w'' \right) = \mathbf{K}\Delta U + \mathbf{M}\Delta U' \tag{91}$$

Now let us recall that $U = u^\eta + \Delta U = u^{\eta + \delta \eta}$ ($u^\eta$ is the temperature field solution of the perturbed direct problem), so that the right-hand side of (91) becomes:

$$\mathbf{K}\Delta U + \mathbf{M}\Delta U' = -\Delta \mathbf{K} U \tag{92}$$

where $\Delta \mathbf{K}$ is the perturbation of $\mathbf{K}$ induced by the conductivity perturbation $\delta \eta$.

Let us now consider tha case $\gamma \gg 1$ in eqn. (91): the measured field $U$ is considered very accurate, and accordingly given a large weight in the functional (75). Then, from eqns. (91) and (92), one has for the solution $w_\Omega$ of the coupled equations (88), (89):

$$\mathbf{K}w_\Omega = -\Delta \mathbf{K} U + O(\frac{1}{\gamma}) \tag{93}$$

which means that $\mathbf{K}w_\Omega$, which is computed without actual knowledge of $\Delta \mathbf{K}$, takes nonzero values only at points (or on elements, in a FEM approach) where $\delta \eta$ is nonzero. This allows, at least in the rather idealized situation considered here where $U$ is known with great confidence over the entire $\Omega$, the geometrical localization of the defect. Similarly, one can see from (78), (87) and (93) that the value of the constitutive equation error is given by:

$$E(\eta) = \frac{1}{2} a_\Omega^\eta (w, w)$$
$$= -\frac{1}{2} a_\Omega^{\delta \eta} (w, U) + O(\frac{1}{\gamma}) \tag{94}$$

Expression (94) is additive with respect to $\Omega$ and can then be split into a sum over a partition of $\Omega$ in subdomains (e.g. finite elements), which can be used to indicate which are the subdomains with nonzero $\delta\eta$.

Then, the subsequent minimization of $J(\eta)$ can be given a reduced size by inspecting the distribution of the density of $E(\eta)$ over $\Omega$ and deciding in advance where $\eta$ has to be corrected. As usual, it can be desirable to use an exact expression of the gradient of $J(\eta)$ with respect to $\eta$:

$$
\begin{aligned}
\delta J(\eta) &= \frac{\partial J}{\partial \eta}\delta\eta \\
&= \delta\mathcal{L}(u_\Omega, q_\Omega, w_\Omega; \eta) \\
&= \frac{\partial \mathcal{L}}{\partial \eta}\delta\eta \\
\frac{\partial \mathcal{L}}{\partial \eta}\delta\eta &= \int_0^T a_\Omega^{\delta\eta}(w_\Omega, 2u_\Omega - w_\Omega)\, dt.
\end{aligned}
\tag{95}
$$

## Comments about the above analysis

1. In this approach, the objective function to be minimized is $J(\eta)$, but the constitutive equation error $E(\eta)$, which contributes to $J$, is used in order to restrict the geometrical area over which a nonzero correction $\delta\eta$ is sought.

2. The error localization property of $E$ have been initially studied and applied in [20], [12] for elastic FEM model updating, allowing substantial reduction of the size of the inversion problem. To our best knowledge, it has not yet been applied to thermal inverse problems.

3. In more realistic situations where $U$ is only known over a subset of $\Omega$ and with small but not infinitesimal $1/\gamma$, the localization property is expected to hold, though obviously in an approximate manner.

4. A similar kind of error functional has also been introduced by ·for transient thermal inverse problems, but with no attempt to geometrically localize the conductivity error.

5. Using (95), any conventional optimization algorithm using gradients can be used. As in other cases discussed in this paper, the computation of the variation $\delta J$ of the error functional $J(\eta)$ uses two temperature fields $u_\Omega$, $w_\Omega$. However, their computation is somewhat more complicated due to their coupling through eqns. (88)–(89).

# References

[1] Balageas D.L., Deom A.A., Boscher D.M. (1987) - Characterization of Non Destructive Testing of Carbon-Epoxy Composites by a pulsed photothermal method, Materials Evaluation (45)4, pp.461-465.

[2] Banks H.T., Kojima F. (1989) - Boundary Shape Identification Problems in Two-dimensional Domains Related to Thermal Testing of Materials. Quart. Appl. Math. 47(2), pp 273-293.

[3] Bonnet M., Bui H.D., Planchard J. (1989) - Problème inverse pour l'équation de la chaleur: applications au contrôle non destructif thermique; Report Electricité de France no. HI-70-6391, pp.1-13, Clamart, France.

[4] Brebbia C.A., Telles J.C.F., Wrobel L.C. (1984) - Boundary Element Techniques. Theory and Application in Engineering. Springer - Verlag, 1984.

[5] Carslaw H.S., Jaeger J.C. (1973) - Conduction of heat in solids; Oxford University Press, Oxford, UK.

[6] Calderon A.P. (1980) - On an inverse boundary value problem; Seminar on Numerical Analysis and its applications to Continuum Physics, Soc. Brasilian de Matematica, Rio de Janeiro, pp.65-73.

[7] Connolly T.J., Wall D.J.N. (1988) - On an Inverse Problem, With Boundary Measurements, for the Steady State Diffusion Equation. Inverse Problems, 4 pp 995-1012.

[8] Friedman A., Vogelius M. (1989) - Determining cracks by boundary measurements; Indiana Univ. Math. J.; 38, pp.527-556.

[9] Kohn R., Vogelius M; (1984) - Determining conductivity by boundary measuremnts; Comm. Pure Appl. Math., 37, pp.289-298.

[10] Isaacson D., Isaacson E.L. (1989) - Comment on Calderon's paper "On an Inverse Boundary Value Problem". Math. Comput. 52, pp.553-559.

[11] Kohn R., McKenney A. - Numerical implementation of a variational method for electric impedance tomography. Inverse Problems, 6 pp 389-414, 1990.

[12] Ladevèze P., Reynier M., Nedjar D. - Parametric correction of finite element models using modal tests. *IUTAM Symposium on Inverse Problems in Engineering Mechanics (Tokyo, 11-15 may 1992)*, H.D. Bui & M. Tanaka, eds., Springer-Verlag.

[13] Lavrentiev M.M. (1967) - Some improperly posed problems of mathematical physics, Springer-Verlag.

[14] Lions J.L. (1968) - Contrôle optimal de systèmes gouvernés par des équations aux dérivées partielles, Dunod, Paris.

[15] Lund J., Vogel C.R. (1990) - A fully-Galerkin approach for the numerical solution of an inverse problem in a parabolic partial differential equation. Inverse Problems, 6 pp. 205-217.

[16] Marchuk G.I. (1982) - Methods of numerical mathematics, chapter 7 of *Numerical methods for some inverse problems*, pp 312-351, Springer Verlag.

[17] Menke W. - Geophysical data analysis : discrete inverse theory. Academic Press, 1984.

[18] H.G. Natke (1992) - On Regularization Methods within System Identification, In *IUTAM Symposium on Inverse Problems in Engineering Mechanics*, H.D. Bui & M. Tanaka, eds., Springer-Verlag.

[19] Petryk H., Mroz Z. (1986) - Time derivatives of integrals and functionals defined on varying volume and surface domains. Arch. Mech. 38(5-6), pp.697-724.

[20] Reynier M. - Sur le contrôle de modélisations éléments finis: recalage à partir d'essais dynamiques. PhD thesis, Ecole Normale Supérieure de Cachan, France, 1990.

[21] Tarantola A. (1987) - Inverse problem theory. Elsevier.

[22] Tikhonov A.N., Arsenin V.Y. (1977) - Solutions to ill-posed problems, Winston Wiley, New York, 1977.

[23] Vogel C.R. (1987) - An overview of numerical methods for nonlinear ill-posed problems, in *Inverse and ill-posed problems*, H.W. Engl and C.W. Groetsch, eds., Academic Press.

# Determination of Void Shapes, Sizes, Numbers and Locations Inside an Object with Known Surface Temperatures and Heat Fluxes

G. S. Dulikravich  and  T. J. Martin

Department of Aerospace Engineering, The Pennsylvania State University
University Park, PA 16802, USA

## Summary

During the past several years we have developed an inverse method that allows a thermal cooling system designer to determine proper sizes, shapes, and locations of coolant fluid passages (holes) in, say, an internally cooled turbine blade. The internally cooled object can be made of materials having distinct thermal diffusivities. Thermal expansion has been neglected although, in principle, it could be included. The result is a simple Laplace's equation for the steady temperature field over a multiply connected domain that is subject to overspecified thermal boundary conditions. This same methodology with minor modifications concerning the type of thermal boundary conditions has been utilized in the nondestructive detection of possible voids in objects with known surface temperature and heat flux distributions.

## Inverse design of coolant flow passages

In the case of an inverse thermal design [1-8], the designer can iteratively enforce a desired heat flux distribution $q_{out}^{des}$ on the hot outer surface of the blade, while simultaneously satisfying the desired temperature distributions $T_{out}^{spec}$ on the hot outer surface and specified temperatures $T_{in}^{spec}$ on the cooled surfaces of each of the holes. This constitutes an over-specified boundary value problem. We can solve the direct Dirichlet problem with the specified temperatures on the inner and outer boundaries for the initially guessed configuration. The temperature field analysis was performed using our boundary integral element code with linearly varying temperature along straight surface panels. Nevertheless, the computed outer surface heat fluxes $q_{out}^{comp}$ corresponding to the initial guess will not be the same as the desired outer surface heat fluxes $q_{out}^{des}$. A properly scaled $L_2$ norm of the difference between the desired outer surface heat fluxes $q_{out}^{des}$ and the computed outer surface heat fluxes $q_{out}^{comp}$ is then minimized by iteratively changing the sizes, shapes, and locations of the holes. Starting with a large number of guessed holes, all unnecessary coolant passages will eventually be eliminated when their sizes reduce below a prespecified minimal allowable value. This procedure is conceptually similar to optimization of the time evolution of thermal boundary conditions for an object having a fixed geometry and specified time rate of change of temperature throughout the object. For example, this time-dependent inverse problem has been developed for possible application in the optimization of freezing protocols for organs intended for transplant surgery [9]. Minimization of the $L_2$ norm was performed automatically using a standard gradient search optimization algorithm of Davidon-Fletcher-Powell [10]. Local minimas in the optimization process were successfully avoided by changing the formulation for the objective function whenever the local minimas were detected [6]. Two definitions of the objective function were used. They represented two different forms of the $L_2$ norm of the difference between the computed and the desired heat fluxes on the outer surface. Thus, the objective function was either a normalized global error

$$F_1(x) = \frac{\sum_{j=1}^{N}\left(q_j{}^{comp} - q_j{}^{des}\right)^2}{\sum_{j=1}^{N}\left(q_j{}^{des}\right)^2} \times 100 \qquad (1)$$

or a normalized local error at each node

$$F_2(x) = \sum_{j=1}^{N} \frac{\left(q_j{}^{comp} - q_j{}^{des}\right)^2}{\left(q_j{}^{des}\right)^2} \times 100 \qquad (2)$$

where the outer boundary was discretized with N surface panels. Besides minimizing the heat flux error on the outer surface, the final configuration has to satisfy two constraints [4]: a) specified minimum allowable distance between any two holes, $d^{hole}$, having radii $r_i$ and $r_k$, and b) specified minimum allowable distance between any hole and the outer boundary, $d^{surf}$. The two constraints were incorporated into the objective function using a barrier function [10]

$$B(g(x),w_b) = \frac{1}{w_b}\sum_{i=1}^{M}\left[\sum_{j=1}^{N}\frac{d^{surf}}{\left(D_j^{surf}-d^{surf}-r_i\right)} + \sum_{k=1}^{M}\frac{d^{hole}}{\left(D_k^{hole}-d^{hole}-r_i\right)}\right] \qquad (3)$$

Here, M is the total number of holes and $w_b$ is the user specified barrier coefficient which was reduced as the process kept converging. Thus, the composite objective function can have two forms depending on whether the global or local objective function is used for its evaluation.

$$F_m(g(x),w_b) = F_m(x) + B(g(x),w_b) \qquad m = 1, 2 \qquad (4)$$

In summary, the inverse design protocol consists of the following steps:
(1)    Specify shape of the outer surface and coating of the turbine blade. Also, specify thermal diffusivities of the blade and coating materials.

(2)    Specify temperatures $T_{in}^{spec}$ and $T_{out}^{spec}$ on the inner and outer surfaces, respectively.

(3)    Specify the desired heat flux $q_{out}^{des}$ values on the outer surface.

(4)    Specify manufacturing constraints:
(i) minimum distance $d^{surf}$ between holes and the outer surface,
(ii) minimum distance $d^{hole}$ between any two neighboring holes.

(5)    Specify initial guess for the number of holes, M, their radii, $r_i$, and locations of the centers of the holes, $x_{ci}$ and $y_{ci}$. Thus, there will be 3×M design variables if we limit ourselves to circular holes only. Recently [8], we have allowed hole shapes to vary according to a Lame curve (superelliptic function) expressed as

$$((x - x_{ci})/a_{ci})^{n_i} + ((y - y_{ci})/b_{ci})^{n_i} = 1 \qquad\qquad (5)$$

Here, $x_{ci}$ and $y_{ci}$ are the Cartesian coordinates of the hole center, $a_{ci}$ and $b_{ci}$ are the two semi-axis of the hole contour, and the exponent $n_i$ can vary from $n_i<1$ (a four-pointed star), to $n_i=1$ (a diamond shape), to $n_i=2$ (a circle or an ellipse), to $n_i>>1$ (a square or a rectangle). Moreover, the local x,y Cartesian coordinate system can be inclined at an arbitrary angle $\alpha_i$ with respect to the global x',y' Cartesian coordinate system. Thus, when utilizing Lame shapes we will have to optimize a total of 6×M design variables.

(6)    Using the boundary element method, Laplace's equation for a given multiply connected domain and temperature boundary conditions is solved and heat fluxes $q_{out}^{comp}$ at the outer boundary are computed.

(7)    Determine relative error between the desired and computed heat fluxes and evaluate the objective function. At the same time the barrier function has to be evaluated to determine the composite objective function $F_m$.

(8)    Each of the three (in the case of circular holes) or each of the six (in the case of Lame shaped holes) design parameters are perturbed by a small amount and steps (6) and (7) are repeated for each perturbation.

(9)    The Davidon-Fletcher-Powell gradient search algorithm [8] is used to find gradient directions for each of the design variables and new updated values of the design variables.

(10)   The optimization procedure is repeated starting from step (6) until the corresponding composite objective function $F_m$ is less than a prespecified value. If the dimension of any hole becomes less than a prespecified value, the hole is explicitly eliminated from further optimization. If the optimization procedure reaches a local minimum, the objective function formulation is automatically changed from Eq. 2 to Eq. 3 while continuing optimization starting with step (6).

## Nondestructive detection of voids

This same methodology and the accompanying software have been adapted to nondestructive determination of the actual number, sizes, shapes, and locations of possible voids inside an arbitrarily shaped solid object which can contain regions with different thermal diffusivities. If the voids are assumed to be free of any material (vacuum voids) than the only heat transfer mode across the voids is by pure thermal radiation. Thus, the thermal boundary condition on the surfaces of the voids should be of a Neumann type in terms of the fourth power of temperature. Nevertheless, since the actual void shapes and their surface radiative properties are not known *a priori* , we have decided to treat the surface of each void as an adiabatic boundary. That is, normal derivative of temperature on the surface of each void was set to zero.

A rudimentary example of a well-posed elliptic problem can now be formulated for the object of a known and fixed size and shape by providing measured temperatures $T_{out}^m$ along the outer surface of the body, while at the same time specifying zero temperature gradient normal to the surfaces (walls) of the voids inside the solid object. Laplace's equation can now be integrated numerically in this multiply connected domain subject to measured Dirichlet boundary condition on the outer surface and the specified Neumann boundary conditions on the surfaces of the guessed voids. Nevertheless, the actual number, sizes, locations, and shapes of the voids are still unknown. Therefore, in addition to the measured values of the outer surface temperatures $T_{out}^m$ we also need to provide measured outer surface heat fluxes $q_{out}^m$. This combination of $T_{out}^m$ and $q_{out}^m$ represents an over-specified elliptic problem meaning that in general Laplace's equation has no solution subject to these conditions, except, possibly for a very special combination of the number, sizes, locations, and shapes of the voids. Starting with a large number of guessed voids, all non-existent voids should be eliminated when their sizes reduce below a prespecified minimal value. The optimization algorithm for detection of possible voids then follows the protocol used in the inverse design of coolant flow passages.

## Results for inverse design

To demonstrate the accuracy and efficiency of the analysis and optimization algorithms a circular coated disk made of a thermally homogeneous material was assigned a constant temperature on its outer boundary and another constant temperature on a centrally located circular hole perimeter. The heat flux on the outer boundary is thus constant and analytically known. Pretending that we do not know the actual number of the holes, their sizes and shapes, we have guessed that there are three circular holes (Fig. 1) asymmetrically positioned [6] within the inner disk each having its surface temperature equal to the temperature that a single centrally located hole had. In only nine iteration cycles, requiring 1162 calls to the boundary element analysis code, two out of three initially guessed holes have reduced to a negligible size (Fig. 1), while the third hole converged to the correct solution of a large centrally located hole. The inverse design concept was then applied with confidence to the same coated disk with ten initially guessed holes [6] all having the same surface temperature corresponding to the only correct solution of a centrally located large hole. After successfully circumventing problems with local minimas by automatically alternating between the two formulations (Eqs.(1) and (2)) for the objective function, nine holes reduced to a negligible size while the tenth hole assumed the only correct centrally located size (Fig. 2). The inverse design was then tested with general Lame (superelliptic shapes) where three initially guessed holes had elliptic, rectangular and square shape, respectively. Each hole had the same constant surface temperature assigned that corresponded to the only correct temperature on a centrally located circular large hole. The convergence history of this test case (Figs. 3 and 4) indicate that it took over 120 optimization cycles to eliminate the initially elliptical hole and the initially square hole and to transform the initially rectangular hole into the correctly sized and centrally located circular hole.

## Results for nondestructive detection of voids and cracks

Again, to test the accuracy of the basic analysis code and the optimization algorithm we chose the circular coated disk with a centrally located circular hole. Ratio of thermal diffusivities of the coating and the core material was 1:5. Outer surface and interface between the coatinf and the core material were discretized with 36 panels each. The correct solution was a centrally located circular hole that had zero surface heat flux, while the outer surface of the disk had variable temperature distribution. This time we guessed that there is one elliptically shaped void eccentrically located in the disk and that it has zero surface heat flux. The surfaceof the void was discretized with 16 panels. This initial guess converged (Figs. 5 and 6) to the correct centrally located circular hole solution in ten optimization cycles requiring 375 calls to the analysis code. During this exercise we varied five out of six design variables associated with Lame curve formulation. Lame exponent was kept at the value n = 2.

To test the capability of the code to detect proper number, sizes, shapes and locations of the possible voids, we took the same coated disk configuration but now with a narrow vertical straight crack having zero surface heat flux and placed to the right of the center of the disk as the correct solution. Outside variable temperature boundary condition was the same as in the previous case. We have decided to guess that there are three straight parallel cracks (Fig. 7), one of which corresponded with the correct crack size, shape and location. Each of the initially guessed cracks had zero surface heat flux assigned. During the optimization, the crack geometric parameters corresponding to a general Lame curve formulation were varied except for the minor semi-axis b and for the Lame exponent, n. Although the coating acted as an effective smoothing device for temperature gradients inside the core material, this test case converged after 18 optimization cycles requiring 252 calls to the analysis code. The left-of-center crack and the centrally located crack reduced in size to zero (Fig.8). Thus, the capability of the code to detect thin simple cracks has been verified.

References

1.   Kennon, S.R., and Dulikravich, G.S.: The Inverse Design of Internally Cooled Turbine Blades, *ASME Journal of Engineering Gas Turbines and Power*, (Jan. 1985) 123-126.

2.   Kennon, S.R. and Dulikravich, G.S.: Inverse Design of Multiholed Internally Cooled Turbine Blades," *Int. Jour. of Numerical Methods in Eng.*, Vol. 22, (1986a) 363-375.

3.   Kennon, S.R. and Dulikravich, G.S.: Inverse Design of Coolant Flow Passage Shapes With Partially Fixed Internal Geometries, *International Journal of Turbo & Jet Engines*, Vol. 3, 1, (1986b) 13-20.

4.   Chiang, T.L. and Dulikravich, G.S.: Inverse Design of Composite Turbine Blade Circular Coolant Flow Passages, *ASME Journal of Turbomachinery*, Vol. 108 (1986) 275-282.

5.   Dulikravich, G.S.: Inverse Design and Active Control Concepts in Strong Unsteady Heat Conduction, *Applied Mechanics Reviews*, Vol. 41, No. 6 (June 1988) 270-277.

6.   Dulikravich, G.S. and Kosovic,B.: Minimization of the Number of Cooling Holes in Internally Cooled Turbine Blades, ASME paper 91-GT-103, ASME Gas Turbine Conf., Orlando, FL, (June 1991); also to appear in *Int. J. of Turbo & Jet Engines*, (1992).

7.   Dulikravich, G.S.: Inverse Design of Proper Number, Shapes, Sizes and Locations of Coolant Flow Passages, Proc. 10th Annual Workshop on CFD Appl., NASA MSFC, Huntsville, AL, (April 1992).

8.   Dulikravich, G.S. and Martin, T.J.: Determination of the Proper Number, Locations, Sizes and Shapes of Superelliptic Coolant Flow Passages in Turbine Blades, Proc. of ICHMT Internat. Symp. on Heat Trans. in Turbomachin., Athens, Greece (Aug. 1992).

9.   Dulikravich, G.S. and Hayes, L.J.: Control of Surface Temperatures to Optimize Survival in Cryopreservation, Proc. of ASME WAM'88 Symp. on Comput. Meth. in Bioeng., Ed: R.L. Spilker and B.R. Simon, BED-Vol.9, (Nov.27-Dec.2, 1988) 255-265.

10.  Vanderplaats, G.N.: Numerical Optimization Techniques for Engineering Design, McGraw-Hill, New York, (1984).

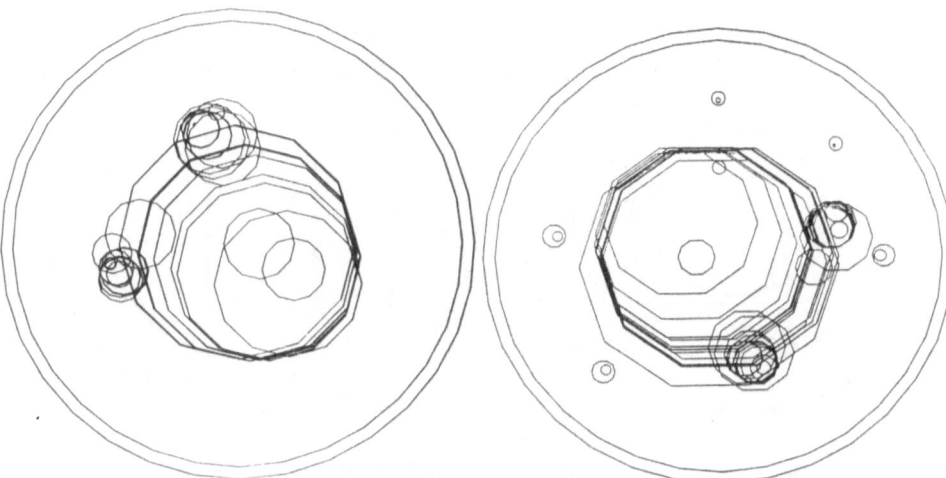

Fig. 1 Inverse design of a centrally located circular hole with an initial guess consisting of three asymmetrically positioned circular holes: configuration evolution history.

Fig.2 Inverse design of a centrally located circular hole with an initial guess consisting of ten asymmetrically positioned circular holes: configuration evolution history.

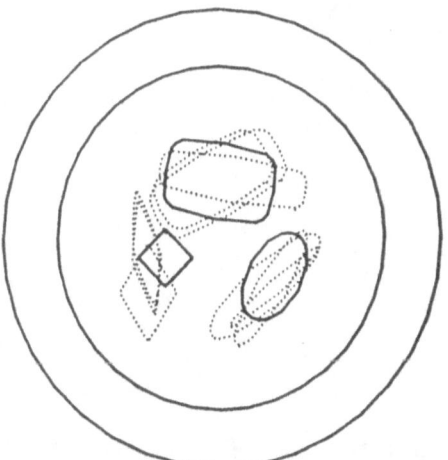 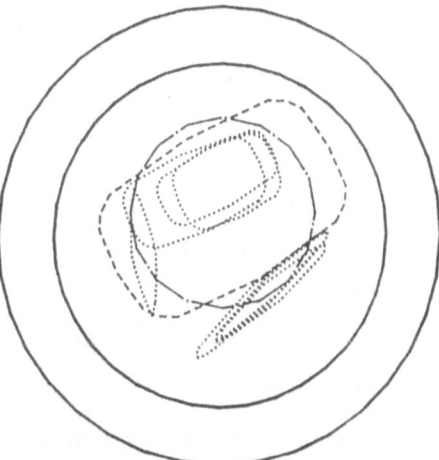

Fig.3 Inverse design of a centrally located circular hole with an initial guess of three Lame-type holes: first 64 optimization cycles.

Fig.4 Inverse design of a centrally located circular hole with an initial guess of three Lame-type holes: optimization cycles 65-120.

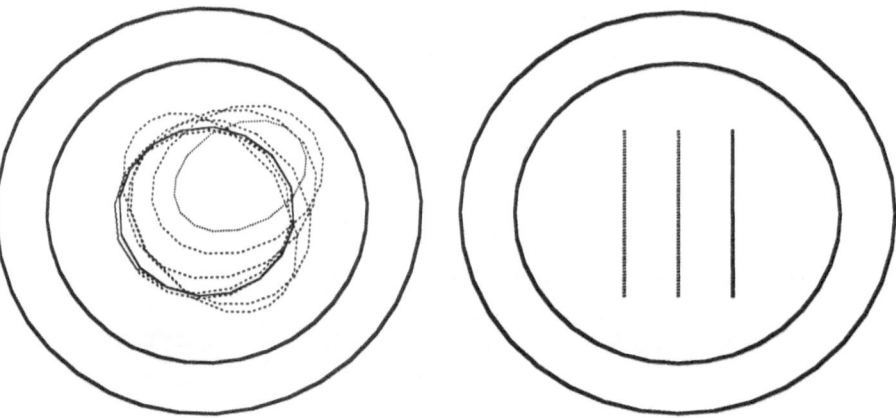

Fig.5 Detection of a centrally located circular void starting with an off-center Lame-type void: evolution history.

Fig.7 Detection of an off-center crack starting with three cracks belonging to a Lame family: initial and final configuration.

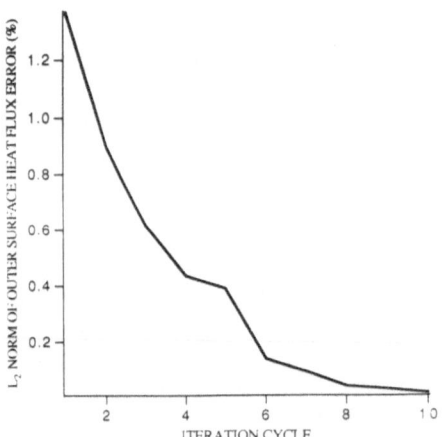 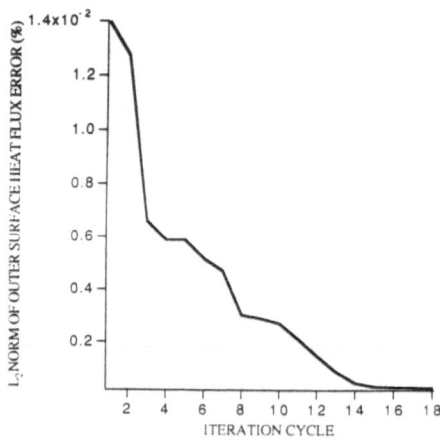

Fig.6 Detection of a centrally located circular void starting with an off-center Lame-type void: convergence history.

Fig.8 Detection of an off-center crack starting with three cracks belonging to a Lame family: convergence history.

# Determination of Time Dependence of Temperature in the Inner Surface of a Cylinder by Measured Information on the Outer Surface

T. Tsuji*, and N. Noda*

* Department of Mechanical Engineering, Shizuoka University
  3-5-1 Jyoohoku , Hamamatsu, 432 Japan

Summary

The dynamic inverse problem to determine the time dependence of the temperature in the inner surface of the cylinder is considered. The indirect boundary method with fictitious surface is used to formulate the present problem. The accuracy of the present method is confirmed by using the experimental data of the hollow cylinder that is subjected to dynamic heat source in the inner surface.

Introduction

In recent years, a considerable effort has been devoted to inverse elastic problems. For example, authors [1]-[3] have discussed the analytical method for the inverse thermoelastic problems. Numerical methods are also used [4]-[6] by many investigators. Oda [4] obtained the contact stress distributions by using the finite element method. Tomishima and Yada[5] obtained the residual stress in the plain plate by the inverse method. However, in these papers, theoretical or numerical data are used instead of the measured data at the boundary. It is important to use experimental data and to consider accuracy of the inverse problem. We [7] investigated an inverse problem to determine inner pressure of a cylinder by using actual experimental data. On the other hand, not only the static problems but also dynamic problems are important. Tanaka and Yamagiwa [8,9] used the boundary element method to obtain the form of the inclusion in the elastic body.

In this paper, the dynamic inverse problem to determine the time dependence of the temperature in the inner surface of the cylinder is considered. The indirect boundary method and the least-squears method are used for the formulation, and the problem is reduced into the system of the algebraic equations with respect to the coefficients of the density functions in the surfaces. The availability of the present method is confirmed by using the experimental data of the hollow cylinder that is subjected to dynamic heat source in the inner surface.

Thermodynamic boundary integral equation

We consider the Fictitious Boundary $\Gamma_f$ as shown in Fig.1 and use the formulation of the indirect boundary method. When the initial temperature is zero, by using density function $\phi(Q,t)$ on the boundary $\Gamma_f$, the temperature $T(P,t)$ at point $P$ and time $t$ can be given as follows:

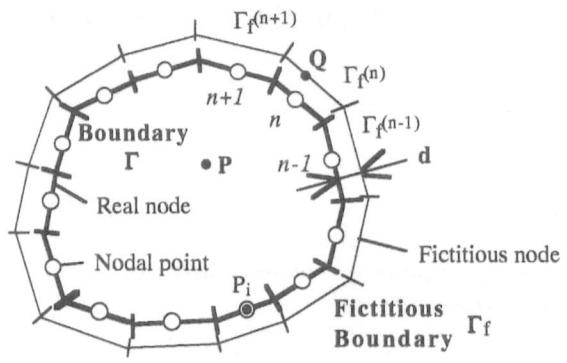

Fig.1 Fictitious Boundary.

$$T(P,t) = \int_0^t \int_{\Gamma_f} \lambda T^*(P,Q,t,\tau)\phi(Q,\tau)d\Gamma_f(Q)d\tau$$

(1)

where the fundamental solution kernel $T^*(P,Q,t,\tau)$ is denoted as :

$$T^*(P,Q,t,\tau) = \begin{cases} \dfrac{1}{4\pi\lambda(t-\tau)} exp[-\dfrac{r^2}{4\kappa(t-\tau)}] & (t>\tau) \\ 0 & (t<\tau) \end{cases}$$

(2)

$\lambda$ and $\kappa$ are thermal conductivity and thermal diffusivity, respectively. We consider the quasi static thermoelastic problem, and denote stress by adding static stress field to the stress fields from thermoelastic potential. In plane strain problem, thermoelastic potential $\Phi(P,t)$ is given by temperature $T(P,t)$ and harmonic function $\phi_1(P)$ as follow:

$$\Phi(P,t) = \frac{1+\nu}{1-\nu}\kappa\alpha\int_0^t T(P,\tau)d\tau + t\phi_1(P)$$

(3)

where $\alpha$ is the coefficient of linear thermal expansion. By substituting Eq.(1) into Eq.(2) and determining $\phi_1(P)$ in order to cancel the singularity, the thermoelastic potential $\Phi(P,t)$ can be shown as follow :

$$\Phi(P,t) = \frac{\kappa\alpha}{4\pi}\frac{1+\nu}{1-\nu}\int_0^t \int_{\Gamma_f} \{E_1[\frac{r^2}{4\kappa(t-\tau)}]-2ln(r)\}\phi(Q,\tau)d\Gamma_f(Q)d\tau$$

(4)

where $E_1(z)$ denotes error function. The displacement $u_i^P$, strain $\varepsilon_{ij}^P$ and stress $\sigma_{ij}^P$ by the thermoelastic potential are given as follows:

$$u_i^P(P,t) = \frac{\partial\Phi(P,t)}{\partial x_i}, \quad \varepsilon_{ij}^P(P,t) = \frac{\partial\Phi(P,t)}{\partial x_i\partial x_j}, \quad \sigma_{ij}^P(P,t) = 2G\{\frac{\partial\Phi(P,t)}{\partial x_i\partial x_j} - \nabla^2\Phi(P,t)\delta_{ij}\}$$

(5)

On the other hand, displacement $u_i{}^s$, strain $\varepsilon_{ij}{}^s$ and stress $\sigma_{ij}{}^s$ for static stress field can be given as follows:

$$u_i^s(P,t)=\int_{\Gamma_f} \psi_k(Q,t)u_{ik}^*(P,Q)d\Gamma_f(Q) \ , \quad \varepsilon_{ij}^s(P,t)=\int_{\Gamma_f} \psi_k(Q,t)\varepsilon_{ijk}^*(P,Q)d\Gamma_f(Q)$$

$$\sigma_{ij}^s(P,t)=\int_{\Gamma_f} \psi_k(Q,t)\sigma_{ijk}^*(P,Q)d\Gamma_f(Q)$$

(6)

where $\phi_k(Q,t)$ is the density function for the static stress field and $u_{ik}^*$, $\varepsilon_{ijk}^*$ and $\sigma_{ijk}^*$ are fundamental solutions as follows :

$$u_{ik}^*(P,Q)=\frac{1}{8\pi G}\frac{1}{1-v}[(3-4v)ln\frac{1}{r}\delta_{ik}+\frac{r_ir_k}{r^2}]$$

$$\varepsilon_{ijk}^*(P,Q)=\frac{-1}{8\pi G(1-v)}\frac{1}{r^2}\{(1-2v)(\delta_{ki}r_j+\delta_{kj}r_i)-\delta_{ij}r_k+2\frac{r_ir_jr_k}{r^2}\}$$

$$\sigma_{ijk}^*(P,Q)=\frac{-1}{4\pi(1-v)}\frac{1}{r^2}\{(1-2v)(\delta_{ki}r_j+\delta_{kj}r_i-\delta_{ij}r_k)+2\frac{r_ir_jr_k}{r^2}\}$$

(7)

$$r_i=x_i(Q^*)-x_i(P) \ , \quad r=\sqrt{r_1^2+r_2^2}$$

$G$, $v$ and $\delta_{ij}$ denote shear modulus, Poisson's ratio and Kronecker delta, respectively. Consequently displacement, strain and stress can be given by the sum of Eq.(6) and (7) as follows :

$$u_i(P,t) = u_i^P(P,t)+u_i^s(P,t), \ \varepsilon_{ij}(P,t) = \varepsilon_{ij}^P(P,t)+\varepsilon_{ij}^s(P,t), \ \sigma_{ij}(P,t) = \sigma_{ij}^P(P,t)+\sigma_{ij}^s(P,t)$$

(8)

The fictitious boundary is divided into $N$-th linear elements and time $t$ is divided by time step $\Delta t$. In each element and time step, the density function is considered as constant. The values of the density functions $\phi(Q,t)$ and $\psi(Q,t)$ for $n$-th element and $M$-th time step is denoted as $\phi^{(n,M)}$ and $\psi_k^{(n,M)}$. Temperature $T$, stress $\sigma_{ij}$ and strain $\varepsilon_{ij}$ at time $t=t_M$ are given as follows :

$$T(P,t_M)=\sum_{m=1}^{M} \sum_{n=1}^{N} A^{(n,M-m+1)}(P) \phi^{(n,m)}$$

(9)

$$u_i(P,t_M)=\sum_{m=1}^{M} \sum_{n=1}^{N} F_i^{(n,M-m+1)}(P)\phi^{(n,m)}+\sum_{n=1}^{N} O_{ik}^{(n)}(P)\psi_k^{(n,M)}$$

(10)

$$\varepsilon_{ij}(P,t_M)=\sum_{m=1}^{M} \sum_{n=1}^{N} H_{ij}^{(n,M-m+1)}(P) \phi^{(n,m)}+\sum_{n=1}^{N} B_{ijk}^{(n)}(P) \psi_k^{(n,M)}$$

(11)

$$\sigma_{ij}(P,t_M)=\sum_{m=1}^{M} \sum_{n=1}^{N} S_{ij}^{(n,M-m+1)}(P) \phi^{(n,m)}+\sum_{n=1}^{N} D_{ijk}^{(n)}(P) \psi_k^{(n,M)}$$

(12)

where

$$A^{(n,m)}(P) = \frac{1}{4\pi}\int_{\Gamma_f^{(n)}} \{E_1(a_m)-E_1(a_{m+1})\} \, d\Gamma_f$$

$$F_i^{(n,m)}(P) = \frac{\alpha(1+v)}{8\pi(1-v)}\int_{\Gamma_f^{(n)}} r_i[\{E_1(a_m)-E_1(a_{m+1})\}+\{\frac{1-e^{-a_m}}{a_m}-\frac{1-e^{-a_{m+1}}}{a_{m+1}}\}]d\Gamma_f$$

$$S_{ij}^{(n,m)}(P) = \frac{\alpha G(1+\nu)}{4\pi(1-\nu)} \int_{\Gamma_f(n)} [-\delta_{ij}\{E_1(a_m) - E_1(a_{m+1})\}$$

$$-\{\frac{1-e^{-a_m}}{a_m} - \frac{1-e^{-a_{m+1}}}{a_{m+1}}\}\{\delta_{ij}\frac{2}{r^2}(\delta_{ij}r_k r_k - r_i r_j)\}] d\Gamma_f$$

$$H_{ij}^{(n,m)}(P) = \frac{\alpha(1+\nu)}{8\pi(1-\nu)} \int_{\Gamma_f(n)} [-\delta_{ij}\{E_1(a_m) - E_1(a_{m+1})\} - \{\frac{1-e^{-a_m}}{a_m} - \frac{1-e^{-a_{m+1}}}{a_{m+1}}\}\{\delta_{ij} - \frac{2}{r^2}r_i r_j\}] d\Gamma_f$$

$$D_{ijk}^{(n)}(P) = \frac{-1}{4\pi(1-\nu)} \int_{\Gamma_f(n)} \frac{1}{r^2}\{(1-2\nu)(\delta_{ki}r_j + \delta_{kj}r_i - \delta_{ij}r_k) + \frac{2}{r^2}r_i r_j r_k\} d\Gamma_f$$

$$B_{ijk}^{(n)}(P) = \frac{-1}{8\pi G(1-\nu)} \int_{\Gamma_f(n)} \frac{1}{r^2}\{(1-2\nu)(\delta_{ki}r_j + \delta_{kj}r_i) - \delta_{ij}r_k + \frac{2}{r^2}r_i r_j r_k\} d\Gamma_f$$

$$a_m = \frac{r^2}{4\kappa m \Delta t} \quad , \quad r_i = x_i(Q) - x_i(P) \quad , \quad r = \sqrt{r_i r_i}$$

## Inverse method for the cylinder

We consider the hollow cylinder with heat source in the inner surface as shown in Fig.2. and determine the temperate changing in the inner surface. The present inverse problem is denoted as shown in Table 1. By using the boundary conditions (iii) to (vi), equations (12) can be shown as follows:

$$\sum_{m=1}^{M} \sum_{n=1}^{N} S_{rr}^{(n,M-m)}(P_i) \phi^{(n,m)} + \sum_{n=1}^{N} D_{rrk}^{(n)}(P_i) \psi_k^{(n,M)} = 0 \qquad (i=1,2,...N)$$

$$\sum_{m=1}^{M} \sum_{n=1}^{N} S_{r\theta}^{(n,M-m)}(P_i) \phi^{(n,m)} + \sum_{n=1}^{N} D_{r\theta k}^{(n)}(P_i) \psi_k^{(n,M)} = 0 \qquad (i=1,2,...N) \qquad (13)$$

where $P_i$ denote the point of $i$-th element and coefficients $S_{rr}^{(n,m)}$, $S_{r\theta}^{(n,m)}$, $D_{rrk}^{(n)}$ and $D_{r\theta k}^{(n)}$ can given by transformation of tensors $S_{ij}^{(n,m)}$ and $D_{ijk}^{(n)}$ into the polar coordinate $(r,\theta)$.

$$S_{i'j'}^{(n,m)} = l_{i'\alpha} l_{j'\beta} S_{\alpha\beta}^{(n,m)} \quad , \quad D_{i'j'k}^{(n)} = l_{i'\alpha} l_{j'\beta} D_{\alpha\beta k}^{(n)} \qquad (14)$$

Table 1  Boundary conditions of the inverse problem.

|  | Outer Surface $i=1,2...N_0$ | Inner Surface $i=1+N_0,...N$ |
|---|---|---|
| Number | $i=1,2...N_0$ | $i=1+N_0,...N$ |
| $T$ | (i) measured (= $\overline{T}$ )) | (ii) unknown |
| $\sigma_{rr}$ | (iii) free (=0) | (iv) free (=0) |
| $\sigma_{r\theta}$ | (v) free (=0) | (vi) free (=0) |
| $\varepsilon_{\theta\theta}$ | (vii) measured (= $\overline{\varepsilon_{\theta\theta}}$ ) | (viii) unknown |

where $i'$ and $j'$ denote $(r, \theta)$ coordinate and $l_{i'\alpha}$ is :

$$l_{i'\alpha} = \begin{bmatrix} \cos(\theta) & \sin(\theta) \\ -\sin(\theta) & \cos(\theta) \end{bmatrix}$$

By using Eq. (13), the density function $\psi_k^{(n,M)}$ for steady stress field can be obtained with respect to the density function $\phi^{(n,m)}$ of temperature field as follows:

$$[\boldsymbol{\Psi}] = -\sum_{m=1}^{M} [\mathbb{D}]^{-1} [\mathbb{S}^{(m)}] [\boldsymbol{\phi}^{(m)}] \tag{15}$$

where

$$[\boldsymbol{\phi}^{(m)}]^t = [\phi^{(1,m)} \cdots \phi^{(N,m)}], \quad [\boldsymbol{\psi}]^t = [\psi_1^{(1,M)} \cdots \psi_1^{(N,M)} \ \psi_2^{(1,M)} \cdots \psi_2^{(N,M)}]$$

$$[\mathbb{D}] = \begin{bmatrix} D_{rr1}^{(1)}(P_1) \cdots D_{rr1}^{(N)}(P_1) & D_{rr2}^{(1)}(P_1) \cdots D_{rr2}^{(N)}(P_1) \\ \vdots \quad \ddots \quad \vdots & \vdots \quad \ddots \quad \vdots \\ D_{rr1}^{(1)}(P_N) \cdots D_{rr1}^{(N)}(P_N) & D_{rr2}^{(1)}(P_N) \cdots D_{rr2}^{(N)}(P_N) \\ D_{r\theta1}^{(1)}(P_1) \cdots D_{r\theta1}^{(N)}(P_1) & D_{r\theta2}^{(1)}(P_1) \cdots D_{r\theta2}^{(N)}(P_1) \\ \vdots \quad \ddots \quad \vdots & \vdots \quad \ddots \quad \vdots \\ D_{r\theta1}^{(1)}(P_N) \cdots D_{r\theta1}^{(N)}(P_N) & D_{r\theta2}^{(1)}(P_N) \cdots D_{r\theta2}^{(N)}(P_N) \end{bmatrix}$$

$$[\mathbb{S}^{(m)}] = \begin{bmatrix} S_{rr}^{(1,m)}(P_1) & \cdots & S_{rr}^{(N,m)}(P_1) \\ \vdots & \ddots & \vdots \\ S_{rr}^{(1,m)}(P_N) & \cdots & S_{rr}^{(N,m)}(P_N) \\ S_{r\theta}^{(1,m)}(P_1) & \cdots & S_{r\theta}^{(N,m)}(P_1) \\ \vdots & \ddots & \vdots \\ S_{r\theta}^{(1,m)}(P_N) & \cdots & S_{r\theta}^{(N,m)}(P_N) \end{bmatrix}$$

The stress and strain can be given with respect of density function $\phi^{(n,m)}$ as follows :

$$\varepsilon_{i'j'}(P, t_M) = \sum_{m=1}^{M} \sum_{n=1}^{N} E_{i'j'}^{(n,M-m)}(P) \phi^{(n,m)} \tag{16}$$

$$\sigma_{i'j'}(P, t_M) = \sum_{m=1}^{M} \sum_{n=1}^{N} G_{i'j'}^{(n,M-m)}(P) \phi^{(n,m)} \tag{17}$$

where $E_{i'j'}^{(n,m)}(P)$ and $G_{i'j'}^{(n,m)}(P)$ are shown as :

$$E_{i'j'}^{(n,m)}(P) = H_{i'j'}^{(n,m)}(P) - \sum_{k=1}^{N} \sum_{j=1}^{N} \{B_{i'j'x}^{(j)}(P) W^{(j,k)} + B_{i'j'y}^{(j)}(P) W^{(j+n,k)}\} S_{rr}^{(n,m)}(P_k)$$

$$- \sum_{k=1}^{N} \sum_{j=1}^{N} \{B_{i'j'x}^{(j)}(P) W^{(j,k+n)} + B_{i'j'y}^{(j)}(P) W^{(j+n,k+n)}\} S_{r\theta}^{(n,m)}(P_k)$$

$$G_{i'j'}^{(n,m)}(P) = S_{i'j'}^{(n,m)}(P) - \sum_{k=1}^{N} \sum_{j=1}^{N} \{D_{i'j'x}^{(j)}(P) W^{(j,k)} + D_{i'j'y}^{(j)}(P) W^{(j+n,k)}\} S_{rr}^{(n,m)}(P_k)$$

$$- \sum_{k=1}^{N} \sum_{j=1}^{N} \{D_{i'j'x}^{(j)}(P) W^{(j,k+n)} + D_{i'j'y}^{(j)}(P) W^{(j+n,k+n)}\} S_{r\theta}^{(n,m)}(P_k)$$

$$W^{(i,j)} = \begin{bmatrix} D_{rrx}^{(n)}(P_i) & D_{rry}^{(n)}(P_i) \\ D_{r\theta x}^{(n)}(P_i) & D_{r\theta y}^{(n)}(P_i) \end{bmatrix}^{-1}$$

The remaining conditions (i) and (vii) are shown by using Eqs. (9) and (16) as follows:

$$\overline{T(P_i, t_M)} = \sum_{m=1}^{M} \sum_{n=1}^{N} A^{(n,M-m)}(P_i)\, \phi^{(n,m)} \quad (i=1,2...N_0) \tag{18}$$

$$\overline{\varepsilon_{\theta\theta}(P_i, t_M)} = \sum_{m=1}^{M} \sum_{n=1}^{N} E_{\theta\theta}^{(n,M-m)}(P_i)\, \phi^{(n,m)} \quad (i=1,2...N_0) \tag{19}$$

Consequently, the present problem is reduced to obtain $\phi^{(i,m)}$ by Eqs. (18) and (19). In the $M$-th time step $t=t_M$, the values of $\phi^{(i,m)}$ $(m=1,...M-1)$ are obtained already, then the number of unknown is $N$ as $\phi^{(i,M)}$ $(i = 1, 2 \cdots , N)$. However, the number of equations given by Eqs. (18) and (19) is $2N_0$, then the least-squears method is used. We determine $\phi^{(i,M)}$ by minimizing following square sum :

$$S^2 = \sum_{i=1}^{N_0} \left[ \overline{T(P_i, t_M)} - \sum_{m=1}^{M} \sum_{n=1}^{N} A^{(n,M-m+1)}(P_i)\phi^{(n,m)} \right]^2$$
$$+ \sum_{i=1}^{N_0} \left[ \overline{\varepsilon_{\theta\theta}(P_i, t_M)} - \sum_{m=1}^{M} \sum_{n=1}^{N} E_{\theta\theta}^{(n,M-m+1)}(P_i)\phi^{(n,m)} \right]^2 \tag{20}$$

The set of $\phi^{(j,M)}$, which minimize the above equations, can be obtained by following equations.

$$\frac{\partial S^2}{\partial \phi^{(j,M)}} = 0 \quad , \quad (j=1,2, \cdots N) \tag{21}$$

It can describe as following sets of algebraic equations.

$$\sum_{i=1}^{N_0} \left\{ \overline{T(P_i, t_M)} A^{(j,1)}(P_i) + \overline{\varepsilon_{\theta\theta}(P_i, t_M)} E_{\theta\theta}^{(j,1)}(P_i) \right\}$$
$$= \sum_{m=1}^{M} \sum_{n=1}^{N} \phi^{(n,m)} \sum_{i=1}^{N_0} \left\{ A^{(n,M-m+1)}(P_i) A^{(j,1)}(P_i) + E_{\theta\theta}^{(n,M-m+1)}(P_i) E_{\theta\theta}^{(j,1)}(P_i) \right\}, \quad (j=1,2, \cdots N) \tag{22}$$

Above equations can be solved at each time step (at time $t=t_M$). Temperature and stress field can be given by Eqs. (9) to (12).

Experimental results

The hollow cylinder made of acrylic resin is used to obtain the experimental data on outer surface of the cylinder, as shown in Fig.2. This cylinder is subjected to two heat sources of nichrome wire at diagonal points $A$ and $B$ on inner surface. The temperature of the heat sources are measured by thermocouples. On the outer surface, temperature and strain data are measured by 5 thermocouples and 22 strain gages, which are putted on quarter region. By using the measuring system as show in Fig.3, all data are automatically measured and calculated.

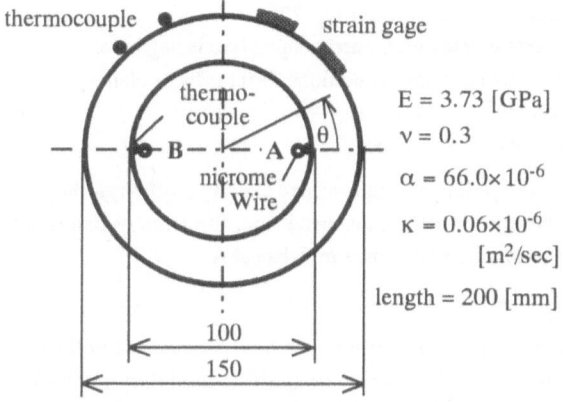

Fig.2 The circular cylinder with heat sources on the inner surface.

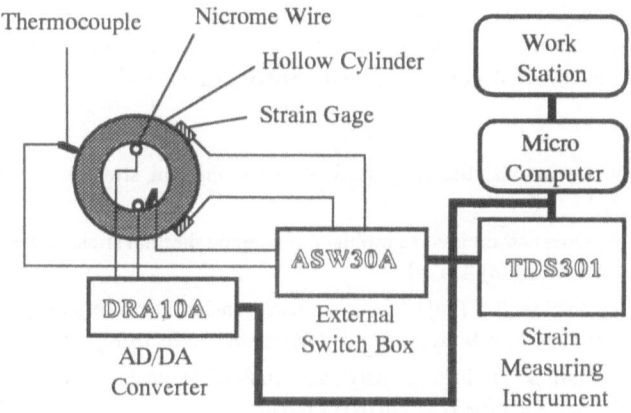

Fig.3 The measuring system for the cylinder with heat sources on the inner surface.

Figures 4 to 7 show the measured temperature and strain data. Temperature at the heat source $A$ in the inner surface is increasing as show in Fig.4. The temperature on the outer surface, as shown in Fig.5, does not change comparing to the temperature at the heat source, but strain $\varepsilon_{\theta\theta}$ changes as shown in Fig. 6. Figure 6 shows the distributions of the strain on the outer surface. These data are interpolated by using the cubic spline and interpolated values are used to the following inverse analysis.

Figure 7 shows the time dependence of the determined inner temperature with time step $\Delta t$ = 10 and 100 [sec]. In this figure, measured temperature of the heat source is also shown. In this determination, we used experiential scheme to determine temperature in each time step. This scheme based on the knowledge that the temperature on the inner surface must increase and shown as follows :

504

1. Determine temperature by using Eq.(22).
2. Search elements where determined temperature is negative.
4. Consider the new boundary conditions $\underline{T}=0$ for these elements.
5. Add these conditions to Eq.(22).
6. Repeat 1 to 5 until convergence.

From Fig.7 the heated points are determined exactly. Although the determined temperature increases slowly compare to the measured one, the time dependence of the determined temperature is almost similar to the measured data at **A**.

Conclusions

The dynamic inverse problem to determine the time dependence of the temperature in the inner surface of the cylinder is considered. The indirect boundary method and the least-squears method are used for the formulation. Using the stress free condition on the surface, the problem is reduced into the system of the algebraic equations with respect to the coefficients of the density functions for temperature. By using the experimental data, which from the hollow cylinder that is subjected to dynamic heat source, time dependence of the inner surface is determined. The determined values are in good agreement with the measured values at inner surface and availability of the present method is confirmed.

References

1. Noda, N.: Optimal heating problem for transient thermal stress in a thick plate, Thermal Stresses, **11-2** (1988) 141-150.
2. Noda, N.: On a certain inverse problem of coupled thermal stress fields in a thick plate, ZAMM, **68-9** (1988) 411-415.
3. Noda, N.; Ashida, F. ; Tsuji, T.: An inverse transient thermoelastic problem for a transversely isotropic body, J. Applied Mechanics, **56** (1989) 791-797.
4. Oda, J. ; Moto, S : On inverse analytical technique to obtain contact stress distributions, Trans. JSME, **55**, (1989) 872-878 (in Japanese).
5. Tomishima, T. ; Yada, T : Study on an identification method of residual stresses in a plate by inverse analysis, JSME Int. J., **32-1** (1989) 31-37.
6. Kubo, S ; Ohnaka, K. ; Ohji, K.: Identification of heat-source and force using boundary integrals, Trans. JSME, **54** (1988) 1329-1334 (in Japanese).
7. Tsuji, T ; Noda, N. ; Tanaka, Y : Determination of Boundary Values in the Inner Surface of a Cylinder by using Boundary Element Method, Applied Stress Analysis, Elsvier Applied Science, (1990) 171-179.
8. Tanaka, M. ; Yamagiwa, K : Application of boundary element method to some inverse problems in elastodynamics, Trans. JSME, **54** (1988) 1054-1059 (in Japanese).
9. Tanaka, M.; Nakamura, M.; Yamagiwa, T. ; Nakayama, C. : Analysis of Some Inverse Problems in Elastodynamics by the Boundary Element Method, Trans. JSME, **55** (1989) 1445-1452 (in Japanese).

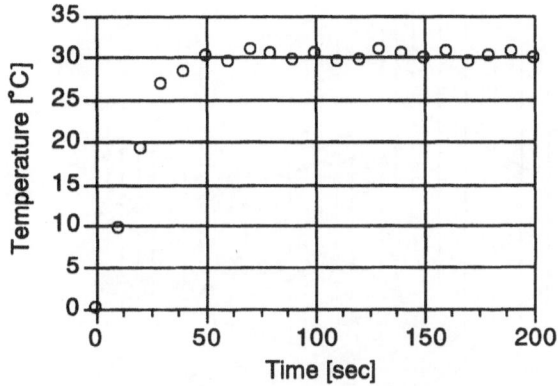

Fig.4 Applied temperature at point $A$.

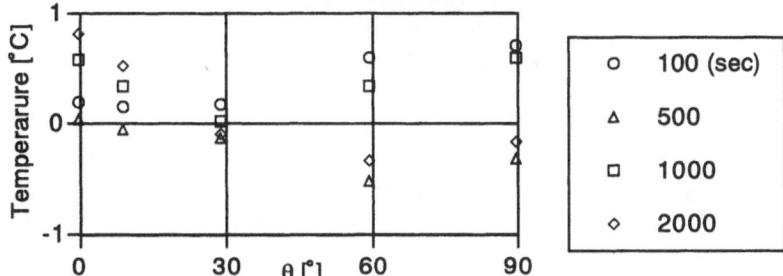

Fig.5 Distribution of measured temperature $\overline{T}$ .

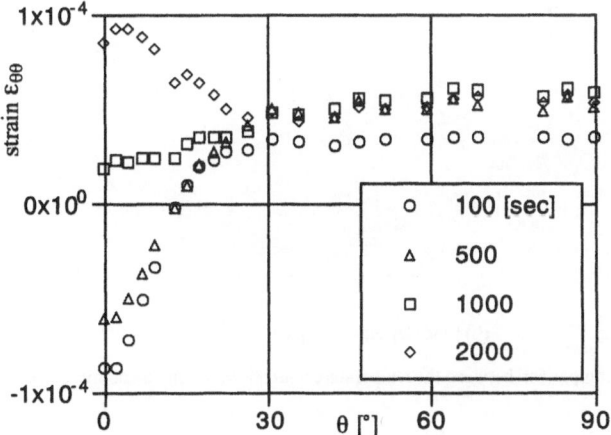

Fig.6 Distribution of measured strain $\overline{\varepsilon_{\theta\theta}}$.

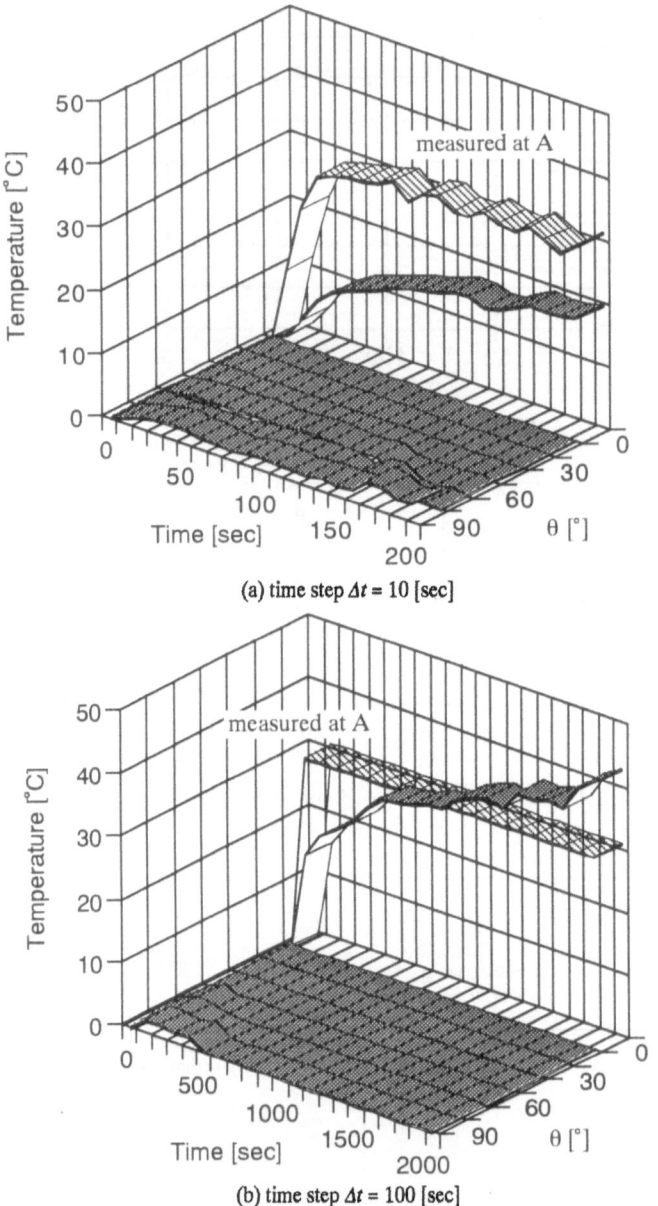

(a) time step $\Delta t = 10$ [sec]

(b) time step $\Delta t = 100$ [sec]

Fig.7. The comparison between the temperature which is determined and measured.

# Transfer Function Approach for Solving Inverse Heat Conduction Problems in Piping Using Experimental Data

T.H. CHAU and P. MORILHAT
Electricité de France, Research and Development Division
25 Allée Privée, Carrefour Pleyel, 93206 St Denis Cedex 1, FRANCE

SUMMARY

To solve inverse heat conduction problem, it is generally assumed that data used for identification are temperatures issued from thermocouples. When complex conduction phenomenon occurs, external temperature measurements are not appropriated.

This paper presents an investigation on the use of external *deformation measurements* for the reconstruction of internal thermal field.

PRESENTATION

Nuclear piping in service are submitted to thermal stresses which need to be evaluated to insure integrity of the structure. For safety purposes, only external wall of pipes can be instrumented while stress maximum is usually located on internal wall. This means we have to solve an inverse heat conduction problem to get temperature field. Stress tensor is then directly computed from this field.

A large number of investigations have been done on the inverse heat conduction problems ([1],[2],[3]). Those methods assumed that data used for identification are external measured temperatures. However in some problems met in piping, we show that use of external temperatures could lead to wrong solution due to complex conduction phenomena, instrumentation limitation and ill-posedness of the mathematical problem.

This paper presents two investigations to solve actual thermal problems using *deformation measurements* as data. The first one is related to a quasi-steady periodic temperature fluctuation (thermal striping) in one-dimensional structure and the second one is related to a steady thermal stratification in a plane section of straight pipe.

For those current studies, a transfer function approach is valid as the phenomenon we work on are located in the thermal linear domain.

## 1. QUASI-STEADY PERIODIC THERMAL STRIPING

Some nuclear piping is submitted to internal temperature fluctuations which are due to level variations of an interface separating two layers of fluids at different temperatures. Those fluctuations may induce *thermal striping* fatigue damage. We have to estimate what type of on-site instrumentation is most appropriated for the survey of this phenomenon.

Following assumptions are made for this investigation :
- thermal fluctuation signal is a quasi-steady periodic one,
- ratio pipe wall thickness/internal radius is small,
- radial heat conduction is locally predominant.

### 1.1 Principles ([4],[5])

Fourier transform of one-dimensional heat conduction equation is given by :

$$\frac{d^2 \, \overline{T}(x,f)}{dx^2} - i \, \frac{2\pi \, f}{a} \, \overline{T}(x,f) = 0 \qquad (1)$$

with an assumed perfect heat insulation on the external wall ($x = e$) as boundary conditions, where f is the frequency of fluctuations signal and a the thermal diffusivity of the material (for ferritic steel A42, $a = 12.65 \ 10^{-6} \ m^2 s^{-1}$).

Solution of (1) which determines temperature field for the quasi-steady periodic state has the following expression :

$$\overline{T}(x,f) = \Theta(x,f) \ . \ \overline{T}_0(f) \qquad (2)$$

where the effect of through thickness heat conduction is totally included in $\Theta(x,f)$ which analytical form is :

$$\Theta(x,f) = \frac{ch \ [(1+i).k.(e-x)]}{ch \ [(1+i).k.e]} \qquad (3)$$

$$\text{with } k = \left[ \frac{\pi \, f}{a} \right]^{1/2}$$

Internal temperature $\overline{T_0}(f)$ is then obtained from (2) using known external temperature $\overline{T}(e,f)$ recorded by thermocouples. Time evolution $T_0(t)$ results from inverse Fourier transform.

In the same way, we can reach internal temperature $\overline{T_0}(f)$ by means of thermal stress. Through thickness temperature gradient induces axial stress which is given by relation:

$$\sigma(x,t) = \frac{E.\alpha}{1-\nu} [Tm(t) - T(x,t)] \qquad (4)$$

where Tm(t) is the average of through thickness temperature

$$Tm(t) = \frac{1}{e} \int_0^e T(x,t).\,dx$$

and the mechanical characteristics of material

| | |
|---|---|
| E : Young modulus | (for A42 steel, E = 202 $10^3$ MPa) |
| $\alpha$ : dilatation coefficient | ( $\alpha$ = 11.87 $10^{-6}$ /°C) |
| $\nu$ : Poisson coefficient. | ($\nu$ = 0.3) |

Using Fourier transform of relation (4) and substituting expression of $\overline{T}(x,f)$ obtained from (2), we get :

$$\overline{\sigma}(x,f) = \Sigma(x,f)\,.\,\overline{T_0}(f) \qquad (5)$$

where $\Sigma(x,f)$ which includes all the through thickness effects has the analytical form :

$$\Sigma(x,f) = \frac{E.\alpha}{1-\nu} \left[ \frac{(1-i).th[(1+i).k.e]}{2.k.e} - \frac{ch[(1+i).k.(e-x)]}{ch[(1+i).k.e]} \right] \qquad (6)$$

Unknown temperature $\overline{T_0}(f)$ is then obtained from (5) where $\overline{\sigma}(e,f)$ is given by strain gauges. Inverse Fourier transform gives the time evolution $T_0(t)$.

## 1.2 Application

Functions $\Theta(x,f)$ and $\Sigma(x,f)$ are computed from relations (3) and (6) for a ferritic steel (A42) pipe with wall thickness e = 21.4 mm.

Evolution of $|\Theta(e,f)|$ and $|\Sigma(e,f)|$ versus frequency f is shown in figure 1. Its represents the attenuation due to wall thickness.

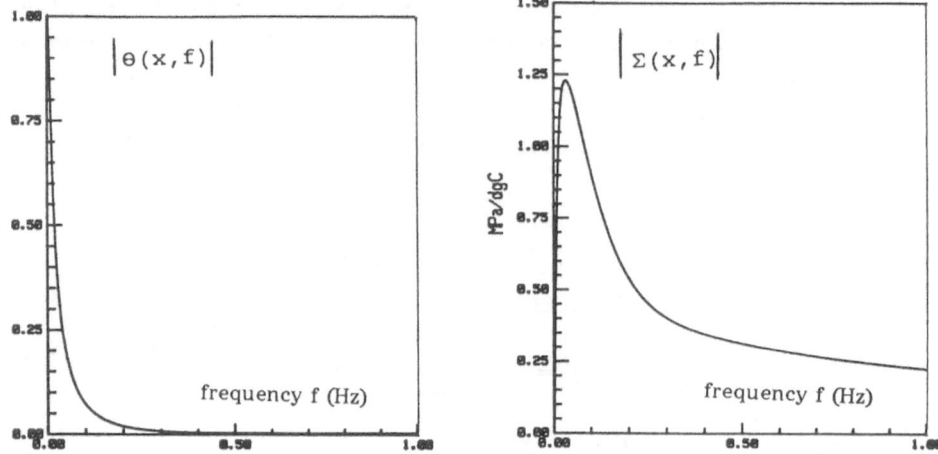

Figure 1 : Evolution of $\Theta(e,f)$ and $\Sigma(e,f)$

We note that beyond the frequency of 0.5 Hz, internal temperature variations cannot be detected by thermocouples ($\Theta(e,f) = 0$) while we can use strain-gauges data up to 25 Hz. Estimation of internal temperature fluctuations in the thermal striping range of 0-10 Hz is carried out better by strain-gauges measurements than by thermocouples ones.

One point we have to evaluate for an experimental instrumentation is an estimation of the error when sensors (thermocouple or strain-gauge) and physical phenomenon are not located at the same position (figure 2).

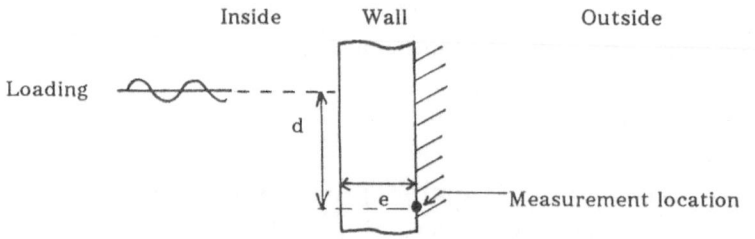

Figure 2 : Sensor and phenomena locations

Figures 3 and 4 presents evolution of error in the estimation of $\Theta(e,f)$ and $\Sigma(e,f)$ for differents locations of sensors.

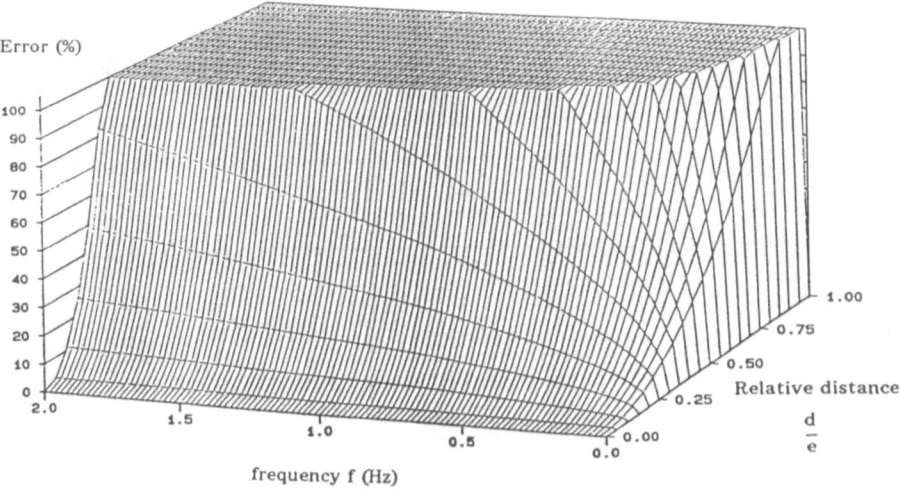

Figure 3 : Error on estimation of $\Theta(e,f)$

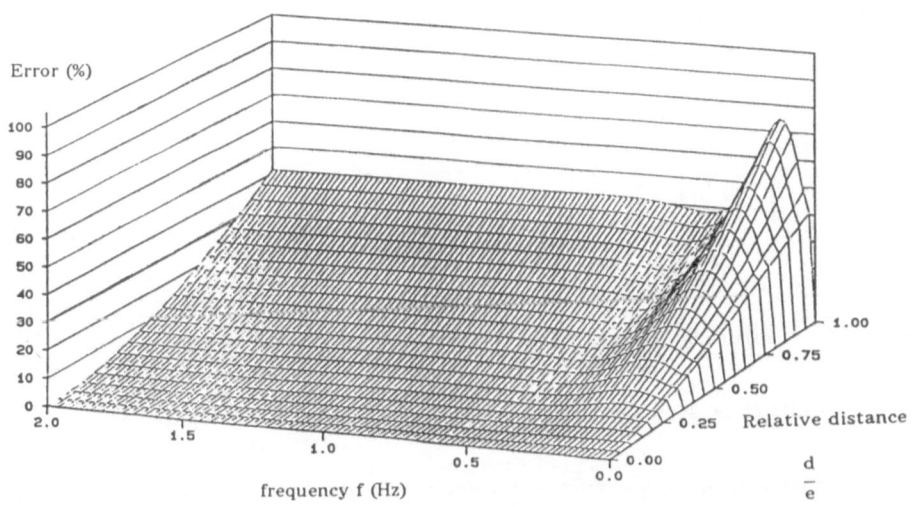

Figure 4 : Error on estimation of $\Sigma(e,f)$

512

Previous figures show that $\Theta(e,f)$ calculation is more concerned with diffe-rence in sensor and phenomenon location than $\Sigma(e,f)$ one. For instance, we find out at $d = e/2$ and $f = 0{,}2$ Hz an 100% error on estimation for $\Theta(e,f)$ and 20% for $\Sigma(e,f)$ .

This evaluation shows another interest of using strain gauges to solve inverse heat conduction problems due to thermal striping.

1.3 Application

Test of previous method has been processed with mock up results.

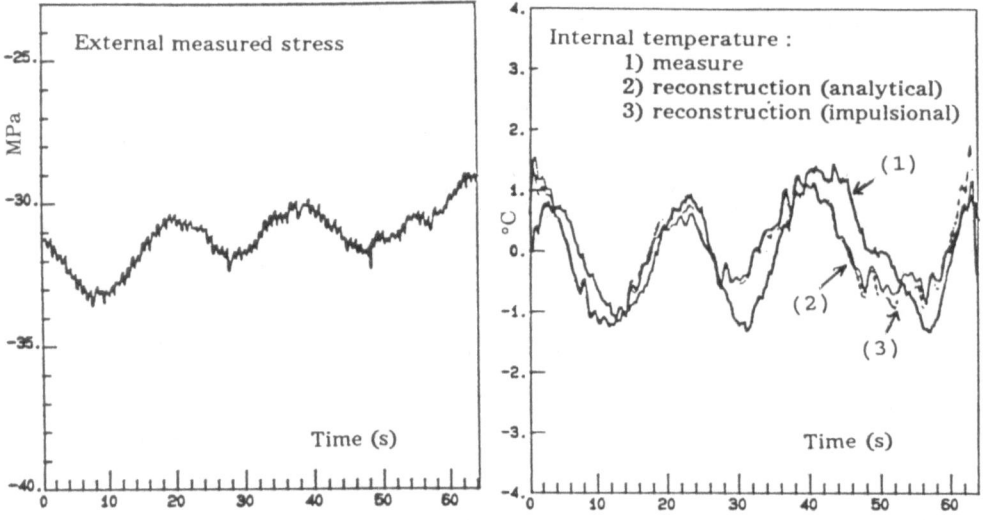

Figure 5 : Reconstruction using deformation measurements

Results show good agreement between experimental and computed results.

Limitations of current method are due to use of one-dimensional heat conduction model. An extension of the method in two-dimensional models is under investigation. It

requires calculation of an impulsional signal (Dirac function) which has been successfully tested with one-dimensional model as shown in figure 5.

## 2. THERMAL STRATIFICATION

Structures in service are usually submitted to non-uniform temperature distribution. Reconstruction of internal thermal field in a plane section of straigth pipe can be
done using transfer function approach.

### 2.1 Principles

Initial internal temperature field is approximated by a linear combination of
basic loadings as shown in figure 6.

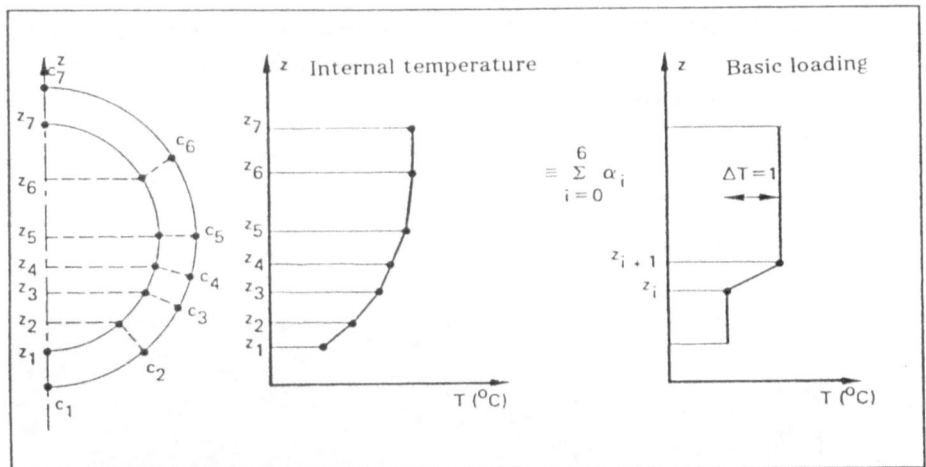

Figure 6 : Principle of internal thermal profile decomposition

Thermal field is computed by means of finite element code for previous internal
basic loadings. It will give for each one of them the temperature $R_{i,k}$ detected at external wall thermocouple $C_k$.

514

For the initial internal thermal field, the principle of superposition gives corresponding external temperature at thermocouple $C_k$ as a linear combination, with the *same coefficients* $\alpha_i$, of those basic loading thermocouple response $R_{i,k}$.

$$\text{Text } (C_k) = \sum_{i=0}^{6} \alpha_i \cdot R_{i,k}$$

Then, knowledge of thermocouples measurements allows to solve the linear equation system (7), to find out coefficients $\alpha_i$ and consequently, the unknown internal temperature.

### 2.2 Limits of method

Different numerical simulations are made for testing this method. Results show good agreements except for low level of stratification.

In this case, influence of circumferential heat conduction is more important than radial heat conduction one and the thermocouple gives a temperature warmer than the expected as shown in figure 7.

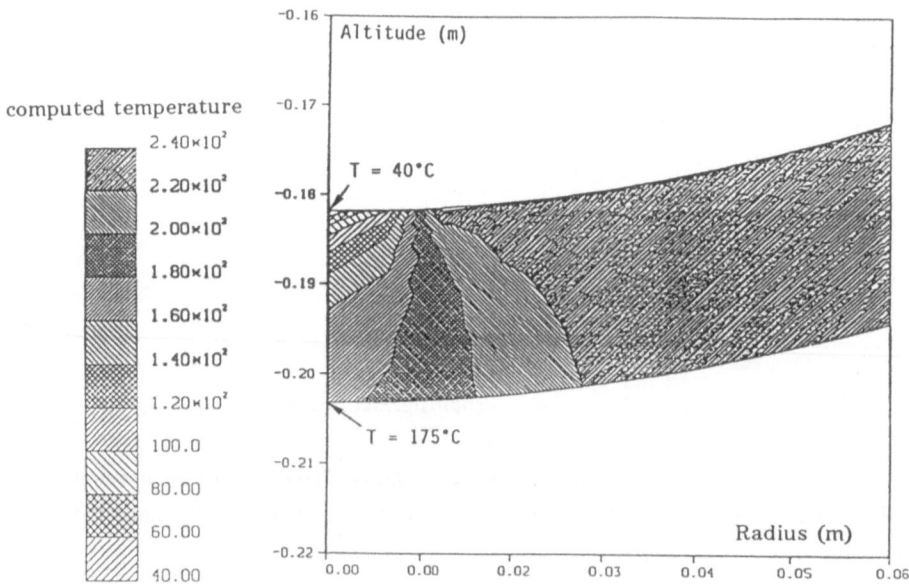

Figure 7 : External temperature detected in case of low-level stratification

Hence, this detected temperature leads to wrong solution for previous reconstruction method based on radial heat conduction.

Investigation has been done to solve this problem using *deformation measurements*.

Indeed, through-thickness temperature gradient will induce a proportional bending stress. It means that in case of low level stratification as shown in figure 7, stress is more or less important depending on internal lower point temperature. Then use of strain-gauges data is a possible way for experimental detection of low-level stratification.

However, computed stress for basic loadings defined previously depends strongly on ideal model such as choice of plane stress element or plane strain element. Those hypothesis are hardly met in actual case. Real boundary conditions are required for application of the transfer function method using external deformation measurements. This means that its implementation for reconstruction of internal temperature will not be easy for a structure in service.

CONCLUSION

The study shows limitations on solving some inverse heat conduction problems with external temperatures. It also investigated on use of deformation measurements in those cases.

For a quasi-steady periodic problem in one-dimensional structure, a method is implemented and checked with experimental results.

For complex heat conduction problem in a plane section, interest of using deformation measurements is shown. But its application in actual problem is difficult as computed stress strongly depends on ideal model choice which is not met for structure in service.

Other methods than transfer function ones are investigated to extend validity domain of using deformation to solve inverse heat conduction problems.

REFERENCES

[1]     Beck, J.V.; Blackwell, B.; St Clair, C.R.: Inverse heat conduction. Ill posed problems, *Wiley Interscience* (1985)

[2]     Reinhardt, H.-J.: A numerical method for the solution of two-dimensional inverse heat conduction problems, *International Journal for Numerical Methods in Engineering*, 32 (1991) 363-383.

[3]     Raynaud, M.; Bransier, J.: A new finite-difference method for the nonlinear inverse heat conduction problem, *Numer. Heat Transfer* (1986)

[4]     Chau, T.H.; Morilhat, P.; Maye, J.P.: Inverse methods in thermomechanics : application to an industrial case, *PLURALIS Calcul des structures et intelligence artificielle Vol 4* (1991)

[5]     Morilhat, P.; Maye, J.P.; Brendle, E.; Hay, B.: Résolution d'un problème thermique inverse par méthodes analytique et impulsionnelle. Application à la détermination des fluctuations thermiques pariétales à l'intérieur d'une conduite, *Int. J. Heat Mass Transfer* (to be published)

# A Method for Disbond Detection in Thermal Tomography by Domain Decomposition Method

Fumio Kojima

Mechanical Engineering Department, Osaka Institute of Technology
5-16-1 Ohmiya, Asahi-ku, Osaka 535, Japan

Summary

A computational method for the disbond detection in aircraft lap joints and in the adhesive joints between aircraft skin and reinforcing doublers is considered using thermographical data. An idea related to method of mappings applies to a parameter estimation problem for a 2-D heat diffusion system. A new computational algorithm for estimating the length of disbond is proposed based on the domain decomposition method.

Introduction

Recently, due to the increasing average of aircraft, the demand has grown for the assessment of structural integrity of aerospace materials. To satisfy this demand, the technique should be noncontacting and allow inspections of a large area of the structure in a short time period. Thermographic inspection of aerospace materials is a good candidate for satisfying this requirement. Figure 1 illustrates this thermal technique using the infrared imaging. As depicted in Fig. 1, the heat is injected to the front surface with lasers or

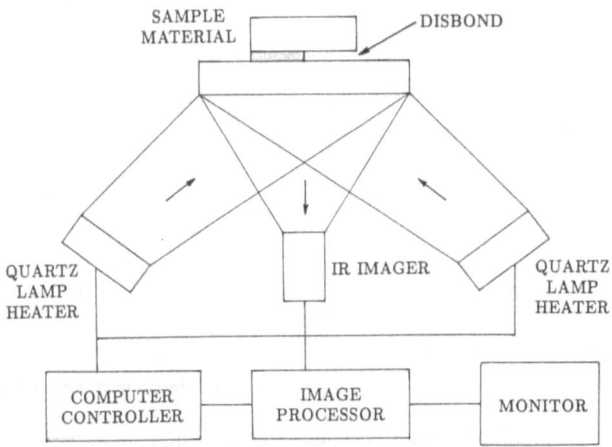

Figure 1: Infrared measurement system in disbond detection

IR lamps and then the subsequent surface temperatures can be recorded by the infrared images without intimate contact with the structure. Mathematically, to estimate the structural information with the thermal data can be formulated as a geometrical heat inverse problem. The inverse formulation of 2-D thermal tomography has been discussed in [1][2] using the idea related to the method of mappings and in [3] using the boundary integral equation method. Application of an artificial neural network to thermal tomography has been studied in [4]. Our efforts on the inverse algorithm arising in thermal tomography is directed to disbond detection of aircraft systems. The presence of disbonds in aircraft lap joints and in the adhesive joints between aircraft skin and reinforcing doublers causes severe damage in passenger airplanes. A new method is proposed using the domain decomposition approach. The domain decomposition method is very classical tool for solving elliptic partial differential equations. Recently, there has been strong revival of the interest in this method due to its potential in highly parallel computing environments ( See [5]). Our aim of this paper is to show the applicability of the inverse approach based on the domain decomposition method to the practical thermal technique.

## Parameter Estimation for Disbond Detection

Sample materials with epoxy bonded area considered here is depicted in Fig. 2. Thermal imaging data can be obtained from a viewed side of material surface in injecting the heat by IR lamp to the same side. The research reported here is aimed at development of a technique for estimating the disbond region that appears at the interface on back side of materials. Thermographic inspection is suitable for checking the quality of bonding at the interface between two materials. If the bonding is poor ( i.e., disbond ) perfect insulation of heat flow occurs at the interface. Accordingly, the corresponding thermal data at the viewed side of materials shows a rapid increasing temperature. For convenience of

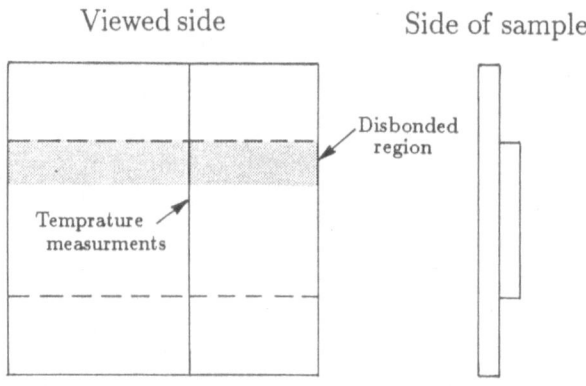

Figure 2: Sample materials

discussions, we restrict our attention in sectional plan of materials. Suppose that one dimensional thermal data are taken on the line as shown in Fig. 2. Then the input and output relations for this nondestructive measurement system can be described by a two dimensional heat diffusion equation defined on 'T-shaped' domain as shown in Fig. 2. Let $G$ denote the 'T-shaped' domain in two-dimensional Euclidean space and let $u = u(t, x)$ be the temperature at time $t$ at location $x = (x_1, x_2)$. The whole boundary of sample materials is denoted by boundary $\partial G$. The viewed side of materials is denoted by $S(\subset \partial G)$. Under our restriction, the detection of disbond region at the interface is converted into a problem for estimating the length of disbond line. In the sequel, the

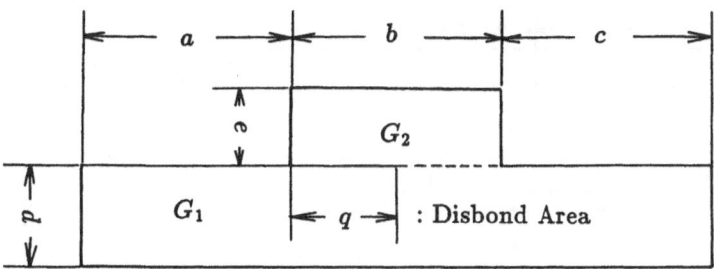

Figure 3: 'T-shaped' domain

disbond line at the inteface is denoted by $\sigma(q)$ where $q$ implies the length of disbond as shown in Fig. 2. In this case, the temperature distribution $u = u(t, \cdot)$ is defined on $G/\sigma(q)$ for each time $t$. The thermographic data from the front surface can be represented as

$$y(t) = \mathcal{H}u(t, q)|_S \quad \text{on } \mathcal{T} \tag{1}$$

where $u(t, q)$ is the solution to the heat diffusion equation,

$$\frac{\partial u(t)}{\partial t} - \kappa \Delta u(t) = 0 \quad \text{in } \mathcal{T} \times G/\sigma(q) \tag{2}$$

with the initial and boundary conditions

$$u(0) = u_0 \quad \text{on } G/\sigma(q) \tag{3}$$

$$\frac{\partial u(t)}{\partial n} = \begin{cases} g/\kappa & \text{on} \quad S \\ 0 & \text{on} \quad \partial(G/\sigma(q))/S \end{cases} \cdot \tag{4}$$

In the above equations, $\mathcal{T}$ denotes the time interval $(0, t_f)$ during which the process is heated and observed. The boundary input $g$ implies heat source level from the front surface $S$. The measurement operator $\mathcal{H}$ in (1) is associated with the data acquisition by infrared imaging technique. The coefficient $\kappa$ is the thermal diffusivity of the material inspected. Our inverse problem is stated as the following output least square problem:

(IDP) Given the thermographic data $\{y_d(t)\}_{t \in \mathcal{T}}$, find an optimal solution $q^* \in Q$ which minimizes the cost

$$J(q) = \frac{1}{2} \int_0^T |\mathcal{H}u(t, q) - y_d(t)|^2 \, dt \tag{5}$$

where $Q$ is an appropriate admissible compact set.

## Method of Mappings

In this section, the method of mapping technique is applied to 'T-shaped' domain $G/\sigma(q)$. Figure 3 illustrates the dimensions of sample materials. In this figure, the unknown parameter $q$ satisfies $0 < q < b$. It is noted that the interface condition on the disbond line is replaced by the Neuman boundary condition. To apply the method of mapping technique, we decompose $T$-shaped domain into two rectangular subdomains, $G = G_1 \cup G_2$

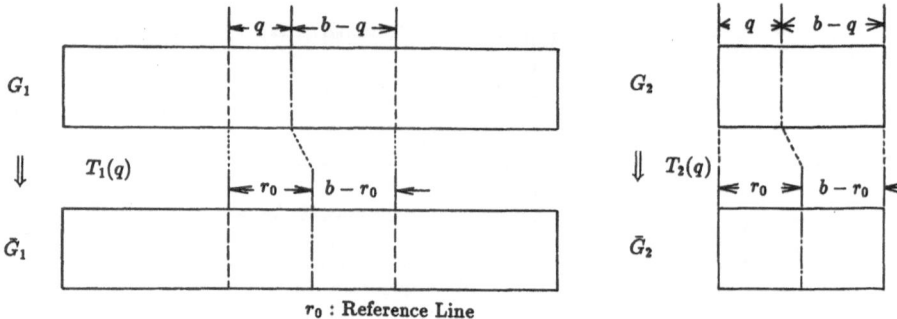

Figure 4: Mapping in subdomain $G_1$ and $G_2$.

as shown in Fig. 3. Let $r_0$ be a reference line corresponding to the unknown parameter $q$. We introduce a mapping operator from $\bar{G}_1$ into $G_1(q)$,

$$T_1(q): \begin{cases} x_1 = f_1(q, z_1) \\ x_2 = z_2 \end{cases} \qquad (6)$$

where

$$f_1(q, z_1) = \begin{cases} z_1 & (0 \leq z_1 \leq a) \\ q(z_1 - a)/r_0 + a & (a < z_1 \leq a + r_0) \\ (b - q)(z_1 - a - b)/(b - r_0) + a + b & (a + r_0 < z_1 \leq a + b) \\ z_1 & (a + b < z_1 \leq a + b + c) \end{cases}$$

Similarly, a mapping operator from subdomain $G_2$ into $\bar{G}_2$ is described as follows:

$$T_2(q): \begin{cases} x_1 = f_2(q, z_1) \\ x_2 = z_2 \end{cases} \qquad (7)$$

where

$$f_2(q, z_1) = \begin{cases} qz_1/r_0 & (0 < z_1 \leq r_0) \\ (b - q)(z_1 - b)/(b - r_0) + b & (r_0 < z_1 \leq b) \end{cases}.$$

Figure 4 illustrates these mappings, $T_1(q)$ and $T_2(q)$. Let us define the system state on the reference domain $\bar{G}_1 \cup \bar{G}_2$,

$$\tilde{u}_i \stackrel{def}{=} u \circ T_i(q) \quad (i = 1, 2).$$

Then the transformed system can be described by

$$\sum_{i \leq 2} \left\{ \left\langle \frac{d\tilde{u}_i}{dt}, \phi \right\rangle_{L^2(\bar{G}_i)} + \sigma_i(q)(\tilde{u}_i(t), \phi) \right\} = L(q)(\phi)$$

$$\text{for } \forall \phi \in H^1(\bar{G}_1 \cup \bar{G}_2) \qquad (8)$$

with

$$\tilde{u}_i(0) = \bar{u}_0 \circ T_i(q)$$

where $\sigma_i(q)(\cdot, \cdot)$ and $L(q)(\cdot)$ denote, respectively, a sesquilinear form on $H^1(\bar{G}_i) \times H^1(\bar{G}_i)$ and a linear form on $H^1(\bar{G}_1)$ such that, for $\phi, \psi \in H^1(\bar{G}_1 \cup \bar{G}_2)$,

$$\sigma_i(q)(\phi, \psi) \stackrel{def}{=} c \int \int_{\bar{G}_i} \left\{ \left( \frac{df_i(q)}{dz_1} \right)^{-2} \frac{\partial \phi}{\partial z_1} \frac{\partial \psi}{\partial z_1} + \frac{\partial \phi}{\partial z_2} \frac{\partial \psi}{\partial z_2} \right\} dz,$$

and

$$L_1(q)(\phi) \stackrel{def}{=} \int_0^{a+b+c} \tilde{g} \frac{df_1(q)}{dz_1} dz_1.$$

Thus our problem is to find an optimal solution of

$$J(q^*) = \min_{q \in Q} \frac{1}{2} \int_0^T |\mathcal{H}\tilde{u}_1(q)|_S - y_d(t)|^2 dt \tag{9}$$

subject to the system (8).

## Algorithm

Since the analytic gradient of cost can be easily computed, various optimization routines are available for solving the problem (9). One popular minimizer for (9) is a quasi-Newton method with BFGS-secant approximation for Hessian ( See e.g. [3] ). At each iteration step, this method searchs a feasible direction and require a line search for choosing an appropriate step size. To implement this method, we need to solve a couple of parabolic partial differential equations recursively at each iteration step. This yields considerably computational amounts for the implementation of this algorithm. Hence we intend to save the execution time of this algorithm using the domain decomposition technique for solving the corresponding parabolic equations. There are two alternative methods for this approach ( see [6] ), i.e., non-overlapping ( Schur method ) and overlapping method ( Schwartz method ). In this paper, due to the geometrical structure of domain, we choose non-overlapping method ( see [5] ). For the convenience of discussions, the finite Galerkin solution can be written by

$$\tilde{u}^N = [\tilde{u}_1^N, \tilde{u}_2^N, \tilde{u}_3^N]^T$$

where the $\tilde{u}_i^N$ corresponds to the solution at the node number in $\bar{G}_i (i = 1, 2)$ and the node number belonging to the interface $I (i = 3)$, respectively. The finite Galerkin approximation of (8) can be represented by

$$C_h^N \frac{d\tilde{u}_h^N(t)}{dt} + A_h^N(q)\tilde{u}_h^N(t) = B_h^N(q)\tilde{g}_h^N(t), \tag{10}$$

with

$$\tilde{u}_h(0) = \bar{u}_{0,h}^N.$$

The implicit numerical scheme for the above equation is to solve a linear system of algebraic equation

$$K^N(q)\tilde{u}^N(t + \Delta) = b^N(t) \tag{11}$$

at each time step $t = k\Delta (k = 0, 1, 2, \cdots, n_t)$, where

$$K^N(q) = C_h^N(q) + \Delta A_h^N(q)$$

and

$$b^N(t) = C_h^N(q)\tilde{u}_h^N(t) + \Delta B_h^N(q)\tilde{g}_h^N(t),$$

respectively. The element matrix can be decomposed into the block matrices,

$$K^N(q) = \begin{bmatrix} K_{11}^N(q) & 0 & K_{13}^N(q) \\ 0 & K_{22}^N(q) & K_{23}^N(q) \\ K_{13}^N(q)^T & K_{23}^N(q)^T & K_{33}^N(q) \end{bmatrix}.$$

In this decomposition, $K_{ii}^N(q)$ represents coupling between the pair of nodes in $\bar{G}_i(i = 1, 2)$ and $I(i = 3)$, and $K_{i3}^N(q)(i = 1, 2)$ imply coupling between the pairs belonging to $\bar{G}_i$ and $I$, respectively. Similarly, $b^N(t, q)$ is decomposed into three parts:

$$b^N = [b_1^N, b_2^N, b_3^N]^T,$$

corresponding to the node points at $\bar{G}_1, \bar{G}_2, \bar{I}$, respectively. By applying block-Gaussian elimination to (11), the Schur complement of the element matrix $K^N(q)$ is given by

$$K_C^N(q) = K_{33}^N(q) - \sum_{i \le 2} K_{i3}^N(q)^T (K_{ii}^N(q))^{-1} K_{i3}^N(q). \tag{12}$$

Using this, we can reduce the system (11) to

$$K_C^N(q)\tilde{u}_3^N = \tilde{b}_3^N(q) \tag{13}$$

where

$$\tilde{b}_3^N(t, q) = b_3^N(t, q) - \sum_{i \le 2} K_{i3}^N(q)^T (K_{ii}^N(q))^{-1} b_i^N(t, q). \tag{14}$$

Then the solutions of $\tilde{u}_1^N(t + \Delta)$ and $\tilde{u}_2^N(t + \Delta)$ can be obtained by solving

$$K_{ii}^N(q)\tilde{u}_i^N(t + \Delta) = \tilde{b}_i^N(t, q) - K_{33}^N(q)\tilde{u}_3^N(t + \Delta) \qquad \text{for} \quad i = 1, 2.$$

The size of the element matrix $K_C^N$ is relatively small, compared with the number of nodes in $\bar{G}_i$. Hence the computational saving can be achieved using an appropriate linear algebraic solver. To solve Eq.(13), we use the iterative method based on the conjugate gradient algorithm. We note that the use of direct method by the Cholesky decomposition in Eq.(13) requires heavily computational volumes due to the computation of inverse matrices of diagonal block $K_{ii}$ in (12)

The proposed algorithm consists of two parts. The first part of our algorithm is devoted to determining the region of the crack to be identified. To get accurate the numerical solution, we select the reference line $r_0$ in Fig. 3 so as to choose it at the neighborhood of the right end of crack as much as possible. Secondly, we apply the quasi-Newton method to solve the minimization problem (9) on that region. Thus our estimation algorithm is stated as follows:

**Step 1:** Determine the inspection interval:

Step 1.1 : Set a finite number of points at the interface,

$$0 < \bar{r}_1 \leq r^1 < \cdots < r^n \leq \bar{r}_2 < b.$$

Step 1.2 : Find an appropriate reference line $\hat{r}^*$ such that

$$J(\hat{r}^*) = \min_{i \leq n} J(r^i).$$

Step 1.3 : Select two inspection intervals at both sides of $\hat{r}^*$. Namely, if $\hat{r}^* = r_j$, set the closed intervals as follows:

$$\bar{I}_L \subset (r_{j-1}, r_j) \quad \text{and} \quad \bar{I}_R \subset (r_j, r_{j+1}).$$

**Step 2 :** Estimate the length of crack:

Step 2.1 : Set the reference lines as

$$r_0^L \in \bar{I}_L \quad r_0^R \in \bar{I}_R.$$

Step 2.2 : Solve the optimization problems,

$$J(\hat{q}_L^*) = \min_{q \in I_L} J(q) \quad \text{and} \quad J(\hat{q}_R^*) = \min_{q \in I_R} J(q).$$

Step 2.3 : Evaluate values of the cost for $J(\hat{q}_L^*)$ and $J(\hat{q}_R^*)$, and determine the estimated length of crack as follows :

$$\hat{q} = \begin{cases} \hat{q}_L^* & \text{if} \quad J(\hat{q}_L^*) + \delta \leq J(\hat{q}_R^*) \\ \hat{q}_R^* & \text{if} \quad J(\hat{q}_L^*) \geq J(\hat{q}_R^*) + \delta \\ \hat{r}^* & \text{if} \quad |J(\hat{q}_L^*) - J(\hat{q}_R^*)| < \delta \end{cases}$$

## Computational Experiments

In a series of experiments, the algorithm proposed was tested using the simulation data. An alminum sample with epoxy bondline was considered as depicted in Fig. 2. A material sample with dimensions $1.0 \times 0.125$ (inch) ( the subdomain $\bar{G}_1$ ) and $0.25 \times 0.125$ (inch) ( the subdomain $\bar{G}_2$ ) was used, i. e.,

$$a = 0.375, \quad b = 0.25 \quad c = 0.375 \quad \text{and} \quad d = e = 0.125$$

The thermal diffusivity coefficient in these experiments was $\kappa = 0.12197(\ (inch)^2/sec\ )$. To discretize the system model by a bilinear finite-element method, each subdomain $\bar{G}_j (j = 1, 2)$ is divided into a finite number of elements $\{e_k\}_{j=1}^{K_j}$ and a number $N_j (\geq K_j)$ of nodes defined by $\{z_i\}_{i=1}^{N_j}$ are selected in $\bar{G}_j$. Each element is preassigned as an axiparallel rectangle with nodes at the vertices. The number of finite elements and nodes in computational experiments were set as

$$K_1 = 16 \times 2 \quad N_1 = 17 \times 3; \quad K_2 = 4 \times 2 \quad N_2 = 5 \times 3$$

Integration in the element matrices and vectors in (10) can be computed through piecewise bilinear basis functions. The discrete state model can be solved with respect to discrete time $t = k\Delta(k = 0, 1, \cdots, n_t)$ where $\Delta = t_f/n_t$. The initial and final time, and number of time divisions were taken as $t_f = 2.4(\ sec\ )$ and $n_t = 8$ in our computations. The initial state and the boundary input were preassigned as $u_0 = 0.$ and $g = 1.0$, respectively. To obtain the model output, the number of observation points was set as $m = 15$ and each sensor is located at

$$\xi_p^i = i/16 \quad (i = 1, 2, \cdots, m).$$

For these test computations, simulated data $\{y_d(t)\}$ were generated by first solving the finite-element model (10) with the same number of finite elements and nodes. Random noise at various levels was then added to the numerical solution, thereby producing simulated noisy data. For the implementation of the proposed algorithm, the number of points related to the reference lines was set as $n = 3$ and those were located at

$$r_i = i/4 \quad (i = 1, 2, \cdots, n).$$

The gradient of the cost functional was computed analytically and the corresponding Hessian was evaluated using the BFGS secant update. Table 1 reports the estimated parameter results for the data without noise and with relative noise. These test computations suggest that the method proposed perform well with reasonable level of noise.

Table 1. Estimated results in computational experiments

|           | Example 1. | Example 2. | Example 3. | Example 4. |
|-----------|-----------|-----------|-----------|-----------|
| True Value | 0.0500    | 0.100     | 0.150     | 0.200     |
| Noise Free | 0.0500    | 0.100     | 0.150     | 0.200     |
| 5% Noise   | 0.00625   | 0.0688    | 0.131     | 0.181     |
| 10% Noise  | 0.0451    | *         | 0.146     | 0.200     |
| 15% Noise  | 0.00625   | 0.0262    | 0.0798    | 0.139     |
| 20% Noise  | 0.0449    | 0.0771    | 0.119     | *         |
| 25% Noise  | 0.00625   | *         | 0821      | 0.131     |
| 50% Noise  | 0.0563    | 0.0688    | *         | 0.131     |

$* = $ No Convergence

Acknowledgements

This research was supported by the National Aeronautics and Space Administration under NASA Contract No. NAS1-18605 while the author was in residence at the Institute for Computer Applications in Sciences and Engineering (ICASE), NASA Langley Research Center, Hampton, VA 23665 USA. The author would like to express his sincere appreciation to Professor H.T. Banks ( North Calolina State University ), Dr. W.P. Winfree ( NASA Langley Research Center ) and Dr. R. G. Carter ( Argonne National Laboratory ) for numerous conversations and encouragements during the course of this study.

References

1. Banks, H.T. ; Kojima, F : Boundary shape identification problems in two
   -dimensional domains related to thermal testing of materials, Quart. Appl. Math.,
   **47** (1989) 273-293.

2. Banks, H.T. ; Kojima, F ; Winfree, W.P. : Boundary estimation problems arising
   in thermal tomography, Inverse Problems, **6** (1990) 897-921.

3. Kojima, F : A computational method for thermal tomography by time dependent
   boundary integral equation method, International Series of Numerical Mathematics,
   Birkhäuser, **100** (1991) 207-217.

4. Kojima, F : Backpropagation learning using the trust region algorithm and application to corrosion detection in thermal testing of materials, Proc. of the International
   Symposium on Nonlinear Phenomena in Electromagnetic Fields, Nagoya, 1992
   ( to appear )

5. Bjørstad, P.E. ; Wildlund, O.B. : Iterative methods for the solution of elliptic
   problems on regions partitioned into substructures, SIAM J. Numer. Anal., **23**
   (1986) 1097-1120.

6. Chan, T.F. ; Goovaerts, D : Schwartz = Schur : Overlapping versus nonoverlapping
   domain decomposition, Submitted to SIAM J. Math. Anal.

# Chapter 9
# Other Engineering Applications

# Inverse Problem Approach Based on the Kalman Filtering and Its Applications

Akira MURAKAMI* and Takashi HASEGAWA*

* Dept. of Agricultural Engineering, Kyoto University
  Sakyo-ku, Kyoto 606-01, Japan

## 1 INTRODUCTION

This paper focuses on the formulation and the numerical performances of the inverse analysis procedure which utilizes prior information in a probabilistic sense. A numerical strategy of Kalman filtering and the extended Kalman filtering technique, in conjunction with the finite element method, is adopted to characterize the material properties of a linear/nonlinear elastic medium considering noisy observations. In applying such a procedure to inverse problem, it is possible to overcome the indeterministic problems which appear in an inverse analysis even when the number of unknowns exceeds the number of measurements. Several hypothetical results of loading tests are given to characterize a distribution of the material properties.

## 2 INVERSE ANALYSIS APPROACH BASED ON KALMAN FILTERING

It is worthwhile to investigate the usefulness and the applicability of the Kalman filtering technique[1], [2] with respect to the inverse problem. A formulation of the inverse analysis procedure is herein furnished. In this article, boldface lower-case and boldface capital letters denote vectors and matrices, respectively, while Italic lower-case and capital letters denote scalar. Superscripts T and $-1$ mean transpose and inverse, and subscripts $t$ and $k$ describe time $t = t$ in continuous time and time $t = k\Delta t$ in discrete time, respectively.

### 2.1 Kalman filtering and extended Kalman filtering

This section describes the structure of the Kalman filtering technique as the mathematical basis for it. The algorithms of Kalman filtering and the extended Kalman filtering technique will be briefly reviewed in order to formulate the following example problems. Details of this are given in References [3], [4] and [5].

## 2.1.1 Kalman filtering

$$y_t = H_t x_t + v_t, \tag{1}$$

$$x_{t+1} = F_t x_t + G_t w_t \tag{2}$$

where $x_t$ is a state vector $(n \times 1)$, $y_t$ is an observation vector $(p \times 1)$, $w_t$ is plant noise $(m \times 1)$, $v_t$ is observation noise $(p \times 1)$, $F_t$ is a state transition matrix $(n \times n)$, $G_t$ is a driving matrix $(n \times m)$ and $H_t$ is an observation matrix $(p \times n)$. Both noises are 'Gaussian' and 'white'.

$$E(w_t) = E(v_t) = \mathbf{o}, \tag{3}$$

$$E\left\{ \begin{pmatrix} w_t \\ v_t \end{pmatrix} \begin{pmatrix} w_s^T v_s^T \end{pmatrix} \right\} = \begin{pmatrix} Q_t & \mathbf{o} \\ \mathbf{o} & R_t \end{pmatrix} \delta_{ts}, \quad R_t > 0, \tag{4}$$

$$E\{w_t x_s^T\} = E\{v_t x_s^T\} = \mathbf{o} \quad \text{for} \ t \geq s \tag{5}$$

where $E\{\cdot\}$ is an average operator and $\delta_{ts}$ is Kronecker's delta.

Now, assume that linear system equations (1) and (2) denote a field equation governing the geomechanical problem at hand. Noises are assumed to comply with the stochastic characteristics listed in equations (3)–(5). Some deductions based on Bayes' theorem introduce the algorithm of Kalman filtering; they are summarized below:

**Filter equation:**

$$\hat{x}_{t/t} = \hat{x}_{t/t-1} + K_t[y_t - H_t \hat{x}_{t/t-1}], \tag{6}$$

$$\hat{x}_{t+1/t} = F_t \hat{x}_{t/t} \tag{7}$$

**Kalman gain:**

$$K_t = P_{t/t-1} H_t^T [H_t P_{t/t-1} H_t^T + R_t]^{-1} \tag{8}$$

**Estimate error covariance matrix:**

$$P_{t+1/t} = F_t P_{t/t} F_t^T + G_t Q_t G_t^T, \tag{9}$$

$$P_{t/t} = P_{t/t-1} - K_t H_t P_{t/t-1} \tag{10}$$

**Initial condition:**

$$\hat{x}_{0/-1} = \bar{x}_0, P_{0/-1} = \Sigma_0. \tag{11}$$

### 2.1.2 Extended Kalman filtering

In cases where the following nonlinear equation dominates an observation equation, on the other hand, the Taylor series expansion is utilized around the estimate of $\hat{x}_{t/t-1}$ in order to avoid the nonlinearity appearing in the observation equation.

$$y_t = h_t(x_t) + v_t \tag{12}$$

where $h_t$ is the nonlinear vector function of state variable $x_t$. Any term beyond the second order is disregarded in this expansion, and the following expression is then obtained as the refined observational equation:

$$
\begin{aligned}
y_t &= h_t(\hat{x}_{t/t-1}) + \left(\frac{\partial h_t}{\partial x_t}\right)_{x_t=\hat{x}_{t/t-1}} (x_t - \hat{x}_{t/t-1}) + v_t \\
&= h_t(\hat{x}_{t/t-1}) + H_t(x_t - \hat{x}_{t/t-1}) + v_t \\
&= H_t x_t + \{h_t(\hat{x}_{t/t-1}) - H_t \hat{x}_{t/t-1}\} + v_t.
\end{aligned}
\tag{13}
$$

Here,

$$\eta_t = y_t - h_t(\hat{x}_{t/t-1}) + H_t \hat{x}_{t/t-1} \tag{14}$$

substituting equation (14) into equation (13), and then

$$\eta_t = H_t x_t + v_t. \tag{15}$$

The second and third terms in equation (13) can be calculated at time $t-1$. Thus, the value of $\eta$ can be 'observed' at time $t$ using observation $y_t$. In turn, equations (2) and (15) constitute a linear system set which can then be applied to the ordinary Kalman filtering algorithm. Such a formulation in the case of nonlinearity is referred to as 'extended Kalman filtering'[4].

## 2.2 Kalman filter in conjunction with the finite element method

When parameter identification is handled as an inverse problem, the unknown parameters should be taken as $x_t$. Then, the following stationary condition of the parameters can be described as the state equation without involving system noise:

$$x_{t+1} = I x_t. \tag{16}$$

Now, consider the inverse problem of characterizing the material properties, Lamé's constants $\lambda_i$ and $\mu_i$, of the elastic medium within the framework of the finite element method. The above condition also holds for the unknown Lamé's constants, because they are constant[6], [7].

### 2.2.1 Extended Kalman filter-FEM

On the other hand, a finite element equation which is written for the equilibrium solution
under the plane strain condition is derived as

$$\boldsymbol{Ku = f}. \tag{17}$$

Through this equation, the observed displacements can be rewritten as the nonlinear
functions of the unknowns, i.e., Lamé's constants,

$$\boldsymbol{u^* = h(x_t) + v_t}. \tag{18}$$

A set of equations (16) and (18) can be applied to the extended Kalman filtering procedure. A direct formulation using equation (17) also allows for the search of parameters
in another way[8] and appears in the next section. Recently, other formulations of the
extended Kalman filtering procedure for an inverse analysis have also been examined in
conjunction with CSM (the Charge Simulation Method)[9] and BEM[10].

Figure 1 shows a flowchart of the computation based on the extended Kalman filter-
FEM.

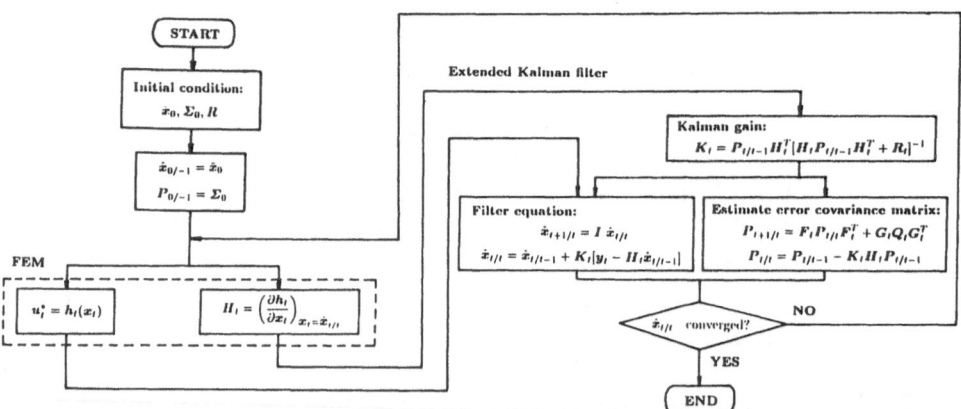

Figure 1: Flowchart of computation based on the extended Kalman filter-FEM (after
Suzuki [11])

### 2.2.2 Kalman filter-FEM

The discrete finite element system, denoted by the observed displacement is sought for
here. The direct strategy is to use the condensation technique to eliminate the unknown
and the boundary displacements.

Equation (17) is block-partitioned, separating the observed displacement, $u^*$, from that to be eliminated, $u$. Thus,

$$\begin{bmatrix} K_{11} & K_{12} \\ K_{21} & K_{22} \end{bmatrix} \begin{Bmatrix} u^* \\ u \end{Bmatrix} = \begin{Bmatrix} f^* \\ f \end{Bmatrix}, \quad K^{ij} = \begin{bmatrix} K_{11}^{ij} & K_{12}^{ij} \\ K_{21}^{ij} & K_{22}^{ij} \end{bmatrix} \tag{19}$$

where $K_{lm}^{ij}$ ($i = 1, \ldots, n$; $j = 1, 2$; $l, m = 1, 2$) is the partitioned stiffness submatrix, and $\lambda_i$ and $\mu_i$ are Lamé's constants in the $i$-th region and are the unknown parameters which correspond to vector $x_t$ in the filtering equation. Equation (19) makes up the condensed system, namely,

$$f^* - Q(\lambda, \mu)f = [(\lambda_j K_{11}^{j1} + \mu_j K_{11}^{j2}) - Q(\lambda_i, \mu_i)(\lambda_j K_{21}^{j1} + \mu_j K_{21}^{j2})]u^* \tag{20}$$

where $Q(\lambda, \mu) = (\lambda_j K_{12}^{j1} + \mu_j K_{12}^{j2})[\lambda_j K_{22}^{j1} + \mu_j K_{22}^{j2}]^{-1}$ ($j$: sum)
and is equivalently rewritten as[12]

$$f^* - Q(\lambda, \mu)f = [(K_{11}^{11} - QK_{21}^{11})u^*, (K_{11}^{12} - QK_{21}^{12})u^*, \ldots,$$
$$(K_{11}^{n1} - QK_{21}^{n1})u^*, (K_{11}^{n2} - QK_{21}^{n2})u^*] \begin{Bmatrix} \lambda_1 \\ \mu_1 \\ \vdots \\ \lambda_n \\ \mu_n \end{Bmatrix} \tag{21}$$

e.g. $y_t = f^* - Q(\lambda, \mu)f$; $\quad x_t = (\lambda_1, \mu_1, \ldots, \lambda_n, \mu_n)^T$;

$H_t = [(K_{11}^{11} - QK_{21}^{11})u^*, (K_{11}^{12} - QK_{21}^{12})u^*, \ldots, (K_{11}^{n1} - QK_{21}^{n1})u^*, (K_{11}^{n2} - QK_{21}^{n2})u^*]$.

By adding the noise term, this equation corresponds to observation equation (1).

As seen in the previous section, the state equation denotes the stationary condition of the parameters. Due to the nonlinearity of the observation equation, an iterative procedure should be adopted under the observed constant displacement until a convergence of the estimated parameters can be achieved. Suffix $t$ is not the time axis, but the 'iterative axis' in this case.

## 3  NUMERICAL PERFORMANCES

Herein, two example problems will be dealt with while observing the hypothetical ground behavior.

*Example-1*  To examine the validity of the foregoing procedure, the accuracy of the present method is assessed by a comparative analysis using some algorithms for the problem appearing in Reference [13]. Figure 2 shows the problem setup and provides a finite element discretization for the system. All geometric conditions and material properties

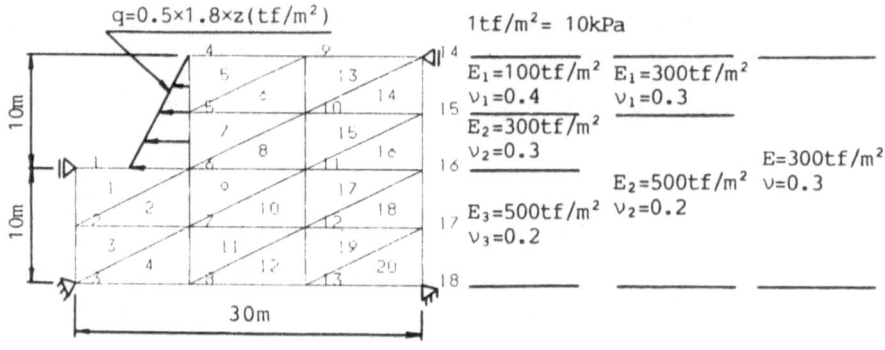

Figure 2: Analytical model (after Arai *et al.* [13])

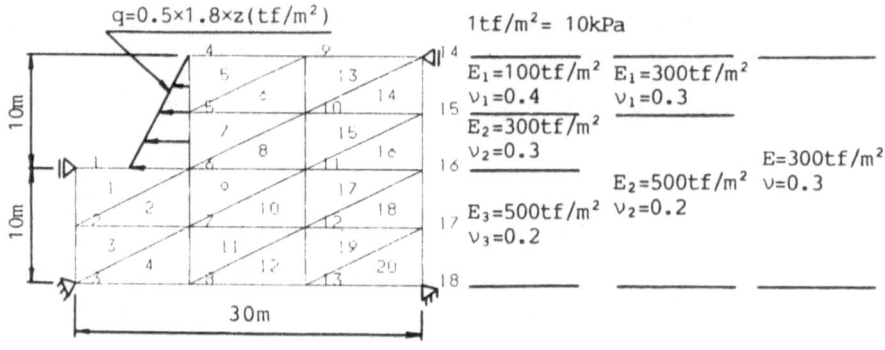

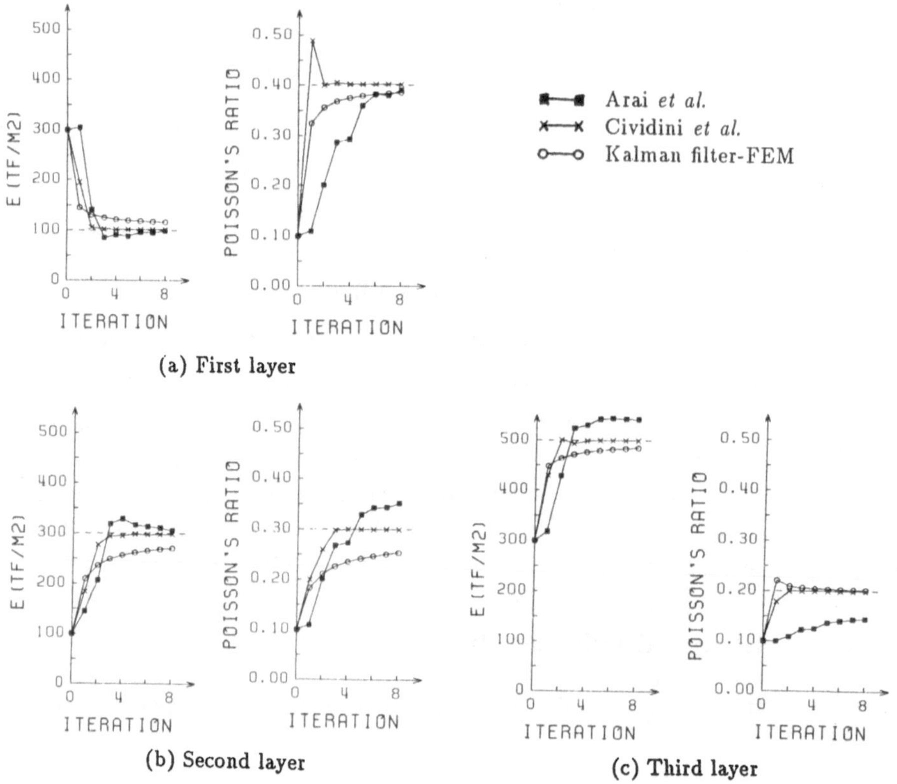

(a) First layer

- ■—■ Arai *et al.*
- ×—× Cividini *et al.*
- o—o Kalman filter-FEM

(b) Second layer  (c) Third layer

Figure 3: Process of parameter identification

used are shown in this diagram. For this problem, Arai *et al.*[14] have compared their own identification with the results obtained through Bayes' approach. In this example, another comparison is made between the solution obtained by the Kalman filter-FEM formulation outlined in the previous section and the results from the other methods.

Figure 3 describes the numerical performances concerning the three-layered soil deposit. Observational nodal points, whose displacements are the numerical input in such cases, are #1, 14 (for settlements) and #4,5,6,9 (for vertical and horizontal displacements) as seen in Figure 2. In this case, however, the discrepancy between the identified results for the second and the third layers is found in Arai's method and the proposed procedure, but not in Bayes' approach. As seen in Figure 2, the number of observed points involved in these layers is 2, whereas the number of observed points in the first layer is 4. It can be recognized that even such poor numerical performances are influenced by the quality and the quantity of the observations. The influence of the observations on the identified results is discussed in References [7] and [8].

*Example-2*  In this example, a more practical problem is examined, i.e., a numerical interpretation of the test embankment. Herein, the distribution of material properties, based on the limited number of measurements is pursued through use of the extended Kalman filter-FEM. A part of the deformations were observed in the hypothetical *in situ* test embankment, as seen in Figure 4. And the distribution of Young's moduli to be estimated is given in Figure 5. The identified distribution of Young's moduli is shown in Figure 6, where it is seen that the existence of a weak layer in the soil deposit can easily be picked up.

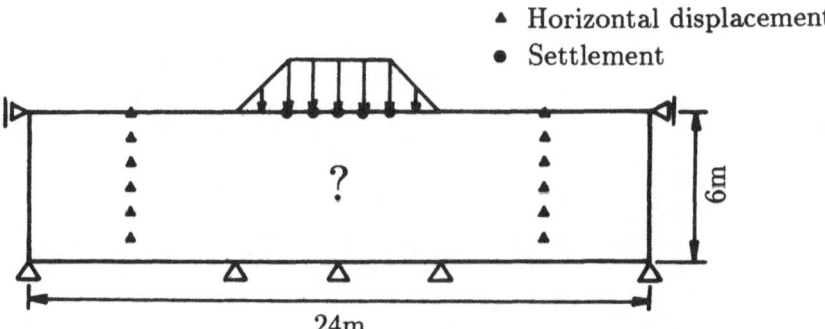

Figure 4: Hypothetical *in situ* test embankment (after Suzuki [11])

536

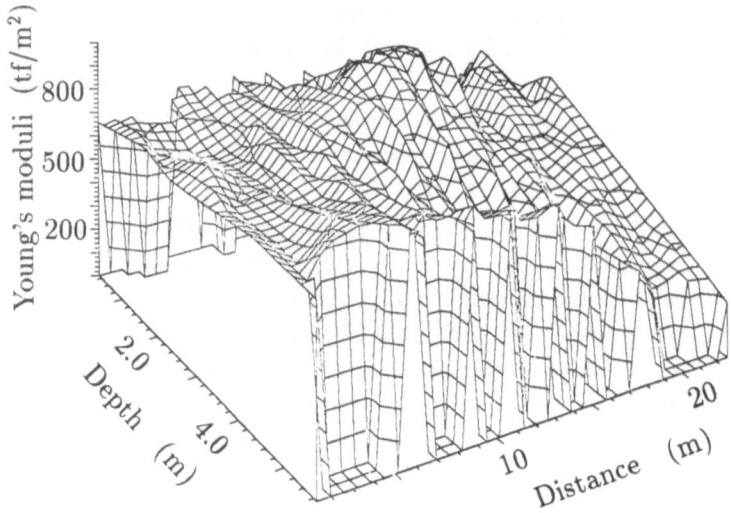

Figure 5: Distribution of Young's moduli to be estimated

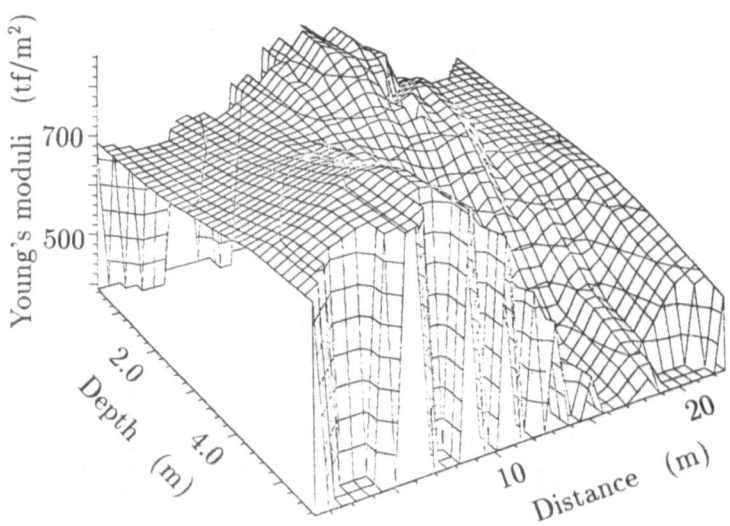

Figure 6: Distribution of identified Young's moduli

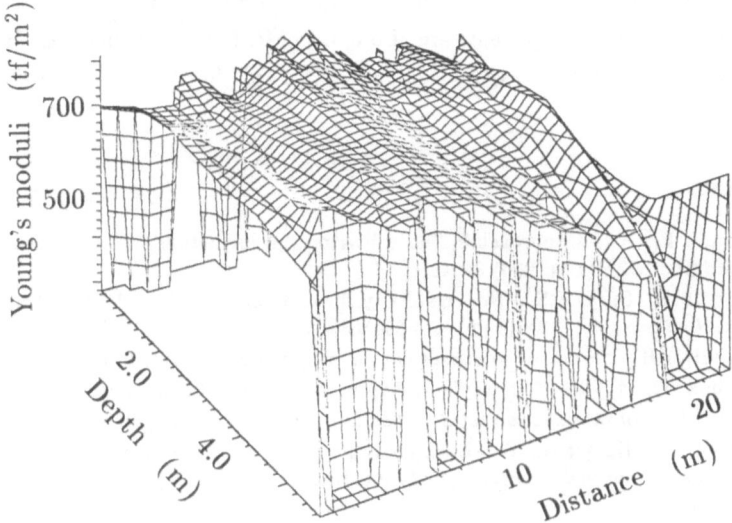

Figure 7: Distribution of identified Young's moduli

In this analysis, the following probabilistic characteristics are given for the initial estimate of the error covariance matrix[11]:

$$\rho(\Delta x, \Delta y) = exp\left[-\left\{\left(\frac{\Delta x}{a}\right)^2 + \left(\frac{\Delta y}{b}\right)^2\right\}\right] \tag{22}$$

where $\rho$ is a correlation factor of the unknowns, $\Delta x$ is the horizontal distance and $\Delta y$ is the vertical distance between two arbitrary points, and $a$ and $b$ are the constants. Suzuki gave the values of 5.0 and 1.0 to $a$ and $b$, respectively. This means that the objective soil stratum is assumed to have the characteristics of a horizontally layered deposit. The initial estimate for the error covariance matrix, especially the non-diagonal terms, has an important role in the identification of the non-homogeneous material properties.

However, an unsuitable estimate for the error covariance matrix may bring about an unreasonable evaluation of the material properties. Figure 7 reveals poor results for the identification when neglecting the non-diagonal terms of the error covariance matrix.

## 4  CONCLUSION

A formulation of the inverse analysis procedure, characterizing material properties, is shown based on Kalman filtering and the extended Kalman filtering technique in conjunction with the finite element method. A numerical evaluation of the hypothetical ground behavior is examined and discussed through the proposed numerical procedure.

538

The authors are indebted to Dr. Makoto Suzuki of Shimizu Corporation for his help and discussions. Acknowledgements are also due to Mr. Kentaro Yamaguchi for his help in the numerical computations and Ms. Heather Griswold for proofreading this article.

# References

[1] Kalman, R.E.: A new approach to linear filtering and prediction problems, *Trans. ASME, J. Basic Eng.*, **82D(1)** (1960) 34–45.

[2] Kalman, R.E.; Bucy, R.S.: New results in linear filtering and prediction theory, *Trans. ASME, J. Basic Eng.*, **83D(1)** (1961) 95–108.

[3] Jazwinski, A.H.: *Stochastic processes and filtering theory*, Academic Press (1970).

[4] Katayama, T.: *Applied Kalman filter*, Asakura-shoten (1983) (in Japanese).

[5] Norton, J.P.: *An introduction to identification*, Academic Press (1986).

[6] Murakami, A.; Hasegawa, T.: Back analysis by Kalman filter-finite elements and a determination of optimal observed points location, *Proc. JSCE* **(388)** (1987) 227–235 (in Japanese).

[7] Murakami, A.; Hasegawa,T.: Back analysis by Kalman filter-finite elements and optimal location of observed points, *Numerical Methods in Geomechanics* (Swoboda ed.), **2** (1988) 2051–2058.

[8] Murakami, A.: *Studies on the application of Kalman filtering to some geomechanical problems related to safety assessment*, Doctoral thesis, Kyoto University (1991).

[9] Murase, H.; Koyama, S.; Honami, N: Kalman filter charge simulation methods for the solution of inverse problems of two- or three-dimensional elasticity, *Trans. of JSME*, **56A(531)** (1990) 106–112 (in Japanese).

[10] Utani, A.; Tosaka, N.: Identification analysis of elastic constants by extended Kalman filter-boundary element method, *Proc. 8th Japan National Symp. Boundary Element Methods*, **8** (1991) 23–28 (in Japanese).

[11] Suzuki, M.: *Basic study upon probabilistic representation of spatial variation of soil properties and its applications to geotechnical problems*, Doctoral thesis, Nagoya Institute of Technology (1990) (in Japanese).

[12] Cividini, A.; Maier, G.; Nappi, A.: Parameter estimation of a static geotechnical model using a Bayes' approach, *Int. J. Rock Mech. Min. Sci. & Geomech. Abstr.*, **20(5)** (1983) 215–226.

[13] Arai, K.; Ohta, H.; Yasui, T.: Simple optimization techniques for evaluating deformation moduli from field observation, *Soils and Foundations*, **23(1)** (1983) 107–113.

[14] Arai, K.; Ohta, H.; Miyata, M.: Comparison of static and statistical methods for back-analysis of elastic consolidation problems, *Computer Methods and Advances in Geomechanics* (Beer, Booker & Carter (eds)), **2** (1991) 949–954.

# The Thin Shell Approach for Some 3D Engineering Inverse Problems

S. ANDRIEUX

Electricité de France. Direction des Etudes et Recherches
1, Avenue du général de Gaulle 92 141 Clamart, France

Number of inverse or identification problems in engineering mechanics are severely ill-posed when formulated in the initial three-dimensional framework. We propose to show in this paper how, for thin or moderately thin domains, the alternative approach of reformuling the problem via shell theories can give rise to some progresses both from the theoretical and practical point of view, and lead to efficient solution procedures of the inverse problems.

A carefull inspection of the shell approach shows that it works like a Tikhonov regularization of the 3D problem [1], but the regularizing norm and parameter arise from the physics and are not arbitrarily choosed as they are in the usual regularization methods.

The central question when dealing with such an approach is to determine how distorted is the resulting Shell Inverse Problem (SIP) with respect to the original three-dimensional one (3DIP). Of course, the answer to this question depends on the problem and on the nature of applications aimed and cannot be universal. It will be nevertheless addressed in some details in the last part of the paper.

The two problems solved via SIP are met in piping. The first one is the determination of stresses in an elastoplastic pipe from profile measurements. In the second one, the inner temperature of the pipe is reconstructed from external temperature measurements. In this last case, a original thermal shell theory has been designed.

Stress determination in an elastoplastic pipe from profile measurements

The problem is to determine the residual stresses in elastoplastic pipes when external or internal profiles w(z) are given (see figure 1). The history of loading (normal pressures distributions) is not known but the principal steps of the deformation of the pipe are available (profile measurements at different instants). In the application (steam generator tubes of PWR power plants near the roll transition zone), the loading path is very complex and consists in five quasi-radial distinct phases at least.

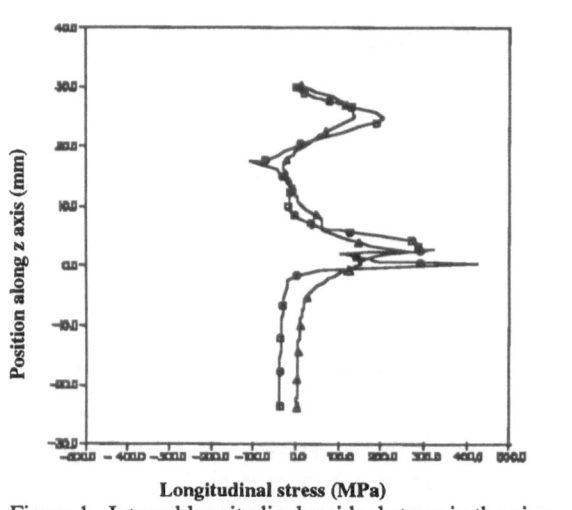

Figure 1:  Internal longitudinal residual stress in the pipe          Sketch of the problem
                    o          Direct computations
                    Δ          Inverse computations

The formulation of this identification problem through the axisymmetric model of thin shell offers three major advantages:

   i)   it allows to reduce it to independant local minimization problems defined on the thickness of the shell at each point of the given profile. These local (constrained) minimizations can be achieved with very low computational costs compared with the 3D problem where the minimization problem is defined over the whole structure.

   ii)  it indicates the kind of regularization of the data needed to stabilize the results with respect to the unavoidable noise in the measurements. Namely, the profile have to be regularized with control of the derivatives up to second order. Direct 2D axisymmetric computations shows the dramatic instabilities encountered with no regularized data.

   iii)  it makes the most of the *a priori* information available in this problem.

The determination of the residual stresses in the pipe comprises two steps : the first one is the regularization of the profile before applying the kinematic displacements to strains relations. It is achieved by a Tikhonov method in which an improved version of Marti's algorithm to choose the optimum regularization parameter is used ([2],[3]). This step is optimized by introduction of Fourier techniques. The second step is the elastoplastic stresses computation in the shell. The constrained minimization form of the problem is solved by a standard radial-return algorithm combined with a secant method dealing with the overall equilibrium condition.

Evaluation of the method has been made by comparing the stresses given by direct 2D axisymmetric elastoplastic computations of a pipe under representative loading pathes and the identified stresses. The results are given on the figure 1 showing a very good agreement even for complicated loading history (unilateral conditions, unloadings...). The discrepancy on the top part of the curve can be attributed to 3D local effects (singularities near the end on contact zones between pipe and roll). On the bottom, the difference is caused by the shell theory used : plane stress condition is assumed but is violated in this part of the pipe. A more refined elastoplastic shell theory can be used to remedies this last difficulty [4].

## Internal temperatures in pipes from external measurements

In situ detection of stratified inner fluid temperature in pipes leads to a kind of inverse problem named the harmonic extension problem in the mathematical literature (see equations (1)). It consists in finding the temperature in a domain from the knowledge of temperature and normal flux on a part of its boundary, provided the thermal equilibrium is reached. From the practical point of view, thermocouples are located at the external boundary of the pipe, and exchange conditions with environment are known (see figure 2).

## The stratification detection problem

Find $(T_c, T_h, \Theta_e)$ from the measured temperatures at points ■■

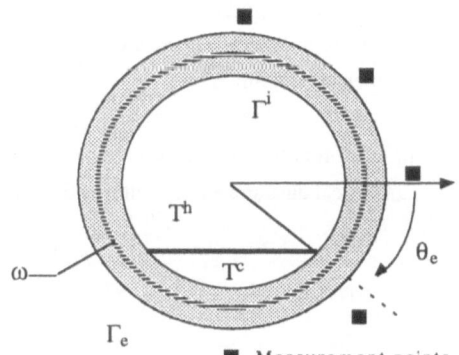

Figure 2

■ Measurement points

542

The uniqueness of the solution of the 3D problem is a consequence (for regular domains) of the Cauchy-Kowaleski theorem, but existence is in general not assured because of compatibility conditions which are to be satisfied between the data (temperature and flux). Furthermore, no kind of physically reasonable continuity of the solution (if it exists) with respect to the data can be achieved. The mostly used method of resolution is the least-square output method, but besides its problematic properties of convexity and convergence, it leads to considerable amount of computations because each iteration needs the resolution of a 3D direct problem.

In order to formulate the thin shell version of the harmonic extension problem, we begin with the derivation of an appropriate shell theory. It can be shown by asymptotic analysis [5] that the through-the-thickness repartition of temperature must be parabolic in order to ensure convergence of the shell equations to the same limit that the 3D one's when the thickness goes to zero. Unlike the mechanical theories of shells, the order of the equations obtained on the middle surface is not increased with regard to the 3D ones. On the other hand, the number of fields extends from one to three .

$$T(x_\alpha,z) = T^0(x_\alpha) + T^1(x_\alpha) z / h + T^2(x_\alpha) 3((z / h)^2 - 1/3)/2$$

z is the thickness space variable, h the half-thickness

This shell theory compares well with 3D calculations even on moderately thick structures (up to 0.2 thickness to radius of curvature ratio) and for sharp boundary conditions as the thermal stratification ones.

Coming back to the identification problem, we can now formulate the shell identification problem: *find the inner surface temperature of the shell when the outer surface temperature and outer normal flux is given.* The main results of the analysis can be summed up as follows [6]:

I.      The shell identification problem has an unique solution.
II.     This solution can be calculated through the successive resolution of two linear (but non symmetric) problems.
III.    The regularization needed concerns the tangential derivative of first order of the measured temperature.

Examples of identification can be found on the figures 3.a and 3.b, where a reconstruction test from analytical data and a real identification are displayed.

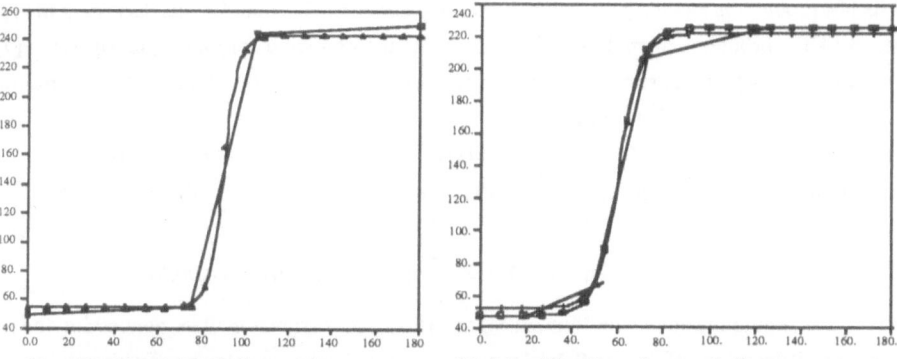

3.a.Identification with 4 observations
   0  Observed temperature
   Δ  Restitued temperature on external
       boundary
     Abscissa : polar angle from the bottom of the pipe

3.b.Identification of an analytic temperature
   0  Analytic direct solution
   Δ  "Observed " temperature
   +  Restitued temperature on ext.bound.
     Ordinate : temperatures in °C

## Regularizations and shells models

We shall show here, on the very simple analytical example of a plate, which distorsion with respect to the 3D problem has to be expected when simplifying it by the shell approach.

Let us consider the following problem: an infinitely long plate, the temperature of wich is fixed (say to zero) on three of its sides, is heated by a unknown top flux. The bottom heat flux is then given (measured). The problem is now to reconstruct the top flux provided the thermal equilibrium is reached throughout the plate (see figure 4).

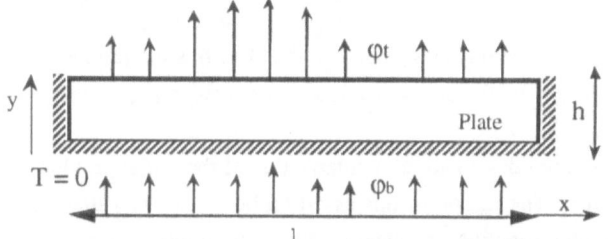

T = 0 on the bottom
and lateral sides.

Bottom heat flux
prescribed:
$\varphi b\,(x) = \sin n\,\pi x\,/\,1$

Figure 4 . The Hadamard problem : Heated plate with zero bottom temperature.

The equations of this two dimensional problem are:

$$
\begin{cases}
\Delta T = 0 \ \ \text{in } [0,l] \times [0,h] \\[2mm]
\\[2mm]
T = 0 \qquad \text{on } [0,l] \times \{0\} \text{ , and } \{0\} \times [0,h] \ \cup \ \{l'_l\} \times [0,h] \\[2mm]
-\dfrac{\partial T}{\partial y} = \varphi^b \ \ \text{on } [0,l] \times \{0\}
\end{cases}
\tag{1}
$$

This last problem is precisely the counter-example given by Hadamard in his definition of well posed PDE's problems, namely the Cauchy problem formulated for an elliptic operator [7]. When the bottom flux is simply a sinus fonction, the solution of the problem throughout the plate has been given by Hadamard :

$$\varphi b\ (x) = \sin n\ \pi x\ /\ l \quad \text{then} \quad T(x,y) = \sin\ (n\ \pi x\ /\ l)\ \sinh\ (n\ \pi y\ /\ l)$$

so that the solution of the inverse problem is, after having decomposed the bottom flux in its Fourier serie :

$$\varphi^t(x) = \sum_n \varphi^t_n\ \sin\ (n\pi x/l)) = \sum_n \varphi^b_n\ \cosh\ (n\pi h/l)\ \sin\ (n\pi x/l))$$

It is convenient to introduce the "bottom to top" flux transfer function $\Phi_{3D}$, connecting the $n^{th}$ Fourier coefficient $\varphi^t_n$ of the top flux $\varphi^t$ to the corresponding bottom flux coefficient $\varphi^b_n$ :

$$\varphi^t_n = \Phi_{3D}\ (\zeta)\ \varphi^b_n \qquad \text{with} \qquad \Phi_{3D}\ (\zeta) = \cosh\sqrt{\zeta} \qquad (3)$$
$$\text{where } \zeta = (n\ \pi h\ /\ l\ )^2.$$

The ill-posedness of the problem appears now through the the fact that the transfer function is not bounded : an arbitrary small deviation in the measured flux $\varphi^b$ can causes an arbitrary large variation on the identified flux $\varphi^t$, provided that the deviation as sufficiently high "mode" non zero components.

The resolution of the same problem via the shell model, which details are omitted here, leads to the following transfer function denoted by $\Phi_S$ :

$$\varphi^t_n = \Phi_S\ (\zeta)\ \varphi^b_n, \qquad \text{with} \qquad \Phi_S\ (\zeta) = \frac{15 + 26\ \zeta\ + 3\zeta^2}{15\ -\ 4\ \zeta +\ \zeta^2} \qquad (4)$$

It is easy to see now that the tranfer function being bounded, the identification problem is no more ill-posed. This feature of the shell approach allows to consider it properly as a regularization of the 3D problem.

It is believed that the shell model provides a natural regularization of the initial problem but keeps some essential features of it. The reason of this assertion lies in the fact that, in the example mentionned, the "transfer fonction" connecting the data (imposed flux at zero temperature at the bottom of the plate) to the output (top temperature) in the shell model exhibits two important properties:

i)  first, it is bounded in the Fourier domain so that continuity with respect to data is restored,

ii)  but twice, this function do not vanish in the Fourier domain even for large frequencies and is not simply a truncature of the 3D transfer function.

This last property allows the shell model to mimic in a realistic way the 3D one : the amplification of high frequencies which causes the ill-posedness of the initial problem is suppressed but these frequencies do not disappear in the shell model. It is then able to describe discontinuous situations and so is not "too much" smoothing.

To complete the analysis of this simple example, let us now derive quickly a Tikhonov regularization of the top flux identification problem. The starting point is a reformulation of the problem via the minimization of a least-square output error functional :

$$\varphi^t \quad \text{minimizes} \quad J(\varphi) = \int_0^1 \left| \varphi_m^b - \varphi^b(\varphi) \right|^2 dx$$

$\varphi^b(\varphi)$ is the bottom flux calculated through the resolution of the direct problem with zero temperature on the lateral and bottom sides and flux $\varphi$ on the top side, $\varphi_m^b$ is the measure.

The Tikhonov-Phillips regularization of the preceeding functional is obtained by adding a term limiting in the solution the oscillations caused by the overwhelming influence of the "high" modes of the measure $\varphi_m^b$. The weight of the added term is controlled via a regularization parameter traditionaly denoted by $\alpha$ :

$$\varphi_\alpha^t \quad \text{minimizes} \quad J_\alpha(\varphi) = \int_0^1 \left| \varphi_m^b - \varphi^b(\varphi) \right|^2 dx + \alpha \int_0^1 \left| \varphi' \right|^2 dx$$

Exact minimization can be achieved using again Fourier decomposition and leads here to the following transfer function $\Phi_\alpha$ :

$$\varphi_n^t = \Phi_\alpha(\zeta) \, \varphi_n^b, \qquad \text{with} \qquad \Phi_\alpha(\zeta) = \frac{\cosh\sqrt{\zeta}}{1 + \alpha \, \xi \left(\frac{1}{h}\right)^2 \cosh^2\sqrt{\zeta}} \qquad (5)$$

Here again the regularization operation becomes apparent through the boundness of the transfer function. On the figure 5 are plotted the tranfer functions defined hereabove, namely the original 3D one, the shell model one and two regularized ones (for three distinct values of regularization parameter $\alpha$).

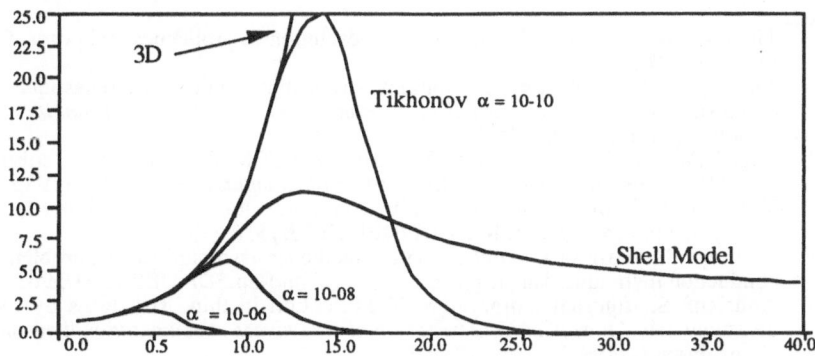

Figure 5 : Bottom to top flux tranfer functions $\Phi(\zeta)$ in the Fourier domain

Three major conclusions can be drawn from this figure, some of them enforcing the remarks already made on the interest of the shell model approach :

1 - For low modes, the three functions give the same bottom to top response, so that great precision can be achieved with any of the methods in this domain,

2 - The Tikhonov-Phillips regularization leads to zero transfer function for high modes whereas the shell transfer function do not vanish,

3 - This troncating feature of the Tikhonov-Phillips regularization is highly dependent on the regularization parameter $\alpha$. The optimal choice of it is a well known (and difficult) problem as it has been quoted in the first part of the paper.

Conclusion

For thin or moderately thick domains, the shell approach of some engineering identification problems has been designed. Together with evident computational costs reduction compared with 3D ones, it allows to get some insights into the problems in hand. Of particular interest are the theoretical results it permits to obtain, but also the informations gained for the regularization needed in any practical application. From this last point of view, it has been shown that the shell model can bring a kind of "natural" regularization on the 3D problem, the quality of which depends obviously on the sophistication of the shell theory used. This physically based regularization is believed to work successfully in number of situations because of its "energetical" content : the estimation of the energy of the 3D solution (or some analogous quantity) is approximated in a very good way by the shell solution energy.

From the mathematical point of view, the link with Tikhonov regularization is perhaps not surprising when recalling that the shell models can often be justified by asymptotic developpment techniques, where precisely the high oscillations terms energy is dropped out in high order terms in the developpment. The Tikhonov parameter plays in some kind the role of the "small parameter" whereas the higher energy terms replace the regularizing semi-norm.

References
1.  Tikhonov A., Arsenine, V., Méthodes de résolution de problèmes mal posés, Ed Mir, Moscou (1976).
2.  Andrieux S., Identification des contraintes dans un tube mince élastoplastique à partir de mesures de profil, in *"Calcul des structures et intelligence artificielle"*, Vol 4, P.Ladeveze Ed, Pluralis (1990)
3.  Engl H.W., Neubauer A. An improved version of Marti's method for solving ill-posed linear integral equations. *Mathematics of Computation.*, **45**, 172, (1985)
4.  Voldoire F., Formulation et évaluation numérique d'un modèle de coque élastoplastique enrichie. Int Rep.EDF/DER HI-73/7518, (1992).
5.  Andrieux S., Marigo J.J., Application des méthodes asymptotiques au problème de la conduction thermique dans les plaques minces, Int Rep.EDF.DER HI71/5963 (1987).
6.  Andrieux S., Internal temperature identification in thin structures by external measurements, *"European conference on new advances in computational mechanics"*, Giens, France, (1991)
7.  Lavrentiev M.M., Some improperly posed problems of mathematical physics, Springer Verlag, Berlin (1967).

# Inverse Analysis for Estimating Galvanic Corrosion Rate

Shigeru AOKI* and Yoshihiro URAI**

*  Department of Mechanical Engineering Science
   Tokyo Institute of Technology
   Ookayama, Meguro-ku, Tokyo 152, Japan
** Undergraduate of Tokyo Institute of Technology
   Ookayama, Meguro-ku, Tokyo 152, Japan

Summary

An inverse problem in which the density of current across the surface of the metal surfaces (this is proportional to the galvanic corrosion rate) is estimated from the measured potential values at some points far from the metal surfaces is discussed. This inverse problem is formulated by employing the boundary element method. Since the system of linear equations obtained is ill-conditioned, some regularization methods, including singular value decomposition and Tikhonov's method (smoothing method and spline method), are applied and the accuracies of these estimation methods are compared. In order to improve the accuracy, a new method using fuzzy clustering techniques is developed. An example problem is solved by the new method in order to demonstrate the accuracy.

## Introduction

It is well known that the economic loss due to galvanic corrosion in engineering structures, such as underground pipeline, off-shore structures and chemical plants, is estimated to be tremendously great. A precise prediction of the corrosion rate is essential to reduce the loss. In case where the electrochemical polarization curves of metal in the electrolyte considered are available, the galvanic corrosion rates can be estimated by

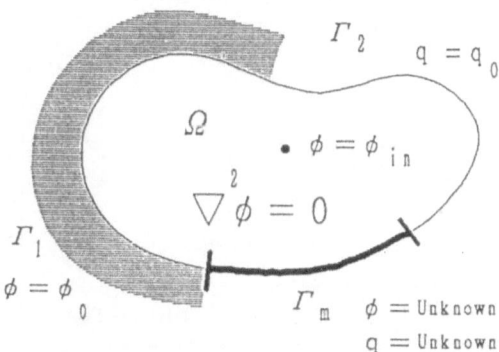

Fig.1 Governing equation and boundary conditions

solving the Laplace's equation with boundary conditions based on the polarization curves[1-3].

The polarization curves, however, are not always available. In this case, it is necessary to estimate the density of the current across the metal surface (the current density is proportional to the corrosion rate) from the potential values measured at some points far from the metal surface. This inverse problem is solved in this paper.

## Basic equations

Let us assume that the surface of the electrolyte domain $\Omega$ is surrounded by $\Gamma$ ($\equiv \Gamma_1 + \Gamma_2 + \Gamma_m$) as shown in Fig.1, where the potential values and current densities are prescribed on $\Gamma_1$ and $\Gamma_2$, respectively while $\Gamma_m$ is the metal surface. If we assume that the electrolyte is homogeneous and that there is no accumulation or loss of ions in the bulk of the electrolyte, the electropotential within the electrolyte, $\phi$, obeys the Laplace's equation:

$$\kappa \nabla^2 \phi = 0 \qquad \text{in} \quad \Omega \qquad\qquad (1)$$

where $\kappa$ denotes the conductivity of the electrolyte. The density of current across the boundary, $q$, is related to the potential as

$$q = \kappa \frac{\partial \phi}{\partial n} \qquad\qquad (2)$$

where $\partial / \partial n$ is the outward normal derivative.

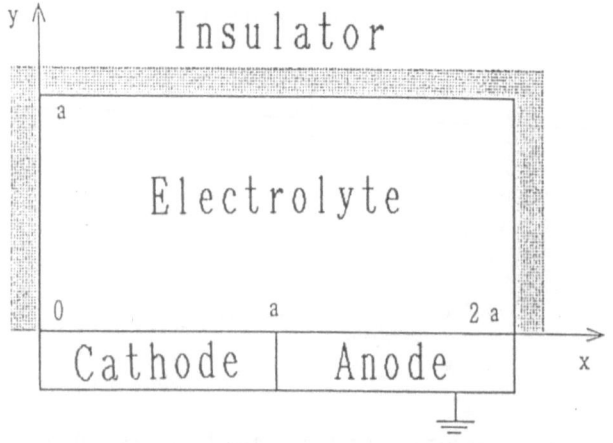

Fig.2 Numerical example problem

The boundary and supplementary conditions are given by

$$\phi = \phi_0 \qquad\qquad \text{on } \Gamma_1 \qquad\qquad (3)$$

$$q = q_0 \qquad\qquad \text{on } \Gamma_2 \qquad\qquad (4)$$

$$\phi, q : \text{unkown} \qquad \text{on } \Gamma_m \qquad\qquad (5)$$

where $\phi_0$ and $q_0$ are the prescribed values of potential and current density, respectively. For an ordinary problem the nonlinear functions $(-\phi = f(q))$ representing polarization curves are given on $\Gamma_m$, while for an inverse problem no condition is given on $\Gamma_m$ as shown in Eq.5. Instead of this the potentials measured at some points far from the metal surfaces, $\phi_{in}$, are given:

$$\phi = \phi_{in} \qquad \text{at internal points} \qquad\qquad (6)$$

Boundary element formulation

Following the ordinary boundary element formulation, we derive the following equations from Eq.(1):

$$\frac{1}{\kappa}[G]\,\underset{\sim}{q}^n - [H]\,\underset{\sim}{\phi}^n = \underset{\sim}{0} \qquad\qquad (7)$$

where $[H]$ and $[G]$ are the known matrices[4], and the superscript n stands for the nodal value. A quantity underlined by $\sim$ indicates a vector.

Eq.(7) can not be solved because neither $\phi$ nor $q$ is given on the boundary $\Gamma_m$ (Eq.(5)). The potential measured at some points in the electrolyte, $\phi_{in}$, is related to $\phi^n$ and $q^n$ by

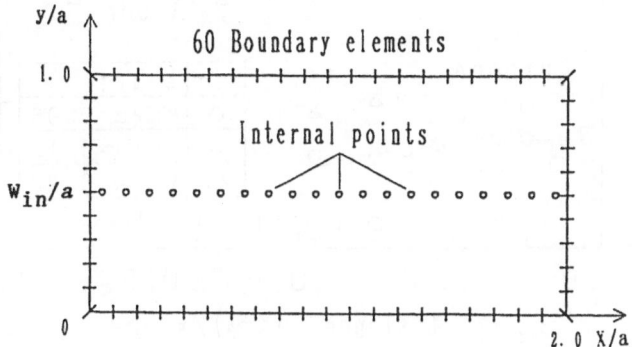

Fig.3 Boundary element discretization

$$\frac{1}{\kappa} [G_{in}] q^n - [H_{in}] \phi^n = \phi_{in} \qquad (8)$$

where $[H_{in}]$ and $[G_{in}]$ are the matrices depending on the shape of the boundary and the location of the measuring points.

Combining Eqs.(7) and (8) and rearranging yield

$$[A] x^n - b^n = 0 \qquad (9)$$

where $x^n$ denotes unknowns including $\phi^n$ and $q^n$ on $\Gamma_m$ and $b^n$ the knowns including $\phi_{in}$. Generally $[A]$ is a non-square matrix. Eqs.(9) are equivalent to the following minimization:

$$\| [A] x^n - b^n \|^2 \longrightarrow min \qquad (10)$$

Numerical Example

Let us consider an example problem of a galvanic corrosion couple consisting of two coplanar strip elements, shown in Fig.2. The electrolyte is bounded by insulators at $x = 0$, $x = 2a$ and $y = a$. We discretize the boundary into 60 constant boundary elements as shown in Fig.3, where $W_{in}$ denotes the distance from the metal surface to the measuring points.

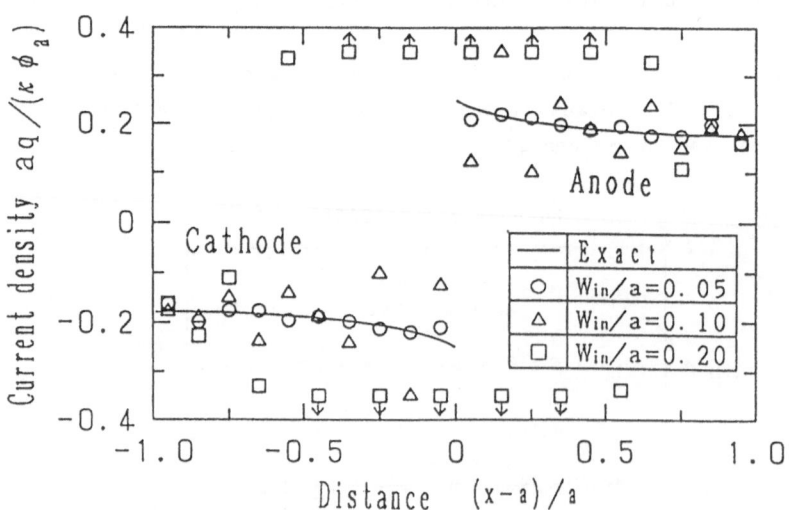

Fig.4 Current density distribution on the surface of galvanic couple (estimated by Eq.(10))

At first we solve an ordinary problem by assuming that the polarization curves of anode and cathode are given by the following equations, and obtain the potential values at measuring points in the electrolyte.

$$\frac{\phi}{\phi_a} = \begin{cases} \log_{10}\left(-\frac{q}{q_a}\right) + 1.1 & \left(\frac{q}{q_a} \leq -1\right) \\ -0.1\frac{q}{q_a} + 1 & \left(-1 \leq \frac{q}{q_a} \leq 1\right) \\ -\log_{10}\left(\frac{q}{q_a}\right) + 0.9 & \left(\frac{q}{q_a} \geq 1\right) \end{cases} \qquad \text{Anode}$$

$$\frac{\phi}{\phi_a} = \begin{cases} \log_{10}\left(-\frac{q}{q_a}\right) + 0.1 & \left(\frac{q}{q_a} \leq -1\right) \\ -0.1\frac{q}{q_a} & \left(-1 \leq \frac{q}{q_a} \leq 1\right) \\ -\log_{10}\left(\frac{q}{q_a}\right) - 0.1 & \left(\frac{q}{q_a} \geq 1\right) \end{cases} \qquad \text{Cathode}$$

where $\phi_a$ and $q_a$ are the known numbers for normalization. The value of conductivity $\kappa$ is chosen as $\kappa/(aq_a/\phi_a) = 10$ where $a$ is the depth of the electrolyte (Fig.2). The obtained potential values are rounded off to three significant figures to take account of the measurement errors, and are used as the values of $\phi_{in}$ in the inverse problem.

Then, we solve the inverse problem by the above-mentioned method. The results for the current density on the metal surfaces are shown in Fig.4 where the

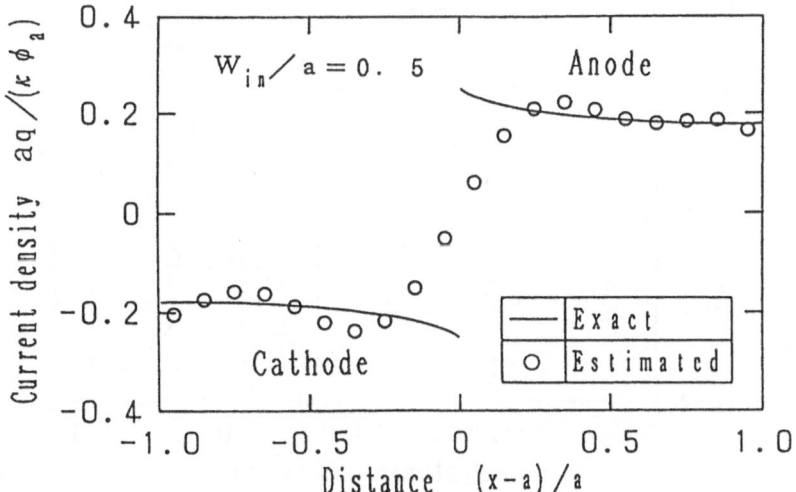

Fig.5 Current density distribution on the surface of galvanic couple (estimated by singular value decomposition )

solid curves indicate the solution of the ordinary problem, i.e., the exact solution for the inverse problem. It is found that estimated values oscillate intensely as the distance of the measuring points from the metal surfaces, $W_{in}$, increases, and that the estimation becomes completely impossible for $W_{in}= 0.2a$ by this method. This suggests that some regularization is needed.

## Regularization
### Singular value decomposition
The matrix $[A]$ in Eq.(10) is decomposed as

$$[A] = [U] [D] [V]^T \tag{11}$$

where $[U]$ and $[V]$ are unitary matrices, $[D]$ is a "diagonal" matrix. The components of $[A]$ are the singular values, $\sigma_\lambda$ ($\lambda = 1,2,\cdots,i+m$), where i is the total number of boundary elements, m is the number of elements on $\Gamma_m$, and

$$\sigma_1 \geq \sigma_2 \geq \sigma_3 \geq \cdots \geq \sigma_R > \sigma_{R+1} = \cdots = \sigma_{i+m} = 0 \tag{12}$$

The number R in Eq.(12) is referred to as the rank of $[A]$. It is possible to reduce the oscillation in the solution by replacing some of small singular values with zeros, i.e., by reducing the rank of $[A]$ [3]. A method for determining the effective rank objectively has been proposed in literature[3].

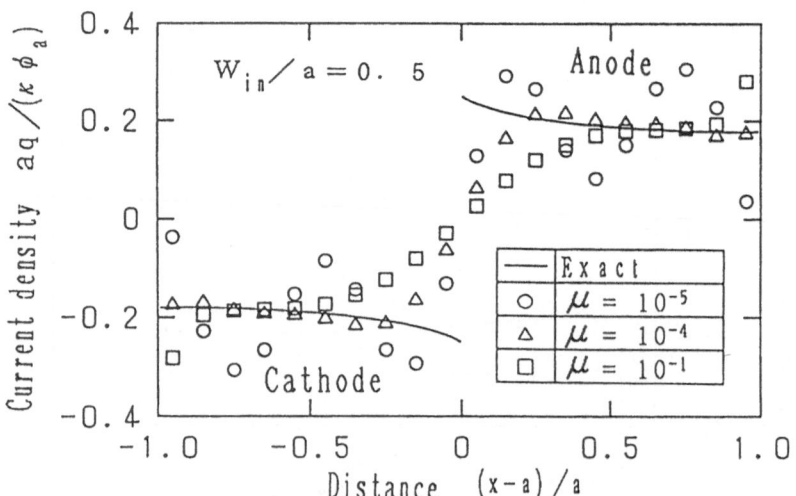

Fig.6 Current density distribution on the surface of galvanic couple (estimated by smoothing method, i.e., Eq.(13))

We apply this regularization method to the above numerical example problem. The result is shown in Fig.5 where the rank was reduced from 80 to 66. It is found that the extraordinary oscillation in Fig.4 has been removed even for $W_{in}/a = 0.5$.

*Smoothing method*

Following Tikhonov[5], we modify Eq.(10) by adding a smoothing term, i.e.,

$$\| [A]\,\underline{x}^n - \underline{b}^n \|^2 + \mu \| \underline{\phi}_m^{''n} \|^2 \longrightarrow \min \qquad (13)$$

where $\phi_m^{''n}$ denotes the second derivative of the potential on metal surfaces with respect to the tangential distance, and $\mu$ is an appropriate positive number. The number $\mu$ is effectively determined by minimizing the residual, Q, which is defined by $Q = \Sigma \{ (\phi_{in} - \phi_{in}^{est}) / \phi_{in} \}$. The $\phi_{in}^{est}$ is the potential (at the measuring points) obtained by Eq.(13) and a function of $\mu$.

Figure 6 shows the result obtained by applying this method to the example problem. It is found that the extraordinary oscillation observed in Fig.4 has been removed, and an estimation becomes possible for $W_{in}/a = 0.5$ if $\mu$ is appropriately chosen (the number $\mu = 10^{-4}$ was determined by minimizing the residual Q). The CPU time was short compared with the singular decomposition method.

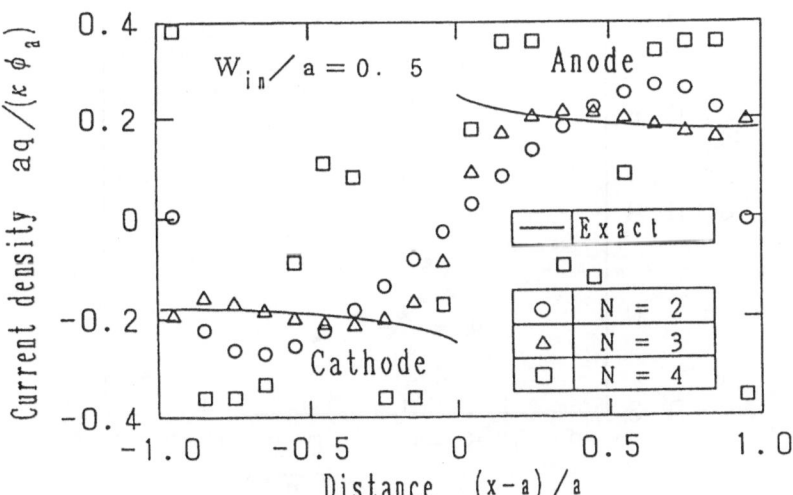

Fig.7 Current density distribution on the surface of galvanic couple (estimated by spline method, i.e., Eq.(14))

*Spline method*

Following Tikhonov again, we modify Eq.(10) in the following form:

$$\| [A] \underline{x}^n - \underline{b}^n \|^2 + \nu \| \underline{g}_m^n - \underline{g}^n \|^2 \quad \longrightarrow \min \qquad (14)$$

where $g$ is a spline function of x and $\nu$ is an appropriate positive number. The variables to minimize Eq.(14) are $\phi^n$ , $q^n$ and the parameters of the spline function. The number $\nu$ is determined in the same way as $\mu$ in Eq.(13).

The result (for $W_{in}/a = 0.5$) obtained by applying this method to the example problem is shown in Fig.7 where the metal surface was divided into $1\sim3$ subregions, i.e., the number of nodal points N was chosen as $2\sim4$, and a cubic spline function was used. It is found that a good result is obtained for N = 3. The CPU time was nearly the same as that of the smoothing method (Eq.(13)).

The accuracy of estimation by these three methods is nearly the same, and it becomes possible to obtain an appropriate solution for $W_{in}/a \leqq 0.5$ by using one of these methods. For $W_{in}/a > 0.5$, however, the estimation became impossible again even if one of these methods is applied. In the next section we will develop a new method for obtaining a good estimation for $W_{in}/a > 0.5$.

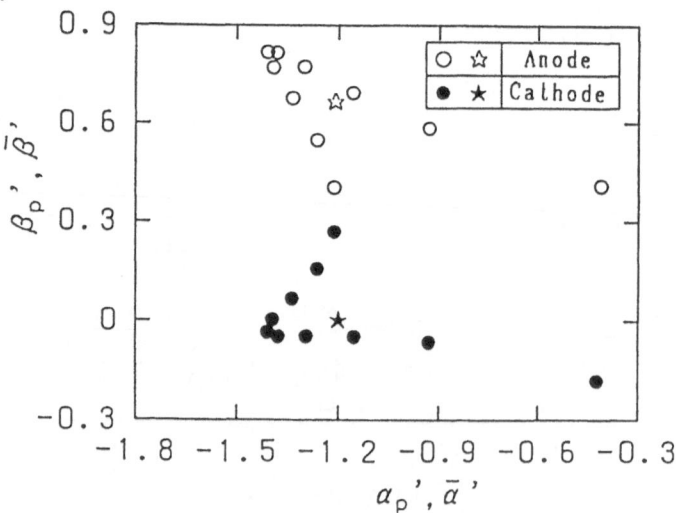

Fig.8 Feature vectors and cluster center in $\alpha_p'$- $\beta_p'$ plane

## Application of clustering technique

### Estimation of the polarization curve

Let us estimate the polarization curve at first. For simplicity the polarization curves for each element are assumed to be linear;

$$\frac{q}{q_a} = \alpha_p \frac{\phi}{\phi_a} + \beta_p \quad (p=1,\cdots,m) \tag{15}$$

where $\phi_a$ and $q_a$ are the known numbers, and $\alpha_p$ and $\beta_p$ $(p=1,\cdots,m$ ; $m =$ number of elements on metal surfaces) are the parameters to be estimated.

More than two different fields are created by impressing current from an external power source, and more than two sets of solutions for $\phi_p$ and $q_p$ (nodal values of $\phi$ and $q$ for element p, respectively) are obtained by applying one of the above mentioned regularization methods. The values of $\alpha_p$ and $\beta_p$ are estimated from the sets of solutions $\phi_p$ and $q_p$. The estimated values of $\alpha_p$ and $\beta_p$ show a considerable scatter.

By applying the fuzzy k-means clustering technique[6], the elements on the metal surfaces are classified into two groups, anode and cathode. As components of a feature vector, the properly scaled values of $\alpha_p$, $\beta_p$, $q_p^j$ and $\phi_p^j$ $(j=1,2,\ldots,h$; h = the number of fifferent fields) are chosen. From the cluster center of each group the average values of $\alpha_p$ and $\beta_p$ for

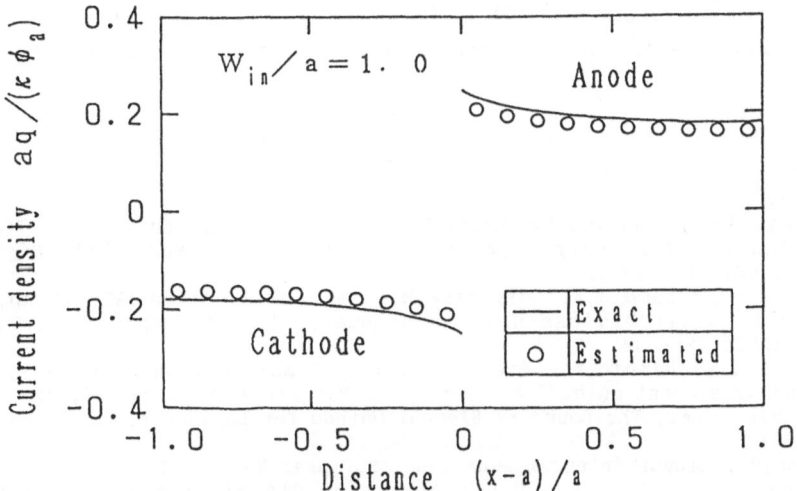

Fig.9 Current density distribution on the surface of galvanic couple (estimated by clustering technique)

556

anode and cathode are determined. Let $\bar{\alpha}_r$ and $\bar{\beta}_r$(r=1,2) denote the average values. Thus, the polarization curves of anode and cathode are estimated as

$$\frac{q}{q_a} = \bar{\alpha}_r \frac{\phi}{\phi_a} + \bar{\beta}_r \qquad (r=1,2) \qquad (16)$$

*Estimation of current density*

The current density distribution on metal surfaces is obtained by performing an ordinary analysis using the estimated polarization curve, Eq.(16). We apply this method to the example problem. Five different fields (h = 5) are created by impressing current from the center of the top surface of the electrolyte ( x = y = 0.5a in Fig.2). The sets of $\phi_p$ and $q_p$ are obtained by the smoothing method (Eq.(13)) and then $\bar{\alpha}_r$ and $\bar{\beta}_r$(r=1,2) are determined.

The feature vectors and cluster center in $\alpha_p'$- $\beta_p'$ plane are shown in Fig.8 where

$$\alpha_p' = \alpha_p / \triangle\alpha, \qquad \beta_p' = \beta_p / \triangle\beta \qquad (17)$$

where

$$\triangle\alpha = \max(\alpha_1, \cdots, \alpha_m) - \min(\alpha_1, \cdots, \alpha_m)$$
$$\triangle\beta = \max(\beta_1, \cdots, \beta_m) - \min(\beta_1, \cdots, \beta_m)$$

Figure 9 shows the current density distribution on the metal surfaces estimated by performing an ordinary analysis using Eq.(16) with the obtained values of $\bar{\alpha}_r$ and $\bar{\beta}_r$(r=1,2). It is found that the estimated current density distributions are in good agreement with the exact values even for $W_{in}/a$ = 1.0.

References
 1. Zamani,N.G.; Porter,G.F.; Mufti,E.J.: A survey of computational efforts in the field of corrosion engineering, *Int. J. Num. Meth. Engng.*, 23 (1986) 1295-1311.
 2. Aoki,S.; Kishimoto,K.; Miyasaka,M.: Analysis of potential and current density distributions using a boundary element method, *Corrosion*, 44 (1988) 926-932.
 3. Aoki,S.; Kishimoto,K.: Prediction of galvanic corrosion rates by the boundary element method, *Mathl Comput. Modelling*, 15 (1991) 11-22.
 4. Brebbia, C.A.; The Boundary Element Method for Engineers, Pentech Press (1978).
 5. Kress,R.; Linear integral equations, Springer-Verlag (1989).
 6. Bezdek,J.C.; Kastelaz,P.S.: Prototype Classification and Feature Selection with Fuzzy, *IEEE Trans. Systems, Man and Cybernetics*, SMC-7 (1977) 87-92.

# Estimation of an Electron Orbit Using Super-Potentials of Liénard-Wiechert Potentials

H. Kawaguchi  and  T. Honma

Department of Electrical Engineering,
Faculty of Engineering, Hokkaido University
Kita 13, Nishi 8, Kita-ku, Sapporo   Japan

## Abstract

In this paper, a estimation method of an electron orbit using far electric fields is presented. Then, it is indicated that any iterative calculations are not necessary to estimate the orbit, using the super-potentials which result in Liénard-Wiechert potentials. Moreover, this estimation method is applied to two examples ( circular and betatron motion ). One can see good agreement between true orbit and estimatied one in each case.

## 1. Introduction

Liénard-Wiechert potentials are solutions to an inhomogeneous *wave* equation which describes electromagnetic fields produced by a moving point charged *particle*. This means that Liénard-Wiechert potentials possess duality of wave and particle and that the potentials describe interactions between electromagnetic fields and a charged particle. A typical phenomenon of them is radiation from an accelerating charged particle. Furthermore, from the duality of wave and particle, another interesting result is obtained. That is to say, one can indicate that a charged particle orbit is estimated from far electric fields produced by the charged particle *without any iterative calculation*, when the motion is periodic and non-relativistic. In this paper, formulation of the estimation method and the numerical calculation are presented.

## 2. New representation of Liénard-Wiechert potentials and their Fourier Expansion

In this section, the new representation of Liénard-Wiechert potentials and the super-potentials for the Liénard-Wiechert potentials, which were presented by one of the authors, are summarized.

Liénard-Wiechert potentials $A^i$ are solutions to inhomogeneous wave equation which describes electromagnetic fields produced by a moving point charged particle[1].

$$\Box A^{i}(ct, \mathbf{x}) = \frac{e}{\varepsilon_0 c^2} u^{i}(t)\delta[\mathbf{x} - \mathbf{y}(t)]$$

(1)

where $c$ is the velocity of light, $\Box$ is D'Alembertian, $u^i = (c, \mathbf{v}(t))$ is the four velocity of the particle, $y(t)$ is trajectory of the particle with a parameter $t$, $e$ is elementary charge and $\varepsilon_0$ is dielectric constant. Then, Liénard-Wiechert potentials are written as follows :

$$A^{i}(ct, \mathbf{x}) = \frac{e}{4\pi\varepsilon_0 c^2} \frac{u^{i}(\tau)}{R_k(\tau) u^{k}(\tau)}$$

(2)

where $R_k(\tau)$ is displacement vector defined by $R_k(\tau) = x^i - y^i(\tau)$ ( $y^i(\tau)$ is the four dimensional position vector of the particle defined by $(c\tau, \mathbf{y}(\tau))$ ). And then, $\tau$ is so-called "retarded time" which satisfies the following causal relation :

$$\tau = t - \frac{|\mathbf{x} - \mathbf{y}(\tau)|}{c}$$

(3)

Liénard-Wiechert potentials depend on the time $t$ and the position $\mathbf{x}$ through this recursive relation.

Here the equation ( 3 ) can be rewritten in the following form

$$\tau = \tau(ct, \mathbf{x})$$

(4)

because retarded time $\tau$ is determined uniquely for any $t$ and $\mathbf{x}$ [1]. Then, one can regard the four dimensional position vector $y^i(\tau)$ as functions of $ct$ and $\mathbf{x}$. That is to say,

$$y^{i}(\tau) = y^{i}[\tau(ct, \mathbf{x})] = y^{i}(ct, \mathbf{x})$$

(5)

These functions are just super-potentials for Liénard-Wiechert potentials, because Liénard-Wiechert potentials can be expressed using these functions $y^i(\tau)$ as follows[2] :

$$A^{i}(ct, \mathbf{x}) = -\frac{e}{8\pi\varepsilon_0 c} \Box y^{i}[\tau(ct, \mathbf{x})]$$

(6)

The equation (6) tells us that *coordinates of the electromagnetic system* $A^i(ct,\mathbf{x})$ are related with *coordinates of the particle* $y^i(\tau)$ by D'Alembertian $\Box$, directly. Using Eq. (6), the Hertzian tensor potentials $\pi^{ij}$ which result in Liénard-Wiechert potentials can be introduced[2,3]. The differential form diagram of electrodynamics[4] containing the super-potential $y^i$ and the Hertzian tensor potential $\pi^{\mu\nu}$ is drown in Fig.1. The notation * denotes a dual operator, the notation d̄ denotes a exterior derivative operator. One can find similarities between the super-potentials and Liénard-Wiechert potentials.

Now, the equation (6) can be derived from the Fourier expansions of Liénard-Wiechert potentials $A^i(ct,x)$ and super-potentials $y^i(\tau)$. If the motion of a particle is periodic, Fourier expansion of $y^i(\tau)$ is written as follows ( see Appendix A )

$$y^i(ct,x) = \frac{1}{2\pi} \sum_{n=0}^{\infty} \frac{C_n}{n} \int_a^b \sin n[\omega t - \sigma - \frac{\omega}{c} R(x;\sigma)]\, dy^i(\sigma) \qquad (7)$$

where $\omega$ is an angular frequency. Then $\Box\, y^i(\tau)$ becomes

$$\Box\, y^i(ct,x) = -\frac{w}{2\pi c} \sum_{n=0}^{\infty} C_n \int_a^b \frac{b\cos n[\omega t - \sigma - \frac{\omega}{c} R(x;\sigma)]}{R(\sigma)}\, dy^i(\sigma) \qquad (8)$$

The Fourier expansion of Liénard-Wiechert potentials is as follows : [1]

$$A^i(ct,x) = \frac{e}{8\pi\varepsilon_0 c^2} \sum_{n=0}^{\infty} C_n \int_a^b \frac{b\cos n[\omega t - \sigma - \frac{\omega}{c} R(x;\sigma)]}{R(\sigma)}\, dy^i(\sigma) \qquad (9)$$

From Eqs. (8) and (9), one can find that Eq. (8) is just Fourier expansion of Liénard-Wiechert potentials except for the factor $-e/8\pi\varepsilon_0 c$. Thus, the equation (6) was proved using the Fourier expansions of $A^i(ct,x)$ and $y^i(\tau)$.

## 3. Estimation method of electron orbit from far electric fields

One can indicate that the orbit of a charged particle such as an electron is estimated from far electric fields. From Eqs. (7) and (9), spatial components of $y^i(\tau)$ and $A^i(ct,x)$ are written as follows :

$$y(ct,x) = y_0(ct,x) + \frac{1}{\pi} \sum_{n=1}^{\infty} \frac{1}{n} \int_a^b \sin n[\omega t - \sigma - \frac{\omega}{c} R(x;\sigma)]\, dy(\sigma) \qquad (10)$$

$$A(ct,x) = -\frac{ew}{16\pi^2\varepsilon_0 c^2} \sum_{n=0}^{\infty} C_n \int_a^b \frac{b\cos n[\omega t - \sigma - \frac{\omega}{c} R(x;\sigma)]}{R(\sigma)}\, dy(\sigma) \qquad (11)$$

where $y_0(ct,x)$ is the term for $n=0$ in the Eq. (10). Taking note that $dy_0/\omega dt = 0$ in Eq. (A7), $y_0$ has to depend on only spatial components x. And then, the following expression of $y_0(x)$ can be obtain from numerical calculation

$$y_0(x) = -\frac{1}{2\pi}\frac{\omega}{c}\int_a^b R(x;\sigma)\,dy(\sigma) \qquad (12)$$

For far fields ( *i.e.* $R(x;\sigma) \simeq R(x) - x \cdot y(\sigma)/R(x)$ )

$$y_0(x) \simeq \frac{1}{2\pi}\frac{\omega}{c}\frac{1}{R(x)}\int_a^b [x \cdot y(\sigma)]\,dy(\sigma)$$

$$= M \times \frac{x}{R(x)} \qquad (13)$$

where $R(x)$ is the distance from the center of a charged particle trajectory to observation point, the vector $M$ is a constant defined by

$$M \equiv \frac{1}{4\pi}\frac{\omega}{c}\int_a^b y(\sigma) \times dy(\sigma) \qquad (14)$$

Here, taking note that far electric fields are calculated by

$$E(ct, x) = -\frac{\partial}{\partial t}A(ct, x)$$

$$= -\frac{e\omega^2}{8\pi^2\varepsilon_0 c^2}\sum_{n=1}^{\infty} n \int_a^b \frac{\sin n[\omega t - \sigma - \frac{\omega}{c}R(x;\sigma)]}{R(x;\sigma)}\,dy(\sigma) \qquad (15)$$

one can find a following relation using Eqs. (10), (13) and (15) :

$$y(ct, x) \simeq M \times \frac{x}{R(x)} - \frac{8\pi\varepsilon_0 c^2}{e\omega^2}R(x)\sum_{n=1}^{\infty}\frac{E_n(ct, x)}{n^2} \qquad (16)$$

The equation (16) reveals the relation between a charged particle trajectory ( or super-potentials ) and the far electric fields produced by the particle. Now, it is found that the first term of Eq. (16) can be neglected, when the motion is non-relativistic, because taking note that $|x/R(x)| \leq 1$ and $|(\omega/c)y(\sigma)| \sim v/c$ ( $v$ is the velocity of the particle ), one can find the following relation :

$$y_0(x) \simeq \frac{1}{2\pi}\frac{\omega}{c}\frac{1}{R(x)}\int_a^b [x \cdot y(\sigma)]\,dy(\sigma)$$

$$\leq \frac{1}{2\pi}\frac{v}{c}\int_a^b dy(\sigma)$$

$$\lll [\text{ the second term of equation (13) }] \qquad (17)$$

Then, the following value $\Psi$, defined by

$$\Psi = -\frac{8\pi\varepsilon_0 c^2}{e\omega^2} \sum_{n=1}^{\infty} \frac{E_n(ct,\mathbf{x})}{n^2} , \qquad (18)$$

represents the trajectory shape divided by distance $R(\mathbf{x})$, when the motion is non-relativistic. If far electric fields are observed at some points, one can calculate their trajectory shapes $\Psi_1$, $\Psi_2$,····· using Eq. (18). Differences between these trajectory shapes are only their amplitudes which are $1/R_1(\mathbf{x})$, $1/R_2(\mathbf{x})$, ·····. Therefore calculating values $\Psi_1$, $\Psi_2$, $\Psi_3$, $\Psi_4$ for any four points and comparing their amplitudes, the center of trajectory can be estimated.( see Appendix B ) Consequently, one can say that a charged particle trajectory $y(ct,\mathbf{x})$ is calculated by the far electric fields, if the motion is periodic and non-relativistic.

Two examples ( circular and betatron motion ) are shown in Fig.2 and Fig.3. A dotted line denotes a true trajectory and a solid line denotes the trajectory calculated by Eq. (16) in each figure. One can find good agreement.

## 4. Summary

The estimation method of electron orbit using the super-potentials of Liénard-Wiechert potentials was presented in this paper. When the electron motion is periodic and non-relativistic, good agreement of true orbit and estimated one was obtained. However, this estimation method can not be applied to relativistic case, because there are not any ways to estimate the first term of Eq. (6). The estimation of electron orbit in the relativistic case is future work.

## 5. References

[1]  Landau, L.D.; Lifshitz, E.M.: *Classical Theory of Fields* (New York: Pergamon) (1971) Sec. 63 and 64.
[2]  Kawaguchi, H.; Murata, S. : Hertzian tensor potential which result in Liénard-Wiechert potential, *J.Phys.Soc.Jpn.*, 58[3] (1989) 848-855.
[3]  Nisbet, A. : Hertzian electromagnetic potentials and associated gauge transformations, *Proc. R. Soc. London* , Ser. A. 132 (1955) 250-263.
[4]  Schutz, B.F. : *Geometrical methods of mathematical physics* ( Cambridge University Press) , (1980) Sec. 4.19
[5]  Whittaker, E. T.:   *A Treatise on the Analytical Dynamics* (Cambridge, Cambridge) (1912) p.89.

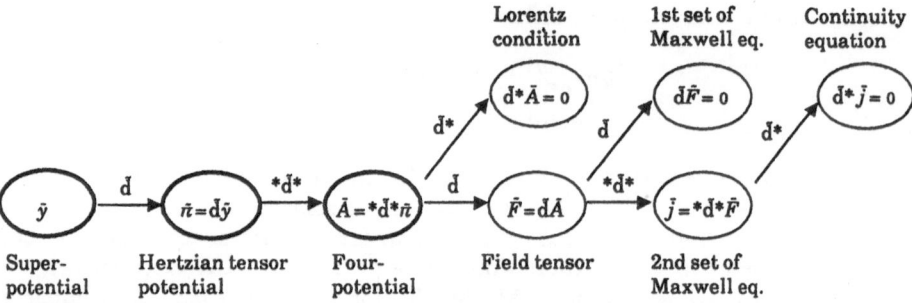

**Fig. 1** Differential form diagram of electrodynamics

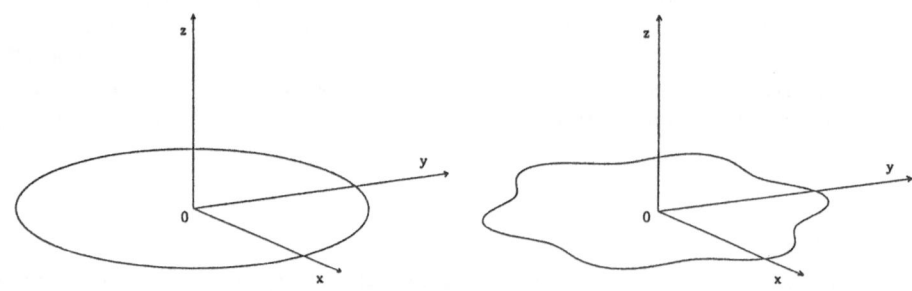

**Fig. 2** True orbit and calculated one using equation (16) for circular motion.

**Fig. 3** True orbit and calculated one using equation (16) for betatron motion.

## Appendix A

Whittaker's procedure is used[5]. The equation (3) is also written as

$$\omega t = \omega \tau + \omega \frac{|\mathbf{x} - \mathbf{y}(\omega \tau)|}{c} \equiv \omega \tau + \frac{\omega}{c} R(\mathbf{x}; \omega \tau) \tag{A1}$$

Differentiating equation (A1)

$$\omega \, dt = \omega \, d\tau + \frac{\omega}{c} \, dR(\mathbf{x}; \omega \tau) \tag{A2}$$

Divided equation (A2) by differentiation of $y^i(\omega \tau)$ $(dy^i(\omega \tau))$ and expanding its inversion in Fourier series

$$\frac{\omega \, dt}{dy^i(\omega \tau)} = \frac{\omega \, d\tau}{dy^i(\omega \tau)} + \frac{\omega \, dR(\mathbf{x}; \omega \tau)}{c \, dy^i(\omega \tau)}$$

$$\frac{dy^i(\omega \tau)}{\omega \, dt} = \frac{1}{\dfrac{\omega \, d\tau}{dy^i(\omega \tau)} + \dfrac{\omega \, dR(\mathbf{x}; \omega \tau)}{c \, dy^i(\omega \tau)}}$$

$$= \frac{1}{2\pi} \sum_{n=-\infty}^{\infty} \int_0^{2\pi} \frac{exp(-inu) \, du}{\dfrac{\omega \, d\tau}{dy^i(\omega \tau)} + \dfrac{\omega \, dR(\mathbf{x}; \omega \tau)}{c \, dy^i(\omega \tau)}} exp(in\omega t) \tag{A3}$$

Transforming the variable $u$ into $\sigma$ defined by

$$u = \sigma + \frac{\omega}{c} R(\mathbf{x}; \sigma) \tag{A4}$$

we obtain

$$\frac{dy^i(\omega \tau)}{\omega \, dt} = \frac{1}{2\pi} \sum_{n=0}^{\infty} C_n \int_a^b \cos(n[\omega t - \sigma - \frac{\omega}{c} R(\mathbf{x}; \sigma)]) \, dy^i(\sigma) \tag{A5}$$

where $C_n = 1$ for $n = 0$, $C_n = 2$ for $n \neq 0$ and $b \, (= a + 2\pi)$ is determined by the following relation

$$2\pi = b + \frac{\omega}{c} R(\mathbf{x}; b) \tag{A6}$$

And then integrating Eq. (A5) with respect to $\omega t$, the equation (7) is derived. Now, it should be noticed that the term for $n = 0$ in equation (A5) $(dy_0^i(\omega \tau)/\omega t)$ is written as follows:

$$\frac{dy_0^i(\omega \tau)}{\omega \, dt} = \frac{1}{2\pi} \int_a^b dy^i(\sigma)$$

$$= (1, 0) \tag{A7}$$

## Appendix B

From far electric fields at four points, one can know the values $\Psi_1$, $\Psi_2$, $\Psi_3$, $\Psi_4$ and calculate ratios $R_2(x)/R_1(x)$, $R_3(x)/R_2(x)$, $R_4(x)/R_3(x)$ and $R_1(x)/R_4(x)$.

On the other hand, it is well-known that if there are two fixed points denoted by $A$ and $B$, a point $P$, which satisfies following condition

$$\frac{\overline{AP}}{\overline{BP}} = const. \qquad (B1)$$

forms a sphere ( see Fig. B1 ). Therefore, one can say that the center of trajectory $C$ exists on the spheres defined by

$$\frac{\overline{X_2 C}}{\overline{X_1 C}} = \frac{R_2(x)}{R_1(x)} \ , \quad \frac{\overline{X_3 C}}{\overline{X_2 C}} = \frac{R_3(x)}{R_2(x)} \ , \quad \frac{\overline{X_4 C}}{\overline{X_3 C}} = \frac{R_4(x)}{R_3(x)} \ , \quad \frac{\overline{X_1 C}}{\overline{X_4 C}} = \frac{R_1(x)}{R_4(x)} \qquad (B2)$$

where $X_1$, $X_2$, $X_3$ and $X_4$ denote the four observation points. Then, the center $C$ can be determined by four spheres uniquely. If the center is determined, the distances $R_1(x)$, $R_2(x)$, $R_3(x)$ and $R_4(x)$ are obtained easily. Consequently, the electron orbit y ( ct, x ) can be found.

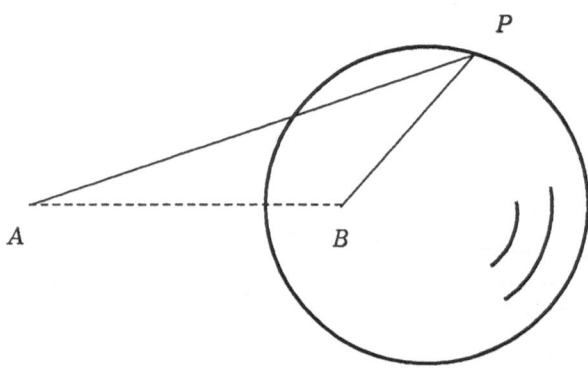

Fig.B1  sphere defined by Eq. (B1)